ADVANCES IN CHEMICAL PHYSICS

VOLUME XLVI

EDITORIAL BOARD

ADVANCES IN CHEMICAL PHYSICS

EDITED BY

I. PRIGOGINE

University of Brussels
Brussels, Belgium
and
University of Texas
Austin, Texas

AND

STUART A. RICE

Department of Chemistry
and
The James Franck Institute
The University of Chicago
Chicago, Illinois

VOLUME XLVI

AN INTERSCIENCE® PUBLICATION

JOHN WILEY & SONS

NEW YORK • CHICHESTER • BRISBANE • TORONTO

An Interscience® Publication

Library of Congress Catalog Card Number: 58-9935
ISBN 0-471-08295-3

Printed in the United States of America

10 9 8 7 6 5 4 3 2 1

CONTRIBUTORS TO VOLUME XLVI

JOHN S. DAHLER, Departments of Chemical Engineering and Chemistry, University of Minnesota, Minneapolis, Minnesota

RASHMI C. DESAI, Department of Physics, University of Toronto, Toronto, Ontario, Canada

LOTHAR FROMMHOLD, Department of Physics, University of Texas, Austin, Texas

AKIRA IKEGAMI, The Institute of Physical and Chemical Research, Hirosawa, Wako-shi, Saitama, Japan

MYUNG S. JHON, Departments of Chemical Engineering and Chemistry, University of Minnesota, Minneapolis, Minnesota

RONNIE KOSLOFF, The James Franck Institute, The University of Chicago, Chicago, Illinois

MASUO SUZUKI, Department of Physics, University of Tokyo, Hongo, Bunkyo-ku, Tokyo

MICHAEL TABOR, La Jolla Institute, P.O. Box 1434, La Jolla, California

INTRODUCTION

Few of us can any longer keep up with the flood of scientific literature, even in specialized subfields. Any attempt to do more, and be broadly educated with respect to a large domain of science, has the appearance of tilting at windmills. Yet the synthesis of ideas drawn from different subjects into new, powerful, general concepts is as valuable as ever, and the desire to remain educated persists in all scientists. This series, *Advances in Chemical Physics*, is devoted to helping the reader obtain general information about a wide variety of topics in chemical physics, which field we interpret very broadly. Our intent is to have experts present comprehensive analyses of subjects of interest and to encourage the expression of individual points of view. We hope that this approach to the presentation of an overview of a subject will both stimulate new research and serve as a personalized learning text for beginners in a field.

ILYA PRIGOGINE
STUART A. RICE

CONTENTS

COLLISION-INDUCED SCATTERING OF LIGHT AND THE DIATOM POLARIZABILITIES

LOTHAR FROMMHOLD

Department of Physics, University of Texas, Austin, Texas

CONTENTS

This work was made possible through the support of the Robert A. Welch Foundation and the Joint Services, Electronics Program.

I. INTRODUCTION

The subject matter of this chapter is the determination of diatom polarizabilities of the simple gases, in particular from new measurements of their collision-induced Raman spectra. The term "diatom" is meant to describe the quasimolecular complex of two interacting atoms (or monomers), which may either be free and in a collisional encounter or else lightly bound together as a so-called dimer or van der Waals molecule.

The polarizability of matter is responsible for the dielectric, refractive, and light scattering properties of gases. Specifically, the diatomic polarizability is related to the second coefficients of the virial expansions of such properties of imperfect gases. In a collisional encounter, atomic polarizabilities are known to vary with the changing internuclear separations (see Section II.A). The diatom polarizability is defined to be an excess polarizability, that is, the sum of the polarizabilities of two interacting atoms, minus the sum of the polarizabilities of the unperturbed (noninteracting) atoms. It is sometimes referred to as the collision-induced (part of the total pair) polarizability. As in the case of diatomic molecules, the diatom polarizability is a tensor. Its components vanish as the internuclear separation approaches infinity. The physical quantities that are accessible to measurement are the invariants of that tensor, *trace* and *anisotropy*, which may be considered orientational averages of it. The trace is the spherically symmetric part of the diatom polarizability and is thus related to polarized scattering of light, and to refraction at moderate to high gas densities. The anisotropy causes depolarization and is also related to field-induced birefringence (Kerr effect).

Collision-induced scattering of light (CIS) was demonstrated about 10 years ago by McTague and Birnbaum,[1,2] who used a conventional 90°-scattering Raman apparatus and densities of various rare gas samples from 10 to 150 amagat. Whereas at low densities (< 1 amagat) the light scattering of the rare gases consists of only the well-known Rayleigh-Brillouin process,[3] at higher densities nearly exponential Stokes and anti-Stokes wings appear[1,2] with intensities that vary as the square of the gas density.[1,2] These collision-induced Raman spectra of the diatoms are highly depolarized in most gases. Hence these are generated mainly by the anisotropy of the diatom polarizability tensor. At even higher densities, new spectral contributions appear that feature intensities proportional to the third, fourth, or higher power of the gas density,[2,4] which describes what we loosely refer to as the "virial expansion of light scattering." We note that one may find this term objectionable, because a rigorous virial treatment of collision-induced lineshapes has apparently not been given. Practically speaking, however, it is not hard to avoid serious confusion and to treat the case of the two-body interactions (the "first-order term" of the virial expansion) with the desirable rigor, both experimentally and theoretically.

A 1974 paper by Gelbart,[5] entitled "Depolarized Light Scattering by Simple Fluids," reviews the field of collision-induced scattering of the simple gases. Since then this field has advanced in various ways. For example, first measurements of CIS in helium were recently reported. Helium is of a particular significance since in this case more or less sophisticated *ab initio* calculations of the diatom polarizability were communicated. Such measurements can be considered a test of the fundamental theory. The first *polarized* collision-induced Raman spectra could be obtained and thus the square of the trace could be measured. An alternative to the "method of moments" was recently developed in which the spectral distribution function is computed from wavemechanics, for a direct comparison with experiment. All the information contained in the spectra is thus utilized, not just some integral properties of the distribution. The major uncertainties of the measurement of diatom polarizabilities, from which the earlier method of moments suffered, are thus completely eliminated. This chapter is an attempt to update those parts of Gelbart's article that relate to the binary collisions, dimers, and measurements of the diatom polarizability. It is more specific in these areas. At the same time, we try to keep the chapter self-contained for the reader's convenience.

II. THE POLARIZABILITY OF THE DIATOM

A. Classical Treatment

An external electric field $\mathbf{F}_0$ induces in an isolated atom a dipole moment $\mathbf{p}=\alpha_0\mathbf{F}_0$. Two atoms (1 and 2) acquire a total induced dipole moment of $\mathbf{P}=\alpha_1\mathbf{F}_1+\alpha_2\mathbf{F}_2$. If the two atoms are collisionally interacting, the α_1, α_2 may be different from the noninteracting atoms' polarizabilities; they are functions of the internuclear separation r. Furthermore, the local fields $\mathbf{F}_1, \mathbf{F}_2$ are distorted by the induced dipole fields of the perturber, according to

$$\mathbf{F}_1=\mathbf{F}_0+\mathbf{F}_{12} \qquad \text{and} \qquad \mathbf{F}_2=\mathbf{F}_0+\mathbf{F}_{21} \tag{1}$$

The dipole field $\mathbf{F}_{12}$ due to the second atom, at the location of the first, is given by

$$\mathbf{F}_{12}=\frac{3\mathbf{n}_{21}(\mathbf{p}_2\cdot\mathbf{n}_{21})-\mathbf{p}_2}{r^3} \tag{2}$$

The unit vector $\mathbf{n}_{21}$ points from the second atom to the first. Consequently, the induced moment in the second atom, $\mathbf{p}_2=\alpha_2\mathbf{F}_2$, depends on $\mathbf{F}_{21}$ (the field of the first) according to (1). An analogous equation can be written for $\mathbf{F}_{21}$, which depends similarly on $\mathbf{F}_{12}$. These two equations can be solved simultaneously for $\mathbf{F}_{12}$ and $\mathbf{F}_{21}$, and one finds that their components are all proportional to $\mathbf{F}_0$ (the external field). Consequently, the dipole moment of the pair can be written in tensor form:

$$\mathbf{P}=(\alpha_{jk})\mathbf{F}_0 \tag{3}$$

Elements α_{jk} are specified below. The diatom polarizability, which is defined as the pair polarizability (3) minus the sum of the polarizabilities of the two separated (noninteracting) atoms, is thus given by

$$\mathbf{P}=\left[\alpha_{jk}-(\alpha_{10}+\alpha_{20})\,\delta_{jk}\right]\mathbf{F}_0 \tag{4}$$

where δ_{jk} is the Kroneker delta. The subscript "0" serves as a reminder that the atomic polarizabilities of the unperturbed (or infinitely separated) atoms are to be subtracted.

From the above, the diatom polarizability tensor can be readily computed. In the frame of the diatom, it is a diagonal tensor of the form

$$\left[\alpha_{jk}-(\alpha_{10}+\alpha_{20})\,\delta_{jk}\right]=\begin{pmatrix}\alpha_{\parallel} & 0 & 0\\ 0 & \alpha_{\perp} & 0\\ 0 & 0 & \alpha_{\perp}\end{pmatrix} \tag{5}$$

with components $\alpha_{\parallel}$ and $\alpha_{\perp}$, the diatom polarizabilities in directions parallel and perpendicular to the internuclear axis.

To compute $\alpha_{\parallel}$, we choose $\mathbf{n}_{21}$ and $\mathbf{n}_{12}$ parallel to $\mathbf{F}_0$ and obtain from (2) the field components in that direction as

$$F_{12}=\frac{2p_2}{r^3} \qquad \text{and} \qquad F_{21}=\frac{2p_1}{r^3} \tag{6}$$

with $p_2=\alpha_2\cdot(F_0+F_{21})$ and $p_1=\alpha_1\cdot(F_0+F_{12})$. This system of equations can be solved, with the result

$$F_{12}=\frac{(2\alpha_2/r^3)(1+2\alpha_1/r^3)F_0}{(1-4\alpha_1\alpha_2/r^6)} \tag{7}$$

$$F_{21}=\frac{(2\alpha_1/r^3)(1+2\alpha_2/r^3)F_0}{(1-4\alpha_2\alpha_1/r^6)} \tag{8}$$

This allows one to put P in the form of (3). With $\alpha_{\parallel}=P/F_0$, we obtain

$$\alpha_{\parallel}=(\Delta\alpha_1+\Delta\alpha_2)+\frac{4\alpha_1\alpha_2(r^3+\alpha_1+\alpha_2)}{(r^6-4\alpha_1\alpha_2)} \tag{9}$$

The polarizability of the pair is thus seen to consist of two terms. The first one is determined by the variation of the atomic polarizabilities with internuclear separation: $\Delta\alpha_i=\alpha_i(r)-\alpha_{i0}$ for $i=1,2$, which usually is a not very well known function of r. (For crude estimates, this term is often ignored for convenience.) The second term may be considered to be a correction, which allows one to use the external field F_0 in place of the local fields [as in (3)]. In essence, it describes the effect of the distortion of the external field in the collisional encounter of two atoms ("field fluctuation term"). The $\alpha_i=\alpha_i(r)$ are the perturbed atomic polarizabilities. However, at not too small separations r, one may substitute the α_{i0} and thus obtain an asymptotically useful description.

Similarly, the computation of $\alpha_{\perp}$ assumes $\mathbf{n}\perp\mathbf{F}_0$. Equation 6 is thus replaced by

$$F_{12}=-\frac{p_2}{r^3} \quad \text{and} \quad F_{21}=-\frac{p_1}{r^3} \tag{10}$$

with p_1, p_2 as above. Solving this system, one obtains

$$F_{12}=-\frac{(\alpha_2/r^3)(1-\alpha_1/r^3)F_0}{(1-\alpha_1\alpha_2/r^6)} \tag{11}$$

$$F_{21}=-\frac{(\alpha_1/r^3)(1-\alpha_2/r^3)F_0}{(1-\alpha_2\alpha_1/r^6)} \tag{12}$$

which leads as above to

$$\alpha_{\perp}=(\Delta\alpha_1+\Delta\alpha_2)-\frac{\alpha_1\alpha_2(2r^3-\alpha_1-\alpha_2)}{(r^6-\alpha_1\alpha_2)} \tag{13}$$

Here, too, we have the two contributions due to (1) the variation of the atomic polarizability, and (2) the accounting for the fluctuating field.

The diatom polarizability tensor, (5), has two invariants: the spherically symmetric part, conveniently referred to as "trace,"

$$a(r)=\frac{\alpha_{\parallel}+2\alpha_{\perp}}{3} \tag{14}$$

[even though (16) represents only one-third of the tensor's trace], and the anisotropy,

$$\gamma(r) = \alpha_{\parallel} - \alpha_{\perp} \tag{15}$$

In the experiments discussed below, quantities related to these invariants (which are orientational averages of the polarizabilities) are measured.

Classical approximations of trace and anisotropy can be obtained from (9) and (13) to (15). For easy reference below we mention that the trace is much more affected by the rather unknown $\Delta\alpha_i$ (the variation of the atomic polarizabilities with separation) than the anisotropy. This is so because the $\Delta\alpha$ of (9) and (13) are nearly the same and cancel in the expression of the anisotropy, (15), whereas they add to become the dominant term for the trace (14).

The most common approximation of the diatom polarizability is the so-called point-dipole, or dipole-induced dipole (DID) model, which assumes that the atomic polarizabilities do not vary with varying proximity of the perturber ($\Delta\alpha_i = 0$ for $i = 1, 2$). Furthermore, one usually assumes like atoms ($\alpha_1 = \alpha_2$) and obtains for this model

$$\alpha_{\parallel} = \frac{4\alpha_0^2}{r^3 - 2\alpha_0} \tag{16}$$

$$\alpha_{\perp} = -\frac{2\alpha_0^2}{r^3 + \alpha_0} \tag{17}$$

The DID models of the invariants are given by (14) to (17) as

$$a(r) = \frac{4\alpha_0^3}{r^6 - \alpha_0 r^3 - 2\alpha_0^2} \tag{18}$$

$$\gamma(r) = \frac{6\alpha_0^2 r^3}{r^6 - \alpha_0 r^3 - 2\alpha_0^2} \tag{19}$$

For reasons discussed below, the asymptotic forms (for large r) are often preferred:

$$a(r) = \frac{4\alpha_0^3}{r^6} \tag{20}$$

$$\gamma(r) = \frac{6\alpha_0^2}{r^3} + \frac{6\alpha_0^3}{r^6} \tag{21}$$

We see below that the DID model of the anisotropy (21) is usually a fairly good approximation, with deviations from the actual anisotropy typically less than about 10% for all physically accessible separations. Equation 20, however, is not only of the wrong sign in important cases, the actual absolute magnitude of the trace is typically three to five times as great, owing to the neglected $\Delta\alpha$ term.

The reasoning given above is not new. Classical arguments similar in spirit and substance were advanced in many tractats as quoted below. Silberstein[6] is usually given credit for having introduced the DID model in 1917. It is shown below that an account of the atomic hyperpolarizabilities must be made for a more accurate description of the field-fluctuation terms, (9) and (13).

B. *Ab initio* Calculations

A discussion of the various methods of computing atomic and molecular polarizabilities can be found in a recent review.[7] We simply mention here that second-order perturbation theory gives the dynamic polarizability tensor components of a diatom (in atomic units) in response to a photon as

$$\alpha_{\|,\perp}(\omega)=\sum_b \frac{(f_{\|,\perp})_{nb}}{(E_b-E_n)^2-\omega^2} \tag{22}$$

The circular frequency ω equals 2π times the frequency of the incident photon. The static polarizabilities are given by (22), with $\omega=0$. The sum extends over all states b. The oscillator strengths are defined as usual by

$$(f_{\|})_{nb}=\tfrac{2}{3}\left(\frac{g_b}{g_n}\right)(E_b-E_n)\langle n|z|b\rangle\langle b|z|n\rangle \tag{23}$$

$$(f_{\perp})_{nb}=\tfrac{4}{3}\left(\frac{g_b}{g_n}\right)(E_b-E_n)\langle n|x|b\rangle\langle b|x|n\rangle \tag{24}$$

The letter E designates diatom energies, with a subscript n or b for the ground or excited state, respectively, and the g_b, g_n represent the rotational degeneracies. The computation of $\alpha_{\|}, \alpha_{\perp}$ involves determining the first-order correction $\Phi_n^{(1)}$ to the wave function in the presence of the external field perturbation. Calculations fall into two main categories, those based on correlated (configuration-interaction) wave functions, and those based on uncorrelated Hartree-Fock schemes (SCF). The "direct sum-over-states" method expands the first-order corrections, $\Phi_n^{(1)}$, in terms of the complete set of unperturbed eigenfunctions.[8] Alternatively, the sums like (22) are formally closed in some way and the $\alpha_{\|}, \alpha_{\perp}$ are obtained in a Cauchy expansion. Variational principles are also common ("variation-perturbation method").

A brief historical sketch is given next. Quantum-mechanical calculations of the polarizability of interacting atoms were first carried out by Jansen and Mazur,[9, 10] who showed that wave-mechanical corrections of the DID model are in general needed for the terms in r^{-6} of the asymptotic expansions. Consequently, only the leading term of the DID anisotropy appears to be useful, rendering the DID trace model useless. Certain and Fortune[11] calculated the exact coefficients of r^{-6} for the asymptotic helium diatom polarizabilities, using a variational principle. Buckingham[12] showed that the leading term A_6/r^6 of the symptotic expansion of the trace differs from the DID model, (20), because of the need to apply a dispersion-type correction to the polarizability of one atom in the presence of the other. It depends on the hyperpolarization and is given by[12]

$$A_6 = 4\alpha_0^3 + \tfrac{5}{9}\cdot\Gamma\cdot\frac{C_6}{\alpha_0} \tag{25}$$

The hyperpolarizability is Γ and C_6 is the coefficient of r^{-6} in the dispersion force. For helium, the coefficient A_6 is seen to agree closely with Certain and Fortune's fundamental work ($A_6 = 41.6a_0^9$ versus $39.0a_0^9$ of Ref. 11. Here $a_0 \cong 0.529 \times 10^{-10}$ m = Bohr's radius.)

The polarizability of the hydrogen diatom was evaluated as a function of the internuclear separation r by Kolos and Wolniewicz,[13] Ford and Browne,[14] and others. However, in the case of hydrogen, the interaction is so strong that the results are not relevant to nonbonded diatoms. The repulsive $^3\Sigma_u^+$ state of H_2 is more appropriate for the diatoms of interest here and was considered by DuPre and McTague.[15] The helium diatom was first investigated by Lim et al.[16] using a SCF basis set of eight Gaussian orbitals centered on each atom. However, in that work serious difficulties were observed at large separations. More recently, Buckingham and Watts[17] used SCF and two different basis sets (of 16 and 18 functions) to treat the helium diatom. Their results showed, however, some dependence on the basis set. The authors, therefore, recommended the numerical results be used with caution.

O'Brien et al.,[18] using a 30-function Gaussian basis set, obtained polarizabilities $\alpha_\parallel, \alpha_\perp$ for the helium diatom believed to be accurate to $\pm 5\%$, for nine different separations. The invariants of the polarizability tensor of He_2 are given in Table I. The trace can be approximated by an exponential as

$$a(r) = \frac{A_6}{r^6} - \lambda_t \exp\left(-\frac{r}{r_t}\right) \tag{26}$$

TABLE I.
The Coefficients of (26) and (27) Derived from *ab initio* Helium Diatom Polarizabilities

Ref.	Method	A_6 (a_0^9)	λ_t (a_0^3)	r_t (a_0)	B_ε ($\mathrm{cm}^6/\mathrm{m}^2$)	λ_a (a_0^3)	r_a (a_0)	$B_K \cdot 10^{15}$ [$\mathrm{cm}^9/(\mathrm{erg})(\mathrm{m}^2)$]	Remarks
CF[11]	Var.	38.97	—	—	(+0.012)	—	—	4.02	Long range only
O'al.[18]	SCF	38.97	12.14	0.70	−0.085	15.6	0.74	2.61	
FC[19]	SCF	{ 9.24	0.	— }	−0.094	—	—	—	For $r \geq 6.5a_0$
		{ 0.	12.9	0.69 }	—	—	—	—	For $r < 6.5a_0$
FC[19]	SCF	—	—	—	—	42.5	0.608	2.91	
D[20]	CfI'n	38.97	61.23	0.55	−0.058	19.10	0.71	2.67	$r > 3a_0$
KK[21]	SCF	38.97	17.59	0.67	−0.093	6.76	0.90	2.26	$r > 3.5a_0$
CMB[27]	Semiclassical	(0)	(16.22)	(0.56)		(0)	(0)		$4a_0 \leq r$

Also given are the second virial dielectric and Kerr coefficients, B_ε and B_K, obtained from these *ab initio* data. Semiclassical results are also quoted for comparison. (Bohr's radius $a_0 = 0.52917706 \times 10^{-10}$ m.) Methods used: Var.: variational; SCF: self-consistent field; CfI'n: configuration interaction; Semiclassical: as explained in the text.

Similarly, the anisotropy is consistent with the expression

$$\gamma(r) = \frac{6\alpha_0^2}{r^3} - \lambda_a \exp\left(-\frac{r}{r_a}\right) \tag{27}$$

with constants A_6, λ_t, r_t, λ_a, r_a, α_0 given in Table I. We remark that the trace is negative, except for large separations, which is attributed to electronic overlap. For the first time, a second virial dielectric coefficient B_ε (discussed below) having the proper sign and magnitude of the experimental value could thus be obtained from the fundamental theory.

Fortune and Certain[19] used an uncorrelated SCF approach to compute the polarizability of the helium diatom. A Slater basis set is used that was reoptimized at each internuclear separation. At the large separations ($r \geq 6.5a_0$), the calculations reduce to the DID approximation. Results in the form of a table can again be approximated by analytical functions. For the trace, we have

$$a(r) = \begin{cases} \dfrac{A_6}{r^6} & \text{for} \quad r \geq 6.5a_0 \\ -\lambda_t \exp\left(-\dfrac{r}{r_t}\right) + 0.00116a_0^3 & \text{elsewhere} \end{cases} \tag{28}$$

For the anisotropy, we get an expression like (27), with constants as in Table I.

A recent computation of the helium diatom polarizability by Dacre[20] accounts for electron correlation. A limited basis set $(5s; 3p; 2d)$ of Gaussian expansions of Slater orbitals is used. The inclusion of electron correlation causes an increase of γ by 7 to 10%.

Figures 1 and 2 compare the various *ab initio* results for helium. It is unfortunate that at the time of this writing no *ab initio dynamic* (that is, frequency-dependent) polarizability data exist. (All theoretical data mentioned are *static* polarizability data.) Our empirical models presented below are all *dynamic* models and must be compared with caution with the theoretical work. Asymptotic forms of the helium diatom trace[11] and the DID anisotropy are also displayed for comparison in Figs. 1 and 2. The asymptotic helium trace (which describes in essence the field fluctuations, but neglects the rapid variation of the atomic polarizabilities at close range), differs substantially from the computed values at near-range. Summing up the discussions concerning the helium diatom polarizability invariants (Figs. 1 and 2), we note that the various computational efforts have resulted at least in a qualitatively consistent description of the r dependence, over the range of at thermal collisions accessible separations r. However, quantitatively, the variations among the various works amount to $\lesssim$20% for the anisotropy and to factors of $\approx$2 for the trace, which may be considered the present uncertainty of the theoretical work for helium.

Recently, the first *ab initio* computation of the neon diatom polarizability appeared.[21] Similarly, a calculation of the argon diatom polarizability was given.[22] However, one must bear in mind that these systems consist of

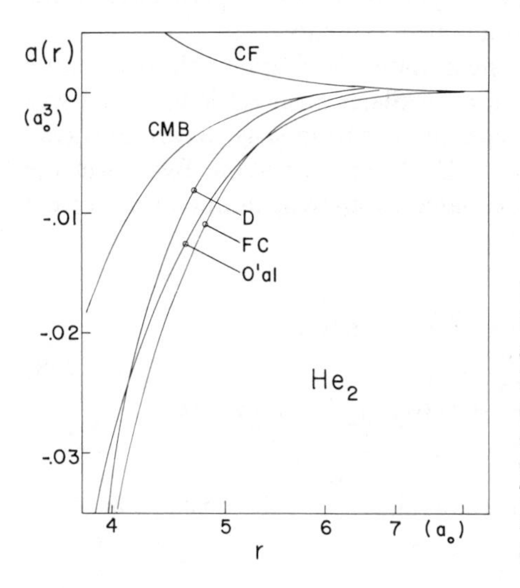

Fig. 1. The *ab initio* calculations of the trace, $a(r)$, of the helium diatom polarizability (references as in Table I: 0′a1,[18] FC,[19] D,[20] asymptotic tract CF.[11]). Kress and Kozak's[21] results are in near coincidence with curve 0′a1 for $r > 5a_0$, and only $\approx$10% above curve FC for $r \leq 5a_0$. The latest semiclassical computations[27] are also given (curve CMB) for comparison.

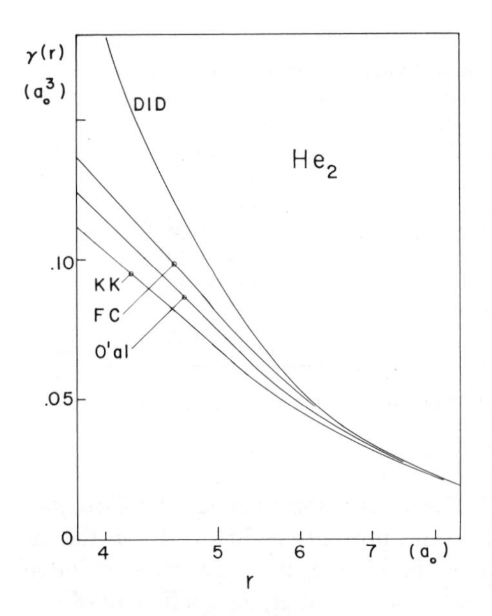

Fig. 2. The *ab initio* calculations of the anisotropy, $\gamma(r)$, of the helium diatom polarizability (references as in Table I: 0′a1,[18] FC,[19] KK[21]). In near coincidence with curve 0′a1 are the data by Dacre,[20] at small $r \lesssim 7a_0$. The DID model is also plotted for comparison.

very many electrons. Consequently, the numerical accuracy of such *ab initio* efforts is at present still limited.

C. Other Computational Results

Three alternative, semiclassical methods have been used for the computations of diatom polarizabilities. These are of interest particularly in the case of the heavier rare gases, for which the fundamental theory has been of a limited value.

Oxtoby and Gelbart[23, 24] assign a polarizability density to each atom and then treat the diatom on the basis of classical electrodynamics. The resulting equation can be solved iteratively. The first iteration gives the correct r^{-3} term of the DID asymptotic expansion of the anisotropy (21), while giving an exponential overlap contribution that reduces the anisotropy at short range.[23] Exchange and correlation effects can be included in the extension to second order.[24] Results are given for helium, neon, and argon diatoms. The spherically symmetric part of the diatom polarizability is again of the form of (26), with $A_6 = 41.6a_0^9$, $224a_0^9$, $9665a_0^9$, [after (25), see Ref. 12]; $\lambda_t = 64.4a_0^3$, $37.9a_0^3$, $629a_0^3$; and $r_t = 0.573a_0$, $0.626a_0$, $0.763a_0$, respectively, for these gases. The anisotropy is given by (27), with $\lambda_a = 42a_0^3$; $243a_0^3$, $1162a_0^3$; and $r_a = 0.632a_0$, $0.644a_0$, $0.901a_0$, for helium, neon, argon diatoms. (Here $a_0 = 0.52917706 \times 10^{-8}$ cm = Bohr's radius.)

An alternative method of computing diatom polarizabilities, a "density functional formalism," is given by Harris et al.[25, 26] It treats the atoms using

TABLE II.
Measurements of the Dielectric Second Virial Coefficient B_ε (cm^6/mol^2)

	From Ref. 35 (49°C)	From Ref. 36 (25°C)
Helium	-0.060 ± 0.040	-0.11 ± 0.02
Neon	-0.30 ± 0.10	-0.24 ± 0.04
Argon	$+0.72 \pm 0.12$[a]	$+0.79 \pm 0.10$
Krypton	5.6 ± 0.3	4.83 ± 0.40
Hydrogen	0.03 ± 0.10	—
Nitrogen	—	1.12 ± 0.30

[a] This value is taken from T. K. Bose and R. H. Cole, *J. Chem. Phys.*, **52**, 140 (1970).

an "electron gas approximation." The calculation includes the Coulomb and exchange effects between the atoms, but only to first order in the interatomic interaction. Correlation of the charge distributions is .ncluded only where they overlap. Clarke et al.[27] point out that, the coefficient of r^{-6} therefore is not correct. Results are given in Ref. 27 for helium and argon diatom polarizabilities.

Clarke et al.[27] discuss these treatments[23–26] most thoroughly and introduce a number of ideas for further improvement. They point out various difficulties of both methods and proceed to develop a third alternative, which they apply to the helium and argon diatom. They argue that the anisotropies of their best computations are very close indeed to the DID model. Surprisingly, Oxtoby and Gelbart's data are not very well reproduced in their calculations, even though the physical effects accounted for are virtually identical to those of Oxtoby and Gelbart.[24] Overlap is shown not to contribute to any experimentally significant reduction of the DID anisotropy to first order in the interatomic interaction.

For comparison with the *ab initio* data, we have plotted the trace data of Clarke et al.[27] in Fig. 1. We note that the semiclassical data deviate much more from any one of the *ab initio* results than the *ab initio* data differ among themselves. The semiclassical anisotropy agrees essentially with the DID curve of Fig. 2 and is not plotted. (The polarizability of the helium atom is taken from Ref. 28.)

D. Measurement of the Trace

The isotropic part of the diatom polarizability (or rather some thermodynamic average of it) can be measured from a determination of the second virial dielectric (or refractive) coefficient $B_\varepsilon(T)$ and, most recently, from recordings of the collision-induced, polarized Raman spectra.

The relationship between the static dielectric constant ε and the atomic polarizability α_0 of nonpolar gases is given by the Clausius-Mosotti function[29, 30]

$$\frac{\varepsilon-1}{\varepsilon+2}=\frac{4\pi N_0\alpha_0}{3v}+\frac{N_0^2B_\varepsilon}{v^2}+\cdots \tag{29}$$

where $N_0=6.022\times10^{23}$ (Avogadro's number) and v is the mole volume. (At the frequencies of visible light, the dynamic ε is commonly replaced by the square of the refractive index, $\varepsilon=n^2$, and dynamic α_0's can thus be determined by what is now called the Lorenz-Lorentz formula.) For ideal (or for dilute imperfect) gases, the second term on the right-hand side of (29) is zero (or vanishingly small). At densities roughly comparable to atmospheric densities, the "second virial dielectric coefficient" B_ε becomes important. It is related to the diatom trace according to[10, 12, 18, 31–36]

$$B_\varepsilon(T)=\frac{8\pi^2}{3}\cdot\int_0^\infty a(r)\exp[-\beta V(r)]r^2\,dr \tag{30}$$

The temperature T is given by $\beta=1/\kappa T$, where κ is the Boltzmann factor, and $V(r)$ designates the interatomic potential, which we consider a known quantity.[37] If $B_\varepsilon(T)$ is measured over a wide range of temperatures, the trace $a(r)$ can conceivably be obtained for all physically accessible separations r given by $V(r)\lesssim\kappa T$. At temperatures near absolute zero, the classical expression, (30), must be replaced by a quantum mechanical one, in which we substitute the density-independent term of the quantum pair distribution function (the two-body Slater sum)[38–40] for the Boltzmann factor.

Orcutt and Cole,[35] as well as Vidal and Lallemand,[36] have measured the dielectric virial coefficient B_ε of a number of gases (see Table II). For helium and neon, B_ε is negative, which is inconsistent with the DID model of the trace [(18) or (20)]. The *ab initio* diatom polarizability computations for helium, which are compiled in Table I, suggest that the negativity of $a(r)$ is a direct consequence of the electronic overlap, which was neglected in the DID model. From (30), second dielectric coefficients B_ε have been computed for all *ab initio* trace models. The results are given in Table I (sixth column) and are seen to be consistent with the measurements[35, 36] of Table II, (except, of course, the asymptotic expression,[11] which is of the wrong sign and magnitude). In the computations of B_ε, we have used several of the most highly refined, semiempirical interaction potentials presently known,[14, 42] with virtually identical results.

Kerr and Sherman[43] measured the second dielectric coefficient of helium at a temperature of 4°K, but under conditions of very high density. They obtained $B_\varepsilon = -0.030 \pm 0.004$ and -0.023 cm^6/mol^2 for the isotopes ^{3}He and ^{4}He, respectively. Using Fortune and Certain's trace,[19] Bruch et al.[38, 40] computed a slightly inconsistent $B_\varepsilon = -0.025$ and -0.030 cm^6/mol^2, respectively.

Summarizing, it is seen that measurements of B_ε exist for most gases, for a few selected temperatures. Consequently, their value for a determination of the diatom trace is presently limited to a simple consistency test: any realistic model of the trace used in (30) must be in agreement with the measurements of B_ε (Table II).

Recently, an alternative method for measuring the dynamic trace was communicated and used for two gases: helium[44, 45] and neon[46, 47]. The new method is based on the collision-induced Raman spectra and is discussed in Section III. Here we will briefly look at the results. Figure 3 presents a measurement and, for comparison, wave-mechanical computations of the polarized Raman spectrum of helium.[44, 45] Absolute scattering intensities were determined relative to Raman lines of known intensity. As is shown below, the polarized Raman spectra can be computed from wave mechanics if a model of the trace and the interaction potential are given.[48] We use the so-called HFD-C potential,[42] and the (static) *ab initio* results of the helium diatom trace as presented in Table I, to compute the spectra displayed in Fig. 3. The agreement is satisfactory, although not perfect. We note that the polarized spectrum of the trace amounts to only $\approx 1\%$ of the total intensity at low frequency shifts, but dominates the signals at high frequencies. As a consequence, the low-frequency falloff of the experimental spectrum is somewhat uncertain and may not be significant at all. The spectra computed on the basis of Certain and Fortune's work[19], and of Kress and Kozak's work[21], are consistent with the measurement[77], which, however, mirrors the slower high-frequency falloff characteristic of Dacre's

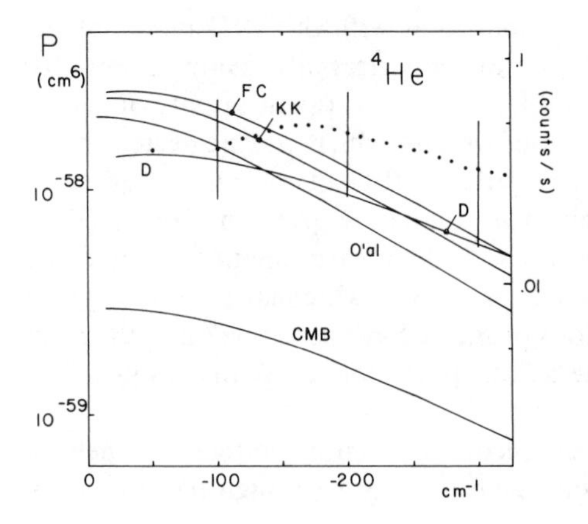

Fig. 3. Wave-mechanical computations of the polarized Raman spectrum (Stokes side) of the helium diatom are compared with the experimental results[45, 77] (●) on an absolute intensity scale [defined in (68) to (71)]. Trace models as in Table I are used as input: O'al,[18] FC,[19] D,[20] KK,[21] and CMB,[27] together with the HFDHE2 potential.[42] We note that other refined, semiempirical potential functions give virtually identical results.

data[20]. This behavior suggests a small range r_t of the trace, as is implied in the works of Dacre[20] and Clarke et al.[27]. If Dacre's trace coefficient λ_t is increased slightly from 61 $a_0{}^3$ to 70 $a_0{}^3$, with the range r_t unaltered, a good fit of the high-frequency portion of the experimental spectrum is obtained. At the same time, the second dielectric virial coefficient B_ε thus computed is consistent with the measurements[35,36]. The absolute magnitude of Clarke et al.'s trace near $r \leqq 5\ a_0$ appears to be too small.

The data presented in Fig. 3 demonstrate a point that is made somewhat more forcefully below. The spectroscopic method is quite powerful, because we do not simply compare two numbers (like a computed and measured B_ε), but rather we compare the shape of a spectral distribution on an absolute intensity scale. A trace varying slowly with r, for example, gives rise to a more "monochromatic" (i.e., narrow-banded) shape, and vice versa. (The reason becomes quite plausible below, when we discuss the classical theory of collision-induced lineshapes.) Hence a measurement of a spectral distribution, along with the absolute intensity, is roughly comparable with a measurement of a whole set of B_ε coefficients at different temperatures. Unfortunately, in the case of the polarized spectrum, the signals obtainable are extremely weak, and relatively large error bars must be accepted as a consequence. Therefore, the shapes of the spectral distributions are not as well defined as one would like. Even so, the spectrum is consistent with only some of the *ab initio* trace models and points out the nature of the weakness of the others.

It is interesting to note that the helium trace models, whose computed Raman spectra are consistent with the new experimental spectra,[44, 45] are associated with B_ε values between -0.085 and -0.095 cm^6/mol^2 (see the sixth column of Table I). The overlap of the two direct measurements of B_ε[35, 36] for helium is an almost identical -0.090 to -0.1000 cm^6/mol^2. Under the assumption that the differences between the static[35, 36] and dynamic (i.e., at the laser frequency) diatom polarizabilities are rather negligible, we conclude that the spectroscopic and virial experimental methods are in agreement for helium. Similarly, as is seen below, the spectroscopically acceptable neon trace models are associated with B_ε values from about -0.22 to -0.30 cm^6/mol^2 (see Refs. 46 and 47). The overlap of the direct measurements of B_ε[35, 36] for neon is a very satisfactory, consistent -0.20 to -0.28 cm^6/mol^2 (Table II). Two very different experimental techniques have thus generated data concerning the trace of helium and neon, which are in close agreement.

We mention that the spectroscopic method of determining diatom polarizabilities, as it is discussed here, is based on the full information obtainable from the observed lineshapes. Historically, however, the lineshapes of collision-induced Raman spectra were almost ignored, in favor of the "moments" (i.e., integrals) of the spectral distribution functions, in the belief that the zeroth and second moments (sometimes supported by the fourth) would in essence contain all the useful information. Powerful theoretical arguments have been advanced in favor of the method of moments, on the basis of the fact that a rigorous virial expansion of the moments appears to be feasible. (No rigorous virial expansion can be given for the lineshapes.) On the other hand, for the purposes of this chapter, a virial expansion is dispensable, provided one uses gas densities small enough that the diatom spectra are the only ones contributing to the observed signals. More important is the use of a wave-mechanical formalism, as it exists for the computation of the lineshape. (Moment expressions are based on classical physics; wave-mechanical "corrections" are known, but rarely used.) One should, furthermore, consider the fact that measurements of particularly the zeroth moment are usually seriously affected by the intense Rayleigh line.[49, 50] Even modern double monochromators feature an instrumental Rayleigh profile of approximately ± 5 slitwidths, which affects the low-frequency collision-induced spectra where they contribute the most to the zeroth moment.[51] Furthermore, particularly at higher pressures, the intercollisional process[52] is likely to affect the observed intensities to an unknown extent. At high frequencies, particularly for the higher moments, the precision suffers once more from the truncation of the spectrum. These uncertainties can all be completely avoided if lineshapes are used directly, where these are known to be purely of diatomic origin. We simply mention here the possibility of measuring the diatom trace spectroscopically using the method of moments, which we, however, consider inferior to the direct study of the lineshape. Furthermore, no trace data have as yet been obtained by the method of moments.

E. Measurement of the Anisotropy

The anisotropy of the diatom polarizability (or some thermodynamic average of it) can be measured (*1*) by determining the second virial Kerr coefficient $B_K(T)$ as a function of temperature, (*2*) from a recording of the collision-induced depolarized Raman spectra, and (*3*) from measurements of the depolarization ratio.

Like all matter, gases become birefringent in the presence of an electric field (electrooptic Kerr effect).[53–56] For nonpolar gases, the molar Kerr constant is given by[56]

$$ {}_mK(T) = \frac{4\pi N_0 \Gamma}{81} + \frac{B_K(T)}{v} \tag{31} $$

where Γ is the atomic hyperpolarizability[56, 57] and B_K is the second virial

Kerr coefficient given by[56]

$$B_K(T) = \frac{8\pi^2 N_0^2 \beta}{405} \cdot \int_0^\infty \gamma(r)\gamma_0(r) \exp[-\beta V(r)] r^2 \, dr \tag{32}$$

As above, we use $\beta = 1/\kappa T$, N_0 = Avogadro's number, and an interatomic potential $V(r)$ assumed to be known. At optical frequencies, the anisotropy is called $\gamma(r)$; the subscript "0" refers to the electrostatic (zero-frequency) case. If $B_K(T)$ is measured over a wide range of temperatures, one could conceivably obtain $\gamma(r) \cdot \gamma_0(r)$ from such data.

At very low temperatures, a wave-mechanical expression[38] must be substituted for (32). Measurements of B_K at room temperatures were communicated[56] for argon, krypton, xenon, and SF_6, which are discussed below. For helium, the constant B_K is very small and no measurements exist. Table I gives the computed values of $\langle \gamma^2 \rangle$ for helium:

$$\langle \gamma^2 \rangle = 4\pi \int_0^\infty [\gamma_0(r)]^2 \exp[-\beta V(r)] r^2 \, dr \tag{33}$$

from which the second virial Kerr coefficient can easily be computed (with the assumption $\gamma \equiv \gamma_0$).

As an illustration of the use of collision-induced Raman spectra, we show in Fig. 4 the experimental depolarized helium spectrum,[45] along with wave-mechanical computations of it, using the various anisotropy models of Table I as input. The work of Fortune and Certain[19] reproduces very nearly the experimental depolarized spectrum, but the other anisotropy data appear to generate rather similar spectra. The asymptotic anisotropy[11] leads to spectral intensities slightly higher than the experiment. The other models[18–21] give rise to up to 50% less intense spectra. The experimental uncertainties are as indicated, that is, much less than the deviations among the various trace models. (The semiclassical calculation[27] agrees almost exactly with the curve labeled CF, Fig. 4.)

The experiment[45] is bracketed by the curves CF[11] (which is based on essentially the leading term of the anisotropy, $6\alpha_0^2/r^3$) and FC[19] [using (27)]. This suggests that the λ_a value of FC[19] (Table I) must be slightly reduced for a better fit. Indeed, Dacre[20] points out that the inclusion of electron correlation, which is neglected in FC[19], will increase γ by a few percent as needed. A perfect fit[45, 58] (shown below, Fig. 6) can be obtained using a dynamic anisotropy with $\lambda_a = 39a_0^3$ (instead of the $42.5a_0^3$ of Ref. 19). Similar measurements of $\gamma(r)$ are described for other gases below.

Historically, spectroscopic measurements of anisotropies have been attempted most often using the method of moments. However, much of what

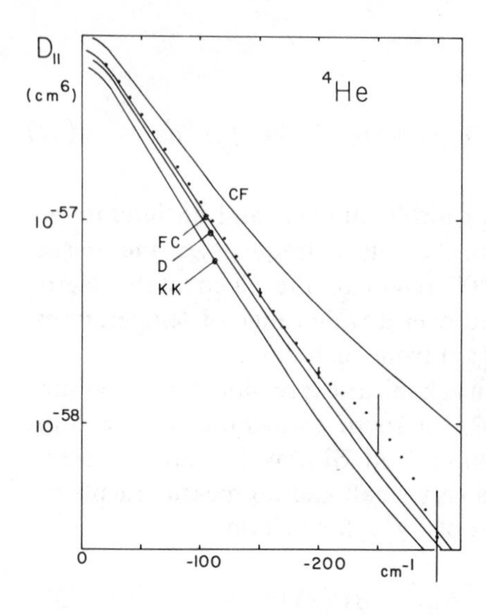

Fig. 4. Wave-mechanical computations of the depolarized Raman spectrum (Stokes side) of the helium diatom are compared with the experimental results[45, 77] (●) on an absolute intensity scale [defined in (68) to (71)]. Anisotropy models as in Table I are used as input: CF,[11] FC,[19] D,[20] KK,[21] along with the HFDHE2 potential.[42] Nearly coincident with curve D is the spectrum computed on the basis of O'Brien et al.'s[18] anisotropy data. Similarly, the data by Clarke et al.[27] give rise to a spectrum that nearly coincides with the one based on the asymptotic anisotropy (CF).

is said at the end of the preceding paragraph applies to that case as well. Particularly, the zeroth moment cannot be determined experimentally with great precision because of the presence of the Rayleigh-Brillouin line,[49, 50] the instrumental profile of the monochromator,[51] and the intercollisional process.[52] The accuracy of the higher moments suffers from the truncations of the spectrum at high frequencies. And finally, two (or three) spectral moments do not nearly carry all the information of the spectral distribution. Hence only rarely do the empirical models of the anisotropy thus obtained reproduce the lineshapes of the measurements from which they were derived![59, 60] Therefore, we prefer to base our conclusions directly on the observed lineshapes and to avoid the method of moments completely. (We look below at some empirical models of the anisotropy that were derived by the method of moments; computed lineshapes are shown for comparison with the experiment.)

F. Remarks

1. Frequency Dependence

The theoretical diatom polarizabilities quoted are all obtained for zero frequencies ("static polarizabilities"). Most measurements, however, are taken at frequencies of the visible. Consequently, a "dynamic" value of the polarizability results from such work. It has been argued that, for the rare gases, dynamic and static polarizabilities are likely to differ by just a few percent.[61] However, as we see below, the best measurements presently possible are of $\pm 4\%$ precision, or thereabout, and the differences between static and dynamic polarizability are beginning to matter. It is most desirable that in the future theorists consider the frequency dependence and provide dynamic data, at least for some of the most commonly used laser frequencies (4880 and 5145 Å).

2. *Range*

The internuclear separations accessible to measurement at room temperature range from a lower limit, given roughly by $V(r_{lo}) \approx 5\kappa T$, to some upper limit, r_{up}, determined by the range of the polarizability. Estimates can easily be obtained from a study of the integrand of the total depolarized intensity, which is given by an expression proportional to (33). (A similar expression exists for the total intensity of the polarized collision-induced spectrum, with the square of the anisotropy replaced by the square of the trace.) The integrand is always positive, and r_{lo}, r_{up} are readily determined, say as the 8 and 92% bounds of the total integral, (33).

For the case of helium, it thus appears that a range of internuclear separations from about $3.6a_0$ to $5.1a_0$ matters as far as the polarized spectrum is concerned. Trace values outside this range do not affect the total intensity significantly. Similarly, for depolarized scattering, a range from $4.2a_0$ to $12a_0$ matters, which reflects the long-range behavior of the anisotropy. (For comparison, we mention that the root of the interatomic potential, $V(\sigma)=0$, is given by $\sigma=5.09a_0$.) Similar numbers are seen below to hold for the alternative experiments mentioned above at room temperature. If such experiments are conducted at higher temperatures, r_{lo} is reduced by only a few tenths of a_0; at low temperatures, r_{lo} is slightly larger. The range of the polarizability measurements is specified below for the other gases.

3. *Models*

The fundamental theory has not provided us with a closed, analytical form of trace and anisotropy as a function of r. It has been argued[62] that perturbation theory will give the polarizabilities in the form of a series such as

$$a(r)=\sum_{i=0}^{\infty}\frac{\tilde{a}_i}{r^{2i+6}} \tag{34}$$

$$\gamma(r)=\frac{6\alpha_0^2}{r^3}+\sum_{i=0}^{\infty}\frac{\gamma_i}{r^{2i+6}} \tag{35}$$

with $\tilde{a}_0=A_6$ (25), and all other coefficients $\tilde{a}_i$, γ_i unknown. A suitably truncated series should, therefore, provide a reasonable "model" of the invariants of the diatom polarizability. On various occasions, models of the form

$$\gamma(r) = \frac{6\alpha_0^2}{r^3}(1 - b_m/r^m) \tag{36}$$

with constants b_m and $m > 0$ (not necessarily integer) to be determined, were also proposed. However, our experience indicates that near the root of the interatomic potential, $r = \sigma$, the variation of the polarizability with the internuclear separation r is most rapid and is best approximated by an exponential, as in (26) and (27); a single inverse power term [as in (36)] is not a very useful approximation.[24] Alternatives appear to be either to use a fairly large number of terms in the sums, (34) and (35) (which we consider impractical and which was indeed never seriously attempted), or else ignore the series of inverse powers and look for other "correction terms" that reproduce the rapid variations with r [like the exponential, (26) and (27)].

For the purposes of this chapter, we adopt practical models with exponentials [(26) and (27), and also (28)], as if these were the only choices. In this way, only two parameters (λ_0, r_0) are to be determined. Of these, the characteristic length r_0 determines the shape of the distribution, and λ_0 largely determines the absolute intensity. With the spectroscopic method recommended here, these parameters can be determined well. In almost every case the theoretical polarizability data could be fitted closely to such forms for the limited range of r values that is significant for the experiment. On occasion, we have used an additional inverse sixth (or eighth) power term, such as Certain and Fortune's long-range coefficient for the anisotropy[11] $(\gamma_0 = 12.07\alpha_0^3)$, without real gain or noticeable improvement of the fit. (This term amounts to a correction of less than 2% at all physically accessible separations, $\sigma \lesssim r$.) The introduction of a third term (such as γ_0/r^6 or a_1/r^8), in addition to the leading term and the exponential, appears to be at present rather unwarranted.

4. *Nonlinear Response*

Throughout this chapter we assume that the electric fields of the experiment are small enough so that the hyperpolarizabilities of the diatom can be ignored. In the experiments discussed, we see that typical laser fields amount to less than 10^4 V/cm. Under such conditions, the linear response to the laser field has been directly demonstrated in our laboratory.

5. *Three-Body Contributions*

The diatom processes discussed in this chapter are typically weak, but give signals proportional to the squares of the gas density. Experimenters, therefore, will work at the highest densities possible. In every case the quadratic density dependence should be verified by suitable measurements before an evaluation of the diatom polarizabilities is attempted. For binary collision-induced Raman spectra, the permissible range of gas densities (which is different for different frequency shifts) has been reported for many gases.[2,4]

The intercollisional effect mentioned above[52] is a very special kind of three-body contribution. Whereas the intracollisional (collision-induced) light scattering process features a density-independent spectral bandwidth of roughly one reciprocal duration of a collision,[1,2] the intercollisional process (which can never be separated experimentally and is superimposed with the collision-induced contribution at low frequencies) is of a bandwidth of about one reciprocal mean free time between collisions.[52] At low densities (near atmospheric density), this means a negligible bandwidth of about $\pm 0.2\ \mathrm{cm}^{-1}$. At the hundreds of atmospheres commonly used in some laboratories, however, the intercollisional spectrum may extend over a band of almost $\pm 100\ \mathrm{cm}^{-1}$.

Collision-induced spectra must be obtained under conditions of vanishingly small interference from many-body effects if these are to be used to measure diatom polarizabilities. In more practical terms, this requirement calls for a careful experimental demonstration that, at all frequency shifts that enter the measurement, the intensities vary as the square of the gas density. Unfortunately, this condition has not always been verified in the past, and some of the existing confusion concerning diatom polarizabilities can be traced to this neglect. Particularly at the low-frequency shifts, the intercollisional effect is apt to affect the observed intensities, and thus mainly the zeroth moment of the distribution. This can give rise to results that are at variance with other data. Similarly, in several laboratories, three-body and higher effects are routinely accounted for by using a virial expansion of the spectral distribution, $I(\omega)=I_2(\omega)\rho^2+I_3(\omega)\rho^3+\cdots$, which series is typically truncated after the cubic or, sometimes, the quartic term. We point out, however, that in fact this virial expansion of the distribution has no rigorous basis. The density dependence at low frequencies, for example, which the intercollisional effect generates, is certainly not described by any virial expansion of a distribution function. Uncertainties may thus enter the evaluation of the diatom polarizabilities, which can be completely avoided by other methods.

In conclusion, we want to emphasize here again that the consideration of the lineshape, taken at shifts where the intensities vary demonstrably as the square of the gas density, avoids all these uncertainties of the method of moments and of the virial expansion of the spectra and so on, and should be used preferentially. The measurement of the part of the lineshape that is clearly the diatomic spectrum free from interference is all that is needed for an evaluation of diatom polarizabilities, if such work is supported by the usual calibration of the absolute intensities, and by wave-mechanical computations of the lineshape.

G. Tentative Summary

The various existing *ab initio* computations of the helium diatom polarizabilities are not in perfect agreement among themselves, although in general their features are quite similar. In particular, from Table I and a look at experimental results (Fig. 3), one would conclude that the characteristic length r_t of the trace is very nearly given by $0.55a_0$, with error bars possibly as small as $\pm 0.06a_0$ or so. Similarly, as far as the helium anisotropy is concerned, the true (dynamic) numbers are slightly above Fortune and Certain's[19] values, as suggested by Fig. 4. The computation of the trace that best reproduces the polarized spectrum is the most successful one in reproducing the depolarized spectrum (FC[19]) of the anisotropy, although a smaller range r_t will improve the agreement with the experiment. However, the observed variances are often only of the order of ~10% and are not very serious. Since the collision-induced spectra are proportional to the square of the invariants, small differences near $r \cong \sigma$ in the invariants appear amplified in the spectral distributions. This very fact renders the spectroscopic method most suitable for measuring the diatom polarizability.

The quality of the results of *ab initio* computations for gases heavier than helium, which are immensely more complex, will likely be somewhat less than that of the presently best helium calculations.

Significantly, the most critical semiempirical computations are now in support of the experimental findings[58–60] that the leading term of the anisotropy, $6\alpha_0^2/r^3$, approximates very closely the true anisotropy at the physically accessible separations (CMB[27]). However, for helium, this theory appears to slightly underestimate the small deviations from $6\alpha_0^2/r^3$. It is, furthermore, most unfortunate that the semiclassical calculations of the helium *trace* deviate most substantially from both the *ab initio* results and the experimentally inferred trace (Fig. 4). We take this as an emphatic warning to accept the results of such calculations with some caution. There is at present no "good" theory of this or any other approximate nature. Consequently, the theoretical data available for gases heavier than helium may presently not be of very high accuracy.

Virial coefficients B_ε, B_K should be known over a range of temperatures if one wants to obtain good measurements of the polarizability tensor invariants. In the absence of such knowledge of the temperature variation, the spectroscopic method (but not the method of moments) appears to be superior. Collision-induced Raman spectra must be obtained on an absolute intensity scale. Their shapes contain much useful information concerning the rate of variation of the invariants with separation r, that can readily be obtained (r_t, r_a). Theoretical and experimental intensities can be matched by adjusting the amplitudes λ_t, λ_a. Much of this information appears to be lost if only two or three moments are used instead of the full lineshape. Furthermore, absolute calibrations of the total intensity (as needed for the method of moments) suffer from a number of interferences, which can all be avoided with rigor if lineshapes are considered directly.

III. BINARY COLLISION-INDUCED SCATTERING OF LIGHT

A. Experimental Arrangements

The common 90°-scattering geometry, in which the laser beam and the direction of the observation form a right angle, is typically used for the recording of collision-induced Raman spectra. Double monochromators of about 1 m focal length and F/8 relative aperture, preferably with holographic gratings, are usually needed to suppress the very intense Rayleigh line. In one case, a Michelson interferometer was employed.[63] The spectra are excited with about 1 W of laser power, usually with the blue (4800 Å) or green (5145 Å) line of the argon ion laser. The laser beam is focused onto the pressurized gas in the sample cell; laser focal points of ~25 μ diameter are magnified about 10 times and are imaged on the entrance slit of the monochromator for maximum efficiency.[48, 64] It is advisable to use a polarization scrambler (a quartz wedge) in front of the entrance slit for uniform monochromator transmission of both polarizations. Apertures of the light collecting lens up to F/0.7 have been reported.[48] Photon counting techniques are preferred; gate times from seconds to hours are used in combination with signal-averaging techniques. With internally masked bialkali photomultiplier tubes, dark counts as low as 0.3 counts/sec are possible, along with quantum efficiencies in excess of 10%. Pressures above 1 atmosphere are best determined with a Bourdon-type quartz manometer.

For most applications, a polarization of the incident beam parallel to the direction of observation is desired. This reduces somewhat the possibility of interference due to the polarized Rayleigh line, which may be many orders of magnitude more intense than the signals to be recorded. This is most readily done with a half-wave plate in the incident beam, which is set for a minimum Rayleigh signal. Under such conditions, a collision-induced scattered signal, $S_{\parallel}$, is obtained; the subscript indicates that the polarization of the incident beam is parallel to the direction of observation.

If the depolarization ratio is to be measured, or polarized spectra are to be recorded, the half-wave plate is best removed. The polarization is then perpendicular to the direction of observation, and signals $S_{\perp}$ are recorded.

If the observed spectra are fully depolarized, the ratio $S_\perp / S_\parallel$ is very nearly equal to[65] $b=7:6$. (A slightly more precise value can be determined in an actual experiment, using a purely rotational Raman transition, for example, of hydrogen, with the two polarizations of the incident beam.)

If the spectra are not fully depolarized, we have both a polarized (P) and depolarized (D) component present, and the signals are superpositions of these, according to

$$S_\perp = P + bD \tag{37}$$

$$S_\parallel = a^* P + D \tag{38}$$

(provided, a good polarization scrambler is used). An unpolarized detector is assumed. The attenuation of the polarized component for the unfavorable polarization (||) is called a^* and can easily be measured, studying intensity ratios of the Rayleigh line, preferably with a low-pressure rare gas. Typically, it is of the order $a^* \cong \frac{1}{10}$. For any given frequency shift ν, signals $S_\perp$, $S_\parallel$ are determined. Since a^*, b are known, frequency-independent constants, we can solve the system (37), (38) for P and D, as follows:

$$P_\perp = \frac{(S_\perp - bS_\parallel)}{(1 - a^* b)} \tag{39}$$

$$D_\parallel = \frac{(S_\parallel - a^* S_\perp)}{(1 - a^* b)} \tag{40}$$

Unfortunately, the terms $S_\perp$ and $bS_\parallel$ are often very nearly equal. As a consequence, good measurements of $S_\perp$, $S_\parallel$ may give a rather substantial uncertainty in P, whereas the error in D is essentially that of $S_\parallel$. A vanishing P means that the spectra are "fully depolarized."

B. Absolute Intensities (Calibration)

The differentially scattered total intensities of depolarized collision-induced Raman spectra, relative to the incident flux, is given by

$$\frac{I_{\mathrm{tot}\,\parallel}}{I_0} = \tfrac{6}{45} \cdot k_0^4 \cdot \left\{ \int_0^\infty [\gamma(r)]^2 \exp[-\beta V(r)] r^2 \, dr \right\} \left(\frac{N^2}{2} \right) \tag{41}$$

with $k_0 = 2\pi/\text{laser wavelength}$. $N^2/2$ is the number of diatoms in the sample volume. Since this can be written in terms of the second Kerr virial coefficient, B_K (32), which is known for many gases,[53] an absolute intensity calibration was often considered dispensable. In recent years, however, two methods have been developed for a direct calibration of collision-induced spectra, one based on the integral of the spectral distribution (the zeroth moment), and the other on a direct calibration of the continuum intensity at a given frequency shift. Both methods use Raman lines of known intensity (rotational lines of hydrogen or nitrogen) as a standard.

Raman intensities of diatomic molecules are given by[65]

$$\frac{\partial I_{\perp}/\partial\Omega}{I_0}=\frac{N\cdot k^4\left(|\langle n'J'|a|nJ\rangle|^2+\frac{7}{45}b_{J'}^{J}|\langle n'J'|\gamma|nJ\rangle|^2\right)(2J+1)e^{-\beta E}}{Q(T)} \tag{42}$$

The quantum numbers n, J characterize the initial state; a prime symbolizes the final state. The $b_{J'}^{J}$ are the squares of the angular matrix elements, given by[65]

$$b_{J-2}^{J}=\frac{1.5J(J-1)}{(2J-1)(2J+1)} \tag{43}$$

$$b_{J}^{J}=\frac{J(J+1)}{(2J-1)(2J+3)} \tag{44}$$

$$b_{J+2}^{J}=\frac{1.5(J+1)(J+2)}{(2J+1)(2J+3)} \tag{45}$$

Molecular trace and anisotropy are called a and γ, respectively. The selection rules are $J\rightarrow J$ for the polarized scattering, and $J\rightarrow J$, $J\pm2$ for depolarized scattering. $Q(T)$ is the sum-over-states. If photon counting techniques are employed, the factor k^4 in (42) must be replaced by k^3k_0, with k_0 being 2π over the laser wavelength. N is the number of molecules in the volume irradiated. For pure rotational transitions ($n=n'$; $J\neq J'$), the trace part vanishes. The square of the dynamic anisotropy matrix element is known for the hydrogen molecule with high precision;[14] it has also been measured for nitrogen.[66] This explains the calibration standards commonly used in this field.

For a beam polarization parallel to the direction of observation, $\partial I_{\parallel}/\partial\Omega$, we replace the factor $\frac{7}{45}$ in (42) by $\frac{6}{45}$ and suppress the trace matrix elements totally.[65]

We note that the differentially scattered intensities must be obtained by integrating (42) over the receiving cone of the light-collecting lens. Observable Raman line intensities are roughly proportional to the slitwidth if the image of the light source is much wider than the entrance slit. (Under such conditions continuum intensities depend on the square of the slitwidth.) Knowledge of the exact slitwidth is, however, not required for the intensity calibration if for discrete Raman lines the word "signal" is taken to mean the integral of the nearly triangular line profile (to be viewed as the product of slitwidth and line intensity[48]), in units of counts per second times wave numbers. These integrals can be measured with precision and are designated S_H or S_N for a calibration with hydrogen or nitrogen, respectively, at a specified Raman transition. (These have the same dependence on the slitwidth as the continuum intensities, roughly quadratic.) Another subscript ($\perp, \|$) is added to specify beam polarization.

One method of calibration, the method of moments, obtains the integrated intensity of the collision-induced spectrum by integration over the spectral distribution. This integral can be shown to be proportional to $(\text{slitwidth})^2 \cdot I_{\text{tot}, \|}$, as defined by (41). The transmission of the monochromator and the variation of the detector efficiency with frequency must be fairly well known. More importantly, one needs to interpolate the low-frequency spectrum over a range of about ± 5 slitwidths,[51] across the most intense Rayleigh line. Since depolarized spectra are usually nearly exponential at not too small frequencies shifts, one often employs an exponential extrapolation to zero frequency shift. This gives rise to sizable errors, since lineshape computations as discussed below indicate a substantial deviation from the exponential form near zero frequencies. Alternatively, one might try to obtain an idea of the shape of the spectra by scanning the low-frequency spectra with very small slitwidths. However, particularly at the high pressures commonly found in such work, the intercollisional spectrum is likely to interfere to an unknown extent. Furthermore, the envelopes of the Raman spectra of van der Waals dimers also shape the observable low-frequency spectra in peculiar ways.[48, 67] In short, we think that really no solution exists to measure the integrated intensities dependably with a precision of 10% or better, and that this method best be avoided.

The alternative method is a direct calibration of the continuum intensity at a shift that corresponds to a rotational hydrogen or nitrogen Raman transition. Since the continuum intensity, $S(\nu)$ (which also carries a subscript $\perp, \|$ depending on the beam polarization), features the same dependence on the slitwidth as S_H and S_N (by our choice of the signal S_N^{48}), the ratio $S(\nu)/S_N$ gives precisely the ratio of the diatom and nitrogen scattering cross section times the (known) diatom densities over the hydrogen density. From this ratio, the diatom scattering cross section is obtained in a straightforward manner.

For the purpose of calibration, several investigators ingeniously mix a small amount of hydrogen with the gas under study and measure the continuum and line intensities by simply tuning the monochromator from the low-frequency CIS spectrum to the first rotational line near 350 cm^{-1}. Others prefer two completely separate steps of measuring $S(\nu)$ and S_N independently, which requires somewhat more care not to disturb the alignment. However, at not too high pressures, the refractive index, and thus the focal point location and dimensions, is not affected much, and this procedure is fully justified. At the same time, a better density measurement is obtained. More significantly, nitrogen can be used with its many low-frequency rotational lines, making possible calibration directly at a frequency of the collision-induced spectrum, and not ~350 cm^{-1} away from it.

The direct calibration must be done at a frequency shift at which no interference (due to Rayleigh wing intensities, or the intercollisional process) is possible, that is, at frequencies that are not too low. A great many calibrations were thus undertaken in our laboratory, using different laser intensities, slitwidths, pressures, standard lines, and so on, with the results being consistent in every case. The accuracy obtainable is very nearly that of the standard (8% for nitrogen; slightly better for the hydrogen calibration). Diatom polarizabilities can thus be measured with an absolute precision of ±4% or better, since spectroscopic intensities depend on the squares of these quantities.

C. Wave-mechanical Theory of the Lineshape

Collision-induced Raman spectra can be computed from a wave-mechanical[48, 68] and a classical theory,[1, 2, 48, 59, 62] if the interaction potential is known, and models of the diatom trace and anisotropy. On account of the importance of such calculations, we review these theories here. For realistic potentials,[37] digital computers and more or less involved numerical methods must be used for the computation of such lineshapes. Nevertheless, rigorous computer codes have long been used successfully and it is felt that future work in this field should be based on the comparison of computed wave-mechanical and spectroscopic lineshapes. This suggestion is based on the fact that often empirical, "improved" models of the anisotropy obtained by the method of moments (which attempts to avoid the discussion of the lineshape completely, using instead only two or three moments of the spectral distribution function) have been shown to be *inconsistent* with the measured spectral distribution, on whose lowest moments the models were founded![59, 60] Spectral moments have to be used with care, must be known with precision to be useful and, probably, must be known to higher orders. The consideration of the spectral distribution directly appears to be indispensable.

Classical electrodynamics gives the power radiated by an induced dipole, $(\alpha_{jk})\cdot\mathbf{F}(t)$, per unit solid angle, of polarization $\boldsymbol{\varepsilon}$[69, 70] as

$$\frac{dI_\varepsilon}{d\Omega} = k^4 I_0\{\boldsymbol{\varepsilon}\cdot[\mathbf{n}\times(\mathbf{n}\times\{\alpha_{xz};\alpha_{yz};\alpha_{zz}\})]\}^2 \tag{46}$$

$$= \left(\frac{d\sigma_\varepsilon}{d\Omega}\right) I_0 \tag{47}$$

Equations 46 and 47 define the differential Raman scattering cross section. Here $\mathbf{n}$ and $\boldsymbol{\varepsilon}$ are unit vectors in the direction of observation and of the polarization of the scattered signal, respectively. If unpolarized detectors are used (i.e., most of the time), the scattered power is the sum of two such signals with orthogonal polarization, $\boldsymbol{\varepsilon}_1 \perp \boldsymbol{\varepsilon}_2$. We have assumed a linearly polarized incident beam, $\mathbf{F}=(0;0;F_0)\cos\omega_0 t$. The incident power density, I_0, is related to the time-averaged field according to $I_0 = cF_0^2/8\pi$, here c is the speed of light; $\{\alpha_{xz};\alpha_{yz};\alpha_{zz}\}$ is the third column of the polarizability tensor, α_{jk}, expressed in the laboratory frame, with $j, k = x, y, z$. It is related to the tensor in the molecular frame (5) by the direction cosines, Φ_{jk}, according to[70]

$$\alpha_{jk} = \sum_{i,l} \Phi_{ji}\alpha_{il}\Phi_{kl} \tag{48}$$

with $\alpha_{11}=\alpha_\parallel$; $\alpha_{22}=\alpha_{33}=\alpha_\perp$; $\alpha_{il}=0$ for $i\neq l$, for the diatom. Averaging the α_{jk}^2 over all Euler angles has the result

$$\alpha_{xz}^2 = \alpha_{yz}^2 = \tfrac{3}{45}\cdot\gamma^2 \tag{49}$$

$$\alpha_{zz}^2 = a^2 + \tfrac{4}{45}\gamma^2 \tag{50}$$

with a = (one third of the) trace and γ = anisotropy [(14) and (15)]. Using this result in (46), for unpolarized detectors and the usual 90°-scattering geometry, we thus obtain the familiar classical relationship for the spatially averaged scattering cross sections of an induced dipole[70]

$$\frac{d\sigma_\parallel}{d\Omega} = k^4\cdot\tfrac{6}{45}\gamma^2 \tag{51}$$

$$\frac{d\sigma_\perp}{d\Omega} = k^4\left(a^2 + \tfrac{7}{45}\cdot\gamma^2\right) \tag{52}$$

The subscripts $\|, \perp$ refer to the incident beam polarization being parallel or perpendicular to the direction of observation.

On the basis of Kramers and Heisenberg's formula,[71] wave-mechanical expressions analogous to the above can be obtained,[65, 70, 72, 73] by replacing the squares of the invariants by the square of their matrix elements (the so-called Placzek's polarizability approximation):

$$|(a)_{n'J'}^{nJ}|^2 = \left| \int_0^\infty \psi^*(r; n', J') a(r) \cdot \psi(r; n, J)\, dr \right|^2 \tag{53}$$

$$|(\gamma)_{n'J'}^{nJ}|^2 = b_{J'}^J \left| \int_0^\infty \psi^*(r; n', J') \gamma(r) \cdot \psi(r; n, J)\, dr \right|^2 \tag{54}$$

The $b_{J'}^J$ are given by (43) to (45). Selection rules are $J' = J$ for the trace, and $J' = J, J \pm 2$ for the anisotropy component. The radial nuclear wavefunctions $\psi(r; n, J)$ are solutions of the Schrödinger equation of the relative motion of the atoms; the irrelevant center-of-mass coordinates have been suppressed. An asterisk indicates the conjugate complex. One can, however, use real ψ's and the asterisks can usually be ignored. The radial Schrödinger equation is given by

$$-\left(\frac{\hbar^2}{2\mu}\right)\psi'' + \left[V(r) + \hbar^2 \cdot J \cdot \frac{(J+1)}{2\mu r^2} - E_{nJ} \right]\psi = 0 \tag{55}$$

with μ = reduced mass of the diatom. For bound states, the normalization condition is the usual

$$\int_0^\infty \psi^* \psi\, dr = 1 \tag{56}$$

For diatoms in collisional interaction, the quantum number n becomes continuous and is best replaced by the energy E itself. In this case, the normalization condition makes use of Dirac's delta function:

$$\int_0^\infty \psi^* \psi\, dr = \delta(E' - E) \tag{57}$$

("energy-density normalization"). In other words, asymptotically (for $r \to \infty$) the free-state wave function must approach the form

$$\left[\frac{2\mu}{(\pi \hbar^2 \tilde{k})} \right]^{1/2} \cdot \sin\left(\tilde{k} r - \frac{\pi J}{2} + \eta_J \right) \tag{58}$$

where $\hbar^2\tilde{k}^2 = 2\mu E$ and η_J is the phase shift for elastic scattering. The frequency shift is given by

$$\nu = \frac{(E' - E)}{h} \tag{59}$$

with $h = 2\pi\hbar$ = Planck's constant.

For diatomic molecules, the combination of (51) to (54) defines the light scattering cross section for discrete Raman transitions from the bound state n, J to another bound state n', J'. Raman line intensities can be obtained if the cross sections are multiplied by the number of molecules irradiated and then by the probability P_J of finding a molecule in the state n, J:

$$P_J = g_J \cdot \frac{(2J+1)\exp(-\beta E_{nJ})}{Q(T)} \tag{60}$$

Here g_J designates the weight factor related to the nuclear spin ($g_J = 1$ for even J, and $g_J = 0$ for odd J, for most rare gases), and $Q(T)$ is the sum-over-bound states,

$$Q(T) = \sum_{n,J} g_J(2J+1)\exp(-\beta E_{nJ}) \tag{61}$$

In this way, the intensities of the Raman transitions of diatomic molecules are obtained,

$$\frac{dI_{\parallel}/d\Omega}{I_0} = \frac{d\sigma_{\parallel}}{d\Omega} \cdot N \cdot P_J \tag{62}$$

as in (42). We occasionally call the product of P_J and $d\sigma/d\Omega$ the "population adjusted" cross section.

For the case of collisionally interacting diatoms, continuous Raman spectra result. The differential scattering cross section now becomes a "cross section per unit bandwidth," according to

$$\frac{d\sigma}{d\Omega} \to \left(\frac{\partial^2\sigma}{\partial\Omega\,\partial\nu}\right) d\nu \tag{63}$$

Similarly, the matrix elements of trace and anisotropy [(53) and (54)], owing to the energy density normalization chosen (57), now acquire the energy differentials dE, dE' as factors. A frequency shift of ν (59) occurs in many different transitions, and we take the incoherent sum over all of these.

Specifically, for the computations of the continuum intensities at a shift ν, we replace dE' by $hc\,d\nu$, multiply the matrix elements by their relative populations (61), and integrate over dE while keeping the frequency shift constant (59). Then we sum over all allowed ΔJ, and sum over all partial waves (i.e. the angular momenta J). We also note that the sum-over-states must now include all free states:

$$Q(T)=Q_b(T)+\sum_J g_J(2J+1)\int \exp(-\beta E)\left(\frac{dn_J}{d\tilde{k}}\right)d\tilde{k} \tag{64}$$

The density of states, $dn_J/d\tilde{k}$, is given by the container radius R and the scattering phase shift η_J (with $\tilde{k}$ from $\hbar^2\tilde{k}^2=2\mu E$):

$$\frac{dn_J}{d\tilde{k}}=\frac{\left(R+d\eta_J/d\tilde{k}\right)}{\pi} \tag{65}$$

$Q_b(T)$ is the partition function of the bound states only (61). The derivative of the phase shift is usually small compared to R and can be neglected, except at sharp scattering resonances, which may be included in Q_b instead (which eventually can also be neglected). Under these conditions, we get the familiar sum-over-states of free diatoms:

$$Q(T)=\frac{(2I+1)^2\,v}{2\lambda_0^3} \tag{66}$$

The average deBroglie wavelength λ_0 of the relative motion is given by

$$\lambda_0^2=\frac{\beta h^2}{2\pi\mu} \tag{67}$$

The letter I designates here the nuclear spin quantum number, and v is the mole volume, which we transfer to the left-hand side of the "population adjusted" pair scattering cross section to give

$$\frac{v\,\partial^2\sigma_{\parallel}}{\partial\Omega\,\partial\tilde{\nu}}=k^4\cdot\tfrac{6}{45}\cdot G(\tilde{\nu}) \tag{68}$$

$$\frac{v\,\partial^2\sigma_{\perp}}{\partial\Omega\,\partial\tilde{\nu}}=k^4\left[A(\tilde{\nu})+\tfrac{7}{45}\cdot G(\tilde{\nu})\right] \tag{69}$$

where

$$A(\tilde{\nu})=\frac{2hc\lambda_0^3}{(2I+1)^2}\sum_J g_J(2J+1)\int_0^\infty |(a)_{E'J'}^{EJ}|^2\exp(-\beta E)\,dE \tag{70}$$

$$G(\tilde{\nu})=\frac{2hc\lambda_0^3}{(2I+1)^2}\sum_{J,\Delta J} g_J(2J+1)b_{J'}^J\int_0^\infty |(\gamma)_{E'J'}^{EJ}|^2\exp(-\beta E)\,dE \tag{71}$$

These expressions A and G can be readily calculated with the help of a digital computer if an intermolecular potential $V(r)$ and the trace and anisotropy as a function of r are given. Bandwidths are now measured in units of reciprocal centimeters. From these equations,[48] the scattered intensities per unit volume can be obtained by multiplication with $(N/v)^2/2$, where N/v is the density of monomers. (N monomers in the volume v correspond to $N(N-1)/2$ diatoms in the volume.) Intensities are most commonly computed in terms of scattered power per volume irradiated. In other words, in (62) one would replace N by N/v. Similarly, for collision-induced Raman spectra, one would multiply the population adjusted cross section $\partial^2\sigma/\partial\Omega\,\partial\tilde{\nu}$ by the density of diatoms $N^2/2v$ (but not by their number, $N^2/2$). This amounts to multiplying the quantity $v\,\partial^2\sigma/\partial\Omega\,\partial\tilde{\nu}$ (which is easily computed) by the well-defined $(N/v)^2/2$.

We mention that Raman transitions from a bound state to a free state, or from a free to a bound state, also contribute to the continuum intensities. It is a straightforward matter to also include these in our computed cross section. However, we refrain from writing these expressions because of their insignificance. The theoretical spectra shown later do fully account for such transitions, which were seen to contribute usually less than 2% of the total intensity at any given shift.

The Raman cross sections discussed above are all for scattered intensities (in units of power) per incident flux (power/area). If photon counting is employed, the factor k^4 that appears in all these expressions must be replaced by k^3k_0, where $k_0=2\pi$ over laser wavelength, and $k=2\pi$ over wavelength at the shift considered. In this way, "photon scattering cross sections" are obtained.

It is well known that most of the rare gases form dimers[74] (a notable exception is helium). As a consequence, resolved or unresolved molecular bands, given by (62), with $N=N_2=$ number of dimers in v, are superimposed with the collision-induced background. Whereas it is true that the integrated Raman intensities of the dimers are usually quite small (typically, 10%), on account of their small concentrations, it is also true that particularly at the low-frequency shifts, the dimer band envelopes are of a substantial intensity, in comparison with the collision-induced background.

Raman band envelopes of dimers were seen repeatedly.[48, 67, 68] The number of dimers, N_2, can be obtained from the law of mass action according to[74]

$$\frac{N_2}{N_1^2} = \frac{Q_b(T)}{Q(T)^2} \tag{72}$$

where N_1 is the number of monomers in the volume v, $Q_b(T)$ is the dimer partition function (61), and $Q(T)$ is given by (66). It is thus a straightforward matter to include the bound dimer spectra on an absolute scale in the computation of the diatom spectra.[48]. A thorough discussion, based on classical physics, of the role of dimers can be found elsewhere.[75]

D. Classical Theory of the Lineshape

Lineshapes of the collision-induced Raman process can be obtained on the basis of Newtonian mechanics.[1, 2, 48, 59, 62] The classical theory may be expected to provide good accounting of the physical processes involved, as long as the angular momenta contributing substantially to the observed intensities are much greater than $\hbar$ (correspondence principle). Classical computations take only a small fraction of the computer time required for wave-mechanical calculations of comparable precision. For these reasons, we present a brief sketch of the classical theory.[48]

Classical electrodynamics[69] gives the power radiated per unit solid angle due to the accelerating charge e as

$$\frac{dI}{d\Omega} = \left(\frac{e^2}{4\pi c^3}\right)\{\boldsymbol{\varepsilon}\cdot[\mathbf{n}\times(\mathbf{n}\times\dot{\mathbf{s}})]\}^2 \tag{73}$$

As above, the unit vectors $\boldsymbol{\varepsilon}$ and $\mathbf{n}$ give the directions of the polarization of the scattered intensity and the direction of observation, respectively. The acceleration $\dot{\mathbf{s}}$ (time derivative of the velocity vector $\mathbf{s}$) is replaced by the second time derivative of the dipole moment, $\mathbf{p}(t) = e\mathbf{r} = (\alpha_{ik})\mathbf{F}_0 \cos\omega_0 t$, to give $e\dot{\mathbf{s}} = d^2\mathbf{p}/dt^2$. For ordinary molecules, the motion is periodic and (73) converts directly to (46) above. However, for the aperiodic motion of the collisional encounter, by Parseval's theorem,[69]

$$\int_{-\infty}^{+\infty} \left| \int_{-\infty}^{+\infty} f(t) e^{-2\pi i\nu t}\, dt \right|^2 d\nu = \int_{-\infty}^{\infty} [f(t)]^2\, dt \tag{74}$$

with $[f(t)]^2$ given by (73), we find that the total energy of polarization $\boldsymbol{\varepsilon}$

radiated per solid angle, per collisional encounter, is given by

$$\frac{dE_\varepsilon}{d\Omega}=\int_{-\infty}^{+\infty}\frac{\partial^2 E_\varepsilon}{\partial\Omega\,\partial\nu}\,d\nu \tag{75}$$

where

$$\frac{\partial^2 E_\varepsilon}{\partial\Omega\,\partial\nu}=\frac{1}{4\pi c^3}\left|\int_{-\infty}^{+\infty}\boldsymbol{\varepsilon}\cdot\left[\mathbf{n}\times\left(\mathbf{n}\times\frac{d^2\mathbf{p}}{dt^2}\right)\right]e^{-i\omega t}\,dt\right|^2 \tag{76}$$

Since

$$\boldsymbol{\varepsilon}\cdot\left[\mathbf{n}\times\left(\mathbf{n}\times\frac{d^2\mathbf{p}}{dt^2}\right)\right]=\frac{d^2}{dt^2}\{\boldsymbol{\varepsilon}\cdot[\mathbf{n}\times(\mathbf{n}\times\mathbf{p})]\} \tag{77}$$

by the well-known property of Fourier transform, we get

$$\frac{(\partial^2 E_\varepsilon/\partial\Omega\,\partial\nu)}{I_0}=2k^4\left|\int_{-\infty}^{+\infty}\boldsymbol{\varepsilon}\cdot\left[\mathbf{n}\times(\mathbf{n}\times\{\alpha_{xz};\alpha_{yz};\alpha_{zz}\})\right]\cos\omega_0 t\,e^{-i\omega t}\,dt\right|^2 \tag{78}$$

At this point, the circular frequencies $\omega=2\pi\nu$ range from $-\infty$ to ∞. To deal with the physically more meaningful positive frequencies, we add to (78) an analogous term, in which ω is replaced by $-\omega$, rewrite $\cos\omega_0 t$ as $(e^{i\omega_0 t}+e^{-i\omega_0 t})/2$, and drop all terms with an exponent of $(\omega+\omega_0)$, which are physically irrelevant for the task at hand. In this way we get

$$\left(\frac{\partial^2 E_\varepsilon}{\partial\Omega\,\partial\nu}\right)I_0^{-1}=k^4\left|\int_{-\infty}^{+\infty}\boldsymbol{\varepsilon}\cdot\left[\mathbf{n}\times(\mathbf{n}\times\{\alpha_{xz};\alpha_{yz};\alpha_{zz}\})\right]e^{i(\omega_0-\omega)t}\,dt\right|^2 \tag{79}$$

as the final result. Only positive frequencies are to be considered. Equation 79 is the aperiodic analogue of (46) above; the units are, however, different (energy per encounter per incident flux). The classical scattered power per unit frequency band is obtained by multiplying (79) by the (*density of diatoms*)$\cdot s\cdot b\cdot db\cdot d\Phi$, after averaging over all Euler angles, describing the possible collision planes ($s=$ relative speed of the encounter, $b=$ impact parameter, $\Phi=$ azimuth).

One can show that the right-hand side of (79) can be written as a sum over three squares, which can be expressed as

$$\left|\int_{-\infty}^{+\infty}\alpha_{zz}(t)e^{-i\tilde{\omega}t}\,dt\right|^2=4\left[\int_0^{\infty}a(t)\cos\tilde{\omega}t\,dt\right]^2$$
$$+\tfrac{4}{15}\left[\int_0^{\infty}\gamma(t)\cos 2\Phi(t)\cos\tilde{\omega}t\,dt\right]^2$$
$$+\tfrac{4}{15}\left[\int_0^{\infty}\gamma(t)\sin 2\Phi(t)\sin\tilde{\omega}t\,dt\right]^2$$
$$+\tfrac{4}{15}\left[\int_0^{\infty}\gamma(t)\cos\tilde{\omega}t\,dt\right]^2 \tag{80}$$

with $\tilde{\omega}=\omega_0-\omega=2\pi$ times the frequency shift. Similarly, we get

$$\left|\int_{-\infty}^{+\infty}\alpha_{xz}(t)e^{-i\tilde{\omega}t}\,dt\right|^2=\tfrac{1}{15}\left[\int_0^{\infty}\gamma(t)\cos\tilde{\omega}t\,dt\right]^2$$
$$+\tfrac{1}{5}\left[\int_0^{\infty}\gamma(t)\cos 2\Phi(t)\cos\tilde{\omega}t\,dt\right]^2$$
$$+\tfrac{1}{5}\left[\int_0^{\infty}\gamma(t)\sin 2\Phi(t)\sin\tilde{\omega}t\,dt\right]^2 \tag{81}$$

and an identical expression if the subscript x (to the left) is replaced by y. The trajectory of a collision can be obtained in polar coordinates $r(t)$, $\Phi(t)$, if the potential $V(r)$ is known and if a value of the impact parameter b and of the relative speed s is chosen. The functions $a(t)$ and $\gamma(t)$ are trace and anisotropy, with $r=r(t)$.

Hence for a given impact parameter b and relative speed s and for an unpolarized detector, we have

$$\frac{v\,\partial^2\sigma_{\perp}}{\partial\Omega\,\partial\nu}=k^4\left[A(\nu)+\tfrac{7}{45}\Gamma(\nu)\right]2\pi sb\,db \tag{82}$$

$$\frac{v\,\partial^2\sigma_{\parallel}}{\partial\Omega\,\partial\nu}=k^4\tfrac{6}{45}\Gamma(\nu)2\pi sb\,db \tag{83}$$

where

$$A(\nu)=4\left[\int_0^\infty a(t)\cos(2\pi\nu t)\,dt\right]^2 \tag{84}$$

$$\Gamma(\nu)=1.5\left\{\int_0^\infty \gamma(t)\cos[2\Phi(t)+2\pi\nu t]\,dr\right\}^2+\left[\int_0^\infty \gamma(t)\cos(2\pi\nu t)\,dt\right]^2$$
$$+1.5\left\{\int_0^\infty \gamma(t)\cos[2\Phi(t)-2\pi\nu t]\,dt\right\}^2 \tag{85}$$

where $A(\nu)$ and $\Gamma(\nu)$ are functions of s and b. These expressions, (82) and (83), have to be integrated over all values of the impact parameter $b \geq 0$ and averaged over a Maxwellian speed distribution to obtain the classical analogue of (68) and (69). We note that with increasing b, the functions $A(\tilde{\nu})$ and $\Gamma(\tilde{\nu})$ go to zero very rapidly and no convergence problems are encountered.

We note that the classical formula above gives equal anti-Stokes and Stokes intensities, which is not correct. The experiments, as well as the wave-mechanical theory, indicate instead a Boltzmann factor for the ratio of anti-Stokes and Stokes intensities at any given shift. By artificially introducing the "principle of detailed balancing" into the classical treatment,[76] attempts have been made to compensate for this defect. Usually, one applies some function of the Boltzmann factor on the anti-Stokes side and its inverse on the Stokes side. However, different and inconsistent ways of introducing detailed balancing are being used in different laboratories. No rigorous justification can be given for any particular form of such corrections. Fortunately, this fact need not concern us here as long as we use the rigorous wave-mechanical theory.

E. Remarks Concerning the Lineshape Calculations

Classical and wave-mechanical formulae for the computation of lineshapes can both be readily computer programmed. For classical computations the integration over the impact parameter b is to be done with great caution. For a fixed relative speed that is not too large, for collisions with small impact parameters b, the polar angle Φ rotates through only a small fraction of the full circle, and a nearly exponential spectrum results. With increasing b, the changes of the polar angle Φ.increase to infinity at the critical value b_0. For such "orbiting collisions," the Fourier spectra approach monochromaticity at twice the frequency of rotation, giving rise to a totally different kind of spectrum. For even larger b, the spectra convert back to the exponential form and usually decrease rapidly in intensity.

Large numerical uncertainties may result from the integration over b, unless it is done in small enough steps near the critical b_0.[78] Another problem of the classical lineshape calculations concerns the high-frequency part of the spectrum, which is easily affected by rounding errors. At the high frequencies, very many terms with rapidly changing signs have to be summed in the Fourier transform. It is best to completely separate the time increments used for the integration of Newton's equation of motion from those used for the Fourier transforms, using spline interpolation for the latter. Furthermore, double precision may be called for on most computers for the Fourier transform, especially at high frequencies. A fourth-order Runge-Kutta scheme was adopted in our laboratories for the computations of the trajectories, with a carefully controlled "estimated local truncation error."[79] Classical lineshapes of typically better than 1% numerical precision require only about 200 sec of computer time on the Cyber 170/750 computer.

Wave-mechanical computations are efficient if, for example, a fourth-order Numerov method[80] is used for the integration of the radial Schrödinger equation (55) of the relative motion of the atoms.[81] Those scattering resonances at energies close to the peak of the centrifugal barrier affect the computed spectra much like the orbiting collisions of the classical treatment. It is therefore important to know for each gas, for each model potential used, the positions of these resonances; the integration over these resonance energies must be done most carefully, with small enough energy increments.[48] Wave-mechanical lineshapes for the heavier diatoms of krypton and xenon are prohibitively expensive, unless one computes only a small number of widely spaced partial waves and resorts to interpolations. In this way, the free-to-free transitions can be dealt with at energies well above the barrier, but not at the resonances, which must be treated individually. Wave-mechanical lineshapes for the depolarized helium spectrum require about 1000 sec of computer time for an overall numerical precision of better than 1%. For neon and argon, three to six times as much computer time is required, and much more is needed for krypton and xenon. Fortunately, the free-to-free transitions of the heavier rare gas diatoms can be obtained with precision from the classical formalism. Polarized Raman spectra involve fewer partial waves and are not nearly as expensive to compute.

F. Moments of Spectral Distributions

The mth moment of a spectral distribution $I(\tilde{\nu})$ [as in (68) or (69)] is given by an expression of the form

$$M_m = \int_{-\infty}^{+\infty} \tilde{\nu}^m I(\tilde{\nu})\, d\tilde{\nu} \tag{86}$$

The zeroth moment equals the total intensity of the spectrum.

On the basis of the classical theory, collision-induced spectra are symmetric. As a consequence, using classical theory, only the even moments are nonvanishing. In reality, however, collision-induced spectra are not symmetric. Rather, their anti-Stokes and Stokes intensities, at any given shift, differ by a Boltzmann factor. Nevertheless, numerous attempts have been made to work with even moments only, either by symmetrizing the real distributions in some way or else ignoring the asymmetry. In the past, for example, at shifts $\tilde{\nu}$, the square roots of the Boltzmann factors, $A(\tilde{\nu}) = \exp(-\beta\tilde{\nu}/2)$, with $1/\tilde{\beta} = \kappa T$ in proper frequency units, were computed, and Stokes intensities were multiplied and anti-Stokes intensities divided by this factor, thereby symmetrizing the experimental distribution function for the determination of classical moments.[82] This procedure is artificial and without a theoretical basis. Whereas one does get a symmetric distribution in this way, the resulting lineshapes are not consistent with the classical theory.[77]

Since collision-induced, *polarized* scattering of light was unknown until recently,[44–47], moments of such spectra found in the literature refer commonly to the depolarized part only. In that approximation, the zeroth moment was shown to be related to the second virial Kerr coefficient, (32), and to the thermodynamic average of the squared anisotropy (33). Classical physics has also been used to relate the moments to certain integrals over the anisotropy,[2, 83, 84] but quantum "corrections" have been reported.[85] For the helium diatom, rigorous *ab initio* computations of several moments at cryogenic and room temperatures were communicated.[40]

Moments can be determined from measurements of the spectral distribution. However, as we mention on several occasions, the determination of, particularly, the zeroth moment suffers from two interfering influences, which limit severely the accuracy attainable. (*1*) The Rayleigh line,[49, 50] which is typically from 5 to 7 orders of magnitude more intense than the collision-induced spectra, masks the low-frequency portion of the latter completely over a band of about ± 5 slitwidths.[51] (*2*) The intercollisional effect modifies the low-frequency collision-induced spectrum, possibly by as much as 25% at low frequencies, over a band of ± 1 collision frequency.[52] This band amounts to ± 20 cm^{-1} at 100 atm and proportionally more at higher densities. As a consequence, the low-frequency portion of the spectra, where the collision-induced spectra are most intense and contribute the most to the zeroth moment, is in general not known very well from experiment. More or less dubious extrapolation schemes must be used for an estimate of the unknown intensities, which are so critical for the determination of the zeroth moment. The most commonly employed exponential extrapolation scheme was found to be wrong for all gases for which wave-mechanical lineshapes were computed. Such computation in fact suggests that exponential extrapolation overestimates the zeroth moment typically by as much as 10 or 20%. As far as the higher moments are concerned, the high-frequency cutoff poses another limit to the precision of determining moments.

Knowledge of all the moments M_m (for $0 \leq m \leq \infty$) is theoretically equivalent to the complete knowledge of the distribution function, and vice versa. Usually, however, only two or three of the lowest-order moments are assumed to characterize the distribution function sufficiently. Higher moments have rarely been used. Whereas this may be justified for Gaussian or simple exponential distributions, or when moments are known with little uncertainty, we feel that there is alarming evidence that this is totally inadequate for the kind of distribution functions and/or experimental uncertainties that are of interest here. It has been shown repeatedly that empirical models of the anisotropy, which were chosen in a way to render the theoretical lowest-order moments consistent with their experimentally determined moments, usually do *not* reproduce the lineshape.[59, 60, 67] Thus moments must be used with caution, or be avoided altogether, for the measurement of diatom polarizabilities. Such work should instead be based on the full information available from the lineshapes directly and need not suffer form any of the shortcomings mentioned.

IV. DIATOM POLARIZABILITY DATA

A. Results for Helium

Obviously, a most interesting case for the study of collision-induced scattering of light is helium, since for this gas *ab initio* computations of the diatom polarizabilities exist. However, theoretical estimates of the intensities of the collision-induced helium spectra indicate that these are very weak, roughly 4 orders of magnitude weaker than the analogous, not very intense, argon spectrum. Consequently, it was often believed that useful helium spectra are not likely to be measurable. Recently, however, within a time span of only 6 months, three such measurements were reported independently in the literature.[44, 45, 58, 86–88] Photon counting techniques with gate times of hours were employed. Highly purified gases were used to prevent interference from the more highly polarizable impurities. Typical impurity levels should be below 1 ppm if pressures near 30 atm are used; at 300 atm, the impurities should be even less (0.1 ppm).

One experiment[45] makes use of a half-wave plate in the laser beam to excite spectra with beam polarizations parallel and perpendicular to the direction of observation. Figure 5 shows spectra thus obtained. At 30 amagat, signals from 0.01 to 1 counts/sec are obtained after subtraction of the dark count (~0.3 counts/sec). Measured intensities could be calibrated and put

TABLE III

Wave-mechanical Computation of the Collison-Induced Spectra of Helium (Stokes Side) and Derived Best Diatom Polarizability Model.

a) Spectra at 4880 Å excitation:

	shifts (cm^{-1})							
	−12	−20	−28	−50	−100	−170	−250	−350
$D_{\parallel}(cm^6)$	6.43(−57)	5.80(−57)	5.07(−57)	3.32(−57)	1.26(−57)	3.16(−58)	0.78(−58)	1.58(−59)
$P_{\perp}(cm^6)$	2.00(−58)	2.00(−58)	2.04(−58)	2.09(−58)	1.94(−58)	1.53(−58)	1.05(−58)	6.12(−59)

b) Trace: (26) with $A_6 = 39a_0^9$; $\lambda_t = 70a_0^3$; $r_t = 0.55a_0$. Determined for $1.9\,\text{Å} \le r \le 2.6\,\text{Å}$.

c) Anisotropy: (27) with $\alpha_0 = 0.207\,\text{Å}^3$ (from Ref. 97); $\lambda_a = 39a_0^3$; $r_a = 0.6079a_0$. For $2.2\,\text{Å} \le r \le 5\,\text{Å}$.

d) $B_\epsilon = -0.092\ cm^6/mol^2$ at 49 °C; $\langle a^2 \rangle = 3.34 \times 10^{-76}\ cm^9 \pm 20\%$; $\langle \gamma^2 \rangle = 2.18 \times 10^{-74}\ cm^9 \pm 8\%$

Also listed are the resulting second virial coefficients and thermodynamic averages (33) of trace and anisotropy, using the potential by Aziz et al.[42]. Note that a rage of r_t-values, from about $0.55a_0$ to $.70a_0$, will also give acceptable lineshapes provided an associated λ_t is chosen such that $\langle a^2 \rangle$ remains constant as specified above. Similarly, r_a may vary from roughly $0.58a_0$ to $0.90a_0$ and resulting lineshapes will still fit the experimental spectra, provided a λ_a is chosen such that $\langle \gamma^2 \rangle$ remains constant as specified.

Note added in proof: Since the completion of this chapter new work indicated that our earlier helium calibration[45] was in error, owing to the previously unaccounted for monochromator leadscrew error. We were, however, able to correct the data for this review. We also mention that new measurements of the ^3He diatom spectra recently obtained in our laboratory support the above conclusions concerning the helium diatom polarizabilities (to be published). Furthermore, we note that the other measurements of the depolarized helium spectra[86, 88] were now shown to be affected by many-body contributions[87]. A two-body spectrum could be extracted[87], which is in a much better (although not perfect) agreement with our work, and with Ref. 88. (See F. Barocchi, M. Zoppi, talk presented at the International Conference on Collision-Induced Phenomena, held in Florence/Italy from Sept. 2–5, 1980. Conference proceedings will be published in a 1981 issue of the Canadian Journal of Physics.)

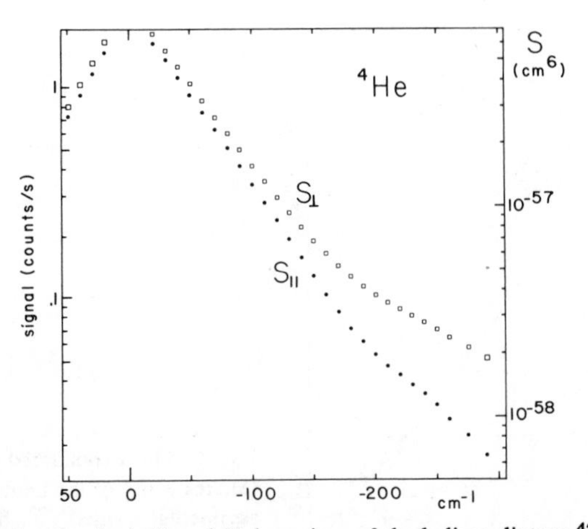

Fig. 5. The anti-Stokes (to the left) and Stokes wings of the helium diatom,[45, 77] taken at ≈30 amagat, with the polarization of the incident beam parallel ($S_{\|}$, ●) and perpendicular ($S_{\perp}$, ○) to the direction of observation. An absolute intensity scale is given to the right, which is explained in text near (68) to (71). From these signals $S_{\|}$, $S_{\perp}$, the polarized and depolarized Raman spectra are obtained, using (39) and (40).

on an absolute scale for both spectra. Tests showed that the intensities varied as the square of the gas density, demonstrating the diatomic origin. Furthermore, the ratios of the anti-Stokes and Stokes intensities were consistent with the usual Boltzmann factor, as required for any Raman spectrum. These experimental tests support the interpretation of these spectra as true binary collision-induced Raman spectra.

As is pointed out above (Section III.A), a parallel beam polarization induces an almost pure depolarized spectrum, whereas perpendicular beam polarization generates a superposition of a polarized and a depolarized component at the detector. Since the two spectra $S_{\|}$, $S_{\perp}$ of Fig. 5 do not agree, a non-negligible polarized component is seen to exist that can be separated (along with the depolarized component) on the basis of (39) and (40). The polarized spectrum, P, is plotted in Fig. 3, and Figs. 4 and 6 present the depolarized spectrum, D. We note that since our work in helium was published[44,45], a small experimental error (≈10%) due to the leadscrew of our monochromator was discovered, which affected mainly the calibration of the absolute intensities. The data presented here in figures 3 through 6 are new data and fully corrected. A detailed discussion of the revisions is in preparation[77].

Figure 3 presents the first collision-induced polarized spectrum ever seen in any gas.[44, 45] The error bars are relatively large, because error is obtained as the difference $S_{\perp} - S_{\|}$ of two signals S, which are nearly the same

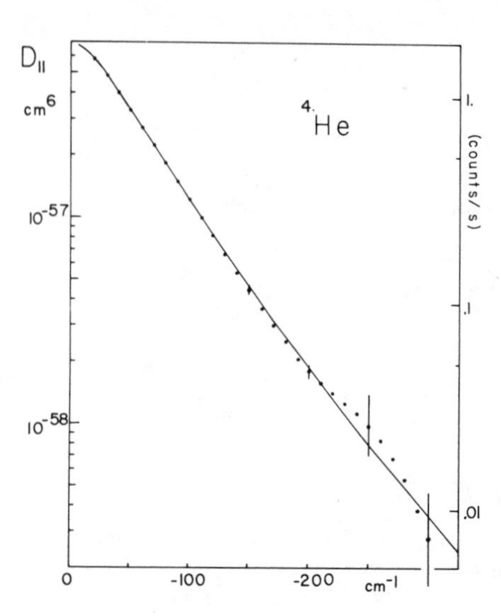

Fig. 6. The depolarized Raman spectrum (Stokes wing) of the helium diatom. (●) experimental results[45, 77]; best theoretical fit, based on the anisotropy model, (27), with $r_a = 0.608a_0$, $\lambda_a = 39a_0{}^3$.

(39). Wave-mechanical lineshapes are computed on the basis of the *ab initio* trace computations[18–21] and (68) to (71). The most accurate of the currently available helium interaction potentials are employed in these calculations,[42, 84] with no discernible differences in the computational results. (If, however, the older and less accurate Lennard-Jones 6-12 potential functions, a critical selection of which was compiled in Ref. 89, are substituted for the more refined potentials[41, 42, 84] frequency-dependent deviations from 6 to almost 20% are observed. The old Lennard-Jones potentials should not be used for the computation of the collision-induced spectra.)

The results displayed in Fig. 3 suggest that the *ab initio* computations of the trace due to Dacre[20], Fortune and Certain,[19] and Kress and Kozak[21] are all consistent with the polarized spectrum,[44, 45] with best fits given by the data of Ref. 19 and, maybe, of Ref. 20. From Table I, we see that Kress and Kozak's model requires a second dielectric virial coefficient of -0.093 cm^6/mol^2; Fortune and Certain's trace requires $B_\varepsilon = -0.094$ cm^6/mol^2 for either potential.[42, 84] We recall that these are consistent with Orcutt and Cole's[35] measurement of $B_\varepsilon = -0.06 \pm 0.04$ cm^6/mol^2, as well as with Vidal and Lallemand's[36] $B_\varepsilon = -0.11 \pm 0.02$ cm^6/mol^2. If the trace constant λ_t derived from Dacre's work is increased by about 20% for a better fit of the experimental polarized spectrum, a second dielectric virial coefficient results which is also consistent with the measurements[35,36].

On the basis of the data considered, and if the analytical model, (26), of the trace is accepted, the characteristic falloff length r_t is seen to be likely somewhere near $0.55a_0 \pm .06a_0$, as required by the fundamental theory[20, 27] (see the fifth column of Table I). Interestingly, the recent computations of the trace by Dacre[20] appear to be slightly inconsistent with the polarized spectrum.[41, 42] The computed polarized spectrum based on the trace of Ref. 20 is somewhat less intense at the lower frequencies. Similarly, the semiclassical theory[27] results in an inconsistent λ_t value, which is too small.

We note that the asymptotic expression of the trace[11] not only leads to unacceptable computed spectra (not shown), but also requires a B_ε of $+0.012\ \text{cm}^6/\text{mol}^2$, which is inconsistent with the measurement.[36] The exponential term of (26) is significant.

We remind the reader that we are comparing measured *dynamic* polarizabilities with computed *static* data. At present, it is not known whether a relationship like (26) holds for the dynamic trace, and what the frequency dependence of the coefficients A_6, r_t and λ_t would be. Our (possibly poorly supported) assumption is that the difference between static and dynamic helium polarizabilities amounts to a few percent only. A wave-mechanical theory of interacting dipoles suggests such a small frequency dependence.[61]

Turning our attention next to the depolarized spectrum of helium, Fig. 4, we note that the asymptotic expression of the anisotropy[11] gives rise to computed spectra, which are only slightly (by ~25%) more intense than the observed spectrum.[44, 45] This observation is in a striking contrast to the trace spectrum, where the asymptotic expression of the trace led to rather unrealistic spectra, and corroborates a remark made above concerning the variation of the atomic polarizabilities of colliding atoms, which nearly cancels in the case of the anisotropy, but generates very substantial changes of the trace [see the discussion following (9)].

We note that for helium, for all physically accessible internuclear separations ($r \geq \sigma$), the asymptotic anisotropy is very nearly in agreement with the DID model. As a consequence, the depolarized spectrum one computes for the DID model (not given) is practically indistinguishable from the curve marked CF in Fig. 4.

The *ab initio* calculations of the anisotropy give rise to the computed spectra shown in Fig. 4 (labels FC[19]; D[20]; KK[21]). The work of O'Brien et al.[18] results in a depolarized spectrum that is everywhere about 4% less intense than the curve labeled D and is not given in the figure on account of this near-identity. Similarly, the anisotropy of Clarke et al.[27] is nearly indistinguishable from the DID model and hence its spectrum practically coincides with the one labeled CF. We conclude that the *ab initio* anisotropy computations all slightly *over*estimate the deviations from DID, whereas the semiclassical theory[27] *under*estimates this. In other words, the exponential correction term (27), appears to be slightly less pronounced than the fundamental theory predicts.

In Fig. 6, we show a "best fit" based on the analytical model of the anisotropy, (27), which we obtain by adjusting the λ_a values of the Table I for different r_a values. The curve of Fig. 6 is obtained using Fortune and Certain's r_a value[19] of $0.6079a_0$, but a λ_a of $39a_0^3$ instead of their $42.5a_0^3$. This gives rise to a perfect fit of the spectroscopic data.

A quick look at the eighth column of Table I shows that the four *ab initio* computations of the anisotropy do not agree very closely among themselves as far as the r_a value is concerned. The smallest r_a is the one just mentioned. The r_a values of O'Brien et al. and Dacre are intermediate (0.74 and $0.71a_0$, respectively), and Kress and Kozak's value is the largest ($0.90a_0$). We have also computed spectra with an intermediate $r_a = 0.725a_0$, and the large $r_a = 0.90a_0$ (not shown). Particularly the intermediate $r_a = 0.725a_0$ gives rise to a quite acceptable fit of the experimental data. The large r_a value is still marginally acceptable. We see that the experimental data available do not at present allow us to determine the r_a value any closer than the various theories: all these values are consistent with the experiment. However, if a r_a value is once chosen, the fit of the experimental spectra requires a certain λ_a, which we found can be obtained from the condition $\langle\gamma^2\rangle = 0.022$ Å$^9 \pm 8\%$ (33) at room temperature. This applies without doubt even to slightly larger or smaller r_a than we have used here. But whatever r_a is chosen, the resulting associated λ_a appear to be substantially smaller (but non-negligible) than the fundamental theory predicts. The theoretical anisotropies are at present all too small. Again, we remark that we compare static theoretical with dynamic experimental polarizabilities as if there was not much difference. More theoretical work is to be done to more accurately estimate the differences between static and dynamic diatom polarizabilities.

An explanatory word concerning the analytical models of the diatom polarizabilities, (26) and (27), which we use in the above discussions nearly exclusively, may be in order. Several of the λ_a, r_a, λ_t, r_t parameters given in Table I are our own inferences from the work shown, which originally consists of a numerical table of polarizabilities at a number of internuclear separations. By no means do we want to imply that models like (26) and (27) are necessary for the computation of the spectra. In fact, we also have used the numerical data in the form of a table and third-order spline interpolation[90] [supplemented by the asymptotic forms, (21) and (25)] for the computation of the spectra, with no discernible difference in the results. It is a straightforward matter to show that, for the case of the helium polarized spectrum, 86% of the total observed intensity is due to the values of the trace over a range of internuclear separations from only $r = 3.6a_0$ to $4.9a_0$. Similarly, for the anisotropy, the significant range of internuclear separations was seen to be from $r \cong 4.2a_0$ to $12a_0$. Accordingly, the analytical models of Table I were chosen in such a way that particularly for these r values a close fit of the diatom polarizability invariants is obtained. Some *ab initio* polarizability data that are outside this range may not be very well represented by such simple analytical forms, (26) and (27); however, this does not affect the computational results in any significant way. The analytical models describe the numerical data with precision, usually even for r values outside the specified ranges.

Since experimental data useful for the determination of the helium diatom polarizability are scarce and mostly of recent origin, it is important to attempt a critical comparison of such data. It is pointed out above that the trace can be measured (a) from the polarized, collision-induced, Raman spectrum, and (b) from the virial expansions of the dielectric Clausius-Mosotti function. Only one measurement of the polarized Raman spectrum of helium[44, 45] exists at present, but the trace models inferred form that work are shown above to be associated with values of the second dielectric virial coefficient in precise agreement with the two existing measurements[35, 36] of B_ε. We also mention above three recent measurements of depolarized Raman spectra of helium that can be used to determine the anisotropy. One of these,[45, 58] taken at 30 amagats, was shown to be free from three-body interference and is reproduced in Figs. 4 to 6. The other two experiments were conducted at much higher densities (from 60 to 255 amagats,[88] and from 60 to 350 amagats,[86, 87] respectively). In these works, three-body contributions were observed and evaluation procedures had to be employed to artificially separate a more or less well defined diatom spectrum from the measurements. LeDuff's published, uncorrected spectrum,[88] taken at 217 amagats and somewhat affected by three-body processes, nearly agrees with the diatom spectra reproduced in Figs. 4 and 6 on an absolute intensity scale. We note that the reciprocal logarithmic slope, measured at 35 cm^{-1} on the Stokes side, amounts to 55 cm^{-1} for the data from Ref. 45, which is in agreement with the extrapolations of the slopes of Ref. 88 to low gas densities (see Fig. 6 of Ref. 88). (Similarly, agreement is observed for the slopes on the anti-Stokes side.) The agreement of the experiments with the other spectrum[86] is at present less satisfactory for some unknown reason.

We note that new work with the rare isotope of helium, ^3He, is now also feasible and most desirable. The common isotope, ^4He, has zero nuclear spin. Consequently, the partial waves of the ^4He diatom occur with even angular momentum quantum numbers only; odd-numbered partial waves are forbidden by symmetry considerations. The rare isotope, on the other hand, features half-integer spin. Accordingly, the odd-numbered partial waves of the ^3He diatom each occur with a statistical weight of three times that of the even-numbered ones, much like the angular momentum states of the ortho and para modifications of molecular hydrogen. This gives rise to rather different collision-induced spectra of the isotopes, even though the interaction potentials, as well as the diatom polarizabilities, are essentially without isotopic distinction. (At the same time, the different mass will further modify the observable spectra in the usual ways.) If measurements of the collision-induced spectra of the rare isotope diatom were obtained, a totally independent, new experimental test of the conclusions presented in this chapter would become available. Such measurements are presently underway in our laboratory.

1. Summary

The results concerning the polarizability of the helium diatom can be summed up as follows (consult Table I for references). For the narrow range of internuclear separations that is significant for the experiments described, we have no difficulties defining analytical models for trace and anisotropy, (26) and (27). These describe the computed data with precision and show the proper asymptotic behavior. Three of the four *ab initio* computations indicate for the trace a characteristic falloff length r_t of between $0.67a_0$ and $0.70a_0$. The experiment is consistent with this value, but gives preference to the smaller $r_a = 0.55a_0$ implied by both Dacre and Clarke et al.[20,27]. Measurements of the dielectric virial coefficient B_ε and the absolute intensity calibration of the spectrum support these *ab initio* trace data. The trace of the semiempirical computations is inconsistent with the observed polarized spectrum. The currently best trace data, over a range of internuclear separations from about $3.6a_0$ to $4.9a_0$, are given by the entries marked D (but with $\lambda_t = 75a_0^3$ and FC in Table I. As far as the anisotropy is concerned, it is seen above that a much broader range of characteristic lengths r_a results from the computations. The resulting spectral lineshapes do, however, barely reflect these differences, owing to the near insignificance of the exponential correction term, that is, to the smallness of λ_a. The fundamental theory is seen to slightly overestimate the effect of overlapping wavefunctions on the anisotropy, by about as much as the semiclassical theory underestimates it. The currently best anisotropy model, over a range of internuclear separations from $4.2a_0$ to $10a_0$, is given by r_a values between about $0.6a_0$ and $0.9a_0$, and a λ_a value chosen such that the thermodynamic average of the squared anisotropy (33) is given by $\langle \gamma^2 \rangle = 0.022$ Å$^9 \pm 8\%$, at $T = 27°$C and for realistic potentials.[42] (The old Lennard-Jones potentials should be avoided.)

B. Results for Neon

Collision-induced Raman spectra of neon were reported from two laboratories.[47, 91] The spectra are only slightly more intense than their helium counterparts, and a special sample cell with internal focusing mirrors was designed as a part of the laser cavity to enhance the weak signals.[91] Gas densities of 30 amagats,[47] and up to 60 amagats[91] were used. Three-body contributions were shown to amount on the average to not more than -5% at the higher density[91] and even less at lower densities.[47] Intensities were seen to be proportional to the square of the gas density, indicating the diatomic origin. Ratios of anti-Stokes and Stokes intensities are consistent with the Boltzmann factor. Signals amounted to ~10 counts/sec or less. Beam polarizations parallel and perpendicular to the direction of observation were employed.[46, 47] The resulting spectra (Fig. 7) are seen to be dissimilar, which is attributed to the presence of a polarized component.[46, 47] On the basis of (39) and (40), the polarized (P) and depolarized (D) components can be separated (see Figs. 8 and 9). Absolute intensity calibrations have been made for all neon spectra.[47, 91]

It is very satisfying to note that the lineshapes of the two different depolarized measurements[47, 91] match where the data overlap (i.e., for shifts of less than 85 cm^{-1}). This is most readily seen when we compare the reciprocal logarithmic slopes on the Stokes sides of the depolarized spectra: one work[47] reports 18.5 cm^{-1} and the other[91] 18.8 cm^{-1}. The absolute intensity scales, which were obtained in one work[46, 47] by a direct calibration of the continuum intensities at a given shift, and in the other[88] via the integrated intensity, appear to be also consistent as is shown below.

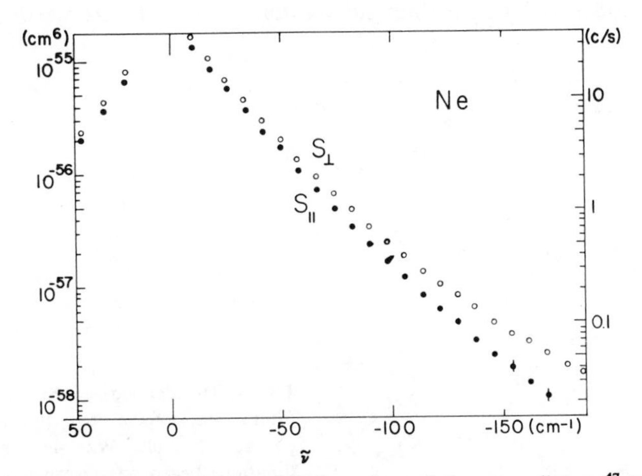

Fig. 7. The anti-Stokes (to the left) and Stokes wings of the neon diatom,[47] taken at ≈30 amagat, with the polarization of the incident beam parallel ($S_{\parallel}$, ●) and perpendicular ($S_{\perp}$, O) to the direction of observation. An absolute intensity scale is given to the left, which is explained in text near (68) to (71). From these signals $S_{\parallel}$, $S_{\perp}$, the polarized and depolarized Raman spectra are obtained using (39) and (40).

Wave-mechanical computations of the lineshape are based as usual on (68) to (71). The so-called MSV potential by Siska et al.[92] is used throughout, but other refined potentials[93] were also considered for trial purposes, with no discernible difference in the end results. We note once more that the use of the old Lennard-Jones 6-12 potentials, which we took from a critical compilation,[89] resulted in spectra that differed from the ones given here by substantial amounts, up to 12%. We consider, therefore, these older potential functions as too inaccurate for our purpose and do not recommend their use.

For neon, bound van der Waals dimers exist: $^{20}Ne_2$, $^{20}Ne^{22}Ne$, and $^{22}Ne_2$. However, their weak spectral contributions consist of the pure rotational transitions[75, 94–96] at frequency shifts below $10 cm^{-1}$, which we ignore. Bound-to-free transitions and free-to-bound transitions are accounted for and are seen to be negligible.

The DID model of the anisotropy gives rise to a spectrum like the curve labeled DID in Fig. 8. (We take the atomic polarizability $\alpha_0 = 0.3992 \times 10^{-24}$ cm^3 at a 4880 Å wavelength, from Ref. 97). At the low frequencies, the deviations relative to the experiment are seen to be quite small, of the order of 10%, but at the higher frequency shifts the depolarized DID spectrum shows too much concavity and deviates increasingly from the experiment. If the asymptotic form of the trace, $a(r) = A_6/r^6$ with $A_6 = 224\ a_0^3$, is used for a wave-mechanical computation of the polarized neon spectrum, curve A of Fig. 9 is obtained. The asymptotic trace does not appear to be a suitable trace model by itself. Rather substantial corrections are needed at close

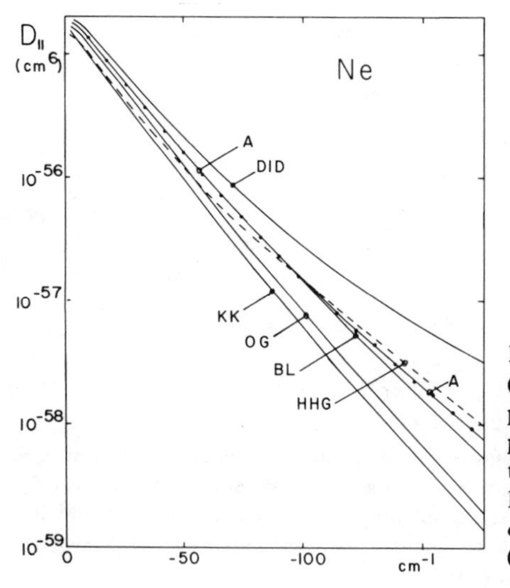

Fig. 8. The depolarized Raman spectrum (Stokes side) of the neon diatom[47]. (●) experimental results. Wave-mechanical computations based on the various models of the anisotropy: DID (19); KK[21]; OG[24]; BL[91]; HHG.[26] Curve A is a best fit, based on the anisotropy model, (27), with $r_a = 0.509 a_0$, $\lambda_a = 900 a_0^3$.

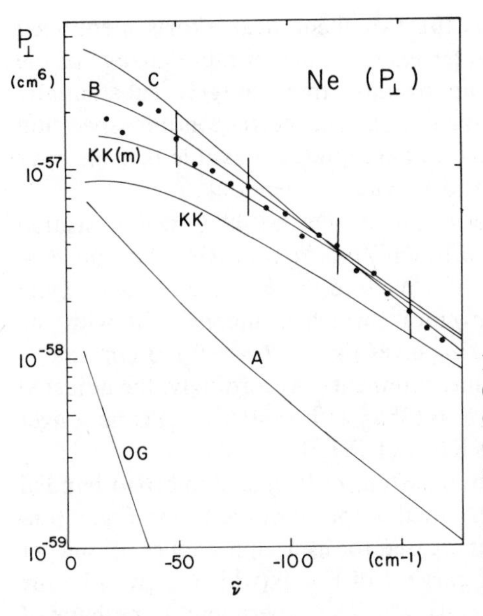

Fig. 9. The polarized Raman spectrum (Stokes side) of the neon diatom[47]. (●) experimental results. Wave-mechanical computations based on the various trace models are also given: OG[24]; KK[21]; KK(m): $\lambda_t = 135 a_0{}^3$, $r_t = 0.626 a_0$; (*A*) asymptotic form ($\lambda_t = 0$); (*B*) $r_t = 0.79 a_0$, $\lambda_t = 65 a_0{}^3$; (*C*) $r_t = 0.80 a_0$, $\lambda_t = 30 a_0{}^3$. [Parameters r_t, λ_t are defined in (26).]

range to account for electronic overlap, just as is seen above for the helium diatom.

Neon diatom polarizabilities were computed from first principles by Kress and Kozak.[21] However, for some time it was known that the computed anisotropy data[21] are not consistent with the depolarized neon spectrum[91] (see the curve KK of Fig. 8). Similarly, the computed trace data[21] are not quite consistent with recent measurements of the polarized spectrum[47] (curve KK of Fig. 9). The observed disagreements are, however, not much more severe than those seen above for the *ab initio* helium work (Figs. 3 and 4), and small "adjustments" (described below) of the analytical models [(26) or (27)] based on the data[21] do lead to acceptable lineshapes. Kress and Kozak's trace, for example, can be fitted to the analytical expression (26) in the critical region of internuclear separations from about $4.5a_0$ to $6.5a_0$, using $A_6 = 224a_0^9$ (from Refs. 12 and 24), $\lambda_t = 115a_0^3$ and $r_t = 0.626a_0$. (The lineshape associated with this analytical model is virtually identical with curve KK of Fig. 9). Increasing λ_t by only 17% to $\lambda_t = 135a_0^3$ leads to curve KK(m) of Fig. 9, which we consider marginally acceptable. (However, we see below that the virial dielectric coefficient associated with the adjusted trace is not consistent with the measurement, indicating a weakness of this particular choice.)

Interestingly, the semiclassical work by Oxtoby and Gelbart[23, 24] has led to a trace model that is also well approximated by (26), with $\lambda_t = 37.9a_0^3$ and the same (!) $r_t = 0.626a_0$. Whereas the two theoretical efforts mentioned agree in their r_t value, which characterizes the range of rapid change of the atomic polarizabilities at beginning overlap, they disagree substantially about the magnitude of that change (λ_t). The corresponding spectrum (curve OG of Fig. 9) is, accordingly, not acceptable. Possible reasons have been discussed by Clarke et al.[27] and by van Kranendonk.[98]

The second virial dielectric coefficient of neon at 49°C was measured twice, with the result $B_\varepsilon = -0.30 \pm 0.10$ cm^6/mol^2 (from Ref. 35); and $B_\varepsilon = -0.24 \pm 0.04$ cm^6/mol^2 (from Ref. 36). We find the B_ε value of the *ab initio* trace[21] to be given by -0.10 cm^6/mol^2, which is inconsistent with the measurements. An adjusted $\lambda_t = 135a_0^3$ gives rise to $B_\varepsilon = -0.143$ cm^6/mol^2, which is still inconsistent with the measurements. Accordingly, the adjusted *ab initio* model introduced above ($\lambda_t = 135a_0^3$ with $r_t = 0.626a_0$) is no longer considered to be of merit (curve KK(m) of Fig. 9).

We note that small r_t values (such as $0.626a_0$ or less) lead to broad-banded polarized spectra featuring relatively small slopes if plotted as in Fig. 9 (unless λ_t is negligibly small). As r_t is increased (or as λ_t approaches 0), steeper slopes result, approaching those of curve A of Fig. 9 (which is given by the asymptotic form of the trace, $a(r) = A_6/r^6$). The experimental lineshape of the polarized spectrum is more nearly described by the steeper slopes of curve A than by those of curves KK or KK(m). Consequently, we have tried various other trace models, (26), with larger r_t values, in an attempt to obtain a model that would fit all existing measurements of B_ε and the polarized spectrum. A best fit is obtained using $r_t = 0.70a_0$, $\lambda_t = 65a_0^3$ (see the curve B of Fig. 9), which is associated with an acceptable B_ε value of -0.022 cm^6/mol^2. Larger values such as $r_t = 0.80a_0$ with $\lambda_t = 30a_0^3$ lead to the unacceptable $B_\varepsilon = -0.31$ cm^6/mol^2 and a somewhat inferior fit of the spectrum (curve C of Fig. 9). The currently available best trace model makes use of $r_t = 0.70a_0$, but a small range of r_t from about $0.65a_0$ to $0.75a_0$ is also consistent with the two existing measurements, if an associated λ_t is obtained for each different r_t such that the condition $\langle a^2 \rangle = 1.35 \times 10^{-75}$ cm^9 $\pm 20\%$ is satisfied. A realistic potential[92, 93] must be used in (33), where the trace, $a(r)$, is substituted for the anisotropy $\gamma(r)$.

Returning next to the experimental tests of the anisotropy models, we repeat that at low frequencies, the DID model leads to intensities only about 10% greater than the observed ones. This calls for a relatively minor correction of $\sim$5% near $r = \sigma$. Unfortunately, the *ab initio* computations of the anisotropy[21] result in a depolarized spectrum, which deviates too much from the measurement, in the other direction. Other classical estimates of the neon diatom polarizabilities[23-26] give rise to curves labeled OG[24] and HHG[26] and are generally not acceptable, see Fig. 8.

TABLE IV
Wave-mechanical Computation of the Collision-Induced Spectra of Neon and Derived Best Diatom Polarizability Model (Stokes Side)

a. Spectra at 4880 Å excitation:

	Shifts (cm^{-1})						
	−15	−30	−50	−80	−120	−160	−200
$D_{\parallel}$ (cm^6)	1.05 (−55)	4.62 (−56)	1.58 (−56)	3.63 (−57)	6.49 (−58)	1.42 (−58)	—
$P_{\perp}$ (cm^6)	2.28 (−57)	2.01 (−57)	1.51 (−57)	8.82 (−58)	4.04 (−58)	1.84 (−58)	8.58 (−59)

b. Trace: (26), with $A_6 = 224 a_0^9$; $\lambda_t = 65\ a_0^3$; $r_t = 0.70\ a_0$. Determined for $2.3\ Å \leq r \leq 3.0\ Å$

c. Anisotropy: (27), with $\alpha_0 = 0.3992 \times 10^{-24}\ cm^3$; $\lambda_a = 900\ a_0^3$; $r_a = 0.509\ a_0$, for $2.5\ Å \leq r \leq 5.6\ Å$

d. $B_\epsilon = -0.22\ cm^6/mol^2$ at 49°C; $\langle a^2 \rangle = 1.42 \times 10^{-75}\ cm^9 \pm 20\%$; $\langle \gamma^2 \rangle = 0.235\ 10^{-72}\ cm^9 \pm 8\%$

Also listed are the resulting second virial dielectric coefficients and thermodynamic averages (33) of trace and anisotropy. Note that a range of r_t -values between 0.65 a_0 and 0.75 a_0 also reproduces the spectra in an acceptable manner, if an associated λ_t is chosen, as described in the text. The permissible range of r_a is even greater.

Bérard and Lallemand[91] have proposed an empirical anisotropy model as in (27), with $\lambda_a = 1090 a_0^3$ and $r_a = 0.509 a_0$. This model provides a satisfactory fit of the experimental data at the lower frequencies, for which it was obtained (curve BL of Fig. 8). This agreement of the model[91] with the data from Ref. 47 (solid dots, Fig. 8) demonstrates the equivalence of the spectroscopic data from the two laboratories,[47, 91] because Bérard and Lallemand were careful to show that their model is consistent with their spectral distribution. At higher frequencies, we observe an increasing deviation of curve BL from the spectrum. A perfect fit is obtained, if we choose $\lambda_a = 900 a_0^3$ instead of $1090 a_0^3$ (see curve A, Fig. 8). Owing to the smallness of the exponential term in (27) near $r = \sigma$, a fairly broad range of r_a values will give very satisfactory fits of the depolarized spectrum, if the λ_a are adjusted such that $\langle \gamma^2 \rangle = 0.235 \times 10^{-72}\ cm^9 \pm 8\%$ is conserved, just as this is observed in the case of the depolarized helium spectrum discussed above. Table IV summarizes the results for neon.

C. Results for Argon

Collision-Induced Scattering of Light in Argon is probably the most thoroughly investigated of all gases. The pioneers in this field have all studied the scattering of light in this gas.[1, 2, 4, 99–104] Many more measurements and treatments related to argon have also been given.[48, 63, 77, 105–111] Three-body and, most recently, even four-body collision-induced spectra and their moments are reported in the literature,[2, 4, 102, 107, 112, 113] but our interest here is exclusively in the two-body spectra and the direct measurement of diatom polarizabilities. We mention that, in much of the previous work, absolute calibrations of the scattered intensities were often not attempted. Instead, the integrated intensity (the "zeroth moment" of the spectral distribution) was related to the measurements of the second virial Kerr coefficient, B_k, which we now believe to be somewhat inaccurate, as we reason below (Table V). To avoid adding to the existing confusion, we must be careful to base our conclusions only on the low-pressure data that have been demonstrated to feature binary collision-induced scattering. In this way, much of the work involving high pressures (above 10 or 20 amagat, roughly) must be considered with caution. The low-frequency part of the Raman spectra, and thus the measured zeroth moments, is easily affected by high gas density.

A polarized contribution of the collision-induced Raman spectrum was never seen in argon, in spite of a careful search.[2, 135] It is possible that at a high-frequency shift, near 280 cm^{-1}, a polarized contribution exists with an intensity of $\approx$30% of the depolarized spectrum.[77] However, because of the extreme feebleness of these signals, no definite conclusion could be drawn. The depolarized argon spectrum, produced by binary collisions (and, to a small extent, by van der Waals dimers), and the anisotropy of the argon diatom polarizability are thus studied here.

We take a recent measurement from our laboratory,[77] obtained at 27 amagat, as a representative argon spectrum (Figs. 10 and 11). We note that the absolute intensity scale is taken from earlier measurements at 3.5, 4.2, and 4.7 amagats,[48] and the lineshape given in Fig. 11 is virtually identical with those obtained at 3 to 5 amagats and published previously (Fig. 3 of Ref. 48). The lineshape of Fig. 10 is, furthermore, virtually identical with a recent measurement of the two-body spectrum by Barocchi and Zoppi,[114] well within the combined errors of the experiments. The intensity falloff is measured over almost 6 orders of magnitude. The error bars at the low frequencies are smaller than the dot size used, and at the highest frequencies they are as indicated in Fig. 10.

We note that the argon dimer does discernibly contribute to the observable spectrum[48] at shifts below 11 cm^{-1}. Such low frequencies are here suppressed. Bound-to-free or free-to-bound transitions are included in our computations, but are seen to be of negligible intensity.[48]

Wave-mechanical calculations based on (68) to (71) were undertaken, using the MSV III potential by Parson et al.[115] For trial purposes, other

TABLE V

Wave-mechanical Computation of the Depolarized Collision-Induced Raman Spectrum of the Argon Diatom (Stokes side)[a]

a. Spectrum at 4880 Å excitation:

Shift (cm^{-1})

	−20	−35	−50	−70	−100	−140	−200	−300	−400	cm^{-1}
$D_\perp$ (cm^6)	1.29 (−53)	3.52 (−53)	1.20 (−54)	3.69 (−55)	8.39 (−56)	1.57 (−56)	1.84 (−57)	1.34 (−58)	3.97 (−59)	cm^6

b. Anisotropy after (19), with $\alpha_0 = 1.68 \times 10^{-24}$ cm^3 (from Ref. 97) for 3.3 Å $\leq r \leq$ 6.2 Å

c. $\langle \gamma^2 \rangle = 4.78 \times 10^{-71}$ $cm^9 \pm 8\%$ at 297.5°K

[a]Note that in this table $D_\perp$ is listed, whereas in Tables III and IV $D_\parallel$ is given. The relationship is $D_\perp = \frac{7}{6} D_\parallel$.

Note added in proof: Recently, we were able to measure the *polarized* Raman spectrum of the argon diatom: J. Chem. Phys., to appear 1, January 1981. We find $A_6 = 9665 a_0^9$; $\lambda = 850 a_0^3$; $\rho_t = 0.764 a_0$. In that work, very accurate and long depolarized spectra were obtained as a side product, which show small, but significant deviations from DID at high frequencies. The inferred model of the anisotropy is given by (19), with $\lambda_a = 9000 a_0^3$; $\rho_a = 0.60 a_0$. (Ref.: talk given by the author at the International Conference on Collision-Induced Phenomena, held in Florence/Italy from Sept. 2–5, 1980. Proceedings will be published in a 1981 issue of the Canadian Journal of Physics.)

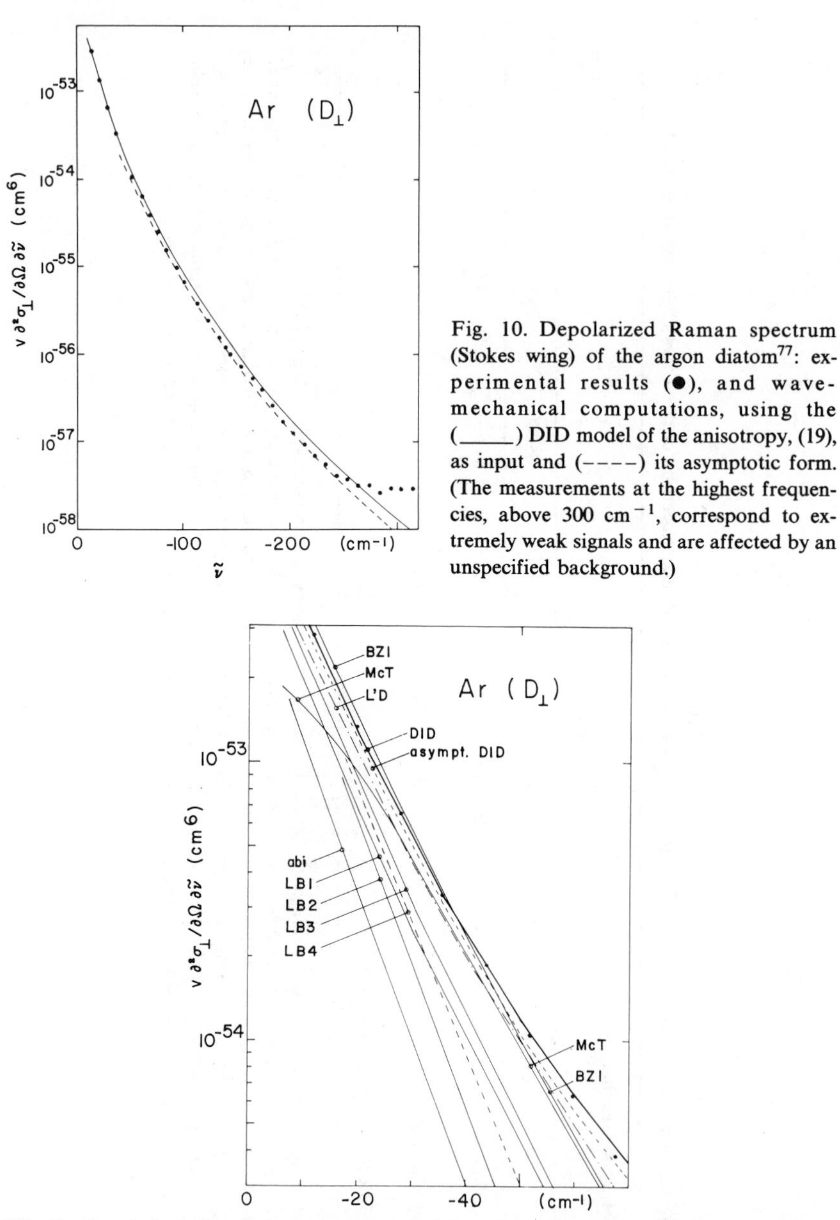

Fig. 10. Depolarized Raman spectrum (Stokes wing) of the argon diatom[77]: experimental results (●), and wave-mechanical computations, using the (_____) DID model of the anisotropy, (19), as input and (– – – –) its asymptotic form. (The measurements at the highest frequencies, above 300 cm^{-1}, correspond to extremely weak signals and are affected by an unspecified background.)

Fig. 11. Depolarized Raman spectrum (Stokes side) of the argon diatom[77]: (●) experimental results. The wave-mechanical computations are based on the various models of the anisotropy: *abi*[22]; DID (19); (– – – –) asymptotic DID LB1 to LB4 (89) to (92); McT[4]; L'D[104]; BZ1[110]. Note that the error bars of the experiment are ±8%, that is, much less than the typical deviations among computations.

argon interaction potentials[116–118] were also used, with negligible differences in the computed spectra.[60] However, the older Lennard-Jones potentials were found to be inadequate,[60] as they give rise to spectra that are about 15% weaker at 60 cm^{-1}, and even more so at higher shifts.

Lineshape calculations for argon at room temperature at low frequency shifts require about 160 partial waves ($0 \leq J \leq 320$). At the highest frequency shifts, only about half as many are needed for a specified precision of $\approx 1\%$ of the numerical results. Furthermore, owing to the massiveness of the argon diatom, the de Broglie wavelengths involved are rather short. Accordingly, the integration of the Schrödinger equation must be done with much smaller step sizes than in the case of helium. As a consequence, the lineshape computations for argon are rather expensive. It is therefore of interest to look for less expensive alternatives, such as the classical lineshape computations we mentioned above. For argon at room temperature, at shifts below 40 cm^{-1}, uncorrected classical and wave-mechanical computations of the Stokes intensities agree within a few percent, except at the lowest frequencies ($\lesssim 12$ cm^{-1}), which we ignore here.[77] (The uncorrected anti-Stokes intensities differ by the usual Boltzmann factor. The term "uncorrected" is used to emphasize that the intensities are not corrected for detailed balancing.) We mention that for the heavier rare gases, the wave-mechanical and classical Stokes intensities differ by even less. For this reason, for shifts $\lesssim 40$ cm^{-1}, classical computations may be considered sufficient. The corresponding savings of computer time are substantial. For argon, a classical lineshape may require less than 200 sec on the Cyber 170/750 but a wave-mechanical computation of comparable accuracy may take 5000 sec or more if all bound-to-bound and bound-to-free transitions are to be included. Furthermore, the required field lengths at execution time are also very different.

Figure 10 displays also two computed wave-mechanical lineshapes, one based on the DID model [(19); solid line of Fig. 10], the other on the asymptotic form, $\gamma(r)=6\alpha_0^2/r^3$ (dashed line). The atomic polarizability at 4880 Å excitation wavelength is taken to be $\alpha_0 = 1.63\ 10^{-24}$ cm^3 (from Ref. 79). At the low frequencies the complete DID model, (19), provides us with a perfect fit of the data (see Fig. 11, heavy line marked DID). From about 70 to 250 cm^{-1} the leading term in the asymptotic expansion gives rise to a perfect fit. Above about 260 cm^{-1}, the observed signals amount to only 0.05 counts/sec (after the dark count of ~ 0.3 counts/sec is subtracted) and a polarized contribution (possibly collision-induced) appears to be superimposed on the depolarized spectrum.[77] (These high-frequency shifts are ignored here.) The deviations from the experimental spectrum relative to the computed DID lineshape (solid line, Fig. 10) amount to 10% at 100 cm^{-1} and up to 30% at 250 cm^{-1}. The statistical uncertainty of the data amounts to only about one-half of these deviations, which are thus seen to be significant. (The numerical uncertainties amount to not more than 2% and can be ignored here.)

It is not hard to find a model of the anisotropy that will fit low- and high-frequency parts equally well. Two such models are of the form

$$\gamma(r)=\frac{6\alpha_0^2 r^3}{(r^6-\alpha_0 r^3-2\alpha_0^2)}-\lambda_a \exp\left(-\frac{r}{r_a}\right) \tag{87}$$

$$\gamma(r)=\frac{6\alpha_0^2}{r^3}+\frac{B_6}{r^6}-\lambda_a \exp\left(-\frac{r}{r_a}\right) \tag{88}$$

The exponential terms amount, however, to only a 1% correction of the DID expression, at separations near $r=3.405$ Å, and B_6 would be very nearly given by the classical value, $B_6=6\alpha_0^3$ [from (21)]. We refrain from recommending such highly refined models, because it is felt that an exponential correction term amounting to only a 1% correction suggests a precision of the data that is at present not justified. We remind the reader that, typically, a spectroscopist records a raw lineshape on an arbitrary intensity scale. Then he corrects it for uniform monochromator transmission and detector sensitivity. Finally, one point of the continuum (or, alternatively, the integrated intensity) is put on an absolute intensity scale. This last process is, at least in our work, the least accurate step, with typical error bars of ± 7 or 8% (see Ref. 48). (For comparison, it is mentioned that the statistical uncertainties amount to typically less than 1% at the low frequencies, where the intensities are high, and to about 10% at the high frequencies.) Small correction terms of the anisotropy [such as the exponentials in (88) and (89)] can be determined well if the uncertainty of the absolute intensity scale can be ignored. However, with the 8% uncertainty of the absolute intensity scale, such corrections appear to be meaningless.

Summarizing these discussions, we conclude that the DID model (19) is a most successful model of the anisotropy of the argon diatom polarizability. Lineshapes computed on its basis agree with the most important part of the experimental spectrum, over a range of intensities of almost 3 orders of magnitude, at frequencies below $\sim 100\ \mathrm{cm}^{-1}$. A small deviation between the DID lineshape and the experiment in the far wing calls for an unspecified, electronic overlap correction of the anisotropy, of the order of only 1% near $r=3.405$ Å. Wave-mechanical depolarized spectra of the argon diatom are also given in Table V.

The main part of this conclusion was more or less forcefully expressed in a number of experimental papers.[60, 99, 104] It is interesting to note that very recently, theoretical work by Clarke et al.[27] led to the same conclusion. Furthermore, we see below that the DID model appears to give rise to spectral lineshapes consistent with the experiments for all of the more highly polarizable gases (Kr, Xe, CH_4 and others). For the sake of completeness, we repeat here that for the less highly polarizable helium and neon, small corrections amounting to roughly -5% near $r=\sigma$ had to be applied to the DID anisotropy to match the computed and observed lineshapes.[44–47]

Nevertheless, on various grounds, many "improved" anisotropy models have been proposed in the past for the argon diatom that usually differ by more than just a few percent at separations $r \cong \sigma$ (corresponding to the root of the interatomic potential, $\sigma = 3.405$ Å) (see Fig. 11 for the associated lineshapes of these models).

An *ab initio* computation of the argon diatom polarizability by Lallemand et al.[22] gives rise to a computed spectrum labeled *abi* in Fig. 11. The computed intensities are only about 30% or less of the observed ones, and the logarithmic slopes are steeper than the experiment permits ($1/8.8\ \text{cm}^{-1}$ versus $1/12.2\ \text{cm}^{-1}$). Obviously, the accuracy of the *ab initio* calculations is at present not sufficient.

Other theoretical work[23, 25] suggests an anisotropy of the form of (27), with $\lambda_a = 1162a_0^3$ and $r_a = 0.901a_0$. The associated spectrum is not shown in Fig. 11, but is almost identical with the dashed curve LB4. Computed intensities are too weak by factors from 0.6 to 0.2, and the computed logarithmic slopes are too steep ($1/8$ versus $1/12.2\ \text{cm}^{-1}$). The "electron gas approximation" was used by Heller et al.[27] for another theoretical estimate of the anisotropy. Using the numerical table[27] and third-order spline interpolation, we get a spectrum that is virtually identical with the one based on the DID asymptotic model. In fact, it can be shown directly that Heller et al.'s anisotropy, for separations of interest here (3.3 Å $\leq r \leq$ 6 Å) is practically equal to the DID values. Clarke et al., similarly, obtain an argon anisotropy, which virtually agrees with the DID model.[27]

Empirical models of the argon diatom polarizability were given by Levine and Birnbaum[119] on the basis of fitting the zeroth, second, and fourth moments of the spectral distribution to simple, two-parameter expressions (setting $x = r/3.405$ Å, with $\sigma = 3.405$ Å):

$$\gamma(r) = \left(\frac{6\alpha_0^2}{\sigma^3}\right)(x^{-3} - 0.289x^{-5.0}) \tag{89, LB1}$$

$$\gamma(r) = \left(\frac{6\alpha_0^2}{\sigma^3}\right)(x^{-3} - 0.473x^{-9.3}) \tag{90, LB2}$$

$$\gamma(r) = \left(\frac{6\alpha_0^2}{\sigma^3}\right)(x^{-3} - 0.226x^{-6.5}) \tag{91, LB3}$$

$$\gamma(r) = \left(\frac{6\alpha_0^2}{\sigma^3}\right)(x^{-3} - 0.330x^{-11.1}) \tag{92, LB4}$$

Unfortunately, no absolute calibration of the scattered intensities was attempted. Instead, the zeroth moment was equated to a value derived from the measurement of the second virial Kerr coefficient.[56] The resulting spectra are too weak. At the same time, the logarithmic slopes are 1/9.7; 1/8.4; 1/9.8; 1/9.0 cm^{-1}. All of these are much steeper than the experiment permits (1/12.2 $cm^{-1} \pm 6\%$). None of these models is consistent with the experiment.

Another empirical model of the anisotropy by McTague et al.,[4] also inferred from moments of the spectral distribution, is of the form of (27), with $\gamma_a = (6\alpha_0^2)0.72$, and $r_a = 1.25a_0$. It gives rise to a lineshape (McT, Fig. 11) that is comparable to the experiment at the higher-frequency shifts. However, at the lower shifts, the intensities are too weak and the computed lineshapes feature a convexity that is not reflected by the experiment.

Lallemand[104] proposed various models by comparing the computed classical correlation functions with the inverse Fourier transforms of the observed spectra. The approach is quite comparable in spirit to our lineshape studies. The conclusions are, as a consequence, often very similar, if not identical, to our conclusions. The reason that correlation functions are used in that work, in preference to the spectral lineshapes themselves, we believe, is a desire to avoid the discussion or implication of detailed balancing (that is, the Boltzmann factor) in an otherwise classical theory. Of the many anisotropy models mentioned,[104] we choose one that was recommended as a good model

$$\gamma(r) = \frac{6\alpha_0^2}{r^3} - \frac{1.28\alpha_0^2}{r^{4.12}} \tag{93}$$

This expression features an inverse 4.12 power of the separation, which by many is considered unphysical. The resulting lineshape (L'D, Fig. 11) is generally a good fit of the spectrum if all intensities are increased by a constant factor (1.2). However, we mention that the model, (93), is not really superior to the DID model, from which it differs for all physically accessible separations by only 5% or less. The success of this model is seen to derive from the insignificance of the correction term.

Recently, on the basis of a new measurement of the depolarized argon diatom spectrum over a wide range of frequencies, Barocchi and Zoppi[110] have argued that realistic models of the anisotropy should be of the form of (88), with three adjustable parameters (B_6; λ_a; r_a). The reasoning is based in part on their sound physical intuition and in part on the desire to have another adjustable parameter available such that more and higher moments can be used with advantage. From their measurements of their zeroth, second, fourth, and sixth moments, combined with an absolute intensity calibration, Barocchi and Zoppi developed two sets of parameters that should, according to those authors, provide a superior fit of the argon spectrum. These sets are

$$B_6=0.313(6\alpha_0^2b^3); \qquad \lambda_a=170\left(\frac{6\alpha_0^2}{b^3}\right); \qquad r_a=0.088b \quad (94;\ \text{BZ1})$$

$$B_6=0.81(6\alpha_0^2b^3); \qquad \lambda_a=11.4\left(\frac{6\alpha_0^2}{b^3}\right); \qquad r_a=0.136b \quad (95;\ \text{BZ2})$$

with $b=7.111a_0$, for use in (88). The spectrum associated with the first set is given in Fig. 11 (curve BZ1); the spectrum obtained for the other model BZ2 is almost identical over the frequency band shown and is, therefore, suppressed. We note that at the low frequencies, the computed intensities are about 8% above the DID lineshape, which describes our experimental data well. (In essence, this is due to the absolute calibrations, which differ by about this amount. The observed differences are within the combined error limits of the experiments and are of no great concern.) We observe, however, steeper logarithmic slopes of these theoretical lineshapes than either experiment permits: $1/10.2\ \text{cm}^{-1}$ for BZ1, and $1/9.8\ \text{cm}^{-1}$ for BZ2 near $20\ \text{cm}^{-1}$ (Stokes side), to be compared with the never disputed experimental value of $1/12.2\pm6\%$, which is supported by all experimental work at low pressures.[48] As a consequence, at the shift of about $80\ \text{cm}^{-1}$, the lineshapes BZ1, and BZ2 feature only an unacceptable 50% of the observed intensities of either experiment. At the higher frequencies (not shown), BZ2 remains roughly at this 50% level below the experiment, and BZ1 drops below the 10% level (relative to the experiment) near $125\ \text{cm}^{-1}$ on the Stokes side. Such behavior is clearly inconsistent with the experiments. By no means are the models BZ1 and BZ2 superior to the DID model, which provides a much closer fit of the spectroscopic data of either experiment, as can be seen in Figs. 10 and 11. This example thus serves as another illustration of what we have come to conclude: for argon, the DID model with no adjustable parameters is as good, if not superior, to all anisotropy models we know featuring two or even three adjustable parameters. This conclusion is firmly founded on the discussion of the full lineshape, not just on a small number of moments. Six orders of intensity variations are thus accounted for with precision, and on an absolute intensity scale of less than 8% uncertainty.

In a strange contrast to this conclusion, several earlier measurements of the thermodynamic average of the anisotropy, $\langle\gamma^2\rangle$, (33), which were based either on the integrated scattered intensities (the zeroth moment of the spectral distribution function)[4, 102, 120, 121] or on the determination of the second virial Kerr coefficient,[56] seemed to indicate that the DID model of the anisotropy is not adequate. Rather, the experimental values of $\langle\gamma^2\rangle$

were found to be from 20 to 40% smaller than the DID value, with one exception of a larger $\langle\gamma^2\rangle$. Table VI gives these measurements, along with some other possibly relevant information.

A quick comparison of the six different determinations of the ratio $\langle\gamma_e^2\rangle/\langle\gamma_{\mathrm{DID}}^2\rangle$ shows that there is not much consistency among any of the data. To a small extent, this is due to differences in the potentials used (the early work made use of Lennard-Jones 6-12 potentials, instead of the more refined models[115–118]), to small differences of temperature, or even to slightly differing values of the atomic polarizability. (We use the polarizability of argon atoms at the green or blue argon ion laser line, using Ref. 97. Other authors use instead the incorrect d.c. value). Also, some authors use the asymptotic form of the DID model and others use (19). All these differences amount to a few percent variation that could easily be corrected. To some other extent, the observed inconsistencies with the DID anisotropy may be related to experimental uncertainties, which are certainly hard to estimate: about 70% of the total intensity is contained in the low-frequency band ranging from a -8 to $+8$ cm^{-1} shift. In this region, the spectra are more or less severely masked by the most intense Rayleigh-Brillouin lines,[49, 50] the associated instrumental wings generated by these structures,[51] the intercollisional light-scattering process,[52] and three-body and higher collision-induced scattering.[2–4] For high-precision work, the zeroth moment is not too useful. Nevertheless, the usual interpolation schemes employed to estimate the unknown low-frequency intensities would almost certainly lead to estimated zeroth moments that are too large, never too small. (This conclusion is based on the knowledge of the wave-

TABLE VI

Ratios of Measured Zeroth Moments in Argon at Room Temperatures, $\langle\gamma^2\rangle$, Relative to the Zeroth Moment Computed on the Basis of the DID Anisotropy Model, $\langle\gamma_{\mathrm{DID}}^2\rangle$

			Somehow accounts for	
Ref.	$\langle\gamma^2\rangle\langle\gamma_{\mathrm{DID}}^2\rangle^{-1}$	Range of pressures (bars)	Three-body	Four-body
FHP[48, 60]	1.00 ± 0.08	3–5	No	No
L[101, 102]	0.81 ± 0.12	10–90	Yes	No
BD[56]	0.73 ± 0.11	20–80	No	No
TTOV[120]	0.67 ± 0.03	200	No	No
McTEH[4]	0.62 ± 0.06	20–150	Yes	No
BNZ[110]	1.13 ± 0.07	10–150	Yes	Yes

mechanical low-frequency lineshapes.) Consequently, one would understand ratios of $\langle \gamma_e^2 \rangle / \langle \gamma_{\rm DID}^2 \rangle$ that are too big. Most values of this ratio given in Table V are, however, too small, less than unity. Moreover, the value inferred from a measurement of the second virial Kerr coefficient[56] is not affected by this latter interference and deviates from unity much like most of the other entries of Table VI. We feel, therefore, compelled to conclude that although all the influences listed do affect the precision of most of these measurements, they are not the decisive ones. The variance of the data (Table VI) is still largely unexplained at this point.

In the search for an acceptable explanation, we note that the experiment done at the lowest gas density (3 to 5 amagat) gives perfect agreement with the DID model.[48, 60] With only a single exception, which is discussed later (the last line of Table VI), the data provided in the table suggest that the higher the gas density of an experiment, the more the measurements deviate from unity, that is, from the DID result. Lallemand,[101, 102] who used relatively small densities (in excess of 10 amagats) obtained a ratio not too much less than unity. Buckingham and Dunmur,[56] in their classical determination of the second virial Kerr coefficient, used higher representative densities (in excess of 20 amagats) to get an even smaller value of 0.73. The work by Thibeau et al.[120] makes use of hundreds of amagat, and the number ratio is still smaller: 0.67 and so forth. Furthermore, the evaluation procedures of the two works just mentioned[56, 110] do not account for three-body collisions, but probably should. (We note that three-body spectra are negative[2, 4] and, if not accounted for, do tend to make zeroth moments appear too small.) None of the experiments mentioned thus far accounts for four-body and higher contributions. We feel that there is some suggestive correlation of otherwise inconsistent data, which should be further investigated.

The data of Barocchi et al.[110] were in part obtained using relatively low densities (10 amagat), and three-body and four-body terms were present and accounted for at the higher densities. Maybe for these reasons, a relatively large ratio $\langle \gamma^2 \rangle / \langle \gamma_{\rm DID}^2 \rangle$ was obtained for the diatom spectrum, which as we show next is consistent with our measurement.[60] Barocchi et al.[110] accept a value of $\langle \gamma_{\rm DID}^2 \rangle = 39.3 \ldots 41.7$ Å^9, which they obtain using the asymptotic expression, $\gamma(r) = 6\alpha_0^2/r^3$. We have normalized instead to $\langle \gamma_{\rm DID}^2 \rangle = 43.1$ Å^9 based on (19). By simply multiplying Barocchi et al.'s entry (Table VI) by their average 40.5 Å^9 and dividing by our value (43.1 Å^9), we obtain the consistent $\langle \gamma_e^2 \rangle / \gamma_{\rm DID}^2 \rangle = 0.94 \pm 0.07$, in support of our conclusion. (We mention above and argue now more directly that Barocchi et al.'s and our calibrations agree to within a few percent. Consequently, their zeroth moment and ours agree to within that error, which becomes evident by renormalization.)

Summarizing the discussion of the data displayed in Table VI, we suggest that there is some evidence that argon pressures that were too high were often used, with the effect of deflating the average of the anisotropy, probably mainly by three-body interactions. However, the low-pressure data do not suffer from this defect, and a proper accounting for three- (and where necessary, four-) body collisions may actually be all that is needed to bring these measurements together. We do not hesitate to base our conclusion on the (uncorrected) low-pressure data, which clearly support the DID anisotropy model for the case of argon, as we see above. The full DID model, (19), is generally preferred over the asymptotic form, which, however, is also quite acceptable. The small differences of full and asymptotic forms of the anisotropy, which differ by merely 4% near $r=\sigma$, may serve as a measure of the electronic overlap corrections one will eventually have to apply in a more realistic anisotropy model. Note that the change of γ by 4% near $r=\sigma$ causes $\langle\gamma^2\rangle$ to change by about 8%.

D. Results for Krypton and Xenon

The heavier rare gases feature Raman spectra that are increasingly affected by the bound dimer contributions.[74] Whereas the helium diatom probably does not have a bound state, a neon dimer exists and possesses about 10 rotational and vibrational states. The approximate number of bound and predissociating ("metastable") diatom states of argon, krypton, and xenon is 120, 400, and 1000, respectively. It is, however, difficult to resolve the dimer spectra. The strongest Raman lines are the pure rotational transitions.[96] Each vibrational state features its own rotational band, with spacings amounting typically to only small fractions of a wave number. All these bands occur at frequencies less than about 8 cm^{-1} and are superimposed. The resulting multiline structure is, furthermore, pressure broadened.[122] What is easily seen at densities of several amagat is the envelope of the rotational bands. Computations of these band envelopes are given in the Figs. 12 (for krypton) and 13 (xenon) (see the curves marked b–b.) A 1 cm^{-1} slitwidth is assumed in these computations. Rotational transitions take place at shifts of <10 cm^{-1}. The shoulders shown at shifts from 10 to 20 cm^{-1} are due to the various vibration–rotation transitions. For simplicity the computations assume an average mass of the atoms and ignore the isotope abundances completely.

The ratios of the anti-Stokes and Stokes intensities of the spectra are given by the usual Boltzmann factor, which differs by only $\approx$15% from unity, on account of the small frequency shifts considered.[67] Measurements at various gas densities indicated that the intensities scale exactly as the squares of the densities up to a maximum density of $\sim$7 amagats for krypton and $\sim$5 amagats for xenon. The spectra given in Figs. 12 and 13,

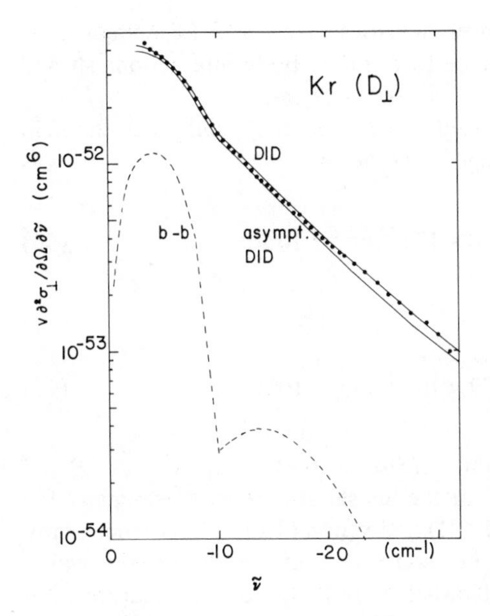

Fig. 12. Depolarized Raman spectrum (Stokes wing) of the krypton diatom[67]: (●) experimental results. Wave-mechanical computations using the DID model of the anisotropy, (19), and its asymptotic form are also shown, along with the envelope of the bound dimer spectrum (b-b).

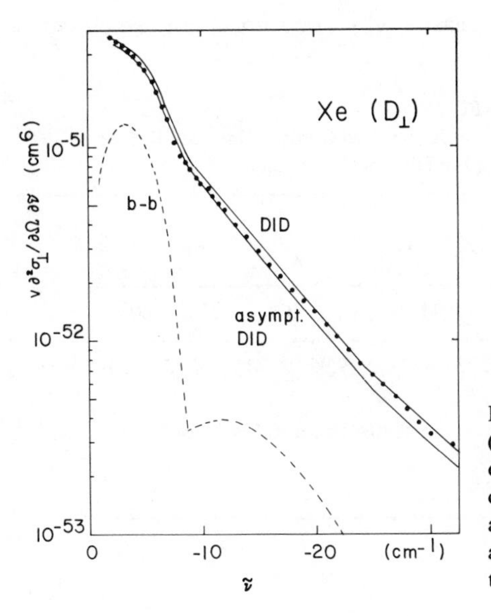

Fig. 13. Depolarized Raman spectrum (Stokes wing) of the xenon diatom[67]: (●) experimental results. Wave-mechanical computations using the DID model of the anisotropy, (19), and its asymptotic form are also shown, along with the envelope of the bound dimer spectrum (b-b).

were, however, taken at much lower densities of 2.7 and 1.8 amagats, respectively.[67] In this way, interference form many-body interactions should be truly minimal.

Absolute intensities were calibrated[67] at a $-12\ \mathrm{cm}^{-1}$ shift with the help of the $S(0)$ rotational line of nitrogen. We obtain

$$v\frac{\partial^2\sigma_\perp}{\partial\Omega\,\partial\tilde{\nu}} = 1.10\times10^{-52}\ \mathrm{cm}^6 \pm 10\% \tag{96}$$

for krypton and

$$v\frac{\partial^2\sigma_\perp}{\partial\Omega\,\partial\tilde{\nu}} = 4.79\times10^{-52}\ \mathrm{cm}^6 \pm 10\% \tag{97}$$

for xenon, both at shifts of $-12\ \mathrm{cm}^{-1}$ (Stokes side).

Wave mechanical computations of the lineshapes are most expensive for gases as massive as krypton (Table VII) and xenon (Table VIII), since many partial wave contributions must be summed, and the de Broglie wavelengths are small, requiring small radial increments for an accurate integration of the Schrödinger equation. Fortunately, wave mechanical calculations are not at all necessary for these gases, except for the bound-state Raman transitions of the dimers. The free state can be dealt with on the basis of the classical trajectories. For the case of argon, the comparison of classical and wave mechanical lineshapes shows negligible differences, amounting to only a few percent. For these heavier gases, these differences should be even less and there is no reason why one should not use classical

TABLE VII.
Wave-mechanical Computation of the Depolarized Collision-Induced Raman Spectrum of the Krypton Diatom (Stokes Side)

a. Spectrum at 4880 Å excitation:

	Shift (cm^{-1})					
	-12	-18	-24	-32	-40	cm^{-1}
$D_\perp$ (cm^6)	1.10 (-52)	4.83 (-53)	2.33 (-53)	9.96 (-54)	4.70 (-55)	cm^6

b. Anisotropy, after (19), with $\alpha_0 = 2.56\times10^{-25}\ \mathrm{cm}^3$ at 4880 Å (from Ref. 97), for 3.6 Å $\leq r \leq$ 6.5 Å

c. $\langle\gamma^2\rangle = 2.22\times10^{-70}\ \mathrm{cm}^9 \pm 8\%$ at 297.5°K

[a] $D_\perp = \frac{7}{6} D_\parallel$

TABLE VIII
Wave-mechanical Computation of the Depolarized Collision-Induced Raman Spectrum of the Xenon Diatom (Stokes side)[a]

a. Spectra at 4880 Å excitation:

	Shift (cm^{-1})					
	−12	−18	−24	−32	−40	cm^{-1}
$D_\perp$ (free-to-free)[a] (cm^6)	4.73 (−52)	1.74 (−52)	7.49 (−53)	2.89 (−53)	1.26 (−53)	cm^6
$D_\perp$ (bound-to-bound)	0.41 (−52)	0.24 (−52)	—	—	—	cm^6

b. Anisotropy after (19), with $\alpha_0 = 4.22 \times 10^{-24}$ cm^3 at 4880 Å (from Ref. 97), for 4.0 Å $\leq r \leq$ 6.8 Å

c. $\langle \gamma^2 \rangle = 1.43 \times 10^{-69}$ cm^9 ± 8% at 297.5°K

[a]Low-frequency dimer contribution (lower line) is also given and should be added to the collision-induced contribution (upper line) before a comparison with the experiment is attempted. $D_\perp = \frac{7}{6} D_\parallel$.

lineshape calculations for the free state (if these are supplemented by wave -mechanical computations for the bound state dimers). We use the so-called MMSV potential by Farrar et al.[123] Use of Lennard-Jones 6-12 potentials leads to lineshapes that deviate by up to 20% at high frequencies from the ones given here, which reflects the inaccuracies of the Lennard-Jones potential.

Two lineshapes are given for each gas. One is computed with the DID model, (19), the other with its asymptotic form, $\gamma(r) = 6\alpha_0^2/r^3$. We use atomic polarizabilities at a wavelength of 4880 Å of 2.56×10^{-24} cm^3 for krypton and 4.22×10^{-24} cm^3 for xenon.[97] The DID model differs from its asymptotic form at $r = \sigma$ by only 6.5 and 8%, respectively. Consequently, the spectra, which depend on the squares of the anisotropy, differ by roughly twice that amount. The precision of the intensity calibration, on the other hand, is about 10%, which means that in the semilogarithmic grid, the experimental points can be shifted as a whole up or down by something slightly less than the differences between these curves. We note that the statistical uncertainties are a negligible 1% or less at the low frequencies, and as large as about 10% at the highest frequencies only.

The DID lineshapes are seen to agree very closely with the experiment (Figs. 12 and 13). For krypton, the experimental data are almost exactly in the middle between the DID and its asymptotic model lineshapes, and no real preference for either model can be given. For xenon, the lineshape associated with the asymptotic form is seen to agree with the data slightly better, but the differences are not very substantial in either case. We note that the logarithmic slopes of the theoretical lineshapes agree exactly with the experimental ones quoted above. As in the case of argon, we can summarize that the DID model reproduces the absolute intensities and spectral distributions of the krypton and xenon diatom Raman spectra. Small modifications of the anisotropy, which change its value relative to the DID model by more than about 8% near $r=\sigma$, are inconsistent with the observed spectra.

Correction terms amounting to much more than just a few percent of the DID anisotropy have been repeatedly reported in the literature. We have shown previously[67] that the anisotropy models of Gersten[62] and Levine and Birnbaum[119] are not consistent with the observed lineshapes at low pressures. Not only are the computed absolute intensities only a small fraction of the observed ones at a shift of $-12\ \text{cm}^{-1}$, but also the shapes are strikingly inconsistent with the observed lineshapes. Shapes are most conveniently compared by the reciprocal logarithmic slopes for frequencies between 10 and 20 cm^{-1} on the Stokes side, which are found to be substantially steeper than the experiment permits.[67] Similarly, the measured second virial Kerr coefficient[56] leads to a value of the thermodynamic average of the squared anisotropy, $\langle\gamma^2\rangle$, which is roughly 70% of the DID value. (A similar situation is discussed above for the case of argon, Table VI.) As in the case of argon, we believe that earlier measurements of $\langle\gamma^2\rangle$, based on either the collision-induced depolarized scattering[1, 2] or the second virial Kerr coefficient,[56] where done at pressures that were too high, where many-body effects unaccounted for did interfere with the measurement. Our spectra given in Figs. 12 and 13 are obtained at the lowest gas densities ever used for such work. Consequently, these data result from almost purely binary interactions and are less affected by many-body interactions than any other measurement known to us. This may possibly explain why our conclusions differ from some other results mentioned in the assessment of the quality of the DID model. For purely binary collisions, the DID model appears to be the only acceptable model of the anisotropy.

E. Results for Selected Molecular Gases

Owing to its cubic symmetry, the methane molecule does not have a low-frequency rotational Raman spectrum. At high enough gas densities, Stokes and anti-Stokes wings appear that are due to the collisional interactions of two methane molecules, just as is seen above for all the monatomic gases.[68, 101, 102, 124, 126] Ratios of the measured average anisotropy, $\langle\gamma^2\rangle$, relative to the theoretical value of $\langle\gamma^2_{\text{DID}}\rangle$, (33), computed with the DID or

TABLE IX.
Measured Ratios of $\langle\gamma^2\rangle$ Relative to the DID Value for Methane

Ref.	Method	$\langle\gamma^2\rangle/\langle\gamma^2_{DID}\rangle$	Density range (amagats)
L[101, 102]	CIS	1.03 ± 0.15	8.5–30
PG[68]	CIS	1.00 ± 0.10	3.3–10
BZST[125]	CIS	1.37	10, 20, ... 100
PF[126]	CIS	1.00 ± 0.10	1.8
TOV[100]	Depol.	0.75 ± 0.03	8–40
WR[128]	Depol.	1.20 ± 0.17	
BO[127]	Kerr	0.89 ± 0.17	

the asymptotic DID model were communicated from measurements of collision-induced Raman spectra, depolarization ratios, and the second virial Kerr coefficient (see Table IX). The results listed are obviously not at all consistent, indicating a certain problem being associated with the determination of this quantity. However, if we restrict our attention to those measurements undertaken at the lowest gas densities (as 1.8 and 3.3 amagats), excellent agreement with the DID model is observed. An identical observation was made above for the case of argon.

This conclusion is supported by the wave mechanical lineshape calculations, based on the DID (19) or asymptotic DID model, see (Fig. 14). Several Lennard-Jones potentials were tried[89] and the differences between the results do not exceed about 4% as long as the roots of the potentials

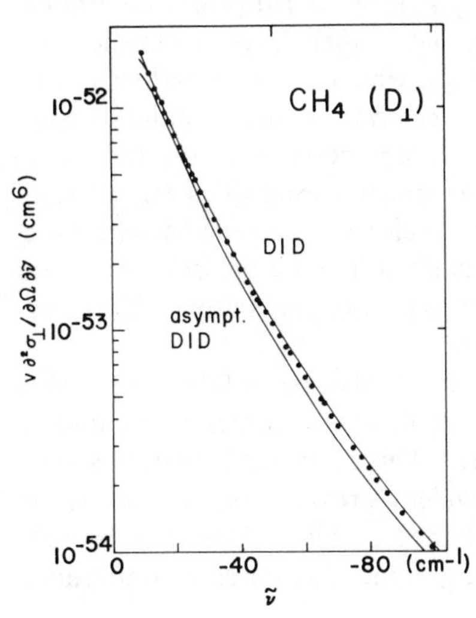

Fig. 14. Depolarized Raman spectrum (Stokes wing) of the methane dimolecule[126]; (●) experimental results. Wave-mechanical computations are based on the DID model of the anisotropy, (19), and its asymptotic form.

used do not differ from $\sigma = 3.817$ Å (from Ref. 129). Alternative forms of interaction potentials were considered in Ref. 102. Other models of the anisotropy, if they differed by more than about $\pm 6\%$ from the DID model near $r \cong 3.8$ Å, were shown to be inconsistent with the observed lineshapes.[68, 126]

Measurements of the collision-induced anisotropy of the polarizability were also reported for some other molecular gases. For hydrogen,[102] the DID model was shown to nearly describe the correct anisotropy. For CF_4, the DID model was also rather successful,[130–132] but a short-range correction was thought to be necessary.[125] More work at very small densities would be useful for a critical assessment of the pair polarizability of molecular gases.

V. CONCLUSION

Direct measurements of the binary collision-induced Raman spectra of gases at low densities were used to obtain new information concerning diatom polarizabilities. In two cases (helium and neon) the *polarized* Raman spectrum of the diatoms could be measured. With the help of wave-mechanical lineshape computations, we were able to formulate accurate trace models for these gases, which were consistent with (a) the two earlier measurements of the second virial dielectric coefficient B_ε, and (b) certain *ab initio* calculations of the diatom polarizabilities. For most other gases, however, a successful separation of polarized diatom Raman spectra of sufficient quality is probably not possible, owing to the superimposed, much stronger *depolarized* spectra. Helium and neon are probably the most favorable cases for the successful separation of the polarized spectrum, owing to their small atomic polarizability [whose fourth power determines the depolarized intensity, (19) and (33)], and to their large electronic overlap correction [i.e., the exponential in (26), which causes negative B_ε and strong polarized intensities]. In any case, *polarized* collision-induced spectra were seen to be negligible in argon, except possibly in the far Stokes wing, where precise measurements of the spectral intensities are difficult. Consequently, the virial expansions of the dielectric Clausius-Mosotti function, or its refractive counterpart,[133] are likely to remain the only source for empirical tests of the spherically symmetric part of the diatom polarizability for gases other than helium and neon.

Depolarized collision-induced spectra could also be obtained for many gases at the lowest densities hitherto reported. These spectra of the diatom are thus the result of a direct measurement. This remarkable feature should be viewed in comparison with the prevalent previous practice of either "correcting" the spectroscopic data obtained at higher densities by subtracting many-body contributions, using some theoretically unfounded virial expansions of the lineshape, or else neglecting such corrections altogether. We note that the maximal gas density that can be used for a recording of diatom spectra varies with the gas used. Whereas, for example, 30 amagats appear to be permissible for helium, on account of the extremely weak intermolecular forces and atomic polarizabilities, 6 amagats was seen to be too much for xenon. The criterion is the demonstrated, accurate density square dependence of the parts of the spectrum, which serve as a basis for evaluating diatom polarizabilities. The results thus obtained often differ from earlier results (which are, at the same time, often mutually somewhat inconsistent), presumably because these were determined with varying amounts of many-body interference present. It is most satisfying to see that other experiments performed at relatively low densities, or experiments that succeeded particularly well to separate many-body interferences, do support this view. Most recently, theory was also able to point out in rather general terms[134] what one might call a major result of this work: The *DID* model of the anisotropy does describe collision-induced, depolarized scattering of light with some precision ($\approx$10% or, in many cases, even better). The wave-mechanical calculations of the lineshapes were an important tool of verification, and with their help for the first time the principle of detailed balancing was naturally and rigorously accounted for in such work. At the same time, the uncertainties of the determination of the zeroth moment could be avoided.

It is hoped that this chapter may help to finalize and reconcile the not always consistent literature concerning diatom polarizabilities.

Acknowledgments

The author wants to thank most cordially his colleagues in all parts of the world who have contributed generously in some form to this endeavor. Very special thanks are due to my dear friend, Dr. Michael H. Proffitt, whose relentless efforts resulted in some of the finest measurements known to date in this field.

References and Notes

1. J. P. C. McTague, and G. Birnbaum, *Phys. Rev. Lett.* **21**, 661 (1968).
2. J. P. McTague, and G. Birnbaum, *Phys. Rev.*, **A3**, 1376 (1971).
3. I. L. Fabelinskii, *Molecular Scattering of Light*, Plenum, New York, 1968.
4. J. P. McTague, W. D. Ellenson, and L. H. Hall, *J. Phys. (Paris)*, **33**, Cl-241 (1972).
5. W. S. Gelbart, *Adv. Chem. Phys.*, **26**, 1 (1974).
6. L. Silberstein, *Philos. Mag.*, **33**, 521 (1917).
7. T. M. Miller, and B. Bederson, *Adv. At. Mol. Phys.*, **13**, 1 (1977).
8. A. Dalgarno, A. L. Ford, and J. C. Browne, *Phys. Rev. Lett.*, **27**, 1033 (1971).
9. L. Jansen, and P. Mazur, *Physica*, **21**, 193 (1955).
10. L. Jansen, and P. Mazur, *Physica*, **21**, 208 (1955).
11. P. R. Certain, and P. J. Fortune, *J. Chem. Phys.*, **55**, 5818 (1971).

12. A. D. Buckingham, *Trans. Faraday Soc.*, **52**, 1035 (1956).
13. W. Kolos, and L. Wolniewicz, *J. Chem. Phys.*, **46**, 1426 (1967).
14. A. L. Ford, and J. C. Browne, *Phys. Rev.*, **A7**, 418 (1973).
15. D. B. DuPre, and J. P. McTague, *J. Chem. Phys.*, **50**, 2024 (1969).
16. T. K. Lim, B. Linder, and R. A. Kromhout, *J. Chem. Phys.*, **52**, 3831 (1970).
17. A. D. Buckingham, and R. S. Watts, *Mol. Phys.*, **26**, 7 (1973).
18. E. F. O'Brien, V. P. Gutschick, V. McKoy, and J. P. McTague, *Phys. Rev.*, **A8**, 690 (1973).
19. P. J. Fortune, and P. R. Certain, *J. Chem. Phys.*, **61**, 2620 (1974).
20. P. D. Dacre, *Mol. Phys.*, **36**, 541 (1978).
21. J. W. Kress, and J. J. Kozak, *J. Chem. Phys.*, **66**, 4516 (1977).
22. P. Lallemand, D. J. David, and B. Bigot, *Mol. Phys.*, **27**, 1029 (1974).
23. D. W. Oxtoby, and W. M. Gelbart, *Mol. Phys.*, **29**, 1569 (1975).
24. D. W. Oxtoby, and W. M. Gelbart, *Mol. Phys.*, **30**, 535 (1975).
25. R. A. Harris, D. F. Heller, and W. M. Gelbart, *J. Chem. Phys.*, **61**, 3854 (1974).
26. D. F. Heller, R. A. Harris, and W. M. Gelbart, *J. Chem. Phys.*, **62**, 1947 (1975).
27. K. L. Clarke, P. A. Madden, and A. D. Buckingham, *Mol. Phys.*, **36**, 301 (1978).
28. R. R. Teachout, and R. T. Pack, *Atomic Data*, **3**, 195 (1971).
29. J. G. Kirkwood, *J. Chem. Phys.*, **4**, 592 (1936).
30. A. N. Kaufman, and K. M. Watson, *Phys. Fluids*, **4**, 931 (1961).
31. A. D. Buckingham, and J. A. Pople, *Trans. Faraday Soc.*, **51**, 1029 (1955).
32. A. D. Buckingham, and J. A. Pople, *J. Chem. Phys.*, **27**, 820 (1956).
33. T. L. Hill, *J. Chem. Phys.*, **28**, 61 (1958).
34. D. A. McQuarrie, and H. B. Levine, *Physica*, **31**, 749 (1965).
35. R. H. Orcutt, and R. H. Cole, *J. Chem. Phys.*, **46**, 697 (1967).
36. D. Vidal, and P. M. Lallemand, *J. Chem. Phys.*, **64**, 4293 (1976).
37. E. B. Smith, *Physica*, **73**, 211 (1974).
38. L. W. Bruch, P. J. Fortune, and D. H. Berman, *J. Chem. Phys.*, **61**, 2626 (1974).
39. J. F. Ely, and D. A. McQuarrie, *J. Chem. Phys.*, **54**, 2885 (1971).
40. P. J. Fortune, P. R. Certain, and L. W. Bruch, *Chem. Phys. Lett.*, **27**, 233 (1974).
41. H. G. Bennewitz, H. Busse, H. D. Dohmann, D. E. Oakes, and W. Schrader, *Z. Phys.* **253**, 435 (1972).
42. R. A. Aziz, V. P. S. Nain, J. S. Carley, W. L. Taylor, and G. T. McConville, *J. Chem. Phys.*, **70**, 4330 (1979).
43. E. C. Kerr and R. H. Sherman, *J. Low Temp. Phys.*, **3**, 451 (1970).
44. M. H. Proffitt, and L. Frommhold, *Phys. Rev. Lett.*, **42**, 1473 (1979).
45. M. H. Proffitt, and L. Frommhold, *J. Chem. Phys.*, **72**, 1377 (1980).
46. L. Frommhold, and M. H. Proffitt, *Chem. Phys. Lett.*, **66**, 210 (1979).
47. L. Frommhold, and M. H. Proffitt, *Phys. Rev.*, **A21**, 1279 (1980).
48. L. Frommhold, K. H. Hong, and M. H. Proffitt, *Mo. Phys.*, **35**, 665 (1978).
49. A. Sugawara, S. Yip, and L. Sirovich, *Phys. Fluids*, **11**, 925 (1968).
50. N. A. Clark, *Phys. Rev.*, **A12**, 232 (1975).
51. M. H. Proffitt, and L. Frommhold, *Rev. Sci. Instr.*, **50**, 666 (1979).
52. J. C. Lewis, and J. van Kranendonk, *Phys. Rev. Lett.*, **24**, 802 (1970).
53. A. D. Buckingham, and J. A. Prople, *Proc. Phys. Soc. (Lond.)*, **A68**, 905 (1955).
54. A. D. Buckingham, *Proc. Phys. Soc. (Lond.)*, **A68**, 910 (1955).
55. A. D. Buckingham, and M. J. Stephen, *Trans. Faraday Soc.*, **53**, 884 (1957).
56. A. D. Buckingham, and D. A. Dunmur, *Trans. Faraday Soc.*, **64**, 1776 (1968).
57. M. N. Grasso, K. T. Chung, and R. P. Hurst, *Phys. Rev.*, **167**, 1 (1968).
58. L. Frommhold, and M. H. Proffitt, *J. Chem. Phys.*, **70**, 4803 (1979).

59. P. Lallemand, *J. Phys. (Paris)*, **33**, Cl-257 (1972).
60. L. Frommhold, and M. H. Proffitt, *Mol. Phys.*, **35**, 681 (1978).
61. P. Mazur, and M. Mandel, *Physica*, **22**, 289 (1956).
62. J. I. Gersten, *Phys. Rev.*, **A4**, 98 (1971).
63. D. J. G. Irwin, and A. D. Mav, *Can. J. Phys.*, **50**, 2174 (1972).
64. J. J. Barrett, and N. I. Adams III, *J. Opt. Soc. Am.*, **58**, 311 (1968).
65. A. Weber, in *The Raman Effect*, A. Anderson, Ed., New York (1973), Chap. 9.
66. C. M. Penny, S. T. Peters, and M. Lapp, *J. Opt. Soc. Am.*, **64**, 712 (1974); see also Bridge and Buckingham, *Proc. Roy. Soc. (Lond.)*, **A295**, 334 (1968); and Powell, Aval, Barrett, *J. Chem. Phys.*, **54**, 1960 (1971).
67. L. Frommhold, K. H. Hong, and M. H. Proffitt, *Mol. Phys.*, **35**, 691 (1978).
68. A. T. Prengel and W. S. Gornall, *Phys. Rev.*, **A13**, 253 (1976).
69. J. D. Jackson, *Classical Electrodynamics*, Wiley New York, 1962.
70. E. B. Wilson, Jr., J. C. Decius, and P. C. Cross, *Molecular Vibrations*, McGraw-Hill, New York 1955.
71. H. A. Kramers, and W. Heisenberg, *Z. Phys.*, **31**, 681 (1925).
72. C. Manneback, *Z. Phys.*, **62**, 224 (1930).
73. G. Placzek and E. Teller, *Z. Phys.*, **81**, 209 (1933).
74. N. Bernardes, and H. Primakoff, *J. Chem. Phys.*, **30**, 691 (1959).
75. H. B. Levine, *J. Chem. Phys.*, **56**, 2455 (1972).
76. P. Schofield, *Phys. Rev. Lett.*, **4**, 239 (1960).
77. L. Frommhold and M. H. Proffitt, in preparation.
78. L. Frommhold, *J. Chem. Phys.*, **63**, 1687 (1975).
79. M. J. Romanelli, *Mathematical Methods for Digital Computers*, Vol. 1, Wiley, New York.
80. J. W. Cooley, *Math. Compt.*, **5**, 363 (1961).
81. The author is indebted to Drs. J. C. Browne and E. Shipsey for their kind help with their computer codes.
82. See, for example, Ref. 88 below, where similar procedures are fully explained. However, many authors use such adjustments without fully specifying the exact details.
83. A. Ben-Reuven and N. D. Gershon, *J. Chem. Phys.*, **51**, 893 (1969).
84. H. B. Levine, *J. Chem. Phys.*, **56**, 2455 (1972).
85. R. W. Hartye, C. G. Cray, J. D. Poll, and M. S. Miller, *Mol. Phys.*, **29**, 825 (1975).
86. F. Barocchi, P. Mazzinghi, and M. Zoppi, *Phys. Rev. Lett.*, **41**, 1785 (1978).
87. F. Barocchi M. Zoppi, to appear.
88. Y. LeDuff, *Phys. Rev.*, **A20**, 48 (1979).
89. J. O. Hirschfelder, C. F. Curtiss, and R. B. Bird, *Molecular Theory of Gases*, Wiley, New York, 1954.
90. T. N. E. Greeville, "Spline Functions, Interpolation, and Numerical Quadrature," in *Mathematical Methods for Digital Computers*, Vol. 2, Wiley, New York, 1968.
91. M. Bérard, and P. Lallemand, *Mol. Phys.*, **34**, 251 (1977).
92. P. E. Siska, J. M. Parson, T. P. Schafer, and Y. T. Lee, *J. Chem. Phys.*, **55**, 5762 (1971).
93. B. Brunetti, F. Pirani, F. Vecchiocatti, and E. Luzzattii, *Chem. Phys. Lett.*, **58**, 504 (1978).
94. Y. Tanaka, and K. Yoshineo, *J. Chem. Phys.*, **53**, 2012 (1953).
95. S. Sengupta, *Chem. Phys. Lett.*, **59**, 423 (1978).
96. L. Frommhold, *J. Chem. Phys.*, **61**, 2996 (1974).
97. A. Dalgarno, and A. E. Kingston, *Proc. Roy. Soc. (Lond.)*, **A259**, 424 (1960).
98. J. von Kranendonk, J. E. Sipe, Mol. Phys. **35**, 1579 (1978).
99. M. Thibeau, B. Oksengorn, and B. Vodar, *J. Phys. (Paris)*, **29**, 287 (1968).
100. M. Thibeau, and B. Oksengorn, *J. Phys. (Paris)*, **33**, Cl-247 (1971).
101. P. Lallemand, *Phys. Rev. Lett.*, **25**, 1079 (1970).

102. P. Lallemand, *J. Phys. (Paris)*, **32**, 119 (1971).
103. P. Lallemand, *C. R.*, **B273**, 89 (1971).
104. P. Lallemand, *J. Chem. Phys.*, **33**, C1-257 (1972).
105. L. Frommhold, *J. Chem. Phys.*, **63**, 1687 (1975).
106. F. Barocchi and M. Zoppi, *Phys. Lett.*, **A66**, 99 (1978).
107. F. Barocchi and M. Zoppi, *Phys. Lett.*, **69A**, 187 (1978).
108. F. Barocchi and M. Zoppi, International School of Physics "Enrico Fermi," Verenna, Italy, 1978.
109. F. Barocchi and M. Zoppi, *J. Chem. Phys.*, **65**, 901 (1976).
110. F. Barocchi, M. Neri, and M. Zoppi, *Chem. Phys. Lett.*, **59**, 537 (1978).
111. D. P. Shelton, and G. C. Tabisz, Molecular Physics, to appear.
112. F. Barocchi, M. Neri, and M. Zoppi, *Mol. Phys.*, **34**, 1391 (1977).
113. F. Barocchi, M. Neri, and M. Zoppi, *J. Chem. Phys.*, **66**, 3308 (1977).
114. The author is grateful to Dr. F. Barocchi for sending us his tabulated intensities of the argon spectrum by Barocchi and Zoppi.
115. J. M. Parson, P. E. Siska, and Y. L. Lee, *J. Chem. Phys.*, **56**, 1511 (1972).
116. M. V. Bobetic and J. A. Barker, *Phys. Rev.*, **B2**, 4169 (1970).
117. J. H. Dymond, and B. J. Alder, *J. Chem. Phys.*, **51**, 309 (1969).
118. R. A. Aziz, and H. H. Chen, *J. Chem. Phys.*, **67**, 5719 (1977).
119. H. B. Levine, and G. Birnbaum, *J. Chem. Phys.*, **55**, 2914 (1971).
120. M. Thibeau, G. C. Tabisz, B. Oksengorn, and B. Vodar, *J. Quant. Spectrosc. Radiat. Transfer*, **10**, 839 (1970).
121. H. B. Levine, and G. Birnbaum, *J. Chem. Phys.*, **55**, 2914 (1971).
122. D. Frenkel, and J. P. McTague, *J. Chem. Phys.*, **70**, 2695 (1979).
123. J. M. Farrar, T. P. Schafer, and Y. T. Lee, in *Transport Phenomena*, J. Kestin, Ed., Proceedings AIP Conference No. 11.
124. F. Barocchi, and J. P. McTague, *Phys. Lett.*, **53A**, 488 (1975).
125. F. Barocchi, M. Zoppi, D. P. Shelton, and G. C. Tabisz, *Can. J. Phys.*, **55**, 1962 (1977).
126. M. H. Proffitt, and L. Frommhold, *Chem. Phys.*, **36**, 197 (1979).
127. A. D. Buckingham, and B. J. Orr, *Trans. Faraday Soc.* **65**, 673 (1969).
128. R. C. Watson, and R. L. Rowell, *J. Chem. Phys.*, **61**, 2666 (1974).
129. A. Michels, and G. W. Nedergragt, *Physica*, **2**, 1000 (1935).
130. D. P. Shelton, M. S. Mathor, and G. C. Tabisz, "Molecular Spectroscopy of Dense Phases," in *Proceedings of the 12th European Congress on Molecular Spectroscopy*, Strasbourg, France, July 1–4, 1975, Elsevier, Amsterdam, 1976.
131. D. P. Shelton, M. S. Mathur, and G. C. Tabisz, *Phys. Rev.*, **A11** (1975) 834.
132. D. P. Shelton, and G. C. Tabisz, *Phys. Rev.*, **A11** (1975) 1471.
133. H. Sutter, *Dielectric and Related Molecular Processes, Spec. Period. Rep., Chem. Soc. Lond.*, **1**, 65 (1978).
134. K. L. Clarke, Ph.D. Thesis, University of Cambridge, England, 1978.
135. See, however, recent work by L. Frommhold, M. H. Proffitt, *J. Chem. Phys.*, to appear January 1, 1981.

Notes added in proof: The reader's attention is directed to the brief notes concerning important new work, which were added in proof on page 40 and 53.

THE ONSET OF CHAOTIC MOTION IN DYNAMICAL SYSTEMS

MICHAEL TABOR

Noyes Laboratory of Chemical Physics, California Institute of Technology, Pasadena, California

CONTENTS

I. INTRODUCTION

Major new results in classical mechanics have provided great insight into the behavior of dynamical systems. The exactly soluble systems of classical mechanics, usually referred to as "integrable" systems, display motion, which is, in a certain sense, well ordered and "regular." By this we mean that the associated trajectories are forever confined to well-defined regions of phase space and show few changes in character when small changes in initial conditions are made. The introduction of even a small perturbation to the system can cause a very dramatic change in the dynamics. Some measure of the regular motion in still preserved but there appears a regime of highly "irregular" motion. In this regime the trajectories are very sensitive to small changes in initial conditions and can wander, in an erratic manner, over large portions of the energetically accessible phase space. The regions of regular and irregular motion are interwoven in a most complicated way and as the perturbation strength is increased the irregular regions grow in size. It must be emphasized that the irregular trajectories are still absolutely deterministic and it is unfortunate that they are often referred to as "stochastic trajectories." Therefore, in this chapter, we make a special effort not to use the term stochastic. Rather we variously use the terms "irregular," "chaotic," and "unstable" to describe such motion. The effect of a small perturbation on an integrable system has been formalized in the celebrated KAM theorem and demonstrated in many numerical experiments.

The presence of an irregular, or chaotic, regime is of great importance in many areas of physics and chemistry. These include, to name but a few, the foundations of statistical mechanics,[1] accelerator[2] and plasma physics,[3] astronomy,[4] and theories of chemical kinetics.[5, 6] There are also important implications for the semiclassical limit of quantum mechanics.[7] Clearly, in all these fields it is of central importance to understand the onset and growth of the chaotic regime. There are a number of standard methods available to distinguish between regular and irregular motion. However, these methods only apply to individual trajectories. Thus of great importance are methods that provide some, hopefully simple, predictions and insights into the appearance of more widespread chaotic motion.

The aim of this chapter is not to give an exhaustive survey of all the many occurrences of irregular motion, but to concentrate on describing some of the methods proposed to predict a "chaotic transition" in dynamical systems, that is, the transition from predominantly regular to predominantly irregular motion. There are now a number of excellent reviews[8–11] available describing the KAM theorem and its implications for dynamical systems. However, to make this chapter self-contained we include a summary of the necessary background before going on to discuss the problems and methods of attempting to predict a "chaotic transition." There have already been some discussions[6, 12] of this matter with particular reference to the problems of energy randomization in molecules. In this chapter we concentrate on the methods themselves rather than on particular applications. We conclude with a section on the implications of the onset of irregular motion for quantum mechanics. There has already been an excellent review by Percival[13] that discusses these matters. Here we examine some of the most recent work in this area.

II. REGULAR AND IRREGULAR MOTION

We begin by considering a conservative Hamiltonian system of N degrees of freedom, that is,

$$E=H(\mathbf{p},\mathbf{q}); \qquad \mathbf{p}=p_1,\cdots p_N$$
$$\mathbf{q}=q_1,\cdots q_N \tag{2.1}$$

for which we have the usual equations of motion

$$\dot{\mathbf{p}}=-\nabla_{\mathbf{q}}H(\mathbf{p},\mathbf{q}) \tag{2.2a}$$

and

$$\dot{\mathbf{q}}=\nabla_{\mathbf{p}}H(\mathbf{p},\mathbf{q}) \tag{2.2b}$$

If the system has N First Integrals in involution, that is, N single-valued, analytic functions of $\mathbf{p}$ and $\mathbf{q}$, $f_i(\mathbf{p},\mathbf{q})$, such that

$$\{f_i,f_j\}=0 \qquad i,j=1,\cdots N \tag{2.3}$$

where { } denotes the Poisson-bracket, then the system can be shown to be integrable by quadrature.[14, 15] Obvious examples of integrable systems are those for which the Hamiltonian can be separated into N independent parts of the form

$$H(\mathbf{p},\mathbf{q})=\sum_{i=1}^{N}H_i(p_i,q_i) \tag{2.4}$$

Here the First Integrals are just the individual "mode energies," that is $f_i = H_i(p_i,q_i)$. However, "separability" of this sort is only a sufficient condition for integrability, not a necessary one. Given the existence of N First Integrals it may be shown that the trajectories are quasiperiodic (see below) and are confined to manifolds with the topology of N-dimensional tori.[14, 15] These manifolds are often referred to as invariant tori. In such cases it is particularly convenient to introduce the canonical variables known as the action-angle variables $(\mathbf{I},\boldsymbol{\theta})$.[16] We then have

$$H=H(\mathbf{I}) \qquad \mathbf{I}=I_1,\cdots I_N \tag{2.5}$$

with equations of motion

$$\dot{\mathbf{I}}=-\nabla_{\boldsymbol{\theta}}H(\mathbf{I})=0 \tag{2.6a}$$

and

$$\dot{\boldsymbol{\theta}} = \nabla_{\mathbf{I}} H(\mathbf{I}) \equiv \boldsymbol{\omega}(\mathbf{I}) \tag{2.6b}$$

which are easily integrated to give

$$\mathbf{I} = \text{constant} \tag{2.7a}$$

and

$$\boldsymbol{\theta} = \boldsymbol{\omega}(\mathbf{I})t + \boldsymbol{\delta} \tag{2.7b}$$

where $\boldsymbol{\delta}$ are a set of arbitrary phases. Each value of the action vector $\mathbf{I}$ specifies the position of a torus in phase space, and the corresponding vector of angles $\boldsymbol{\theta}$ specifies the position of a trajectory on that torus. The angle variables are periodic in 2π (this is a simple consequence of the manifolds having toroidal topology) and we can thus express the original motion in terms of Fourier series, that is,

$$\mathbf{q}(t) = \sum_{\mathbf{m}} \mathbf{a}_{\mathbf{m}}(\mathbf{I}) e^{i\mathbf{m}\cdot\boldsymbol{\theta}(t)} = \sum_{\mathbf{m}} \mathbf{a}_{\mathbf{m}}(\mathbf{I}) e^{i\mathbf{m}\cdot(\boldsymbol{\omega}t+\boldsymbol{\delta})} \tag{2.8a}$$

$$\mathbf{p}(t) = \sum_{\mathbf{m}} \mathbf{b}_{\mathbf{m}}(\mathbf{I}) e^{i\mathbf{m}\cdot\boldsymbol{\theta}(t)} = \sum_{\mathbf{m}} \mathbf{b}_{\mathbf{m}}(\mathbf{I}) e^{i\mathbf{m}\cdot(\boldsymbol{\omega}t+\boldsymbol{\delta})} \tag{2.8b}$$

where the $\mathbf{a}_{\mathbf{m}}(\mathbf{I})$ and $\mathbf{b}_{\mathbf{m}}(\mathbf{I})$ are N-dimensional vectors of Fourier coefficients, each component of which is labeled by an N-dimensional vector, $\mathbf{m}$, of indices. It is in this sense that we refer to the motion as being quasiperiodic. Quasiperiodic trajectories correspond to regular motion.

Notice that in (2.6) we have assigned the frequency vector $\boldsymbol{\omega}$ as a function of the actions $\mathbf{I}$, that is, the frequencies depend on the torus. In this (typical) case the system is said to be nondegenerate and a general nondegeneracy condition can be written as

$$\det\left|\frac{\partial^2 H}{\partial I_i \partial I_j}\right| \neq 0 \tag{2.9}$$

It is instructive to examine the surfaces of constant energy drawn in action space. Thus for a system of two degrees of freedom we plot the contours determined by $E = H(I_1, I_2)$ in the I_1, I_2 plane. In Fig. 1 we plot such a contour for a nondegenerate system. Every point on a given contour corresponds to a different torus in phase space. From (2.6) we can see that the normal at each point gives the frequency vector associated with that torus. Since this is a nondegenerate system the normal has a different direction at

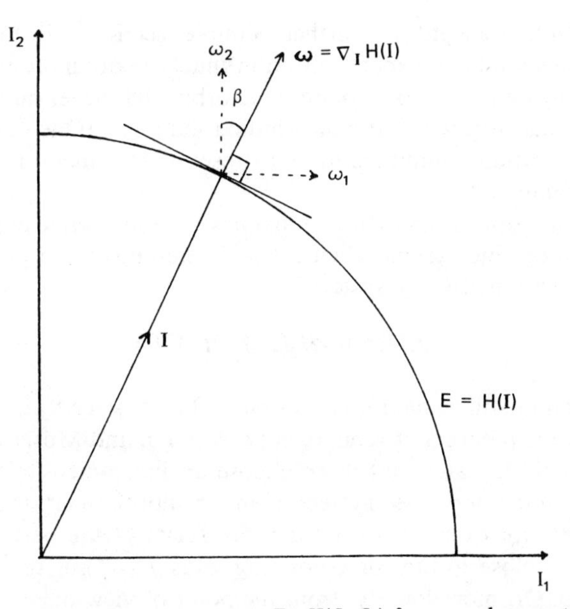

Fig. 1. A typical surface of constant energy $E=H(I_1, I_2)$ for a nondegenerate system. The frequency vector $\boldsymbol{\omega}$ is normal to this surface. The ratio of frequency components is determined by $\tan\beta=(\omega_1/\omega_2)$.

each point along the contour. By way of contrast, consider a pair of uncoupled harmonic oscillators. The Hamiltonian in action variables is given by

$$H(I_1, I_2)=I_1\omega_1+I_2\omega_2 \tag{2.10}$$

and the surfaces of constant energy are just straight lines. In this highly degenerate case all normals have the same direction, since ω_1 and ω_2 are constant frequencies.

Each torus can be characterized by a winding number, α, which is simply the ratio of the two frequencies (given by $\tan\beta$ in Fig. 1), that is

$$\alpha=\frac{\omega_1}{\omega_2} \tag{2.11}$$

When α is rational, that is, $\alpha=P/Q$ where P and Q are integers, the torus is covered by a family of closed orbits. Each member of the family closes after P cycles of θ_1 and Q cycles of θ_2. The integers P and Q completely specify the topology of the closed orbit. For example, if $\alpha=\frac{1}{1}$ the orbit traces out a simple closed circle in configuration space. If $\alpha=\frac{1}{2}$ the orbit has the topology of a figure eight (for further examples see Ref. 17). For typical integrable systems we can expect to find infinitely many more tori with irrational winding numbers, corresponding to orbits that never quite close upon themselves, than tori with rational winding numbers. (One can see this by realizing that rational numbers form a subset of zero measure in the space of all real numbers.)

The key question in analytical dynamics is to ask what happens to the tori of an integrable system when a (small) nonintegrable perturbation is introduced. That is, for the system

$$H(\mathbf{I},\boldsymbol{\theta})=H_0(\mathbf{I})+\varepsilon H_1(\mathbf{I},\boldsymbol{\theta}) \tag{2.12}$$

does the motion still remain quasiperiodic? The answer to this question is provided by the theorem of Kolmogorov, Arnol'd, and Moser (KAM).[14, 15, 18, 19] Essentially it states that if we choose an incommensurate set of frequencies $\boldsymbol{\omega}^*$ in the perturbed system, then for almost all $\boldsymbol{\omega}^*$ and sufficiently small ε, there will exist an invariant torus $T(\boldsymbol{\omega}^*)$ of the perturbed system $H(\mathbf{I},\boldsymbol{\theta})$ that is close to the corresponding torus $T_0(\boldsymbol{\omega}^*)$ of the unperturbed system $H_0(\mathbf{I})$. Or, more loosely, from the point of view of the unperturbed system, KAM tells us that most tori are preserved, albeit in slightly distorted form, for sufficiently small perturbation. The term "most" excludes those tori with commensurable frequencies and their immediate, in a delicate measure-theoretical sense†, neighbors. It is important to note that the theorem applies to nondegenerate systems, that is, (2.9) has to be satisfied. A similar theorem has also been proved for degenerate systems, such as systems of harmonic oscillators, with a condition on the commensurability of the fundamental frequencies.[20] One should also note that formally the "sufficiently small" value of ε is, in fact, incredibly small.[21] However, as we see later, numerical evidence suggests that tori are preserved under far greater perturbation than that suggested by the theorem.

The structure of the phase space of the perturbed system is immensely complicated. In the regions of phase space where tori have been destroyed, a remarkable structure of stable and unstable closed orbits appears. This

†More precisely, a torus with frequencies ω_1, ω_2 is not preserved if the ratio ω_1/ω_2 can be approximated by a rational fraction (see Section V) that falls within the bound

$$\left|\frac{\omega_1}{\omega_2}-\frac{r}{s}\right|<\frac{\kappa(\varepsilon)}{s^{2.5}}$$

The KAM theorem only tells us that $\kappa(\varepsilon)$ is small and that $\kappa(\varepsilon)\to 0$ as $\varepsilon\to 0$. In Section IV we describe a practical method, due to Chirikov, for estimating the "widths" of these zones.

structure is described in more detail below and in Section V. Near the unstable closed orbits one finds trajectories that are no longer confined to smooth manifolds but that behave in a highly erratic or chaotic manner. These are the irregular trajectories that can wander over large portions of the energetically available phase space. However, it should be stressed that for Hamiltonian systems one can generally not assume that these irregular trajectories are ergodic (in the strict sense of the word) over the whole of the energy shell, that is, one cannot assume that time average equals phase space average for these trajectories.

Before we go on to describe various methods of distinguishing between regular and irregular trajectories we make the following digression. The complicated structure of regular and irregular motion is not exclusive to nonintegrable Hamiltonian systems, but is, in fact, the generic[14] behavior for a much wider class of dynamical systems. As an immediate illustration of this consider the pair of difference equations

$$\begin{aligned} x_{n+1} &= x_n + y_n \\ y_{n+1} &= y_n - x_{n+1}^3 \end{aligned} \tag{2.13}$$

This "system" is an example of what is known as an area preserving algebraic mapping. It is area preserving in the sense that the Jacobian of transformation is equal to unity, that is,

$$\left| \frac{\partial(x_{n+1}, y_{n+1})}{\partial(x_n, y_n)} \right| = 1 \tag{2.14}$$

This is exactly analogous to Liouvilles theorem for Hamiltonian systems. For compactness we denote the transformation (2.13) by some operator T such that we can write

$$(x_{n+1}, y_{n+1}) = T(x_n, y_n) = T^n(x_0, y_0) \tag{2.15}$$

In Fig. 2. we plot a number of trajectories for this system in the (x, y) phase plane. Near the origin the orbits lie on smooth curves. These are usually referred to as invariant curves and are the analogues of the invariant tori of integrable Hamiltonian systems. Most striking is the outermost structure. The set of eight small elipses form what is known as a set of "islands." At the center of each island one can find a so-called fixed point of the mapping. These fixed points correspond to a closed orbit; in this case the orbit closes after eight iterations of the mapping, that is, $T^8(x_0, y_0) = (x_0, y_0)$. Because these fixed points are surrounded by smooth curves (the islands) the orbit is stable since a nearby trajectory will stay close to it. Such

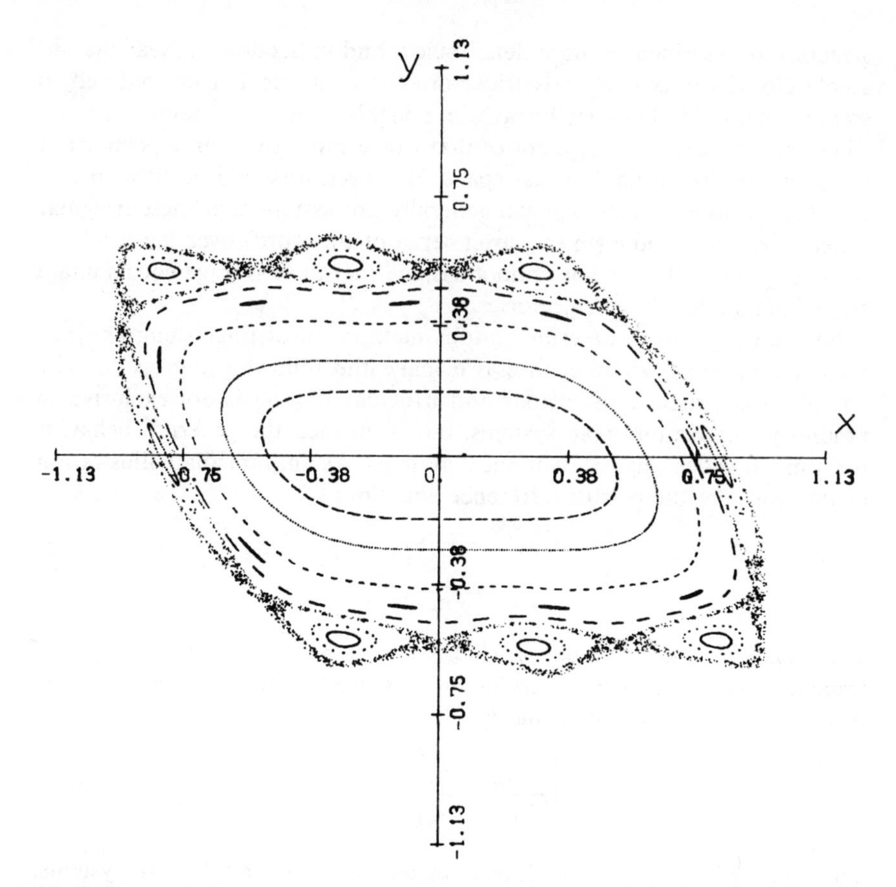

Fig. 2. Some typical phase plane trajectories of the area-preserving mapping given by (2.13). Reproduced by permission from Berry et al.[23]

fixed points are termed "stable" or "elliptic." In between the set of islands we see a completely different structure; in this case the pattern has been generated by 4000 iterations of a single trajectory. This trajectory is irregular, since it no longer lies on an invariant curve but fills up an area of the phase plane. Notice the dense hyperbolic structure that lies between each pair of islands. At the center of this structure one can again find a fixed point of the mapping. As before there is a set of eight fixed points, but these now correspond to an unstable closed orbit. These fixed points are now referred to as "unstable" or "hyperbolic." This structure of alternating stable and unstable fixed points is just like that alluded to above for the destroyed tori of Hamiltonian systems. More detailed discussions of mappings and fixed points can be found elsewhere.[10, 22] The point we are trying to make here is that the dynamics of area preserving mappings and Hamiltonian systems have a lot in common. Hence the study of difficult physical problems can sometimes be greatly simplified, without loss of generality, by the study of mappings. As examples of this we cite recent investigations of the semiclassical limit of quantum mechanics[23] and studies of linear response theory.[24] An obvious problem is to relate a particular mapping to a specific Hamiltonian. This is generally not easy, but for a class of Hamiltonians, corresponding to systems acted on by periodic forces, the Hamiltonian equations of motion take the form of exact difference equations. As we see in Section VII the mapping (2.13) is just such an example.

There are a number of methods at our disposal for distinguishing between regular and irregular trajectories. We discuss three of them below.

A. Surface of Section

This well-known technique is particularly suited for Hamiltonian systems of two degrees of freedom. Here one follows the successive crossings of a trajectory through a surface intersecting the energy shell, for example the (p_y, y) plane at the point $x=0$. The position of the system, at a given energy, on such a surface completely specifies its state to within a sign. (This is, of course, only true for systems with two degrees of freedom and which are quadratic in the momenta.) After a suitable number of crossings a pattern emerges. In the case of quasiperiodic motion the crossing points appear to lie on a smooth curve. This curve, often called an invariant curve, corresponds to the intersection of the torus, on which the trajectory lies, with the surface of section. Should the trajectory under consideration be one that does not close upon itself (irrational winding number), then the successive points eventually fill up the curve densely. This is a manifestation of the fact that such orbits are "ergodic" on the torus.[14] On the other hand, in the case of an orbit being closed, that is, a torus of commensurable frequencies, one only sees a finite number of fixed points; the number being determined by the winding number α (2.11). However, should the trajectory be irregular, no such smooth pattern emerges. Rather, one sees a random splatter of points that fills up some area. There have been many investigations, too numerous to mention here, of various Hamiltonian systems using the surface of section technique. The best known of these is that due to Henon and Heiles,[25] who investigated the Hamiltonian

$$H=\tfrac{1}{2}\left(p_x^2+p_y^2+x^2+y^2\right)+x^2y-\tfrac{1}{3}y^3 \tag{2.16}$$

In Fig. 3 we show surfaces of section obtained for a number of trajectories at energies $E=\frac{1}{12}$, $\frac{1}{8}$, and $\frac{1}{6}$. Below $E=\frac{1}{8}$ nearly all trajectories investigated

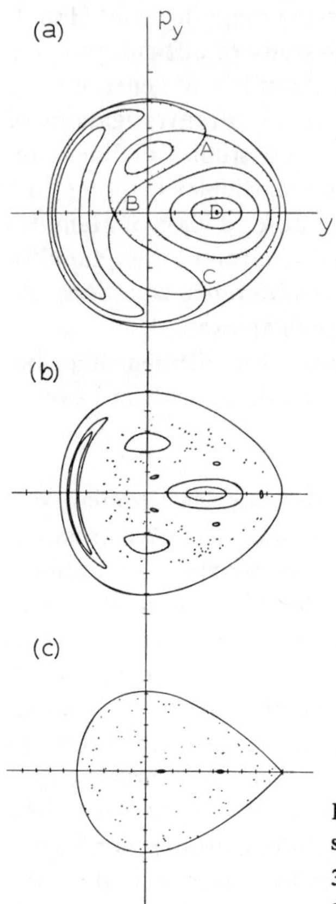

Fig. 3. Surfaces of section obtained for the Henon-Heiles system (2.16) at (*a*) $E=\frac{1}{12}$, (*b*) $E=\frac{1}{8}$, and (*c*) $E=\frac{1}{6}$. In Fig. 3*a* the points A, B, C, D correspond to certain special closed orbits. (Reproduced by permission from Ford[8].)

appear to lie on tori, but above this energy irregular trajectories start to appear in certain regions of the phase space, and by $E=\frac{1}{6}$ (at which the system dissociates) nearly all the trajectories investigated were chaotic. The fact that the surface of section pictures (e.g., Fig. 3*b*) look similar to the phase plane of the mapping (Fig. 2) is no coincidence, since the surface of section is nothing more than an area preserving mapping of the phase plane onto itself.[10, 14]

B. Separation of Trajectories

A characteristic of strongly chaotic systems is that trajectories with similar initial conditions separate exponentially in time. For a certain class of systems known as C-systems this is a rigorous result.[14] Although most Hamiltonian systems have not been proven to be C-systems, the irregular trajectories are also found to display an exponential-like separation, in contrast to quasiperiodic (regular) trajectories, which only separate linearly

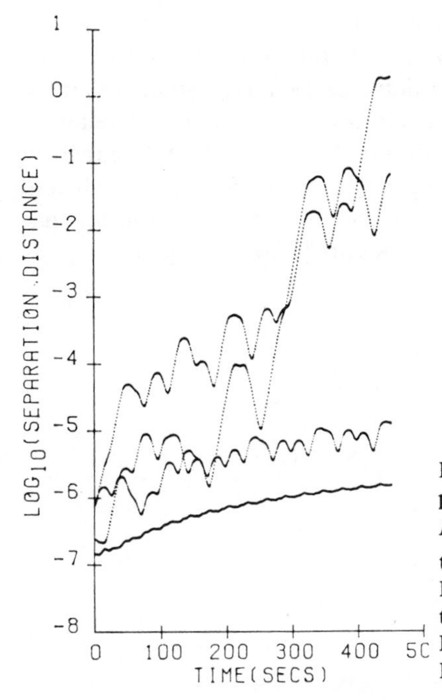

Fig. 4. Separation of four trajectory pairs in phase space for the Henon-Heiles system at $E=\frac{1}{8}$. The two upper curves correspond to trajectory pairs started in the chaotic regions of Fig. 3*b* and the two lower curves correspond to trajectory pairs started in the regular regions of Fig. 3*b*. (Reproduced by permission from Ford[8]).

in time. In Fig. 4 we show some results obtained for the Henon-Heiles system that illustrate this contrasting behavior. The rate of separation of diverging trajectories can be shown to be an "entropy-like"[26] quantity and this can sometimes provide a convenient means of classifying the irregular trajectories.[27] This matter is discussed at greater length by Brumer, who has also applied the technique to collisional trajectories in chemical reactions.[12, 28]

C. Spectral Properties

Interesting results are obtained when one investigates the spectrum (Fourier transform) of trajectories. In the case of quasiperiodic motion it is not difficult to see, using (2.8), that the spectrum will be a finite series of discrete lines. On the other hand, for an irregular trajectory the spectrum changes dramatically; a very "grassy," almost continuous spectrum is observed.[29] Some rigorous results are available. It may be shown that systems that are "weakly mixing" or more strongly irregular* will have an absolutely continuous spectrum. Ergodic systems can at most have a discrete

*There is a well-defined hierarchy of "chaotic" properties, for example, mixing is "more" chaotic than ergodic. More detailed discussions can be found in Refs. 14, 30, 50. It should also be pointed out that the theorems relating to spectra refer to evolution operators rather than individual trajectories.

spectrum.[30] At this stage it is not clear whether, given sufficient resolution,[31] the spectrum of an irregular trajectory of a Hamiltonian system would actually display absolutely continuous portions. One should also note that the spectrum of a family (ensemble) of trajectories will, of course, depend on the particular ensemble chosen (see Section VI). Overall, the change in spectrum as the motion changes from regular to irregular is impressive, and analogous behavior is observed in fluids as the flow changes from laminar to turbulent.[32] Spectra of classical trajectories are shown in Fig. 5.

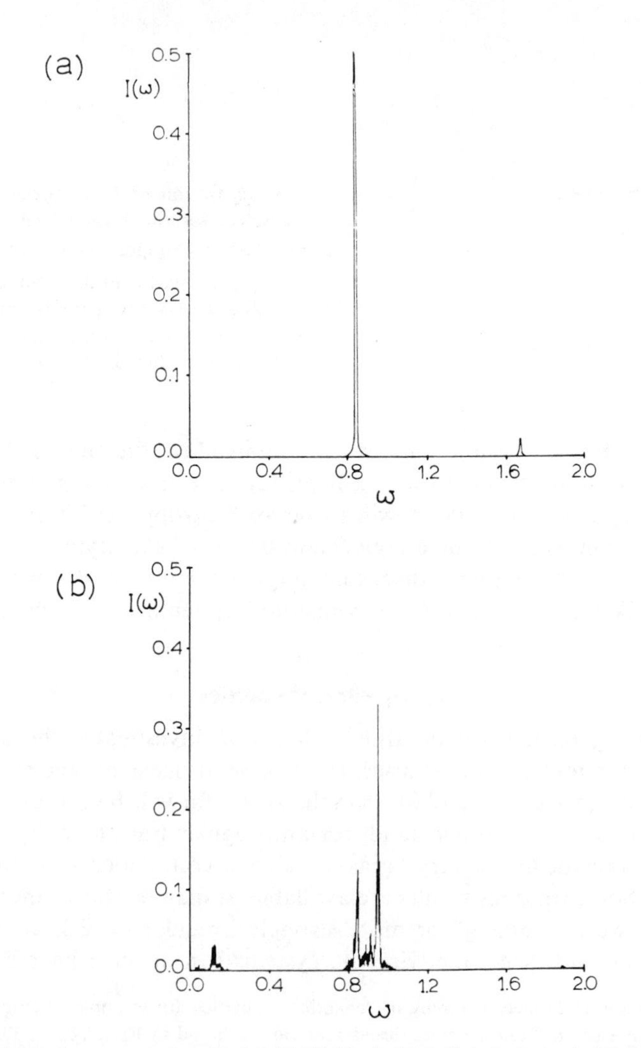

Fig. 5. Classical power spectra, $I(\omega)$, of individual classical trajectories $[x(t)+y(t)]$ of the Henon-Heiles system at $E=\frac{1}{8}$: (a) a regular trajectory and (b) an irregular trajectory.

All the above methods for detecting a transition to chaotic behavior refer, of course, to individual trajectories. If one investigates a sufficient number of trajectories, using one or more of the above techniques, over a wide range of energies one will obtain a picture of the overall dynamical properties of the system. The study carried out by Henon and Heiles using the surface of section technique led them to give an estimate of the measure of irregular trajectories as a function of energy. Their results, which are shown in Fig. 6, indicate a sharp transition to chaotic behavior at about $E=\frac{1}{8}$. Although more detailed investigations later found irregular trajectories at lower energies,[34] the general picture of a fairly dramatic transition, at $E=\frac{1}{8}$, to widespread chaotic motion still remained.

Another study of interest is that carried out by Ford and co-workers on a system derived from the Toda lattice Hamiltonian,[35] namely,

$$H=\frac{p_x^2}{2m_x}+\frac{p_y^2}{2m_y}+\frac{1}{24}\left[e^{(2y+2\sqrt{3}\,x)}+e^{(2y-2\sqrt{3}\,x)}+e^{-4y}\right]-\frac{1}{8} \tag{2.17}$$

Using both the surface of section and trajectory separation techniques it was concluded that for equal masses, $m_x=m_y$, the system was integrable at all energies.[36] Subsequent to this computer analysis Henon proved that the system was indeed integrable, that is, he showed the existence of two first integrals of the motion.[37] On the other hand, when the masses are "split" ($m_x \neq m_y$) irregular trajectories appear. This was investigated by Casati and Ford,[38] again using both trajectory separation and surface of section methods. Their results led them to a crude estimate of the energy, E_c, marking the onset of widespread irregular motion, as a function of mass ratio (see Fig. 7.)

Other simple Hamiltonian systems have also been subjected to detailed dynamical investigations. For example, the Barbanis system,[39] which takes

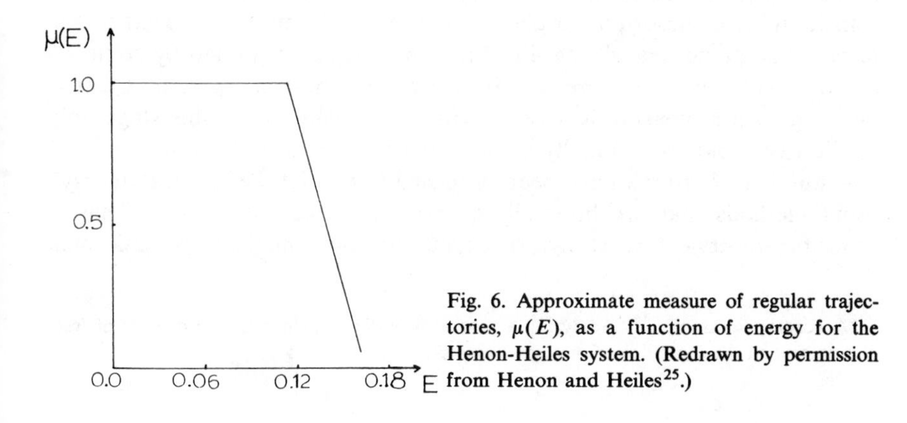

Fig. 6. Approximate measure of regular trajectories, $\mu(E)$, as a function of energy for the Henon-Heiles system. (Redrawn by permission from Henon and Heiles[25].)

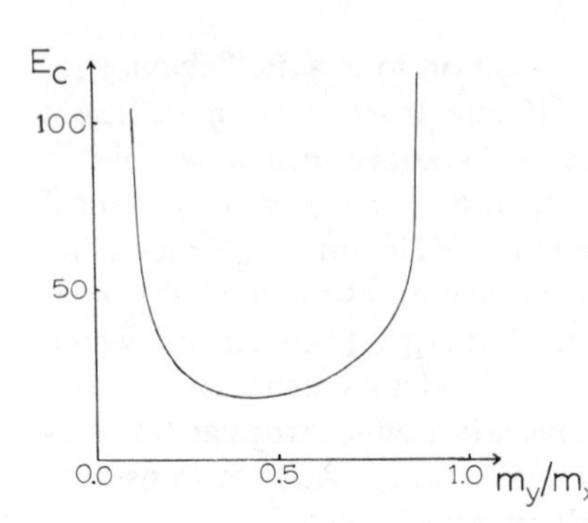

Fig. 7. Approximate "critical energy" as a function of mass ratio m_y/m_x obtained for Hamiltonian (2.17). (Redrawn by permission from Casati and Ford[38].)

the generalized form

$$H=\tfrac{1}{2}\left(p_x^2+p_y^2\right)+\tfrac{1}{2}\left(\omega_x^2x^2+\omega_y^2y^2\right)+\alpha x^2y \tag{2.18}$$

has been studied extensively, both numerically and analytically, by Contopoulos.[40] A number of systems devised with unimolecular kinetics in mind have also been studied in detail.[12, 41, 42] Unfortunately, not many systems of more than two degrees of freedom have been studied in detail.*[43]

In most of the above cases some form of a transition to a predominantly chaotic regime has been observed. This has led to the notion of a "critical energy" marking this transition. Such a notion is very attractive, but it must be treated with caution. The transition, although often quite rapid, is never sharp. Irregular trajectories can be found in regimes of predominantly regular motion and, conversely, regions of regular motion ("islets of stability") can be found well into the chaotic regime. Indeed, it should be emphasized that integrable and near-to-integrable systems are exceptional cases and that for typical (generic) Hamiltonian systems regular and irregular motions are intertwined, in a most complicated manner, right down to the lowest energies. Thus when one talks of a "critical energy" or a "chaotic transition" it is important to give some notion of "measure." That is, the term "widespread chaos" should, if possible, be accompanied by some indication as to what measure is implied by the term "widespread." Clearly, defining such a measure is a very difficult problem and at this stage only qualitative, if not occasionally subjective, estimates have been used.

A number of criteria have been proposed to predict the "critical energy" using methods that are hopefully simpler to implement than a detailed trajectory-by-trajectory study. In examining these methods we ask such

*According to Arnol'd,[15] "Analyzing a general potential system with two degrees of freedom is beyond the capability of modern science."

questions as: (*1*) Does the method have a sound theoretical basis? (*2*) Does it have some sense of measure built in, that is, how widespread is "widespread"?, and (*3*) Does it indicate in which areas of phase space the irregular trajectories first appear?

III. EXPONENTIAL SEPARATION OF TRAJECTORIES

A number of criteria have been proposed to predict the onset of chaotic motion by attempting to find the energy at which adjacent trajectories will display exponential separation. As was discussed in the preceding section, such behavior is a characteristic of the chaotic regime.

We now study a system of N degrees of freedom with a Hamiltonian of the general form

$$H(\mathbf{p},\mathbf{q})=\frac{\mathbf{p}\cdot\mathbf{p}}{2}+V(\mathbf{q}) \tag{3.1}$$

where $\mathbf{p}$ and $\mathbf{q}$ are the N-dimensional vectors of momentum and conjugate coordinate, respectively. The equations of motion are therefore

$$\dot{\mathbf{q}}=\nabla_{\mathbf{p}}H(\mathbf{p},\mathbf{q})=\mathbf{p} \tag{3.2a}$$

$$\dot{\mathbf{p}}=-\nabla_{\mathbf{q}}H(\mathbf{p},\mathbf{q})=-\nabla_{\mathbf{q}}V(\mathbf{q}) \tag{3.2b}$$

Consider some trajectory $\mathbf{q}^{(1)}(t)$. The dynamics are governed by the equations

$$\dot{\mathbf{q}}^{(1)}=\mathbf{p}^{(1)} \tag{3.3a}$$

$$\dot{\mathbf{p}}^{(1)}=-\left[\nabla_{\mathbf{q}}V(\mathbf{q})\right]_{\mathbf{q}=\mathbf{q}^{(1)}(t)} \tag{3.3b}$$

We refer to this trajectory as the reference trajectory. Now take some other (nearby) trajectory $\mathbf{q}^{(2)}(t)$ with corresponding equations of motion

$$\dot{\mathbf{q}}^{(2)}=\mathbf{p}^{(2)} \tag{3.4a}$$

$$\dot{\mathbf{p}}^{(2)}=-\left[\nabla_{\mathbf{q}}V(\mathbf{q})\right]_{\mathbf{q}=\mathbf{q}^{(2)}(t)} \tag{3.4b}$$

We now introduce the variables $\boldsymbol{\xi}$ and $\boldsymbol{\eta}$, which measure the separation of the two trajectories in coordinate and momentum space, respectively.

$$\boldsymbol{\xi}(t)=\mathbf{q}^{(1)}(t)-\mathbf{q}^{(2)}(t) \tag{3.5a}$$

$$\boldsymbol{\eta}(t)=\mathbf{p}^{(1)}(t)-\mathbf{p}^{(2)}(t) \tag{3.5b}$$

For $\mathbf{q}^{(2)}(t)$ sufficiently close to the reference trajectory we may expand the equations for $\dot{\mathbf{q}}^{(2)}$ and $\dot{\mathbf{p}}^{(2)}$ about this trajectory and hence obtain the linearized equations of motion for $\boldsymbol{\xi}$ and $\boldsymbol{\eta}$. Hence

$$\dot{\boldsymbol{\xi}} = \boldsymbol{\eta} \tag{3.6a}$$

$$\dot{\boldsymbol{\eta}} = -\mathbf{V}(t)\cdot\boldsymbol{\xi} \tag{3.6b}$$

where $\mathbf{V}(t)$ is the $N \times N$ matrix of second derivatives of the potential $V(\mathbf{q})$ evaluated along the reference trajectory, that is,

$$[\mathbf{V}(t)]_{ij} = \left(\frac{\partial^2 V(\mathbf{q})}{\partial q_i \partial q_j} \right)_{\mathbf{q}=\mathbf{q}^{(1)}(t)} \tag{3.7}$$

The stability of the motion [as described by (3.6)] is determined by the time-dependent eigenvalues of the $2N \times 2N$ matrix

$$\mathbf{M} = \begin{bmatrix} \mathbf{0} & \mathbf{1} \\ -\mathbf{V}(t) & \mathbf{0} \end{bmatrix} \tag{3.8}$$

where $\mathbf{0}$ and $\mathbf{1}$ are the $N \times N$ null and unit matrices, respectively. Given that one can find the time-dependent transformation $\mathbf{T}$ that diagonalizes $\mathbf{M}$, that is,

$$(\mathbf{T}\mathbf{M}\mathbf{T}^{-1})_{ij} = \lambda_i(t)\,\delta_{ij} \tag{3.9}$$

the motion is governed by equations of the form $\exp[\int_0^t dt'\lambda(t')]$. If any one of the eigenvalues is real, the trajectory separation grows exponentially and the motion is deemed unstable. Conversely, imaginary eigenvalues correspond to stable motion. Clearly though, the eigenvalues, and hence the stability of the motion, can charge character as a function of time. We return to this very important point later.

Solving (3.9) exactly is enormously difficult. The problem can be greatly simplified, however, by taking the following mathematically dubious step. Although the linearized equations (3.6a) and (3.6b) are *only* meaningful when defined with respect to a specific phase space trajectory, that is, the reference trajectory, it has been proposed[44, 45] that the time dependence of $\mathbf{V}(t)$ (and hence $\mathbf{M}$) can be removed by replacing the time-dependent phase point $\mathbf{q}^{(1)}(t)$ by the time-independent phase space coordinate $\mathbf{q}$. This reduces equations (3.6) to a pair of linear autonomous differential equations, that is,

$$\dot{\boldsymbol{\xi}} = \boldsymbol{\eta} \tag{3.10a}$$

$$\dot{\eta} = -\mathbf{V}(\mathbf{q})\cdot\boldsymbol{\xi} \tag{3.10b}$$

with

$$[\mathbf{V}(\mathbf{q})]_{ij} = \frac{\partial^2 V(\mathbf{q})}{\partial q_j \partial q_j} \tag{3.11}$$

where the coordinates $\mathbf{q}$ are regarded as time-independent parameters. The stability of the motion can now be determined by examining the so-called critical points of (3.10a) and (3.10b). These are the equilibrium points of the motion and are determined by the condition $\dot{\boldsymbol{\xi}} = \dot{\boldsymbol{\eta}} = \mathbf{0}$. In this case there is a single critical point at $(\boldsymbol{\eta}, \boldsymbol{\xi}) = (0, 0)$. The stability of motion in the neighborhood of this point is determined by the eigenvalues of the time-independent version of the matrix $\mathbf{M}$, that is, (3.8) with $\mathbf{V}(t)$ replaced by $\mathbf{V}(\mathbf{q})$. As a specific example consider a system of two degrees of freedom. The eigenvalues are determined from

$$\det \begin{vmatrix} -\lambda & 0 & 1 & 0 \\ 0 & -\lambda & 0 & 1 \\ -\mathbf{V}_{11} & -\mathbf{V}_{12} & -\lambda & 0 \\ -\mathbf{V}_{21} & -\mathbf{V}_{22} & 0 & -\lambda \end{vmatrix} = 0 \tag{3.12}$$

which yields the two pairs of roots

$$\lambda_{\pm} = \pm \left[-b \pm \sqrt{b^2 - 4c} \, \right]^{1/2} \tag{3.13}$$

where

$$b = \frac{\partial^2 V}{\partial q_1^2} + \frac{\partial^2 V}{\partial q_2^2} \tag{3.14}$$

and

$$c = \frac{\partial^2 V}{\partial q_1^2} \frac{\partial^2 V}{\partial q_2^2} - \left(\frac{\partial^2 V}{\partial q_1 \partial q_2} \right)^2 \tag{3.15}$$

We assume $b > 0$. Then providing c is positive the roots, $\lambda_{\pm}$, are pure imaginary and the motion, which is now governed by equations of the form $\exp(\lambda_{\pm} t)$, is stable, that is, the critical point is elliptic (stable). However, if c becomes negative a pair of roots will become real, leading to exponential divergence of trajectories, that is, the critical point is hyperbolic (unstable). The quantity c has the same sign as the Gaussian curvature of the potential

$V(\mathbf{q})$ and one is thus led to associate a change from stable to unstable motion with a change in curvature of the potential; that is, one assumes that as a trajectory pair passes into the regions of negative curvature of the potential it will start to separate exponentially. This contention has been tested numerically for the Henon-Heiles system.[46, 47]

For simple potentials it is a fairly straightforward matter to determine the energy at which such a change will occur. This approach was first devised by Toda,[44] who applied it to the Henon-Heiles system. For this potential the Gaussian curvature changes sign at $E=\frac{1}{12}$, which is in quite good agreement with the observed transition to predominantly chaotic motion around $E=\frac{1}{8}$. This method was also derived by Brumer and Duff,[45] who discuss the circumstances under which the time dependence of $\mathbf{V}(t)$ can be neglected (see below). They applied the method to a number of other simple systems and again obtained good results. At first sight, then, it might appear that we have a useful criterion for predicting the "chaotic transition," but closer inspection suggests that this approach is not so satisfactory. One obvious missing ingredient is that the method includes no sense of measure, that is, one has no notion of whether the chaos is widespread or not. Also, the method fails for numerous other systems. Any system that is integrable but whose potential displays an inflection, such as smooth separable potentials with repulsive cores and attractive tails, will be predicted to show a chaotic transition. Systems of the Toda lattice type (2.17) will be deemed stable even when the masses are unequal. On the other hand, the Kepler potential, which has everywhere negative curvature, will be deemed unstable whatever the mass ratios.[48]

However, it should not be too surprising that this method has such serious defects since its derivation has little justification. The key step is to go from the nonautonomous set of equations (3.6a) and (3.6b) to the autonomous set (3.10a) and (3.10b), by replacing the reference trajectory by an arbitrary value of the phase space coordinate. Such a precedure may possibly be valid in a sufficiently small time interval over which the matrix $\mathbf{V}(t)$ can be considered approximately constant. In this sense the stability analysis can only be of a highly "*local*" nature, since the true stability of the motion is determined by the time-dependent eigenvalues (3.9) of the original time-dependent matrix $\mathbf{M}$. This being so, it is not clear that an initial exponential divergence of nearby trajectories will persist "*globally*." In their critique of the Toda method, Benettin et al[47] suggest that there is little connection between local and global instabilities for Hamiltonian systems. The opposite point of view has been taken by Brumer[12, 45] and also by Cerjan and Reinhardt.[46]

The fact that the Toda-Brumer-Duff approach still manages to work in a number of cases has led Cerjan and Reinhardt[46] to consider the method further. In this work the "separation" equations are not linearized, but are left in their exact form, that is,

$$\dot{\boldsymbol{\xi}}=\boldsymbol{\eta} \tag{3.16a}$$

$$\dot{\boldsymbol{\eta}}=f(\mathbf{q}^{(1)}(t),\boldsymbol{\xi}) \tag{3.16b}$$

where f is some (nonlinear) function of the trajectory separation $\boldsymbol{\xi}(t)$ and the reference trajectory $\mathbf{q}^{(1)}(t)$, that is,

$$f(\mathbf{q}^{(1)}(t),\boldsymbol{\xi})=-\left[\nabla_{\mathbf{q}}V(\mathbf{q})\right]_{\mathbf{q}=\mathbf{q}^{(1)}(t)}+\left[\nabla_{\mathbf{q}}V(\mathbf{q})\right]_{\mathbf{q}=\mathbf{q}^{(1)}(t)-\boldsymbol{\xi}(t)} \tag{3.17}$$

As such (3.16a) and (3.16b) are equivalent to the original equations of motion (3.3) and (3.4). If one again makes the same basic approximation of replacing the specific phase space trajectory $\mathbf{q}^{(1)}(t)$ by the phase space coordinate $\mathbf{q}$, one obtains a set of nonlinear autonomous equations, that is,

$$\dot{\boldsymbol{\xi}}=\boldsymbol{\eta} \tag{3.18a}$$

$$\dot{\boldsymbol{\eta}}=f(\mathbf{q},\boldsymbol{\xi}) \tag{3.18b}$$

where now

$$f(\mathbf{q},\boldsymbol{\xi})=-\nabla_{\mathbf{q}}V(\mathbf{q})+\left[\nabla_{\mathbf{q}}V(\mathbf{q})\right]_{\mathbf{q}=\mathbf{q}-\boldsymbol{\xi}} \tag{3.19}$$

where the coordinates $\mathbf{q}$ are again regarded as time-independent parameters. The above equations can again be subjected to a critical point analysis,[49] but now there are other critical points apart from the trivial one at $(\boldsymbol{\eta},\boldsymbol{\xi})=(0,0)$. As a simple example consider the one-dimensional Hamiltonian

$$H=\tfrac{1}{2}(p^2+q^2)+\tfrac{1}{3}\alpha q^3 \tag{3.20}$$

It is not difficult to obtain the separation equations

$$\dot{\xi}=\eta \tag{3.21a}$$

$$\dot{\eta}=-(1+2\alpha q)\xi+\alpha\xi^2 \tag{3.21b}$$

This pair of equations has critical points at

$$(\eta,\xi)=(0,0) \qquad \text{and} \qquad \left[0,\tfrac{1}{\alpha}(1+2\alpha q)\right] \tag{3.22}$$

Linear stability analysis about the trivial critical point (0,0) predicts a transition from stability to instability at $q=-\frac{1}{2}\alpha$. This is the inflection point of the cubic potential in (3.20) and is the result that the Toda-Brumer-Duff method would yield. However, stability analysis of the second critical point $[0, \frac{1}{\alpha}(1+2\alpha q)]$ shows exactly the opposite behavior at $q=-\frac{1}{2}\alpha$, that is, a transition from instability to stability. Cerjan and Reinhardt claim that the presence of the second critical point with stability characteristics opposite to those of the trivial critical point provides some form of stabilizing mechanism. Thus they deduce that the motion is stable, as it must be for a one-dimensional system. This simple example illustrates their general approach. One examines *all* the critical points of the time-independent version of the separation equations (3.18) in order to deduce the overall stability properties of a given system.

A number of systems for which the Toda-Brumer-Duff method incorrectly predicts a chaotic transition were found to be stable on the basis of the above analysis. Apart from the obvious questions about the validity of the time-independent equations (3.18), which, as with the Toda-Brumer-Duff method, reduce the approach to a *local* stability analysis, this approach also suffers from a number of other shortcomings. As before the method has no sense of "measure" built in. Also, although the idea that the simultaneous presence of elliptic and hyperbolic critical points will provide some sort of stabilizing mechanism is intuitively attractive, it is not clear that this should necessarily be the case. As an illustration of some of these problems we present in Appendix A an amusing example (due to Hannay and Berry) of how motion that is always locally stable can be globally unstable.

Nonetheless, the methods described in this section have worked well for a number of examples and more work is required to understand why this should be so. Further studies of the linearized *global* stability equations, that is, (3.6a) and (3.6b), would also be valuable.

IV. OVERLAPPING RESONANCES

The destruction of tori and the onset of widespread chaotic motion can be attributed to what is known as the overlapping of resonances. A very clear description of this concept has been given by Walker and Ford.[34] To understand what is meant by a resonance we take the Hamiltonian (2.12), for two degrees of freedom, and expand the perturbing part in a Fourier series, that is

$$H(\mathbf{I}, \boldsymbol{\theta}) = H_0(\mathbf{I}) + \varepsilon \sum_{m,n} H_{mn}(\mathbf{I}) e^{i(m\theta_1 + n\theta_2)} \tag{4.1}$$

where $\mathbf{I}=(I_1, I_2)$, $\boldsymbol{\theta}=(\theta_1, \theta_2)$, and $H_{mn}(\mathbf{I})$ are the Fourier coefficients. Resonances occur when the phase in (4.1) is stationary, that is, in the vicinity of $m\omega_1 + n\omega_2 = 0$. It is these resonances that are responsible for the famous problem of "small divisiors" in classical perturbation theory.[10] Note, however, that if we have only an isolated resonance, that is,

$$H(\mathbf{I}, \boldsymbol{\theta}) = H_0(\mathbf{I}) + \varepsilon H_{kl}(\mathbf{I}) e^{i(k\theta_1 + l\theta_2)} \tag{4.2}$$

then there are still two constants of the motion, namely, the total energy and $lI_1 - kI_2$. This may be verified by taking the Poisson bracket of H and $lI_1 - kI_2$. Thus (4.2) is still integrable. The effect of the extra term is to cause a gross distortion of the tori in the vicinity of the resonance. Walker and Ford took the integrable Hamiltonian

$$H_0(I_1, I_2) = I_1 + I_2 - I_1^2 - 3I_1I_2 + I_2^2 \tag{4.3}$$

and investigated the effect of adding a 2:2 resonance and a 3:2 resonance, that is,

$$H(\mathbf{I}, \boldsymbol{\theta}) = H_0(\mathbf{I}) + \alpha I_1 I_2 \cos(2\theta_1 - 2\theta_2) + \beta I_1^{3/2} I_2 \cos(2\theta_1 - 3\theta_2) \tag{4.4}$$

The effect of each resonance in isolation and together is shown in Fig. 8. At low energies the two resonant zones are well separated. As the energy of the system is increased the two zones overlap and a "macroscopic zone of instability" appears. By this term Walker and Ford simply meant a clearly visible splatter of points on the surface of section! The size of this zone increases with increasing energy. The structure is further complicated by the appearance of "secondary" resonant zones as the two principle zones approach each other. By means of a numerical investigation the authors were able to predict the energy at which the overlap of the resonances first occurred, that is, they were able to predict (successfully) the onset of widespread chaotic motion. The overlapping of resonances would appear to play a key role in such an onset, as well as give a great deal of physical insight. When principle zones start overlapping, many higher order resonances are also involved and thus one may be moderately confident that fairly large areas of phase space have had (most of) their tori destroyed and that the ensuing chaos will indeed by "widespread."

A means of predicting approximately when resonant overlap will occur has been proposed by Chirikov.[50, 51] This method works best for forced one-dimensional oscillators, a model very useful in the design of accelerators or the study of diatomic molecules subjected to laser radiation.[52] We consider a one-dimensional *nonlinear* oscillator, for example $H_0 = \frac{1}{2}(p^2 + \frac{1}{2}q^4)$, perturbed by an external periodic force, say $V(t) = q\cos(\phi)$, where $\phi = \Omega t + \phi_0$ is the external phase. The unperturbed system, being one dimensional, can always be solved in action angle variables (I, θ) and we can then

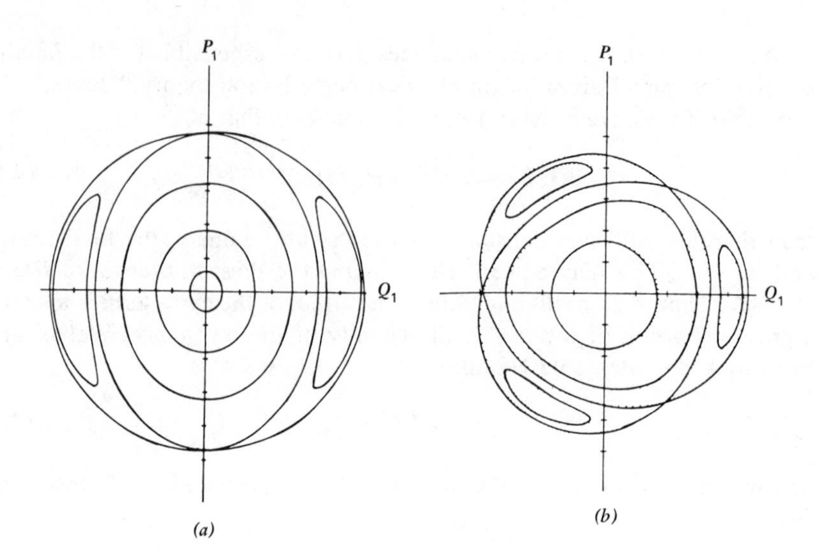

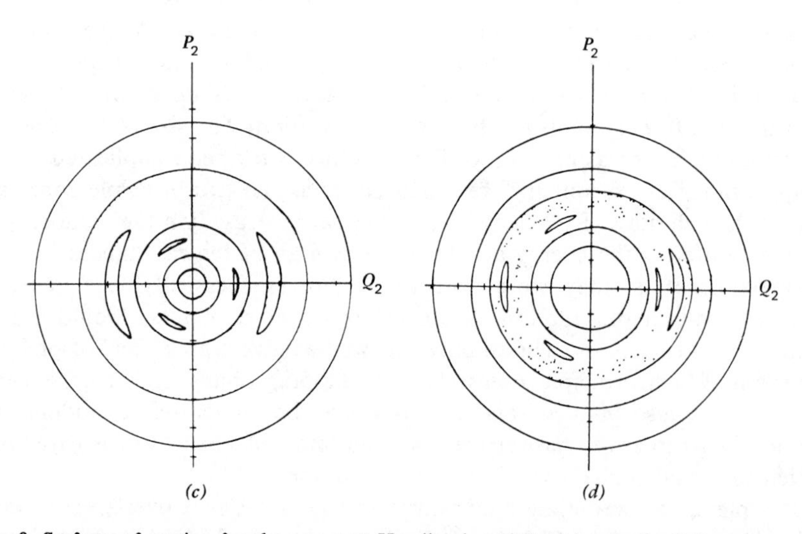

Fig. 8. Surfaces of section for the resonant Hamiltonian (4.4): (*a*) only the 2 : 2 resonance acting ($\beta=0$), (*b*) only the 2 : 3 resonance acting ($\alpha=0$), (*c*) both resonances present ($\alpha=\beta=0.02$) but widely separated at $E=0.18$, and (*d*) the two resonant zones overlapping at $E=0.2095$; the erratic splatter of points is generated by a trajectory started in the region of resonance overlap. Reproduced by permission from Walker and Ford[34].

express the external force as a Fourier series in these variables, that is,

$$H = H_0(I) + \varepsilon \sum_{m,n} V_{mn}(I) e^{i(m\theta + n\phi)} \tag{4.5}$$

In what follows the results are only significant for nonlinear oscillators. In the linear case we know that when the external frequency Ω equals the oscillator frequency ω, the motion "blows up." In the nonlinear case there is also a resonance in the vicinity of $\Omega = \omega$; however, as the amplitude of the oscillator increases, the frequency, which is energy dependent, changes and the system comes out of resonance. Thus nonlinearity stabilizes the resonance. In the above case the oscillator frequency is given by the usual equation, $\omega(I) = \partial H_0 / \partial I$ and there is a resonance at those values of $I = I^r$ such that

$$\frac{\omega(I^r)}{\Omega} = \frac{k}{l} \tag{4.6}$$

In this case the corresponding phase (and its harmonics) $l\theta - k\phi$ are slowly varying compared to other terms in the series (4.5). By virtue of the nonlinearity of H_0 there will of course be other values of I^r giving rise to other resonances and, generally speaking, the set of resonances is everywhere dense. However, to simplify matters we start by considering the resonance (4.6) in isolation and examine the behavior of the Hamiltonian in its vicinity.

We draw on the very clear treatment of resonant Hamiltonians given by Jaeger and Lichtenberg[53] rather than follow the slightly simpler treatment of Chirikov. We introduce the generating function

$$F = F(J, \theta, \phi) = (l\theta - k\phi)J + \theta I^r \tag{4.7}$$

where J is the new momentum and the term θI^r provides, as we see below, a convenient shift in the origin of the new action variable J. From the generating function we obtain the relations

$$I = \frac{\partial F}{\partial \theta} = lJ + I^r \tag{4.8}$$

and

$$\psi = \frac{\partial F}{\partial J} = l\theta - k\phi \tag{4.9}$$

where ψ is the new "resonant" phase conjugate to the new momentum J.

The time derivative of F is also required, that is,

$$\frac{\partial F}{\partial t} = -k\Omega J \tag{4.10}$$

Performing the canonical transformation the Hamiltonian becomes, in terms of the new variables,

$$H = H_0(J) + \varepsilon \sum_{m,n} V_{mn}(J)\exp\left\{ i\frac{1}{l}\left[m\psi + (km+nl)\phi \right] \right\} - k\Omega J \tag{4.11}$$

The transformed Hamiltonian has almost the same form as the original one (4.5). However, by performing this transformation one is effectively putting the observer in a rotating frame in which the rate of change of the new phase measures the slow deviation from resonance. Although it is not explicitly clear that $\dot{\psi} \ll \dot{\phi}$ we assume that near the resonance it is and hence during one complete cycle of ψ, ϕ will have passed through many cycles. The average contribution of these rapidly oscillating terms is zero and this leads us to the next stage, namely, averaging the Hamiltonian over the fast phase variables, that is,

$$\overline{H}(J,\psi) = \frac{1}{2\pi}\int_0^{2\pi} H(J,\psi,\phi)\,d\phi \tag{4.12}$$

The resulting "averaged" Hamiltonian, H, takes the form

$$\overline{H} = H_0(J) = \varepsilon \sum_p V_{pl,-pk} \cos(p\psi) - k\Omega J \tag{4.13}$$

where we have gone over to real arithmetic (assuming $V_{-l,k} = V_{l,-k}$, $V_{00} = 0$, and absorbing a factor of 2 in the Fourier coefficients $V_{pl,-pk}$). There are still all the harmonics to deal with, but at this stage we make the assumption

$$V_{pl,-pk} \ll V_{l,-k} \qquad \text{for} \quad p = 2,3,\ldots \tag{4.14}$$

and (4.13) reduces to

$$\overline{H} = H_0(J) + \varepsilon V_{l,-k} \cos\psi - k\Omega J \tag{4.15}$$

The last stage is as follows. Having worked on the assumption that $\dot{\psi} \ll \dot{\phi}$ in the region of the resonance (4.6), we expand (4.15) about $I = I^r$ to second order, although we assume that the coefficient $V_{l,-k}(J)$ is only a slowly varying function of I. Thus we obtain

$$\bar{H}=H_0(I^r)+lJ\left(\frac{\partial H_0}{\partial I}\right)_{I=I^r}+l^2\frac{J^2}{2}\left(\frac{\partial^2 H_0}{\partial I^2}\right)_{I=I^r}$$
$$+\varepsilon V_{l,-k}(I^r)\cos\psi-k\Omega J \tag{4.16}$$

Since $(\partial H_0/\partial I)_{I=I^r}=\omega(I^r)$ the terms that are linear in J conveniently cancel by the resonance condition (4.6). Dropping the constant term $H_0(I^r)$, we are left with the "resonant" Hamiltonian

$$H_r=\frac{J^2}{2M}+\varepsilon V_{l,-k}\cos\psi \tag{4.17}$$

where the "mass" M is given by

$$M^{-1}=l^2\left(\frac{\partial^2 H_0}{\partial I^2}\right)_{I=I^r} \tag{4.18}$$

The "resonant" Hamiltonian (4.17) has exactly the form of a pendulum Hamiltonian. The boundary between bounded and unbounded motion of a pendulum is termed the separatrix (see Fig. 9) and is determined by the equation

$$J_{sx}=\pm(4M\varepsilon V_{l,-k})^{1/2}\cos\left(\frac{\psi}{2}\right) \tag{4.19}$$

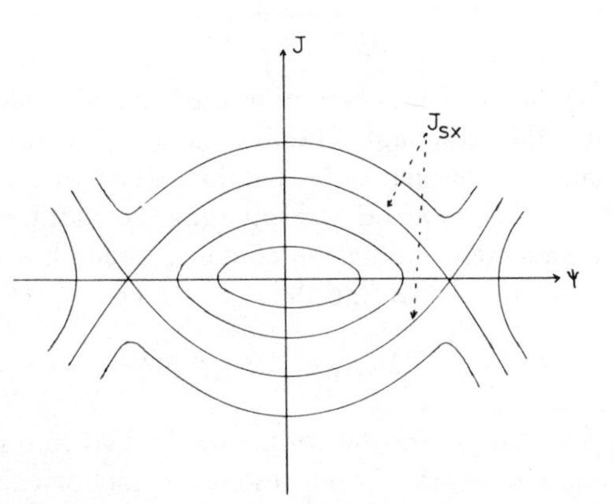

Fig. 9. The phase plane for the pendulum Hamiltonian (4.17). The separatrix J_{sx} separates bounded from unbounded motion.

In the old (I, θ) variables this is simply

$$I_{sx} = I^r \pm (\Delta I^r)\cos\left(\frac{l\theta - k\phi}{2}\right) \tag{4.20}$$

where

$$(\Delta I^r) = 2l(\varepsilon M V_{l,-k})^{1/2} \tag{4.21}$$

The quantity (ΔI^r) is the resonance "half width," which can also be expressed in terms of frequency, that is,

$$(\Delta\omega_r) = \frac{\partial\omega}{\partial I}(\Delta I^r) = \frac{1}{l^2 M} 2l(\varepsilon M V_{l,-k})^{1/2} = \frac{2}{l}\left(\frac{\varepsilon V_{l,-k}}{M}\right)^{1/2} \tag{4.22}$$

Notice that since $M^{-1} = l^2(\partial\omega/\partial I)$ the resonance half width's dependence on the order of the resonance is only in the Fourier coefficients $V_{l,-k}$. From (4.21) and (4.22) we can see that the effect of a resonant perturbation is of $O(\varepsilon^{1/2})$.

Several assumptions have gone into the derivation of the resonant Hamiltonian (4.17). Nonresonant terms in H are neglected by assuming they are rapidly oscillating and hence their average value over a cycle of motion is zero. The resonance harmonics have been neglected by assuming $V_{l,-k} \gg V_{pl,-pk}$ and, furthermore, the higher order terms in the expansion about I^r have been dropped. These assumptions can be summarized in the "moderate nonlinearity condition"

$$\varepsilon \ll \alpha \ll \frac{1}{\varepsilon} \tag{4.23}$$

where $\alpha = (I/\omega)(\partial\omega/\partial I)$. This condition is discussed further by Chirikov.[51]

So far, though, the "resonant" Hamiltonian (4.17) is still integrable, since it consists of only one resonance in isolation. Chirikov's "criterion of overlapping resonances" is obtained by evaluating the width of another (principle) resonance and then finding the coupling strength ε at which the two resonances touch; that is, we find the ε for which

$$(\Delta\omega_r)_1 + (\Delta\omega_r)_2 = \Delta\Omega \tag{4.24}$$

where $(\Delta\omega_r)_1$ and $(\Delta\omega_r)_2$ are the widths of the two resonances and $\Delta\Omega$ is their separation. The width of each resonance zone is calculated independently of all the others; clearly this is a major approximation and one simply hopes that the "moderate nonlinearity condition" (4.23) will ensure that the error is not too great.

Chirikov has tested his method out on a number of simple systems. Of particular interest is the one with model Hamiltonian[50, 51]

$$H(I,\theta,t)=\frac{I^2}{2}+k\sum_{n=-\infty}^{\infty}\cos(\theta-nt) \tag{4.25a}$$

Physically it represents a pendulum being acted on by an infinite series of resonances or, alternatively, by a series of "kicks" at times $t=2\pi m$. Equation (4.25a) can be written in the equivalent form

$$H(I,\theta,t)=\frac{I^2}{2}+2\pi k\cos\theta\sum_{m=-\infty}^{\infty}\delta(2\pi m-t) \tag{4.25b}$$

Each term in the series of resonances provides us with a resonant Hamiltonian

$$\overline{H}^{(n)}=\frac{I^2}{2}+k\cos\psi_n \tag{4.26}$$

where ψ_n is the slowly varying phase $(\theta-nt)$. One can immediately write the resonance half width from (4.22), that is,

$$(\Delta\omega_r)_n=2k^{1/2} \tag{4.27}$$

Resonances occur at every integer value of $\omega=I=I^r=n$ (this is represented schematically in Fig. 10). The spacing between resonances is unity $[\Delta\Omega=I^r_{(n+1)}-I^r_{(n)}=(n+1)-n=1]$ and they touch when

$$(\Delta\omega_r)=\frac{\Delta\Omega}{2}=\tfrac{1}{2} \tag{4.28}$$

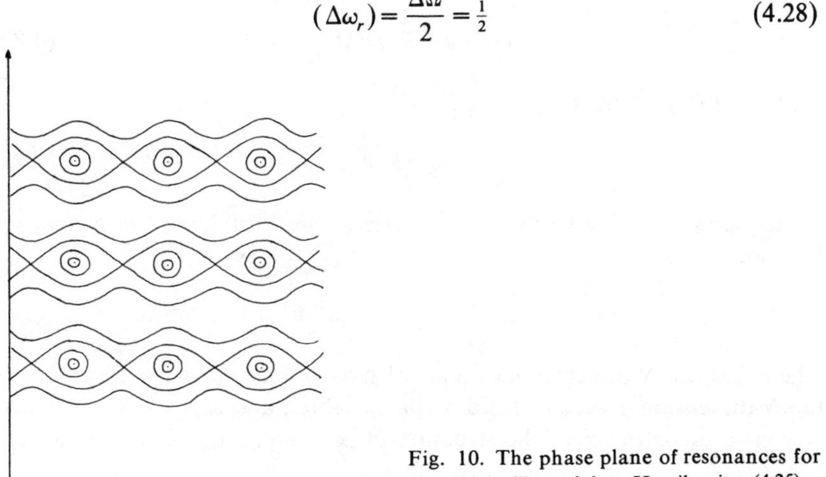

Fig. 10. The phase plane of resonances for the "kicked" pendulum Hamiltonian (4.25).

This enables one to predict the critical value of the perturbation parameter k at which resonance overlap occurs, that is since

$$(\Delta\omega_r)=2k^{1/2}=\tfrac{1}{2} \tag{4.29}$$

this gives

$$k_{\text{crit}}=\tfrac{1}{16} \tag{4.30}$$

Numerical studies of this system suggest that widespread chaos sets in at about $k=\frac{1}{40}$, that is, the overlap criterion is out by about a factor $2\frac{1}{2}$. Further refinements, such as including higher harmonics of the resonant phases, give an improved estimate of $k\simeq\frac{1}{30}$. Even more refinements are possible.[5] In another forced oscillator system examined by Chirikov, the overlap criterion was again found to be accurate to about a factor of 2. Considering all the approximations involved, Chirikov's criterion would seem to be very satisfactory for this class of system. A detailed investigation of systems of the form of (4.25) has been made by Channell[54] and another study of interest is due to Rechester and Stix.[55]

This method can also be applied to conservative systems of two or more degrees of freedom. We now go carefully through the analysis for a system of N degrees of freedom. Starting with the general Hamiltonian

$$H(\mathbf{I},\boldsymbol{\theta})=H_0(\mathbf{I})+\varepsilon\sum_{\mathbf{m}} V_{\mathbf{m}}(\mathbf{I})e^{i\mathbf{m}\cdot\boldsymbol{\theta}} \tag{4.31}$$

where $\mathbf{I}$, $\boldsymbol{\theta}$, and $\mathbf{m}$ are all N-dimensional vectors, there are resonances at those values of $\mathbf{I}=\mathbf{I}^r$ for which

$$\mathbf{k}\cdot\boldsymbol{\omega}(\mathbf{I}^r)=\mathbf{k}\cdot\left[\nabla_{\mathbf{I}}H(\mathbf{I})\right]_{\mathbf{I}=\mathbf{I}^r}=0 \tag{4.32}$$

As before this results in the phase

$$\psi_k=\mathbf{k}\cdot\boldsymbol{\theta} \tag{4.33}$$

being "slow" relative to the other phases. One introduces the generating function

$$F(\mathbf{J},\boldsymbol{\theta})=(\mathbf{I}^r+\mathbf{J}\cdot\boldsymbol{\mu})\cdot\boldsymbol{\theta}^{\dagger} \tag{4.34}$$

where $\mathbf{J}$ is the N-dimensional vector of new actions, $\boldsymbol{\theta}^{\dagger}$ is the transpose of the N-dimensional vector of old angle variables, and $\boldsymbol{\mu}$ is an $N\times N$ matrix of constant coefficients. The structure of $\boldsymbol{\mu}$ is important. The kth element of the product $\boldsymbol{\mu}\cdot\boldsymbol{\theta}^\dagger$ should clearly be the resonant phase ψ_k [see (4.39)]. This still leaves us with $N-1$ other phases that should be linearly independent and, assuming the resonance $\mathbf{k}\cdot\boldsymbol{\omega}=0$ to be isolated, "fast" relative to ψ_k. A simple choice is to take

$$\psi_j=\theta_j \qquad j\neq k \tag{4.35}$$

which results in $\boldsymbol{\mu}$ having the structure

$$\boldsymbol{\mu}=\begin{bmatrix} 1 & 0 & 0 & \dots & 0 \\ 0 & 1 & 0 & & 0 \\ \vdots & & & & \\ k_1 & k_2 & & & k_N \\ \vdots & & & & \\ 0 & & & & 1 \end{bmatrix} \tag{4.36}$$

Other choices for $\boldsymbol{\mu}$ may be possible.[51] The matrix $\boldsymbol{\mu}$ is nonsingular, with $\det(\boldsymbol{\mu})=\mu_{kk}=k_k$ and hence we can define an inverse matrix $\boldsymbol{\mu}^{-1}$. The generating function $F(\mathbf{J},\boldsymbol{\theta})$ gives us the relationship between the old $(\mathbf{I},\boldsymbol{\theta})$ and new $(\mathbf{J},\boldsymbol{\psi})$ variables; that is,

$$\mathbf{I}=\nabla_{\boldsymbol{\theta}}F(\mathbf{J},\boldsymbol{\theta})=\mathbf{I}'+\mathbf{J}\cdot\boldsymbol{\mu} \tag{4.37}$$

from which we have

$$\mathbf{J}=(\mathbf{I}-\mathbf{I}')\cdot\boldsymbol{\mu}^{-1} \tag{4.38}$$

and

$$\boldsymbol{\psi}^\dagger=\nabla_{\mathbf{J}}F(\mathbf{J},\boldsymbol{\theta})=\boldsymbol{\mu}\cdot\boldsymbol{\theta}^\dagger \tag{4.39}$$

from which we have

$$\boldsymbol{\theta}^\dagger=\boldsymbol{\mu}^{-1}\cdot\boldsymbol{\psi}^\dagger \tag{4.40}$$

Transforming the Hamiltonian (4.31) to the new variables gives

$$H(\mathbf{J},\boldsymbol{\psi})=H_0(\mathbf{J})+\varepsilon\sum_{\mathbf{m}}V_{\mathbf{m}}(\mathbf{J})\exp i\left(\psi_k\sum_{j=1}^{N}m_j\mu_{jk}^{-1}+\sum_{j=1}^{N}\sum_{i\neq k}^{N}m_j\mu_{ji}^{-1}\psi_i\right) \tag{4.41}$$

where μ_{ij}^{-1} denotes the (ij)th element of the inverse matrix $\boldsymbol{\mu}^{-1}$. As before, we can average the Hamiltonian over the fast variables $\psi_j (j \neq k)$

$$\bar{H} = \frac{1}{(2\pi)^{N-1}} \int_0^{2\pi} d\psi_1 \ldots \int_0^{2\pi} d\psi_{k-1} \int_0^{2\pi} d\psi_{k+1} \ldots \int_0^{2\pi} d\psi_N H(\mathbf{J}, \psi) \tag{4.42}$$

The averaging process eliminates all terms from the sum over $\mathbf{m}$ in (4.41) except for the term $\mathbf{m} = \mathbf{k}$ and its harmonics. Hence

$$\bar{H} = H_0(\mathbf{J}) + \varepsilon \sum_l V_{l\mathbf{k}} \exp(il\psi_k) \tag{4.43}$$

where we have used

$$\sum_j k_j \mu_{ji}^{-1} \equiv \sum_j \mu_{kj} \mu_{ji}^{-1} = \delta_{ki} \tag{4.44}$$

and the sum over l represents the sum over the harmonics of $\mathbf{k}$. One then proceeds by expanding $H_0(\mathbf{J})$ about the resonant action $\mathbf{I} = \mathbf{I}^r$. This yields

$$\bar{H} \simeq H_0(\mathbf{I}^r) + \mathbf{J} \cdot \boldsymbol{\mu} \cdot \omega(\mathbf{I}^r)^\dagger + \tfrac{1}{2}(\mathbf{J} \cdot \boldsymbol{\mu}) \cdot \mathbf{H}^0 \cdot (\mathbf{J} \cdot \boldsymbol{\mu})^\dagger + \varepsilon \sum_l V_{l\mathbf{k}} \exp(il\psi_k) \tag{4.45}$$

where $\mathbf{H}^0$ is the $N \times N$ matrix of second derivatives of $H_0(\mathbf{I})$ evaluated at $\mathbf{I} = \mathbf{I}^r$

$$(\mathbf{H}^0)_{ij} = \left[\frac{\partial H_0^2(I)}{\partial I_i \partial I_j} \right]_{\mathbf{I} = \mathbf{I}^r} = \frac{\partial \omega_j(\mathbf{I}^r)}{\partial I_i} \tag{4.46}$$

As before we consider just the lowest harmonic and write out $\bar{H}$ explicitly as

$$\bar{H} = H_0(\mathbf{I}^r) + \sum_j \sum_i J_j \mu_{ji} \omega_i + \tfrac{1}{2} \sum_j \sum_i \frac{J_j J_i}{M_{ji}} + \varepsilon V_{\mathbf{k}} \cos \psi_{\mathbf{k}} \tag{4.47}$$

where the "masses" M_{ji} are given by

$$(M_{ji})^{-1} = \sum_l \sum_m \mu_{jl} (\mathbf{H}^0)_{lm} \mu_{im} \tag{4.48}$$

Dropping the constant term $H_0(I^r)$ and using our particular choice of $\boldsymbol{\mu}$

(4.36) gives

$$\overline{H}=\frac{J_k^2}{2M_{kk}}+\varepsilon V_{\mathbf{k}}\cos\psi_k+\sum_{j\neq k} J_j\omega_j+\sum_{j\neq k}\sum_i\frac{J_jJ_i}{2M_{ji}} \tag{4.49}$$

Although the $J_j(j\neq k)$ are constants of the motion (because $\overline{H}$ only involves ψ_k), it does not follow that these momenta are necessarily zero. Ideally one would like to find a transformation to some new momenta such that the last two sets of terms in (4.49) vanish. At this stage it is not clear to the author that such a transformation can be found that would not interfere with the isolation of the original resonance and hence the validity of the above analysis, that is the separation of the "fast" and "slow" phases (cf. Chirikov's orthogonal metric[55a]). Thus we simply hope that near the resonance the net contribution from all these terms is small and that we can again assume the same basic form of "resonant" Hamiltonian

$$\overline{H}_r=\frac{J_k^2}{2M_{kk}}+\varepsilon V_k\cos\psi_k \tag{4.50}$$

Just as with the one-dimensional case we can define a resonance half width

$$\Delta J_k=2(\varepsilon V_{\mathbf{k}}M_{kk})^{1/2} \tag{4.51}$$

From this we easily obtain the vector of widths in the original action variables, that is,

$$\Delta\mathbf{I}=(\mathbf{k})\,\Delta J_k \tag{4.52}$$

Obtaining the resonance width in frequency space requires a little more care. The vector of frequency widths is given by

$$\Delta\omega=\left(\frac{\partial\omega}{\partial J_k}\right)\Delta J_k \tag{4.53}$$

and using the chain rule for $\partial/\partial J_k$ gives

$$\Delta\omega=\sum_i\left(\frac{\partial I_i}{\partial J_k}\cdot\frac{\partial\omega}{\partial I_i}\right)\Delta J_k=\sum_i\left(\mu_{ki}\frac{\partial\omega}{\partial I_i}\right)\Delta J_k \tag{4.54}$$

where we use (4.37). Projecting out the component of $\Delta\omega$ parallel to the **k** vector, that is, in the direction of the resonance, finally yields

$$(\Delta\omega)_k = \frac{\mathbf{k}\cdot\Delta\omega}{|\mathbf{k}|} = \frac{1}{|\mathbf{k}|}\sum_j\sum_i\left(\mu_{kj}\frac{\partial^2 H_0}{\partial I_j \partial I_i}\mu_{ki}\right)\Delta J_k = \frac{2}{|\mathbf{k}|}\left(\frac{\varepsilon V_{\mathbf{k}}}{M_{kk}}\right)^{1/2} \qquad (4.55)$$

which is just the multidimensional analogue of (4.22). The procedure is then as before. One calculates, in isolation, the width of this and a nearby (principal) resonance and then estimates the value of ε at which they will overlap.

As a simple illustration of the method we apply it to the Walker-Ford Hamiltonian (4.4). The details of the calculation are given in Appendix B. A simple graphical approach is taken. We plot (Fig. 11) in the space of the original action variables (I_1, I_2), the 2:2 and 2:3 resonance zones. From this graph we can read the values of I_1 and I_2 at which the two zones first touch. In the present case this corresponds to a zero-order (H_0) energy of approximately 0.227 units, which agrees to within 10% of the observed "critical" energy of 0.2095 units. Another application of the method has been made by Oxtoby and Rice[56] with particular reference to the problem of intramolecular energy transfer. Working with a pair of coupled Morse

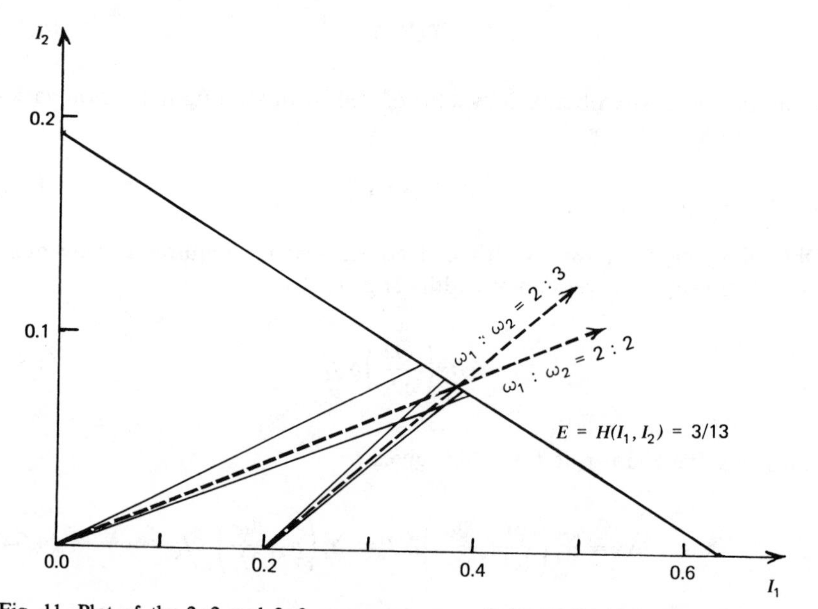

Fig. 11. Plot of the 2:2 and 2:3 resonance zones of the Walker-Ford Hamiltonian (4.4) calculated on the basis of Chirikov's method. The two zones first touch at about $I_1 = 0.338$ and $I_2 = 0.064$, corresponding to a zero-order energy of approximately 0.227.

oscillators they examine, as a function of energy, the regimes of isolated and overlapping resonances. Their results, also presented graphically, led them to suggest that at low energy the predominance of isolated resonance zones can lead to the trapping of vibrational energy and hence slow vibrational relaxation. At higher energies (as might be expected) the dynamics is dominated by overlapping resonances, which can lead to rapid energy exchange. In this regime the statistical theories of unimolecular decay should be applicable. The results of this resonant overlap approach are in qualitative agreement with molecular dynamics calculations. Generally, though, the application of Chirikov's method to multidimensional conservative systems can present a number of practical problems. For example, one requires a nonlinear zero-order Hamiltonian whose solution in action angle variables is known. Furthermore, the results may not be invariant to the choice of the zero-order part.[33] Another problem is that the overlap criterion is usually set up to give the critical value of perturbation parameter ε, whereas frequently one works with fixed ε and seeks the energy at which overlap occurs. Usually there is not a simple scaling to relate a critical ε to a critical energy and in those cases a graphical approach, such as that used here, can be implemented. Despite the success of our simple example the method should generally not be expected to be accurate to more than a factor of 2 or 3, since yet another approximation [the dropping of the additional terms in (4.49)] has been introduced.

Overall, the usefulness of the Chirikov method has been amply demonstrated for forced oscillator systems. Further studies[57] of its applicability to multidimensional conservative systems are now required.

V. CLOSED ORBIT METHODS

An important method for predicting the onset of chaotic motion, based on the stability properties of closed orbits, has been proposed by Green.[58, 59] To understand it we extend our discussion of Section II and examine more clearly the fate of a torus when it is destroyed under perturbation. The KAM theorem tells us that it is those tori with commensurable frequencies that are first destroyed. These are the tori that are covered by families of closed orbits. Such orbits can be characterized by a rational winding number. In a surface of section study of an integrable or near-to-integrable system (of two degrees of freedom) a single one of these orbits would generate a pattern of Q fixed points lying on a smooth curve. Any other member of that family of closed orbits would, of course, generate another set of Q fixed points lying on the same curve. Thus the whole curve is made up of fixed points. (Such fixed points lying on a line are called parabolic fixed points.) On perturbation this family of fixed points is broken up. The Poincare-Birkhoff fixed point theorem[14] tells us that in its place appears $2nQ$ fixed points (n is an arbitrary integer), of which nQ are elliptic fixed points and nQ are hyperbolic fixed points. The elliptic fixed points correspond to stable closed orbits and, as is described earlier, each is surrounded by smooth invariant curves (islands) corresponding to "high-order" tori. On the other hand, the hyperbolic fixed points correspond to unstable closed orbits and an extremely complicated structure develops in their neighborhood (the area becomes dense with so-called homoclinic points[10]). It is in these regions that one finds the chaotic orbits. As we saw earlier, the overall structure in the vicinity of a destroyed invariant curve is quite remarkable. It has been immortalized in a famous picture attributed to Melnikov (see Ref. 51), which we sketch here (Fig. 12). The overall structure is repeated on all scales about the elliptic fixed points.[14]

The method of Greene is based on the hypothesis that the dissolution of an invariant curve (torus) can be associated with the sudden change from stability to instability of nearby closed orbits. To see this more precisely, imagine a weakly perturbed integrable system. According to the KAM theorem, those invariant curves with "sufficiently" irrational winding number are preserved. The neighboring rational (and close-to-rational) curves break up in the manner just described above, that is, a mixture of nQ elliptic (stable) and nQ hyperbolic (unstable) fixed points. Greene's method is based on the observation that when the perturbation is made sufficiently strong (or the energy high enough) the set of nQ stable fixed points also becomes

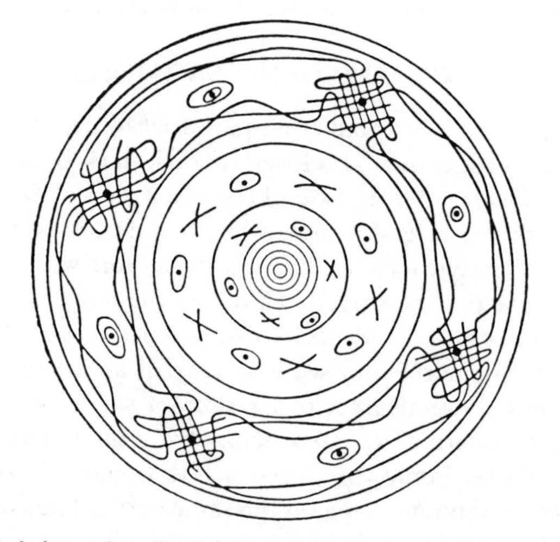

Fig. 12. Structure of alternating elliptic (⊙) and hyperbolic (×) fixed points appearing in the gaps between irrational tori. The "tangled web" structure around the outermost set of hyperbolic fixed points is discussed in Refs. 8 and 10. Reproduced by permission from Ford[8].

unstable (they become "hyperbolic-with-reflection" fixed points). The contention is that this then signals the dissolution of an invariant curve "close" to that set of fixed points. A rather nice way of estimating the closeness of a closed orbit to a given invariant curve is by expressing that curves winding number in the form of a continued fraction, that is,

$$\alpha = a_0 + \cfrac{1}{a_1 + \cfrac{1}{a_2 + \cfrac{1}{a_3 + \ddots}}} \tag{5.1}$$

Where $a_0, a_1, a_2 \ldots$ are positive integers. Continued fractions provide a unique and convergent representation of an irrational number. Successive truncations of (5.1) yield rational fractions with values converging monotonically on the desired irrational. For example, the continued fraction representation of π yields successive truncations 3, 22/7, 333/106, 355/113; the last of which already reproduces π to six decimal places! Thus the successive truncations of the continued fraction representation of an irrational winding number yield the winding numbers of the closed orbits that become ever "closer" to the chosen invariant curve. By following the stability properties of these sequences of closed orbits, as they "close in" on an invariant curve, Greene is able to predict the breakup of that curve.

The two essential ingredients of this method are (*1*) finding the closed orbits and (*2*) determining their stability characteristics. A detailed discussion of the former problem is outside the scope of this chapter. Suffice it to say that there are now well-developed and efficient methods for finding closed orbits of any desired topology (winding number). One approach is described by Greene[59] and another powerful method has been developed by Helleman and Bountis.[60] We describe the stability analysis by example. A system studied in detail by Greene is the "kicked" pendulum system (4.25) introduced in the preceding section. The interesting feature of this system is that when one integrates the corresponding Hamilton's equations for $\dot{I}$ and $\dot{\theta}$ over a period from one "kick" to the next, one can obtain a pair of difference equations of the form

$$I_{n+1} = I_n + \frac{k}{2\pi} \sin 2\pi\theta_n \tag{5.2a}$$

$$\theta_{n+1} = \theta_n + I_{n+1} \tag{5.2b}$$

It is easy to verify that $|\partial(I_{n+1}, \theta_{n+1})/\partial(I_n, \theta_n)| = 1$ and hence that (5.2a) and (5.2b) are another example of an area preserving mapping [see (2.15)]. In this particular case the motion is periodic, with period unity, in both I and θ. The phase space of the system is thus taken to be the unit torus. The parameter k can be regarded as a perturbation parameter. For $k = 0$ the mapping takes the trivial form

$$I_{n+1}=I_n \tag{5.3a}$$

$$\theta_{n+1}=\theta_n+I_{n+1} \qquad [\text{mod } 1] \tag{5.3b}$$

In this case the mapping is clearly "integrable," since all the orbits lie on straight lines. These are just the invariant curves of the unperturbed mapping. In Fig. 13 we show some typical orbits for a value of k, which corresponds to the mapping being strongly perturbed. Here we see some strongly irregular orbits, filling up substantial portions of the phase plane, as well as the typical alternating hyperbolic and elliptic fixed point structure. Notice also that there are still invariant curves remaining that divide the phase space. These curves prevent a trajectory from wandering over the whole phase plane. Clearly, it will not be until these curves are destroyed that the "chaos" will be truly widespread. This point is taken up later.

The stability of a given closed orbit is determined by evaluating the so-called tangent space mapping. This corresponds to linearizing the mapping at each iteration. Thus if we denote the "tangent space" variables as $(\delta I, \delta\theta)$ we have for this space

$$\begin{bmatrix} \delta I_{n+1} \\ \delta\theta_{n+1} \end{bmatrix} = \mathsf{M} \begin{bmatrix} \delta I_n \\ \delta\theta_n \end{bmatrix} \tag{5.4}$$

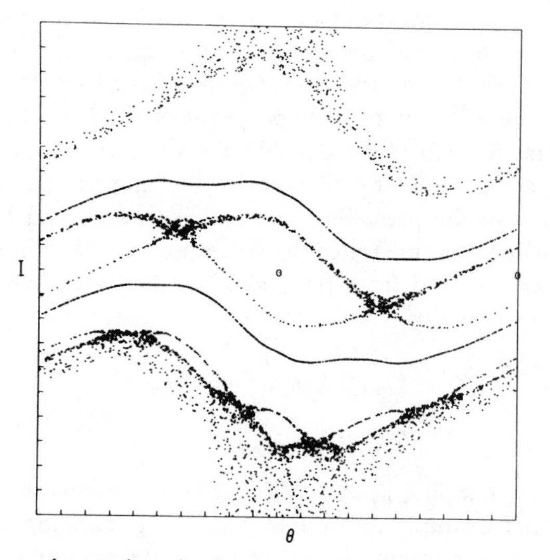

Fig. 13. Five-phase plane trajectories for the mapping (5.2) with $k=0.97$. (Reproduced by permission from Greene[59].)

where

$$\mathsf{M}=\begin{bmatrix} 1 & -k\cos 2\pi\theta_n \\ 1 & 1-k\cos 2\pi\theta_n \end{bmatrix} \tag{5.5}$$

The "tangent space" mapping is also an area preserving mapping, since

$$\det|\mathsf{M}|=1 \tag{5.6}$$

For an orbit that closes after Q iterations of the mapping the eigenvalues, $\lambda_{\pm}$, of the 2×2 matrix

$$\mathsf{M}^{(Q)}=\prod_{n=1}^{Q}\begin{bmatrix} 1 & -k\cos 2\pi\theta_n \\ 1 & 1-k\cos 2\pi\theta_n \end{bmatrix} \tag{5.7}$$

give the stability indices, or Floquet multipliers, of the orbit. Denoting the matrix elements of $\mathsf{M}^{(Q)}$ by $M_{ij}^{(Q)}$ we have explicitly

$$\lambda_{\pm}=\tfrac{1}{2}\left[M_{11}^{(Q)}+M_{22}^{(Q)}\right]\pm\tfrac{1}{2}\left[\left(M_{11}^{(Q)}+M_{22}^{(Q)}\right)^2-4\right]^{1/2} \tag{5.8}$$

where we have made use of condition (5.6), that is,

$$M_{11}^{(Q)}M_{22}^{(Q)}-M_{12}^{(Q)}M_{21}^{(Q)}=1 \tag{5.9}$$

It is a standard result[10] to show that if the eigenvalues are complex the orbits are stable and if they are real the orbits are unstable. Notice that here the stability analysis is of a truly "global" nature, since the eigenvalues are determined from an "integration" around the complete period of the orbit [see (5.7)]. This is in contrast to the local stability analyses of Section III, where the eigenvalues are only evaluated, in effect, over an infinitesimally short period of the trajectories.

Greene introduces a quantity called the "residue," which is defined as

$$R=\tfrac{1}{4}\left[2-\mathrm{Tr}(\mathsf{M}^{(Q)})\right] \tag{5.10}$$

From (5.8) it is easy to see that if $0<R<1$ the eigenvalues are imaginary and hence the orbit is stable, that is, the fixed points are elliptic. If $R<0$ or $R>1$ the eigenvalues are real and hence the orbit is unstable. More precisely, for $R<0$ the fixed points are hyperbolic and for $R>1$ they are "hyperbolic-with-reflection."[8, 10] (For parabolic fixed points $R=0$.) It can be shown that for an orbit of "length" Q, the residue is proportional to k^Q for both large and small k. (Recall that k is the perturbation parameter for the system studied here.) A quantity, called the "mean residue" f, is then introduced that scales away this exponential dependence on Q, that is,

$$f=\left(\frac{R}{\beta}\right)^{1/Q} \tag{5.11}$$

where β is some arbitrary constant. We can now proceed to characterize the stability properties of the sequence of closed orbits converging on a chosen invariant curve. Each successive closed orbit (determined through the successive truncations of the continued fraction representation of the winding number of the chosen curve) has a larger Q, corresponding to increasing topological complexity of that orbit. The remarkable thing is that the corresponding sequence of mean residues is found to converge to some finite value. The rate of convergence seems to be determined by the value of β; for this problem the optimum value was found to be $\beta=\frac{1}{4}$. It is then demonstrated (empirically) that when the converged mean residue becomes greater than unity the invariant curve associated with that sequence of closed orbits is destroyed.

This criterion then enables one to find the value of the perturbation parameter k at which any chosen invariant curve breaks up. For the above system the method has been found to give very accurate results. Furthermore, Greene has made an ingenious extension of his method to predict the onset of widespread chaos. It is based on the conjecture that the more closely an irrational curve can be approximated by a sequence of rationals, the smaller the perturbation (k) required to destroy it. Thus one might reasonably assume that the last invariant curve to be destroyed will be that whose winding number is least closely approximated by a sequence of rationals. This is the invariant curve whose winding number has the continued fraction representation.

$$\alpha=1+\cfrac{1}{1+\cfrac{1}{1+\cfrac{1}{1+\ddots}}}$$

$$=\frac{(\sqrt{5}-1)}{2} \tag{5.12}$$

the famous "golden mean." Thus by the time k is sufficiently large so that this invariant curve breaks up, one may fairly confidently assume that all the other curves have also been destroyed. There will then be no impediment to an irregular trajectory wandering over the whole of the phase plane (see Fig. 13) and widespread chaos will reign. The critical value of k corresponding to the breakup of the golden mean curve was found to be about unity, in close agreement with the observed onset of widespread chaos.

This remarkable method is not restricted to algebraic mappings and can also be applied to Hamiltonian systems such as the Henon-Heiles system. However, for those systems the tangent mapping has to be computed numerically. An early investigation was made by Lunsford and Ford[35] and more recently by Greene.[61] To illustrate some of Green's results we must return to the surface of section illustrations (Fig. 3). In Fig. 3*a* point D, often called a central fixed point, corresponds to a simple closed orbit that is stable.[62] Around this point one finds a family of invariant curves; the associated winding numbers decreasing as one moves further away from the center. Greene considers the invariant curve of this family with irrational winding number $(7+5\sqrt{5})/(41+29\sqrt{5})\simeq 1/5.82$, that is, a curve lying somewhere between the curves with rational winding number $\frac{1}{5}$ and $\frac{1}{6}$. The stability analysis suggests breakup of this irrational curve around $E=0.118$. This seems to correlate quite well with the observed dynamics as manifested, for example, by Fig. 3*b*.

Greene's method represents what is probably the correct approach to predicting the onset of chaotic motion, that is, concentrating on the destruction of individual tori. In this way one avoids the problems and pitfalls associated with predicting the onset of widespread chaos, such as defining the measure of "widespread." Furthermore by following the breakup of a carefully chosen torus or invariant curve, such as the golden mean curve of the kicked pendulum system, one can indeed obtain a good deal of information about the transition to "widespread" instability. Application of the method to multidimensional Hamiltonians may prove to be rather difficult. In these cases my feeling is that here one should concentrate on the stability properties of the orbits of simplest topology. These orbits often seem to form the "cores" of large families of tori and one can envisage that the transition to instability of these orbits might herald a more widespread instability. For example, in the Henon-Heiles system, the central fixed point D of Fig. 3*a* is reported[63] to become unstable around $E=0.148$. This is not too far from the generally accepted onset of widespread chaos around $E=0.125$.

We conclude this section by describing, very briefly, an approach to predicting the onset of widespread chaos, due to Bountis and Helleman,[60, 64] which also utilizes the properties of closed orbits. The method is based on the following ideas. As we described earlier, a closed orbit can be identified by some rational winding number $\sigma=P/Q$, which, for a system of two degrees of freedom, is the ratio of the associated fundamental frequencies of motion ω_1 and ω_2. Clearly, though, for a given ratio P/Q, there can exist an infinite one-parameter family of closed orbits corresponding to the frequencies $\omega_1=\nu_r P$ and $\omega_2=\nu_r Q$, where the parameter ν_r represents a "recurrence frequency."[65] For any given winding number σ, recurrence frequency ν_r, and a pair of initial phases, the initial conditions of the corresponding closed orbit can be found easily.[64] The Henon-Heiles system was investigated in detail. It was found that for those closed orbits with zero initial velocities, the initial conditions $x(0)$, $y(0)$ lay on smooth curves parameterized by the winding number σ, each point along a given "σ-curve" corresponding to a continuous change in ν_r. When plotted out the σ-curves were found to bunch about the $\sigma=1$ curves (see Fig. 14). These curves are associated with the simple (unstable) closed orbits that generate the fixed points A, B, C in Fig. 3*a*. As the energy was increased the bunching of the σ-curves (termed "σ-confluence" by the authors) about the $\sigma=1$ curves was found to increase fairly sharply. The significance of this confluence is as follows. A change in initial conditions perpendicular to a given σ-curve, that is, changing from one curve to another, leads to a change in orbital topology. In the regions of large σ-confluence it becomes possible for a very small

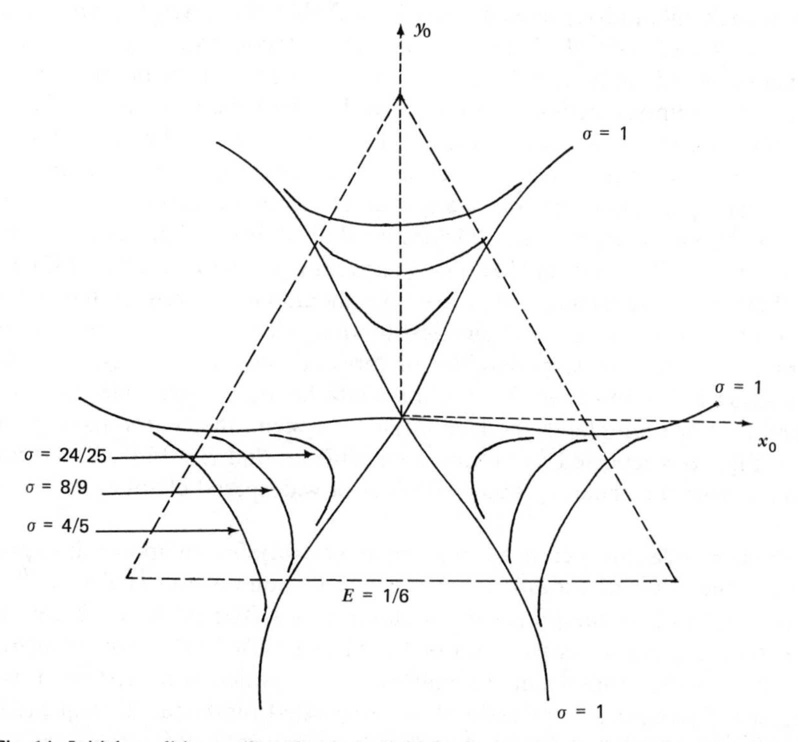

Fig. 14. Initial conditions $x(0)$, $y(0)$ of zero initial velocity closed orbits plotted for different values of $\sigma=\omega_1/\omega_2=P/Q$ (P, Q relatively prime). Notice the "confluence" about the $\sigma=1$ curves. The σ curves are superimposed on the equipotential contour $E=\frac{1}{6}$ (---) of the Henon-Heiles system (2.16). (Redrawn by permission from Helleman and Bountis[60].)

change in initial conditions to lead to a very large change in orbital topology, for example, going from $\sigma = \frac{2}{3}$ to $\sigma = \frac{1}{101}$. The relative periods of motion can then be very different and one could thus expect to see a rapid (initial) divergence of neighboring orbits. This sort of sensitivity to initial conditions is, of course, a characteristic of the chaotic regime. This approach not only indicates how this instability grows with energy but, primarily, locates the regions where it occurs. Indeed, if one looks at the surfaces of section for the Henon-Heiles system, it is precisely in the vicinity of the fixed points associated with the $\sigma = 1$ curves that widespread chaos first appears. An attractive feature of this method is that, to a zeroth order approximation, the σ-curves can be computed in a simple analytical form. Furthermore, these approximate curves agree to within 10 to 20% of the exact ones. It is now highly desirable for there to be further studies of this approach, for example, σ-confluence in nonzero velocity initial condition spaces.

VI. CORRELATION FUNCTION METHODS

We now turn to a very different approach to the onset of chaotic motion. In an intruiging paper by Mo[66] the powerful generalized Langevin equation formalism was used to model a certain correlation function that displayed a change in behavior at the "critical energy." Good agreement was obtained with a number of simple systems such as the Henon-Heiles system. Mo's method raises a number of very interesting questions, which we discuss in some detail below. The approach is based on the following ideas. For a Hamiltonian system the motion of some associated dynamical variable $a(t)$ is given by Liouvilles equation

$$\frac{da(t)}{dt} = iLa(t) \tag{6.1}$$

where iL is the Liouville operator of that system. This motion can also be described by a generalized Langevin equation[68]

$$\frac{da(t)}{dt} = i\Omega a(t) - \int_0^t d\tau K(\tau) a(t-\tau) + F(t) \tag{6.2}$$

providing the proper choice of frequency Ω, memory kernel $K(t)$, and "random force" $F(t)$ is made. The correlation function of $a(t)$ is defined as

$$C(t) = \langle a(0)^*, a(t) \rangle \tag{6.3}$$

where the brackets denote a certain "inner product," which depends on the ensemble chosen. $C(t)$ is related to the memory kernel through the equation

$$\frac{dC(t)}{dt} = i\Omega C(t) - \int_0^t d\tau K(t-\tau)C(\tau) \tag{6.4}$$

When $K(t)$ is taken to be a delta function and the frequency Ω is set to zero, the generalized Langevin equation reduces to the familiar equation for Brownian motion. In this case the correlation function exhibits a pure exponential decay characteristic of a Gaussian Markov process.[69] The introduction of certain projection operators[68] enables the "systematic" and "random" parts of $a(t)$ to be separated and one can then deduce the form of the random force $F(t)$ and the memory kernel $K(t)$. The frequency Ω is related to the initial time derivative of the correlation function; in what follows we take it to be zero.

The Laplace transform of (6.4) yields the simple relationship

$$\tilde{C}(s) = \frac{1}{s + \tilde{K}(s)} \tag{6.5}$$

where the tildes denote the transformed functions and s is the Laplace transform space variable. Mori[70] has shown that $\tilde{K}(s)$ can be represented in the form of a continued fraction, that is,

$$\tilde{K}(s) = \cfrac{D_1}{s + \cfrac{D_2}{s + \cfrac{D_3}{s + \ddots}}} \tag{6.6}$$

where the D_i's are certain functions of the moments, y_n, of the Liouville operator, where $y_n = \langle (iL)^n a, (iL)^n a \rangle$. Substitution of this form of $\tilde{K}(s)$ into (6.5) yields a similar continued fraction representation for $\tilde{C}(s)$. However, rather than work directly with the correlation function of $a(t)$, Mo chooses to work with the correlation function of its time derivative $\dot{a}(t)$, that is, the quantity

$$D(t) = \langle \dot{a}(0)^*, \dot{a}(t) \rangle \tag{6.7}$$

In this case there exists a relationship[71] between this correlation function and the memory kernel, $K(t)$, associated with the original dynamical variable, $a(t)$, namely,

$$\tilde{D}(s) = \frac{\tilde{K}(s)}{1 + \tilde{K}(s)/s} \tag{6.8}$$

If one uses the continued fraction representation of $\tilde{K}(s)$ (6.6), truncated to some finite order, the expression for $\tilde{D}(s)$ can be multiplied out and then transformed back to real time space. It is not difficult to verify that the resulting $D(t)$ can always be cast in the form of various combinations of cosines, the "frequencies" being certain functions of the D_i's. A similar result would also be obtained for the correlation function $C(t)$.

In the work of Mo the dynamical variable $a(t)$ is chosen to be the quantity

$$a(t)=\sum_{i=1}^{n}(p_i^2+q_i^2) \tag{6.9}$$

which is taken to represent the distance of a phase point (on a given trajectory) from the origin. In the regime of regular (quasiperiodic) motion the trajectories are confined to tori. This means [see (2.8)] that $a(t)$ [or $\dot{a}(t)$, or any other function of p and q] can be represented by a Fourier series,

$$a(t)=\sum_{\mathbf{m}} a_{\mathbf{m}} e^{i\mathbf{m}\cdot\boldsymbol{\theta}} \tag{6.10}$$

where $\mathbf{m}$ and $\boldsymbol{\theta}$ are N-dimensional vectors of integers and angle variables, respectively. To compute the correlation function of $\dot{a}(t)$, Mo takes the inner product in (6.7) to be the (normalized) microcanonical average, that is, the quantity $\dot{a}(0)\dot{a}(t)$ is averaged over the whole of the energy shell. Mo then assumes that in the regular regime the quasiperiodic behavior of $a(t)$ [and hence $\dot{a}(t)$] is also displayed by the correlation function (6.7), although since correlation functions are even functions of time, it can only contain cosine terms. Conveniently enough, as is described above, this is just the form taken by $D(t)$ [or $C(t)$] when the continued fraction representation of $K(t)$ is used. That is, it takes the form

$$D(t)=\sum_{i} C_i \cos\Omega_i t \tag{6.11}$$

where the C_i are certain coefficients and the frequencies Ω_i are certain functions of the moments of the Liouville operator. They should in no way be confused with the classical frequencies of the motion. Mo's computations claim that the Ω_i's become complex at an energy close to the observed "critical energy" marking the onset of widespread chaos. At this point (6.11) becomes exponentially diverging and Mo takes this to reflect the exponential separation of nearby trajectories, a characteristic of the chaotic regime. As we now discuss this modeling of the correlation function seems to suffer from some theoretical inconsistencies and its assumed mode of behavior does not agree with the observed behavior.

The physical significance of the generalized Langevin equation formalism is realized once the inner product, that is, ensemble average, has been chosen.[68] Clearly this choice should reflect the dynamics. In the case of quasiperiodic motion the natural choice would appear to be the torus to which a given family of trajectories is confined. A given torus is covered by an infinite one-parameter (the initial phase) family of orbits. The ensemble average is then just a simple phase average. In this case it can be shown [using (2.8)] that

$$D(t)=\langle \dot{a}(0), \dot{a}(t)\rangle = \sum_{\mathbf{m}} |a_{\mathbf{m}}(\mathbf{m}\cdot\boldsymbol{\omega})|^2 e^{i\mathbf{m}\cdot\boldsymbol{\omega}t} \tag{6.12}$$

where $\boldsymbol{\omega}$ is the N-dimensional vector of frequencies associated with the chosen torus. [In the case of the frequencies $\boldsymbol{\omega}$ being incommensurable the orbits cover the torus ergodically and (6.12) can be computed with a single trajectory.] However, if we follow Mo and choose the microcanonical ensemble (rather than a torus), then $D(t)$ can no longer take the simple form (6.12). Rather, it corresponds to contributions of the form (6.12) averaged over all tori. The result will be some nontrivial oscillatory function oscillating about the mean value.

In the chaotic regime we assume that the majority of trajectories are no longer confined to tori, but explore most of the energy shell. Clearly, the most suitable choice of ensemble in this case is the microcanonical one. However, the behavior of the correlation function in this regime is entirely different from its behavior in the integrable regime, despite the use of the same ensemble average. As was mentioned earlier, a characteristic of the chaotic regime is the exponential divergence of nearby trajectories. This means that the correlation between them decreases and hence the microcanonical ensemble averaged correlation function must *decay*, albeit in an oscillatory manner, with increasing time. This is quite easy to see since

$$D(0)=\langle \dot{a}(0), \dot{a}(0)\rangle = \langle [\dot{a}(0)]^2\rangle \tag{6.13}$$

and

$$D(\infty)=\langle \dot{a}(0), \dot{a}(\infty)\rangle = \langle \dot{a}(0)\rangle\langle \dot{a}(\infty)\rangle = \langle \dot{a}(0)\rangle^2 \tag{6.14}$$

Since mean square is greater than square mean, $D(t)$ must decay as t approaches infinity. It certainly cannot diverge exponentially! Notice that we have only been able to obtain the decomposition in (6.14) by assuming that the motion is stochastic (in the strict sense of the word) and hence $\dot{a}(0)$ and $\dot{a}(\infty)$ are uncorrelated. We cannot do this when $a(t)$ is multiply periodic. Generally speaking, in the chaotic regime, we cannot assume that the trajectories are ergodic over the whole energy shell and hence the ensemble average must be obtained by averaging over many trajectories rather than from the time average of a single trajectory. Thus, overall, as the classical motion makes the transition from predominantly regular to predominantly irregular motion, the energy shell averaged (i.e., microcanonical ensemble) correlation function changes from some form of oscillatory behavior to some form of decaying behavior. Although the above arguments have been

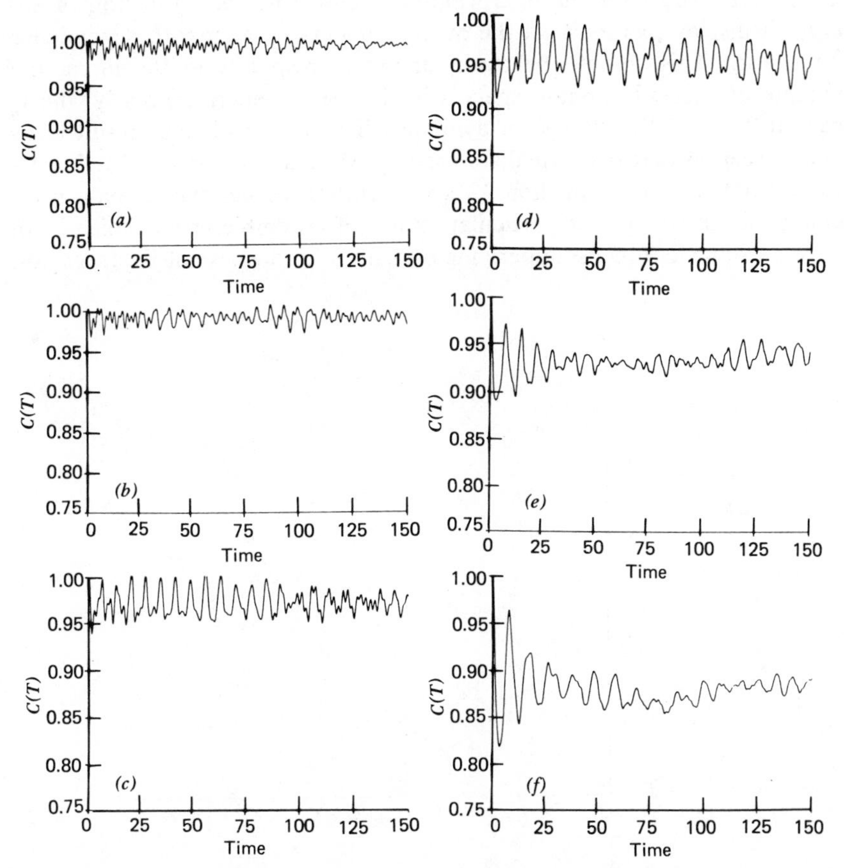

Fig. 15. The (normalized) autocorrelation function $\langle a(0)^*, a(t)\rangle$ where a is given by (6.9) at energies: (a) 0.0367, (b) 0.0767, (c) 0.1167, (d) 0.1367, (e) 0.1567, and (f) 0.1667. Reproduced by permission from Koszykowski et al.[67]

developed for the correlation function $D(t)$ they should, as is indicated earlier, apply equally to the behavior of the correlation function $C(t)$.

To illustrate these ideas we show in Fig. 15 the correlation function $C(t)$ [where $a(t)$ is given by (6.9)] computed at a series of energies for the Henon-Heiles system.[67] At the lowest energies, where the motion is almost entirely quasiperiodic, $C(t)$ displays a regular oscillatory behavior. As the energy increases an initial decay gradually grows and by $E=\frac{1}{6}$ (dissociation threshold) $C(t)$ displays a clearly decaying oscillatory form. The change in $C(t)$ is gradual; there is no noticeably dramatic change about the "critical energy" of $E=\frac{1}{8}$. A somewhat similar trend is observed for the correlation function of $\dot{a}(t)$ and other choices of dynamical variable.[67]

In Fig. 16 we plot the computed values of the first moments of $a(t)$, that is, $y_1=\langle iLa, iLa\rangle$ for the microcanonical ensemble, as a function of energy. It displays, as must all the other moments y_n, a smooth dependence on energy. This smooth behavior simply corresponds to the increasing volume of accessible phase space with increasing energy. Clearly, the y_n cannot "detect" the change in dynamics. This rather obvious result leads one to suspect that the formalism used by Mo may not be suitable for detecting the "chaotic transition." As we mention earlier the formalism requires one to fix upon a particular choice of ensemble. This choice is, in effect, a statement of the underlying dynamics. In the case of the transition

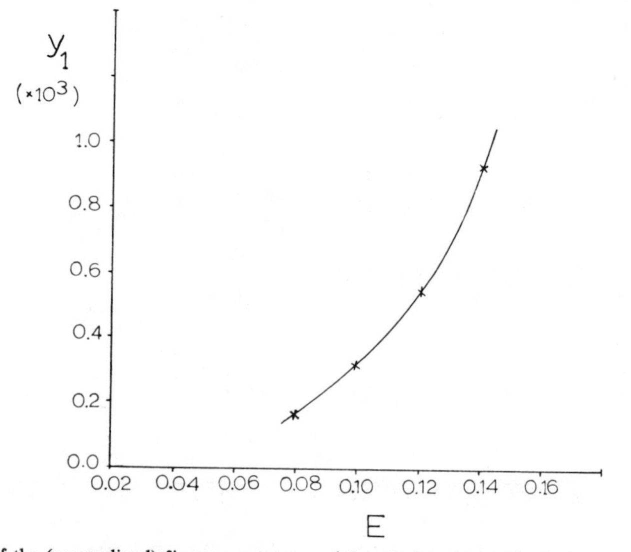

Fig. 16. Plot of the (normalized) first moment, $y_1=\langle iLa, iLa\rangle$, where a is given by (6.9), as a function of energy. Computed by M. L. Koszykowski (private communication).

from regular to irregular motion, one should really change from using the torus as the ensemble to using the energy shell as the ensemble in order to reflect the change in dynamics. This would, for example, lead to a dramatic change in the D_i's in (6.6), with a corresponding effect on $C(t)$ [or $D(t)$]. Otherwise, as we have seen, for a chosen ensemble the moments of the Liouville operator change smoothly, independent of the change in dynamics. The change in the Ω_i's in (6.9) from real to complex at the "critical energy" now seems rather fortuitous since they are only functions of quantities (the y_i's) that are unaffected by the chaotic "transition." It does not seem that there is any simple way of building a change of ensemble into the formalism without a prior knowledge of the critical energy, that is, one must know the answer beforehand. Furthermore, our analysis cannot be the whole story since once a system becomes sufficiently large, correlation functions decay *independently* of the underlying dynamics. As an example of this the reader is referred to the study of the momentum autocorrelation function of a mass defect in a linear chain by Cukier et al.[72] Thus, although the motion is entirely integrable, they are able to observe an almost pure exponential decay for a mass defect ratio 0.1 in a chain of 50 particles. It is interesting to reflect on the nature of the randomness "generated" by the regular trajectories of a sufficiently large integrable system compared with that generated by the irregular trajectories of a smaller, nonintegrable system.

VII. REGULAR AND IRREGULAR REGIMES IN QUANTUM MECHANICS

In the preceding sections we discussed the onset of widespread chaotic motion and possible methods for predicting when this occurs. One of the most fascinating aspects of this "transition" is its implications for the semiclassical limit* of quantum mechanics.

When the motion is integrable the actions provide a set of adiabatic invariants[73] that are suitable for quantization. For a given value of the action vector $\mathbf{I}=(I_1,\ldots,I_N)$ each component I_i is defined by[14, 15, 74]

$$I_i=\frac{1}{2\pi}\oint_{\mathcal{C}_i}\mathbf{p}\cdot d\mathbf{q} \qquad i=1,\ldots N \tag{7.1a}$$

where the $\mathcal{C}_i$ are the N topologically independent contours on the N-dimensional torus associated with $\mathbf{I}$ (see Fig. 17). A bound state can be associated, in the semiclassical limit, with the torus whose actions [defined by

*By this we mean the limit in which parameters of the system, with the same dimensions as $\hbar$, becomes large relative to $\hbar$. We express this as the limit $\hbar\to 0$.

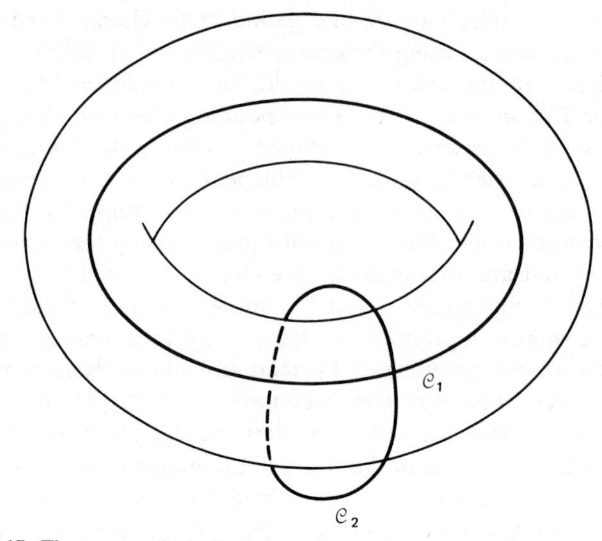

Fig. 17. The two topologically independent paths on a two-dimensional torus.

(7.1a)] satisfy

$$\mathbf{I_m} = \left(\mathbf{m} + \frac{\boldsymbol{\alpha}}{4}\right) \qquad \begin{matrix} \mathbf{m} = m_1, \ldots m_N \\ \boldsymbol{\alpha} = \alpha_1, \ldots \alpha_N \end{matrix} \tag{7.1b}$$

where $\mathbf{m}$ is the vector of quantum numbers and $\boldsymbol{\alpha}$ represents the (fixed) "Maslov indices." These correspond to the number of classical turning points encountered in one cycle around each of the paths $\mathcal{C}_i$. The energy of the $\mathbf{m}$th state is then given by

$$E_{\mathbf{m}} = H(\mathbf{I_m}) \tag{7.2}$$

where the Hamiltonian, which is assumed integrable, is expressed in terms of the (constant) action variables $\mathbf{I}$. The above set of rules was developed by Einstein,[74] Brillouin,[75] and Keller[76] (EBK). They provide a generalization of the old Bohr-Sommerfeld rules for one-dimensional bounded motion. These ideas are discussed in detail by Percival.[13]

In the regime of classically chaotic motion very few tori persist, and for most bound states at these energies the EBK rules are no longer applicable. This problem was appreciated by Einstein[74] and more recently by Percival[6], who has proposed that, in the semiclassical limit, the bound state spectrum might be divided into two parts.

1. A *regular spectrum* corresponding to the regime of predominantly integrable (regular) motion in which all states can be quantized according to the EBK rules.
2. An *irregular spectrum* corresponding to the regime of predominantly chaotic (irregular) motion in which EBK is no longer applicable.

The two different classes of spectrum may have very different properties reflecting the differences in the underlying classical motion.

Our discussion of these properties is divided into two parts. The first deals with eigenvalue-related properties, that is, quantization procedures, sensitivity of energy levels to perturbation, densities of states and so on. This part also includes a discussion of transition probabilities and correspondence principles. The second part deals only with eigenvector-related properties, that is, wave functions, probability densities, Wigner functions, wave packets, and so on and concludes with a few remarks on the vexed issue of "quantum ergodicity."

A. Eigenvalue-Related Properties

The EBK rules provide a practical route to calculating the eigenvalues of the regular spectrum. Operationally it reduces to a problem in classical, rather than semiclassical, mechanics. Many of these methods have been reviewed by Percival.[13] Clearly, though, in the predominantly chaotic regime the EBK rules are no longer applicable. However, Jaffe and Reinhardt[77] found that if, in this regime, one still persists with their algorithm (which ultimately must diverge in this regime) for constructing "good" action variables, that is, one still assumes that tori exist even though, strictly speaking, they do not, the results obtained from EBK quantization still agree quite well with the exact quantal results. This observation touches on a very significant point. The system they examined did not support many (only 66) bound states, that is $\hbar$ is relatively "large" for this system. Clearly, if the regions of phase space occupied by chaotic motion are much smaller than $O[(2\pi\hbar)^N]$ they will have little significance quantum mechanically. Thus in the above case $\hbar$ would appear to be playing some form of "smoothing" role that can "patch up" the otherwise destroyed tori.[78] We return to this very important point later. However, once $\hbar$ becomes sufficiently "small," even this approach is no longer applicable. As yet there is no proven semiclassical quantization condition available for the truly chaotic regime. It has been conjectured that in this case some form of phase space volume quantization might apply,[79, 80] that is,

$$\int d\mathbf{p}\int d\mathbf{q}\,\theta(E-H(\mathbf{p},\mathbf{q}))=m(2\pi\hbar)^N \tag{7.3}$$

where m is a single integer quantum number labeling the energies in order, and θ is the (unit) step function. This condition might also include some additional term analogous to the Maslov index.[79]

Percival has conjectured[7] that the regular and irregular spectra will be distinguishable by means of their transition probabilities. Transitions between states of the regular spectrum will be characterized by strong selection rules, that is, one expects to see a spectrum (in the sense of a spectroscopic observable) of just a few intense lines corresponding to strongly coupled states. On the other hand, states of the irregular (energy) spectrum are expected to be coupled with similar intensity to all those states of a similar energy that correspond to the same "chaotic" regions of phase space, that is, one expects to see a spectrum of many weak lines. There are, as yet, no reported experimental observations that unambiguously indicate the existence of an irregular spectrum.

The correspondence between classical power spectra (as described in Section II) and quantal (transition) spectra is easily understood in the regular regime. Consider the two EBK states $E_{\mathbf{n}}=H(\mathbf{I}=\mathbf{n}\hbar)$ and $E_{\mathbf{m}}=H(\mathbf{I}=\mathbf{m}\hbar)$, where for convenience we drop the Maslov terms. For $E_{\mathbf{m}}$ sufficiently close to $E_{\mathbf{n}}$ we may expand the former about the latter in a Taylor series to obtain, to first order

$$E_{\mathbf{m}}=H(\mathbf{n}\hbar)+\hbar(\mathbf{m}-\mathbf{n})\cdot\left[\nabla_{\mathbf{I}}H(\mathbf{I})\right]_{\mathbf{I}=\mathbf{n}\hbar}+\cdots \tag{7.4}$$

The quantal transition frequency ω_{mn}, is therefore given by

$$\omega_{mn}=\frac{(E_{\mathbf{m}}-E_{\mathbf{n}})}{\hbar}\simeq(\mathbf{m}-\mathbf{n})\cdot\boldsymbol{\omega}(\mathbf{n}\hbar) \tag{7.5}$$

where we use (2.6). Thus the power spectrum of a classical trajectory belonging to the torus with actions $\mathbf{I}=\mathbf{n}\hbar$ will have lines that correspond (approximately) to the $\mathbf{n}\to\mathbf{m}$ quantal transition. In the limit $\hbar\to 0$ or $\mathbf{n}\gg\mathbf{n}-\mathbf{m}$ the classical and quantal frequencies becomes equal.[73] Furthermore, the squared moduli of the classical Fourier coefficients correspond to the quantal transition probabilities. In practice, that is for finite $\hbar$, the best agreement between classical and quantal spectra is often obtained by comparing the quantal spectrum for $\mathbf{n}\to\mathbf{m}$ transition with the classical spectrum of a trajectory lying on the torus with the "mean action" $\mathbf{I}=\hbar(\mathbf{m}+\mathbf{n})/2$ rather than the "initial state action" $\mathbf{I}=\mathbf{n}\hbar$ (see for example Ref. 29).

In contrast to the regular regime the spectra of irregular trajectories are immensely complicated and display, essentially, an infinity of lines. It is not at all clear at this stage how one might compare such spectra with the quantal spectrum. It may well be that a meaningful comparison is only possible when both spectra are averaged over some range of trajectories and states, respectively. Much more consideration of a "correspondence principle" for the irregular regime is now required. (Classical and quantal spectra are compared in Ref. 80.)

Percival has also predicted that regular and irregular states will be distinguishable by their behavior under perturbation. Irregular states will be very sensitive to an external or slowly varying perturbation, whereas regular states will be relatively stable. This conjecture was first tested by Pomphrey,[81] who investigated the eigenstates of the Henon-Heiles type Hamiltonian

$$H=\tfrac{1}{2}\left(p_x^2+p_y^2+x^2+y^2\right)+\lambda\left(x^2y-\tfrac{1}{3}y^3\right) \tag{7.6}$$

using a value of $\lambda=0.088$. The quantity calculated was the "second difference," Δ^2E_i, where

$$\Delta^2E_i=E_i(\lambda-\Delta\lambda)+2E_i(\lambda)-E_i(\lambda+\Delta\lambda) \tag{7.7}$$

which gives a measure of the sensitivity of the ith eigenvalue to small changes ($\Delta\lambda$) in the perturbation. A number of very large second differences were found for states with energies in the predominantly chaotic regime. This would appear to confirm Percival's prediction. More recently we have performed a detailed investigation into second differences,[80] again working with Hamiltonian (7.6) but now with $\lambda=0.1118$. Owing to the symmetry of the potential the eigenvalues can have either A (nondegenerate) or E (doubly degenerate) symmetry. Furthermore, each state can be assigned a principle quantum number and an approximate "angular momentum" quantum number. We found that high angular momentum states all had small Δ^2E_i's, whereas low angular momentum states had consistently larger Δ^2E_i's. This phenomenon has also been noted by McDonald and Kaufman[82] in their study of the eigenstates of the "stadium" (described below). In our study this behavior is consistent with the underlying classical dynamics. The high angular momentum states could all be associated with tori, that is, stable motion, and hence quantized by EBK, even in the predominantly chaotic regime. On the other hand, the low angular momentum states, when they could be computed semiclassically, were found to be associated with those tori that were the first to be destroyed at higher energy. Our study also revealed another interesting feature, namely, the possibility of both level "crossings" and "avoided crossings." At high energies and where symmetry permitted, a number of states were found to cross as a function of λ.[83] If this effect were not taken into account, spuriously large values of Δ^2E_i would have been calculated. We also found one case of what appears to be an "avoided crossing" or "repulsion of levels." This effect yielded large values of $\Delta^2 E_i$.[84]

Investigations into the role played by closed orbits in the semiclassical description of the energy spectrum have provided a number of interesting results. It has been demonstrated that in the case of integrable Hamiltonians[17, 85] or certain types of hollow enclosure,[86] that is, regular energy spectra, a complete description is possible. The results show that there is not a one-to-one correspondence between an individual eigenstate and a particular closed orbit. Rather, each closed orbit makes some oscillatory contribution, corresponding to clusters of levels, to the density of states. As the contributions from more and more topologically distinct closed orbits are added together the clusters gradually resolve themselves into delta functions at the correct (EBK) eigenenergies. Unfortunately, no such simple picture emerges in the case of nonintegrable systems. Here Gutzwiller[86] has proposed quantization condition involving individual closed orbits. Different conditions are obtained depending on whether the orbit is stable or unstable. These notions have proved to be a somewhat contentious issue that has not yet been satisfactorily resolved.

The statistical properties of the regular spectrum have been investigated both theoretically and numerically.[88] For nondegenerate integrable systems, that is, condition (2.9) is satisfied, the distribution of the nearest neighbor level spacing, s, varies as exp(− s). This means that, in order of increasing energy, the levels of the regular spectrum tend to cluster. In the case of degenerate systems, that is, systems of harmonic oscillators, different distributions are obtained. These depend on subtle, number-theoretic properties of the fundamental frequencies.

The statistical properties of the irregular spectrum are not yet well understood. By analogy with the theories of random matrices, used in the study of the statistics of nuclear energy levels,[89] one is tempted to surmise that a "Wigner distribution," such as $P(s)=[(\pi/2)s]\exp[-(\pi/4)s^2]$, might be applicable. Thus in contrast to the regular spectrum one anticipates some form of "repulsion of levels." Some arguments to this effect have been given by Zaslavskii.[90] An interesting numerical study has been carried out by McDonald and Kaufman[91] on the "stadium" system. This system corresponds to an empty, hard-walled enclosure consisting of two semicircles (of radius R) connected by straight walls (of length $2a$). For zero "aspect ratio," $\gamma=a/R$, the enclosure reduces to a simple circle and the motion is integrable. For nonzero values of γ the motion becomes strongly irregular.[92] For $\gamma=0$ the corresponding energy spectrum is regular and, as predicted, a strongly clustered level spacing distribution is found. On the other hand, for $\gamma=1$ (when the spectrum is presumably irregular), a distribution exhibiting a repulsion of levels is observed.

B. Eignevector-Related Properties

It is a well-known result that in the limit $\hbar \to 0$ a solution of Schrodingers equation can be written in the form $\psi = A\exp\{iS/\hbar\}$. A is a certain amplitude and S is the classical, position-dependent action function that satisfies the Hamilton-Jacobi equation.[16] In the case of the motion being integrable this form of ψ can be specified rather completely. The action function S can be written in the form

$$S(\mathbf{q},\mathbf{I}) = \int_{\mathbf{q}_0}^{\mathbf{q}} \mathbf{p}(\mathbf{q}',\mathbf{I})\cdot d\mathbf{q}' \tag{7.8}$$

where $\mathbf{q}_0$ is some arbitrary initial point in the classically allowed region. Notice that in the integrand the momentum $\mathbf{p}$ is expressed as a function of the constant actions $\mathbf{I}$. The variables conjugate to the actions are the angle variables $\boldsymbol{\theta}$. These are defined by the relationship

$$\boldsymbol{\theta} = \nabla_{\mathbf{I}} S(\mathbf{q},\mathbf{I}) \tag{7.9}$$

and are uniformly distributed over the associated torus. The function S plays the role of the generating function that effects the canonical transformation between the $(\mathbf{p},\mathbf{q})$ and $(\mathbf{I},\boldsymbol{\theta})$ variables.[16] This being so we also have the relationship

$$\mathbf{p} = \nabla_{\mathbf{q}} S(\mathbf{q},\mathbf{I}) \tag{7.10}$$

Furthermore, S is a multivalued function of $\mathbf{q}$. This follows from the fact that $\mathbf{p}$ is multivalued, for example, for one-dimensional bounded motion p has two branches (see Fig. 18) determined by

$$p(q,I) = \pm\{2m[H(I) - V(q)]\}^{1/2} \tag{7.11}$$

In fact the condition that $\psi(\mathbf{q})$ be a single-valued function of $\mathbf{q}$, while S is multivalued enables one to deduce the EBK quantization conditions.[75, 76]

The most general form of the amplitude A was deduced by Van Vleck[93] to be

$$A = \det\left(\frac{\partial^2 S(\mathbf{q},\mathbf{I})}{\partial q_j \partial I_k}\right)^{1/2} \qquad j,k = 1,\ldots,N \tag{7.12}$$

which provides a measure of the "classical path density."[94] The semiclassical wave function corresponding to the state with quantum number $\mathbf{m}$ can

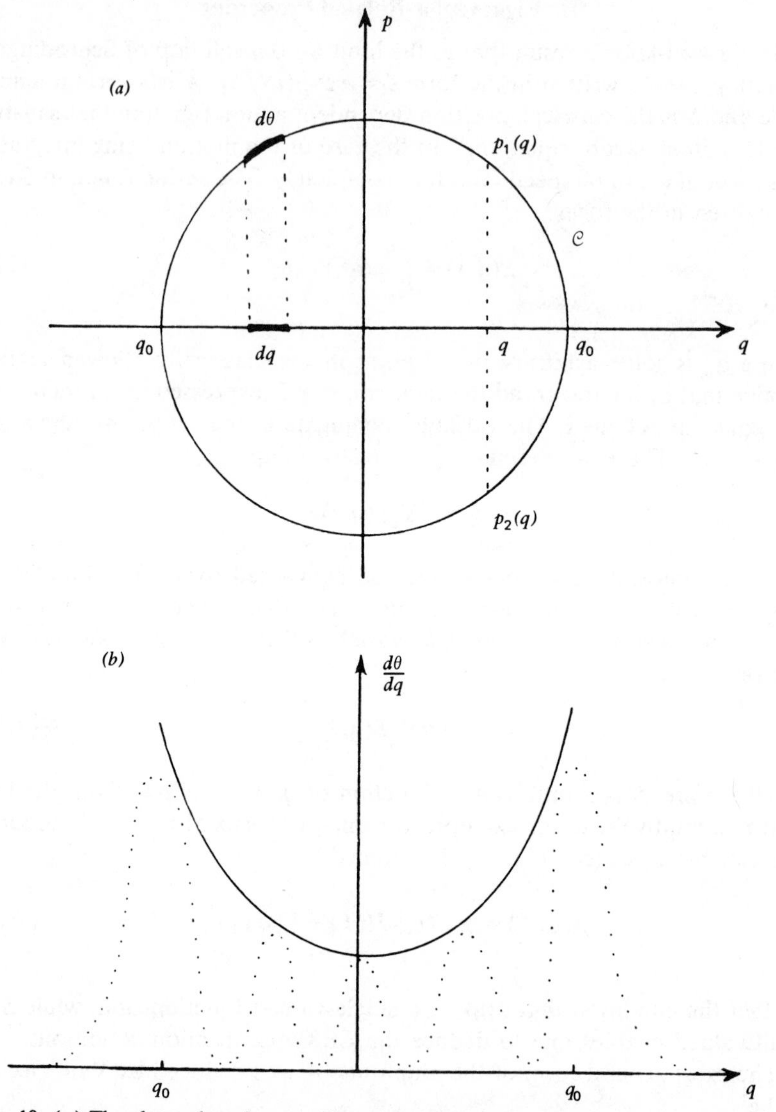

Fig. 18. (a) The phase plane for one-dimensional bounded motion showing a typical curve, $\mathcal{C}$, of constant energy $E=H(p,q)$. The momentum p is a two-valued function of q; the two branches $p_1(q)$ and $p_2(q)$ coalesce at the classical turning points q_0. Turning points are characterized by being those points on $\mathcal{C}$ whose tangents are parallel to the p-axis. (b) The projection of the curve $\mathcal{C}$, given by $d\theta/dq$, onto the q-axis. This projection, which is singular at the classical turning points, gives the smooth envelope of the true, quantal probability density (·····).

now be written apart from certain phase factors, in its most general form, that is,

$$\psi_{\mathbf{m}}(\mathbf{q})=\sum_r \det\left(\frac{\partial^2 S_r}{\partial q_j\,\partial I_k}\right)^{1/2}\exp\left[\frac{iS_r(\mathbf{q},\mathbf{I}_{\mathbf{m}})}{\hbar}\right] \tag{7.13}$$

where the sum over r is the sum over the different branches of S and $\mathbf{I}_{\mathbf{m}}$ satisfies (7.1b), which ensures the single valuedness of ψ.

The probability density $|\psi_{\mathbf{m}}(\mathbf{q})|^2$ evaluated from (7.13) will include oscillatory cross terms corresponding to interference between the different branches of S. These terms can be eliminated by a local averaging over some width Δq,[95] which vanishes slower than $\hbar$ as $\hbar\to 0$. One then obtains a "coarse grained" (classical) probability density with the particularly simple form

$$\overline{|\psi_m(\mathbf{q})|^2}=\sum_r\left|\frac{\partial^2 S_r}{\partial q_i\,\partial I_k}\right| \tag{7.14}$$

This form has a rather interesting geometrical interpretation, namely, it corresponds to the projection of the torus associated with the **m**th state onto the coordinate plane. As a simple example consider one-dimensional bounded motion. Using (7.9) we have

$$\overline{|\psi(q)|^2}=\left|\frac{\partial^2 S}{\partial q\,\partial I}\right|=\left|\frac{d\theta}{dq}\right| \tag{7.15}$$

This limiting form of $|\psi|^2$ (sketched in Fig. 18b) is singular at the classical turning points. These singularities, which are termed caustics, herald the breakdown of the semiclassical wave function (7.12) at these points. None theless, this crude form of $|\psi|^2$ provides, in the semiclassical limit, the envelope of the oscillations of the true quantal probability density.

So far, though, everything we have said only applies to the wave functions of the regular states. Defining a semiclassical form of the wave function for irregular states is much more difficult. The problem is that for irregular trajectories $\mathbf{p}$ is no longer a finitely multivalued function of $\mathbf{q}$; instead it has infinitely many branches. (Alternatively one may say that in the chaotic regime no global solution to the Hamilton-Jacobi equation exists.)

To be able to compare the semiclassical forms of the regular and irregular states it has been suggested[95–97] that the use of the Wigner function[98] may provide a convenient alternative to the study of the wave functions themselves. The Wigner function provides a quantal analogue to classical phase space density. It takes the form

$$W(\mathbf{p},\mathbf{q})=\frac{1}{(\pi\hbar)^N}\int d\mathbf{x}\, e^{-2i\mathbf{p}\cdot\mathbf{x}/\hbar}\psi(\mathbf{q}+\mathbf{x})\psi^*(\mathbf{q}-\mathbf{x}) \tag{7.16}$$

Examination of $W(\mathbf{p},\mathbf{q})$ in a particular phase plane (p_i, q_i) provides one with a quantal analogue to the Poincare surface-of-section.[96] The Wigner function has many interesting properties,[99] but for our purposes the most important is that its projection onto the coordinate plane gives the quantal probability density, that is,

$$\int W(\mathbf{p},\mathbf{q})d\mathbf{p}=|\psi(\mathbf{q})|^2 \tag{7.17}$$

In the case of regular states the semiclassical form of the wave function $\psi_{\mathbf{m}}(\mathbf{q})$ can be used to evaluate the associated pure state Wigner function $W_{\mathbf{m}}(\mathbf{p},\mathbf{q})$. It may then be shown[96, 100] that in the *classical* limit, $\hbar=0$, $W_{\mathbf{m}}(\mathbf{p},\mathbf{q})$ reduces to

$$W_m(\mathbf{p},\mathbf{q})=\frac{1}{(2\pi)^N}\delta(\mathbf{I}(\mathbf{p},\mathbf{q})-\mathbf{I}_{\mathbf{m}}) \tag{7.18}$$

that is, the Wigner function "collapses" onto a delta function on the classical torus associated with the **m**th state. As is described above, projection of this torus onto the coordinate plane gives the limiting form of the probability density. The results of these projections depends on the "orientation" of the torus in phase space. This can lead to a variety of different caustic structures.[95] There are subtle differences between the results obtained for separable and nonseparable integrable systems.[101]

For finite $\hbar$, that is, in the *semiclassical* limit, the Wigner function displays a regular structure of "diffraction fringes." Projection of this form of $W_{\mathbf{m}}(\mathbf{p},\mathbf{q})$ onto the coordinate plane yields the correct oscillatory form of $|\psi_{\mathbf{m}}(\mathbf{q})|^2$. Comparison of the classical and semiclassical limiting forms of $W_{\mathbf{m}}(\mathbf{p},\mathbf{q})$ shows that for regular states the role of $\hbar$ is to add a regular structure (the quantum oscillations) onto a smooth classical background. In Fig. 19 we show a Wigner "surface of section," for a regular state of a Henon-Heiles type system, calculated by Hutchinson and Wyatt.[102]

An investigation of the Wigner function for irregular states might again appear to be frustrated by our lack of knowledge of a semiclassical form for the wave function. However, some progress can be made by adopting a slightly different point of view. For regular states the classical limit of the Wigner function is the phase space manifold (the torus) with which the state

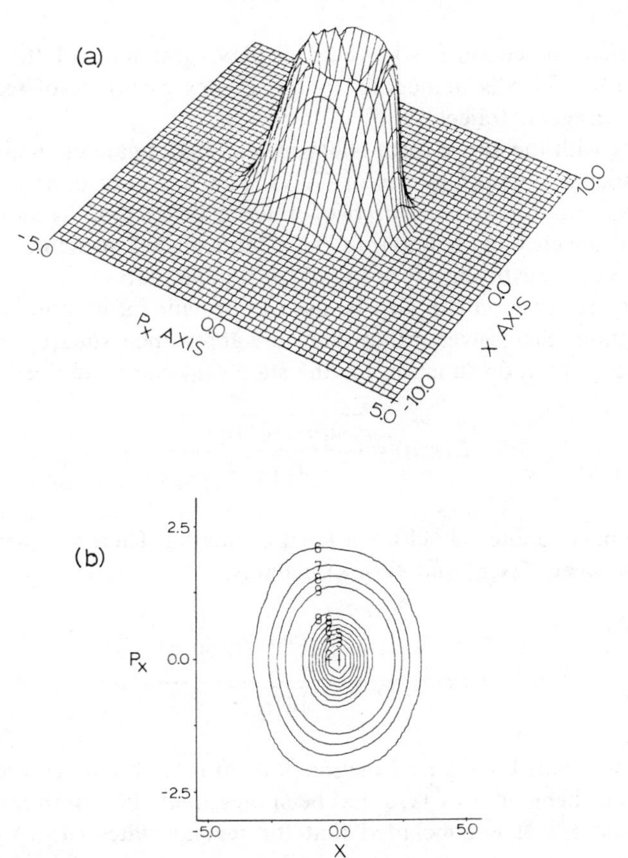

Fig. 19. The Wigner distribution in the (p_x, x) plane of the tenth eigenstate of the Hamiltonian $H=\frac{1}{2}(p_x^2+p_y^2+0.49x^2+1.69y^2)-0.10(xy^2-x^3)$: (*a*) perspective and (*b*) contours. A smooth, concentric pattern of phase space density is clearly demonstrated. (Reproduced by courtesy of Hutchinson and Wyatt.[102])

is associated. In the extreme chaotic regime an irregular state is probably associated with a large portion of the corresponding energy shell. This being so, a reasonable conjecture for the classical limit of the Wigner function is that it be the (normalized) microcanonical distribution,[103] that is,

$$W(\mathbf{p},\mathbf{q})=\frac{\delta(E-H(\mathbf{p},\mathbf{q}))}{\int d\mathbf{p}\int d\mathbf{q}\,\delta(E-H(\mathbf{p},\mathbf{q}))} \tag{7.19}$$

For finite $\hbar$, although the exact form of W is not known, one anticipates that a surface-of-section of W would display a random splatter of phase space density. This is analogous to the Poincare surface-of-section observed for irregular trajectories.[96]

Working with the above form of W one can investigate the limiting form of the associated $|\psi(\mathbf{q})|^2$ by use of (7.17). This has been done by Berry,[95] who shows that for systems of two or more degrees of freedom $|\psi(\mathbf{q})|^2$ vanishes at the classical boundaries. This "anticaustic" structure is in sharp contrast to the caustic structure found for regular states.

The Wigner function can also be used to provide information about the wave functions themselves rather than just about their squared moduli. A spatial autocorrelation function for the state $\psi(\mathbf{q})$ can be defined[95] as

$$C(\mathbf{x},\mathbf{q})=\frac{\overline{\psi(\mathbf{q}+\mathbf{x})\psi^*(\mathbf{q}-\mathbf{x})}}{\overline{|\psi(\mathbf{q})|^2}} \tag{7.20}$$

where the bars denote, as before, a local averaging. There is a simple relationship between $C(\mathbf{x},\mathbf{q})$ and $W(\mathbf{p},\mathbf{q})$, namely,

$$C(\mathbf{x},\mathbf{q})=\frac{\int d\mathbf{p}e^{i2\mathbf{p}\cdot\mathbf{x}/\hbar}\overline{W}(\mathbf{p},\mathbf{q})}{\overline{|\psi(\mathbf{q})|^2}} \tag{7.21}$$

The "coarse grained" Wigner function $\overline{W}(\mathbf{p},\mathbf{q})$ is simply its classical limiting form. The behavior of $C(\mathbf{x},\mathbf{q})$ has been investigated for both regular and irregular states.[95] It is concluded that for regular states $C(\mathbf{x},\mathbf{q})$ is anisotropic, whereas for irregular states it is isotropic, taking the form of a Bessel function for certain forms of potential. Overall, regular states are expected to exhibit strong, anisotropic interference oscillations with just a few scales of oscillations. In the case of irregular states, ψ should exhibit more moderate, spatially isotropic oscillations with a continuous spectrum of wave vectors $(\mathbf{p}/\hbar)$ that is, oscillations on all scales. This would imply that $\psi(\mathbf{q})$ is a Gaussian random function of $\mathbf{q}$. Tests of some of these conjectures are now being carried out.[104]

The role played by $\hbar$ in the chaotic regime is very different from that played in the regular regime.[95, 96] Chaotic classical dynamics displays structure down to arbitrarily fine scales. Here $\hbar$ "smooths away" this fine structure and irregular states display structure only down to scales of order $\hbar$. It is very important to emphasize that the notion of an "irregular state" is a semiclassical, that is, $\hbar \rightarrow 0$, phenomenon. However strong the nonintegrable perturbation, if $\hbar$ is not sufficiently "small," one should not expect to observe irregular states (hence, perhaps, the results of Jaffe and Reinhardt).

Indeed, Berry[96] has predicted that the energy spectrum can go through a number of different semiclassical regimes that depend on the relative sizes of both $\hbar$ and the perturbation parameter ε. The relationship between the limits $\hbar \to 0$ and $\varepsilon \to 0$ is nontrivial.

With these remarks in mind we now turn to some numerical investigations of wave functions in the regular and irregular regimes. Noid et al.[105] have computed the exact quantal probability density for states of a Henon-Heiles type system (with the fundamental frequencies in the ratio 2 : 1 rather than 1 : 1). In Fig. 20 we show their $|\psi(\mathbf{q})|^2$ computed for a regular state. A classical trajectory belonging to the associated (EBK) torus is also shown.

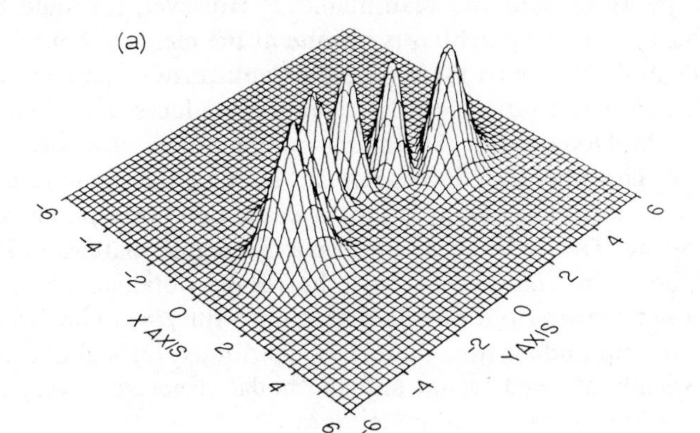

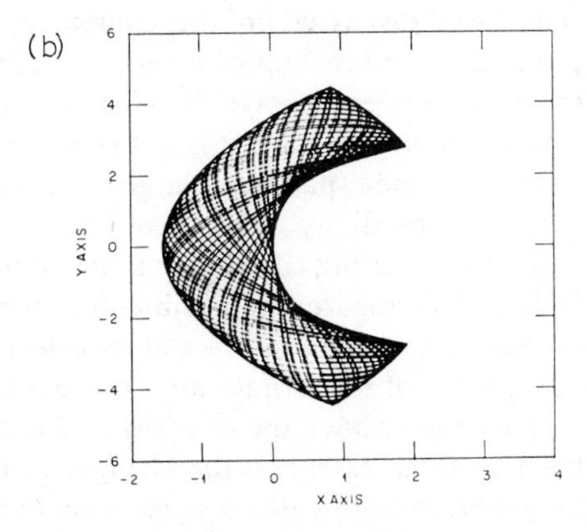

Fig. 20. (*a*) The quantal probability density, $|\psi(\mathbf{q})|^2$, of a state in the regular regime of the Hamiltonian $H = \frac{1}{2}(p_x^2 + p_y^2 + 1.96x^2 + 0.49y^2) - 0.08(xy^2 - 0.08y^3)$ and (*b*) a trajectory belonging to the associated EBK torus. The energy is $E = 4.265$. (Reproduced by permission from Noid et al.[105])

The probability density is clearly confined to the region enclosed by the "boxlike" caustic structure of the associated trajectory. From the regularity of the oscillations of $|\psi(\mathbf{q})|^2$ it is clear that $\psi(\mathbf{q})$ will have a regular pattern of nodal lines. In Fig. 21 we show $|\psi(\mathbf{q})|^2$ computed for a state at a higher energy and for which EBK quantization was not possible. A typical irregular trajectory at that energy is also shown. For this state $|\psi(\mathbf{q})|^2$ appears to have spread, like the trajectory, over most of the classically allowed configuration space. The whole structure of the wave function is now much less regular than before.

Another interesting study is that made of the wave functions of the "stadium" by McDonald and Kaufman.[82, 91] However, it should be noted that for this system the potential is infinite at the classical boundaries and hence $\psi(\mathbf{q})$ must go to zero there (i.e., the "anticaustic" conjecture cannot be tested). For aspect ratio $\gamma=0$ the "stadium" reduces to a circle. In Fig. 22 we show McDonald and Kaufman's computations of a wave function for this case. The amplitude shows regular, strongly directional oscillations and rises to a maximum around an inner circle that corresponds to an underlying caustic. The nodal structure displays a regular pattern of intersecting nodal lines. In Fig. 23 we show the results obtained for a state of almost the same energy but now in a stadium with $\gamma=1$. The difference is striking. The amplitude is now uniformly distributed throughout the whole of the classically allowed region and the nodal structure is very irregular with apparently no crossings of nodal lines.

There has been some discussion about the relationship between the changes in nodal structure and the onset of chaotic motion. The noncrossing of nodal lines has been discussed in the context of nonseparable systems[106] and more recently the breakup of a regular nodal pattern as a possible criterion for "quantum ergodicity."[107] However, although there may well be some connection, at this stage it is not clear how one can distinguish between a change of nodal pattern as a genuine manifestation of some form of underlying "ergodicity" and that due to an increasing "nonseparability" of the system in the coordinate space in which the wave function is plotted. Indeed, it is apparently possible to construct integrable systems whose wave function have arbitrary nodal complexity.[108] More investigations into the properties of nodal lines are now required.

In both of the examples cited above the changes in the wave functions are undoubtably related to some extent to the changes in the underlying classical dynamics. However, in both cases it is not easy to make a direct, "one-to-one," correlation between the changes in the classical dynamics and those in the quantum mechanics. In the regular case we can, since the semiclassical form of the wave function is known. In the irregular case, lack of knowledge of the semiclassical mechanics makes a direct comparison

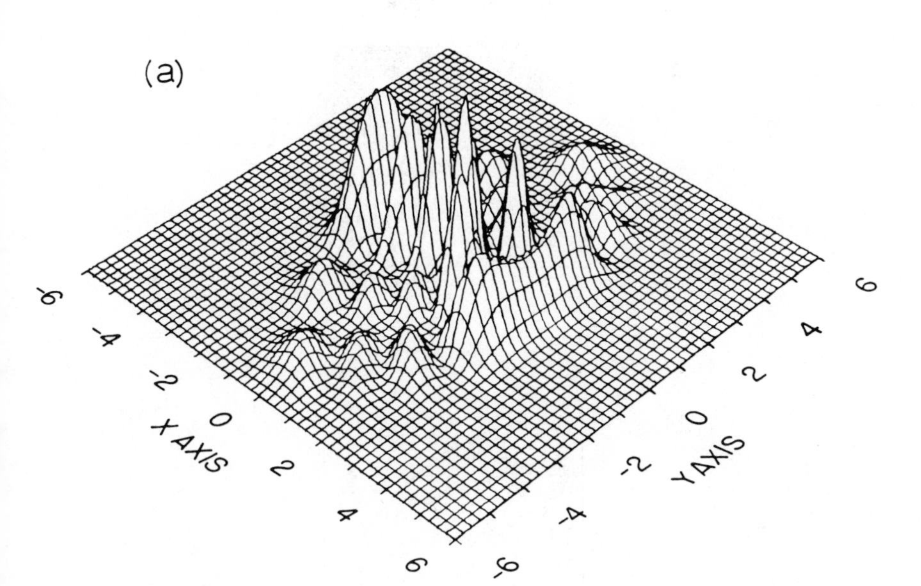

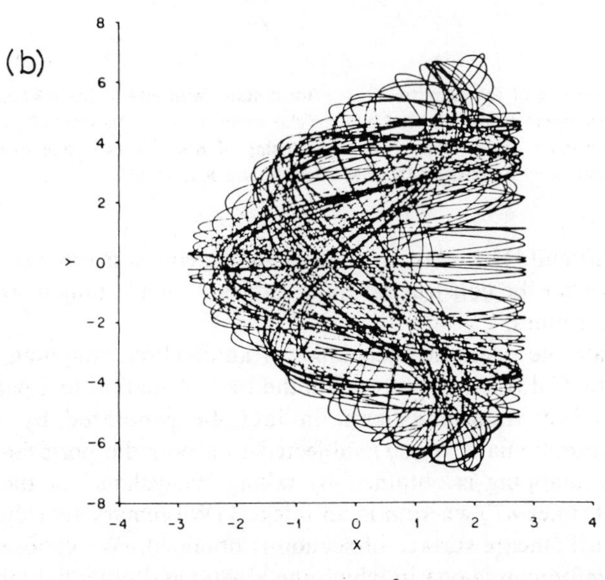

Fig. 21. (*a*) The quantal probability density, $|\psi(\mathbf{q})|^2$, of a state in the irregular regime of the same Hamiltonian used in Fig. 20 and a typical irregular trajectory at the same energy, $E = 8.0$. (Reproduced by permission from Noid et al.[105])

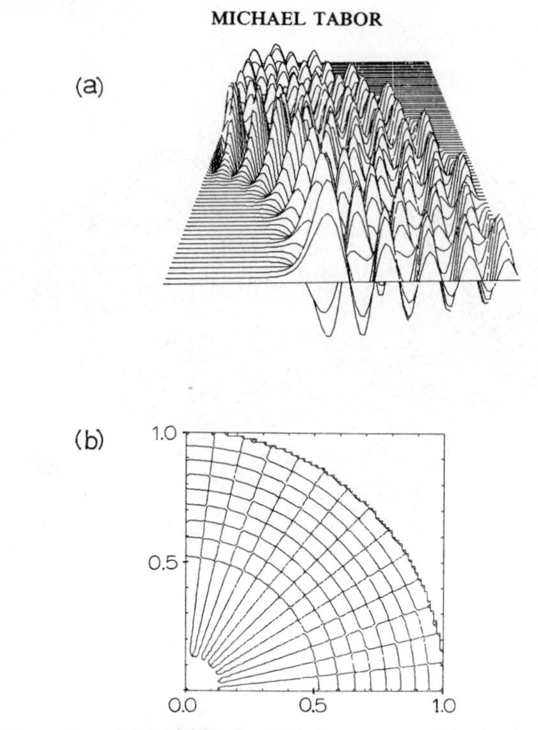

Fig. 22. (*a*) Perspective of the amplitude, $\psi(\mathbf{q})$, for a state, with eigenvalue $k=65.38142$, of the "stadium" with aspect ratio $\gamma=0$, that is, a circular enclosure (one quadrant of ψ is shown). (*b*) The nodal structure of this state. The noncrossing of nodal lines is due to computer graphics. (Reproduced by permission from McDonald and Kaufman[82].)

much more difficult. However, in the case of algebraic mappings a semiclassical theory can be constructed that enables a direct comparison to be made in both regimes.

To investigate the quantum mechanics of an algebraic mapping,[23] such as that given by (2.13), one has to relate the transformation to a particular Hamiltonian. Such mappings can, in fact, be generated by a one-dimensional Hamiltonian $H(p,q)$ subjected to a periodic perturbation of period T. The mapping is obtained by taking "snapshots" of the (p,q) phase plane at times nT, where n is an integer. (We remark that this is essentially how a Poincare surface-of-section is obtained.) We choose a case where the Hamiltonian is one in which the kinetic and potential terms are periodically switched on and off, that is,

$$H(p,q,t)=\frac{p^2}{2\mu\gamma} \qquad 0<t<\gamma T \tag{7.22a}$$

$$=\frac{V(q)}{1-\gamma} \qquad \gamma T<t<T \tag{7.22b}$$

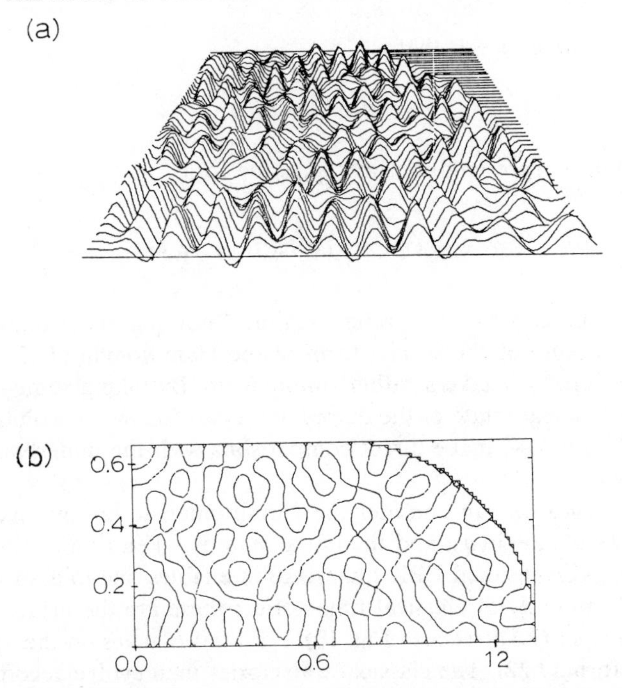

Fig. 23. (*a*) Perspective of the amplitude, $\psi(\mathbf{q})$, for a state, with eigenvalue $k=65.036$, of the stadium with $\gamma=1$ (one quadrant only). (*b*) The nodal structure of this state. Closer analysis revealed no nodal line crossings. (Reproduced by permission from McDonald and Kaufman[82].)

where $0<\gamma<1$. Hamiltonians such as these are akin to those used to describe ray propagation in wave guides. In the case $\gamma=1$ the Hamiltonian (7.22) reduces to that of a free particle subjected to periodic impulses, that is, a Hamiltonian of the form (4.25b) [with $V(q)$ replacing the $\cos\theta$ there]. Integration of Hamilton's equations over a period T yields the mapping (M)

$$M: \quad q_{n+1}=q_n+\frac{p_nT}{\mu} \tag{7.23a}$$

$$p_{n+1}=p_n-T\left(\frac{\partial V}{\partial q}\right)_{q=q_{n+1}} \tag{7.23b}$$

Notice that M is independent of γ. By setting $V(q)=q^4$ and introducing a suitable scaling (7.23a) and (7.23b) reduce exactly to (2.13).

In the same way that M maps the (classical) phase point (p_n, q_n) to another, (p_{n+1}, q_{n+1}), we would like to construct the analogous quantal operator $\hat{U}$ that maps the state of the system $|n\rangle$, at "time" n, to its state

$|n+1\rangle$, at "time" $n+1$, that is,

$$|n+1\rangle = \hat{U}|n\rangle \tag{7.24}$$

In coordinate representation this takes the form

$$\psi_{n+1}(q) = \int dq' \langle q|\hat{U}|q'\rangle \psi_n(q') \tag{7.25}$$

where the subscripts n, $n+1$ refer to "time," not quantum number. It turns out that because of the special form of the Hamiltonian (7.22) the matrix elements $\langle q|\hat{U}|q'\rangle$ take a rather simple form. By taking some initial state that is not an eigenstate of the mapping we can follow its evolution and, as is described below, make direct comparisons with the underlying classical mechanics.

At time $t=0$ we can consider the Hamiltonian to be time independent, that is, $H=p^2/2+V(q)$, and the initial state is taken as a stationary state. Use of one-dimensional EBK quantization enables one to associate such a state with a family of classical trajectories; these are the orbits that lie on the curve $\mathcal{C}_i$ of (7.1) (see also Fig. 18). For $t>0$ H takes on the (perturbed) periodic form (7.22). The classical trajectories then evolve according to the mapping M (7.23). However, rather than follow the time development of a single trajectory we follow the evolution of the *whole family* orbits that, as is described above, can be associated in the semiclassical limit with the initial quantum state.

In Fig. 24 we show how a particular curve $\mathcal{C}$ (family of orbits) evolves under M. After only a few iterations $\mathcal{C}$ develops a remarkable pattern of convolutions. These features can be directly related to the presence of the elliptic and hyperbolic fixed points of the mapping (see Fig. 2), that is, the presence of a chaotic regime. In Fig. 25*a* we show the analogous series of pictures for the evolution of the associated quantum state; here we plot $|\psi(q)|^2$. After only one iteration of the "quantum map," the regular structure of $|\psi(q)|^2$ makes a transition to an entirely different, irregular morphology that exhibits multiple scales of oscillations (there is also a loss of the original, regular nodal pattern). In Fig. 25*b* we show the effect of smoothing these $|\psi(q)|^2$'s over a width of the order of a typical (inverse) deBroglie wavelength, that is, a width of $O(\hbar)$. This is required for the semiclassical analysis.

Comparison of the classical and quantal "maps" is obtained as follows. By projecting the evolving curves onto the coordinate axis one obtains the coarse grained probability density $\overline{|\psi(q)|^2}$. The results of this projection are shown in Fig. 26*a*. After $n=1$ there is a striking proliferation of caustics,

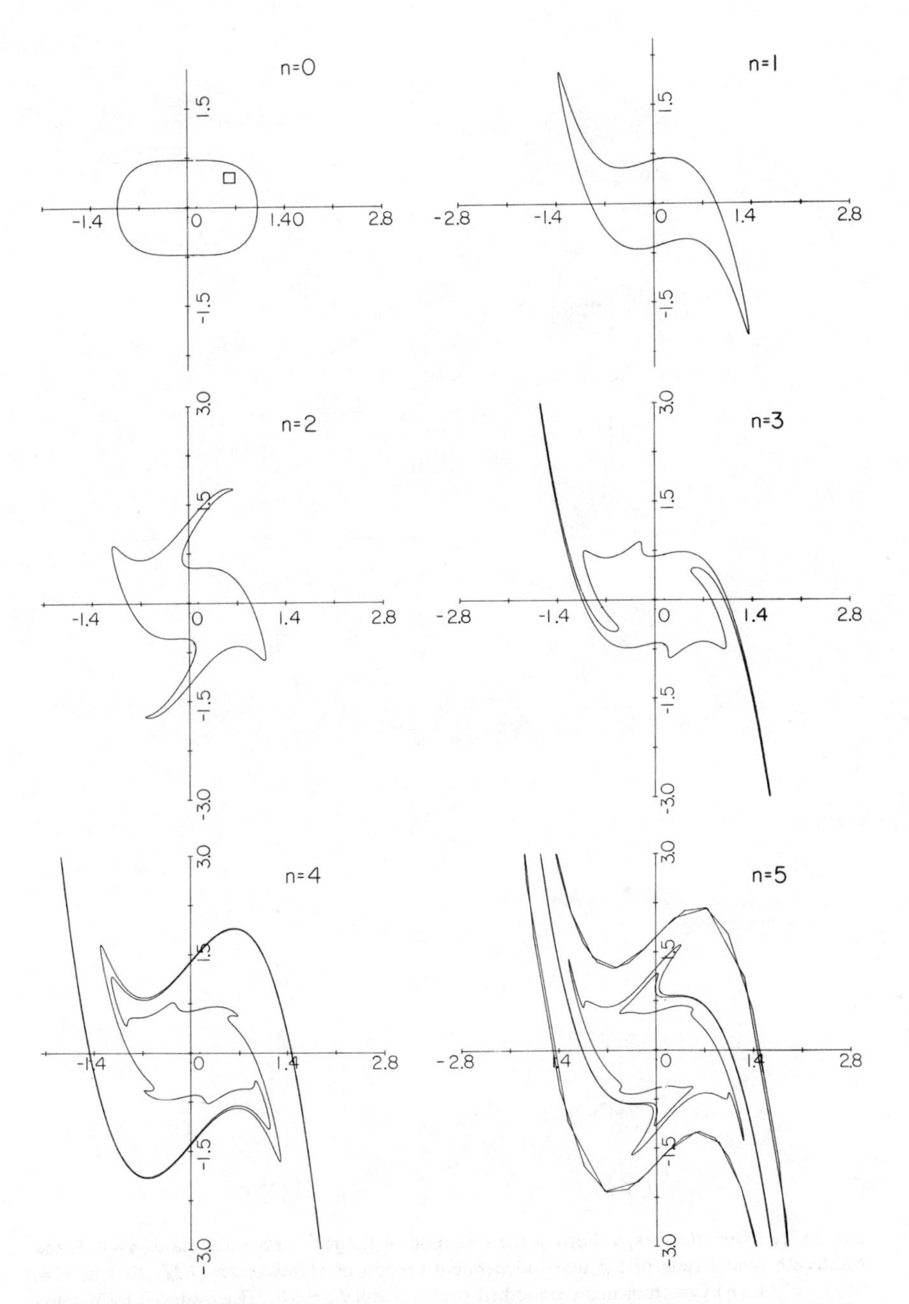

Fig. 24. Classical maps, $\mathcal{C}_n$, of initial family of trajectories $\mathcal{C}_0$. The small square marked in $\mathcal{C}_0$ has "area" $\hbar$. (Reproduced by permission from Berry et al.[23])

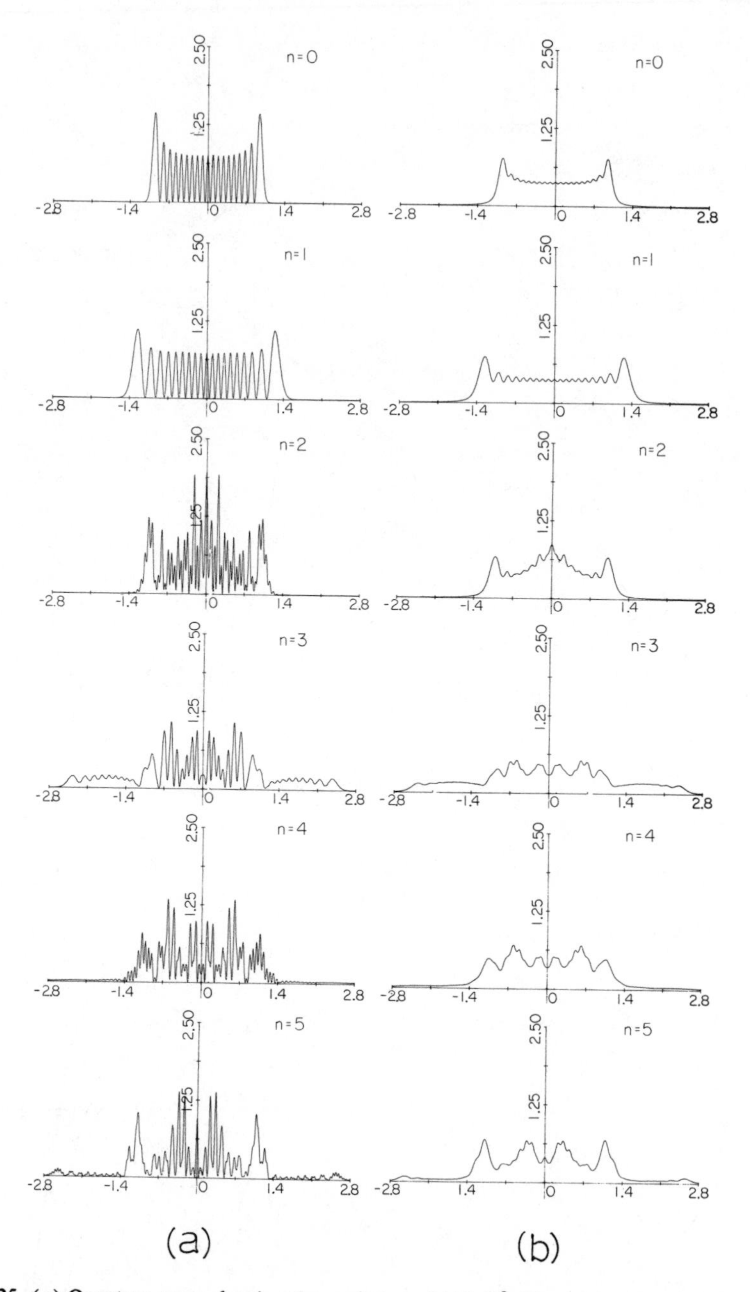

Fig. 25. (*a*) Quantum maps showing the evolution of $|\psi(q)|^2$. The initial state ($n=0$) is the eighteenth bound state of the time-independent version of Hamiltonian (7.22), that is, $H=p^2/2+q^4/4$. (*b*) Quantum maps smoothed over a width $\Delta q=0.05$. (Reproduced by permission from Berry et al.[23])

which seems to be a hallmark of the chaotic regime. At first sight there appears to be little connection between these probability densities and the exact quantal ones. The problem is that many of the caustics are in clusters that cannot be distinguished on a "scale" of $\hbar$. Thus to make a comparison we must smooth the $\overline{|\psi(q)|^2}$ over a width of $O(\hbar)$. This is shown in Fig. 26*b* using the same width as that used in Fig. 25*b*. Comparison of these two figures indicates a clear agreement between the smoothed $\overline{|\psi(q)|^2}$ and the smoothed quantal results. We have thus been able to show a direct relationship between the onset of irregularity in a quantum state and the chaotic nature of the underlying classical dynamics. The details of this work are described in ref. 23. The quantum mechanics of algebraic mappings also has been studied, with slightly different goals in mind, by Casati et al.[109] and by Berman and Zaslavskii.[110]

Some interesting investigations into the evolution of wave packets in multidimensional conservative systems have been made by Heller.[111, 112] His results indicate that symmetry considerations can prevent a wave packet from uniformly exploring all parts of the energetically allowed configuration space. This result is quite independent of the underlying classical mechanics. On the other hand, when symmetry is absent, a wave packet can sample all the allowed space. This is again quite independent of whether the classical mechanics is "ergodic" or not. These studies have been motivated by the desire to find an operationally useful definition of "quantum ergodicity." A very interesting discussion of such definitions has been given by Nordholm and Rice.[97] It is clear that a careful distinction has to be made between those definitions that, in the limit $\hbar \to 0$, reflect some underlying "ergodicity" in the classical mechanics and those results that imply some "statistical" character or behavior (e.g. uniform sampling of allowed configuration space) of the wave functions that is independent of the underlying classical dynamics. Part of the problem seems to be one of language. For example, the term ergodicity implies, strictly speaking, a uniform covering. This can occur for quite regular motion, for example, the covering of a torus by a quasiperiodic trajectory with irrational winding number or the covering of the energy shell by all trajectories in one-dimensional systems, or during the evolution of a nonstationary state in an integrable system. On the other hand, the term "chaotic" (or "irregular" or "stochastic") should be used to imply a definite instability of the motion with the attendent exponential divergence of nearby trajectories or continuous (or near continuous) spectrum. Thus Percival's definition of an "irregular state," which is specifically semiclassical, is firmly tied into the underlying classical dynamics. The current confusion over notions of "quantum ergodicity" could be greatly reduced by a careful definition of terms or even, perhaps, the introduction of new terminology.

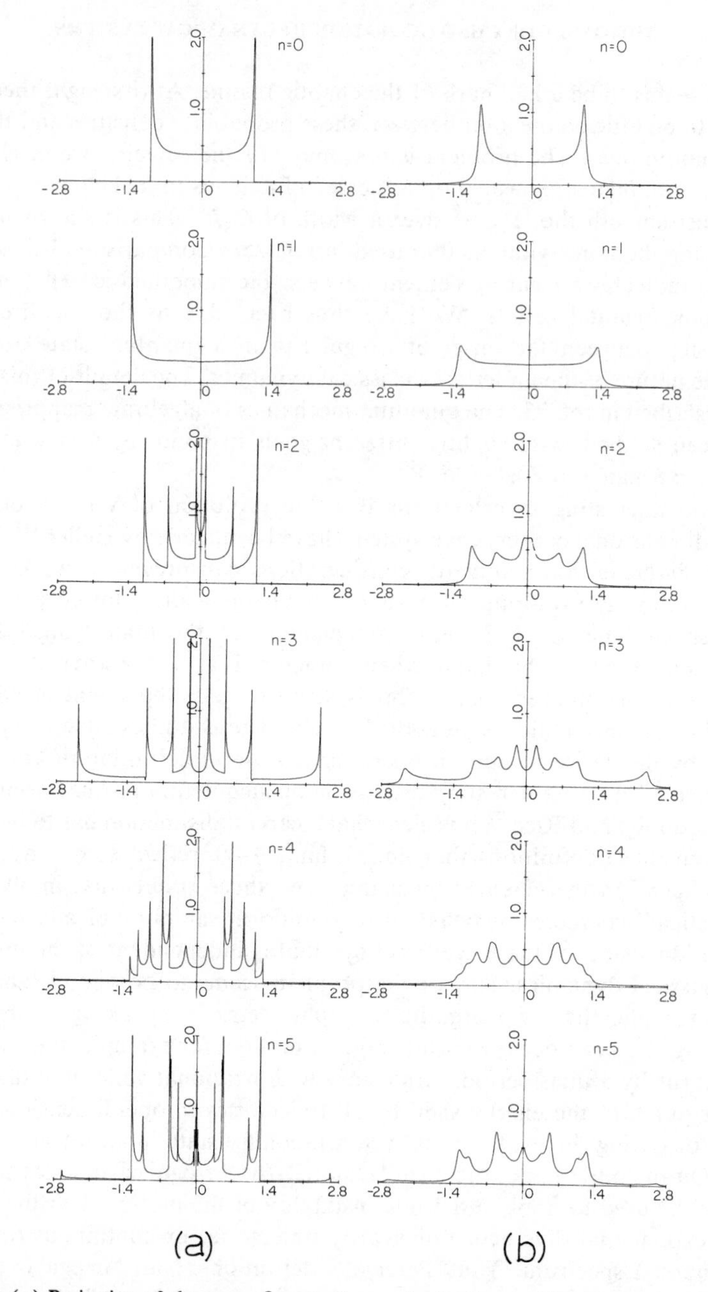

Fig. 26. (*a*) Projection of the maps $\mathcal{C}_n$ onto the *q*-axis showing proliferation of caustics. (*b*) Projections smoothed over the same width as that used in Fig. 25*b*. (Reproduced by permission from Berry et al.[23])

Appendix A

The effect we examine here is that of a child "pumping" a swing.[113] By bobbing up and down the child can create a periodic driving force in resonance with the swing and hence increase his/her swinging amplitude. Normally we would expect the nonlinearity of the swing to stabilize this increase (see Section IV). However, children are generally unaware of the subtleties of nonlinear mechanics and simply adjust their frequency to stay in resonance. This situation can be idealized (with the help of an ideal child) into the motion of a linear oscillator with an "adiabatically switched" frequency. This system, which is an example of parametric resonance, illustrates how motion in the vicinity of a single *stable* critical point (equilibrium point) can be *destabilized*.

We start off by considering a linear oscillator of mass m with (for now) fixed frequency ω. The Hamiltonian is

$$H=\frac{p^2}{2m}+\frac{m\omega^2q^2}{2} \tag{A.1}$$

with the corresponding equations of motion

$$\begin{aligned}\dot{q}&=\frac{p}{m}\\ \dot{p}&=-m\omega^2q\end{aligned} \tag{A.2}$$

This system of equations has the trivial critical point $(p,q)=(0,0)$. For real and positive frequency ω the critical point is easily shown to be stable (elliptic). The equations of motion are easily solved to give

$$\begin{aligned}q(t)&=q(0)\cos\omega t+\frac{p(0)\sin(\omega t)}{m\omega}\\ p(t)&=p(0)\cos\omega t-m\omega q(0)\sin(\omega t)\end{aligned} \tag{A.3}$$

where $q(0)$, $p(0)$ are the initial conditions. These equations can be written in the form of a linear map, that is,

$$\begin{bmatrix} q(t)\\ p(t)\end{bmatrix}=M\begin{bmatrix} q(0)\\ p(0)\end{bmatrix} \tag{A.4}$$

where the mapping M is given by

$$M=\begin{bmatrix}\cos\theta & \frac{1}{\alpha}\sin\theta\\ -\alpha\sin\theta & \cos\theta\end{bmatrix} \tag{A.5}$$

where

$$\theta = \omega t \qquad \text{and} \qquad \alpha = m\omega \tag{A.6}$$

The mapping is easily shown to be area preserving, that is

$$\det(M) = 1 \tag{A.7}$$

We now introduce the "adiabatic switching" of the frequency,

$$\begin{aligned} \omega &= \omega_1 \qquad \text{for a period } t_1 \\ &= \omega_2 \qquad \text{for a period } t_2 \end{aligned} \tag{A.8}$$

The mapping M must now be written as the product of two mappings, namely,

$$M = M_1 M_2 \tag{A.9}$$

where

$$M_1 = \begin{bmatrix} \cos\theta_1 & \dfrac{1}{\alpha_1}\sin\theta_1 \\ -\alpha_1 \sin\theta_1 & \cos\theta_1 \end{bmatrix} \tag{A.10}$$

and

$$M_2 = \begin{bmatrix} \cos\theta_2 & \dfrac{1}{\alpha_2}\sin\theta_2 \\ -\alpha_2 \sin\theta_2 & \cos\theta_2 \end{bmatrix} \tag{A.11}$$

Both M_1 and M_2 are area preserving. Furthermore, in either frequency "mode" the motion shares the same stable critical point at $(p,q) = (0,0)$.

The stability of the combined motion can be determined by examining the eigenvalues of M. These are the roots of

$$\begin{vmatrix} M_{11} - \lambda & M_{12} \\ M_{21} & M_{22} - \lambda \end{vmatrix} = 0 \tag{A.12}$$

where the M_{ij} are the (ij)th elements of M. Using the fact that M is area preserving, that is, $M_{11}M_{22} - M_{12}M_{21} = 1$, we find

$$\lambda_{\pm} = \tfrac{1}{2}\left\{ (M_{11} + M_{22}) \pm \left[(M_{11} + M_{22})^2 - 4 \right]^{1/2} \right\} \tag{A.13}$$

The motion is unstable (i.e., destabilized) if the roots are real. This occurs if

$$|M_{11}+M_{22}|>2 \tag{A.14}$$

The matrix elements M_{11} and M_{22} can be found using (A.9) to (A.11), and with a little algebra the condition (A.14) can be expressed as

$$|\cos\theta_1\cos\theta_2-A\sin\theta_1\sin\theta_2|>1 \tag{A.15}$$

where

$$A=\tfrac{1}{2}\left(\frac{\alpha_2}{\alpha_1}+\frac{\alpha_1}{\alpha_2}\right)\equiv\tfrac{1}{2}\left(\frac{\omega_2}{\omega_1}+\frac{\omega_1}{\omega_2}\right) \tag{A.16}$$

It is easily verified that in the case of $\omega_1=\omega_2$, that is, there is no frequency switching, the condition (A.15) is not satisfied and hence the motion remains (as it must do) stable.

In the case of $\omega_1\neq\omega_2$, destabilization becomes possible. Let

$$A=1+\varepsilon \qquad \varepsilon>0 \tag{A.17}$$

then condition (A.15) becomes

$$|\cos(\theta_1+\theta_2)-\varepsilon\sin\theta_1\sin\theta_2|>1 \tag{A.18}$$

which is satisfied when $(\theta_1+\theta_2)$ is near an integer multiple of π. Now set

$$(\theta_1+\theta_2)=n\pi+\delta \tag{A.19}$$

Hence

$$\cos(\theta_1+\theta_2)=(-1)^n\cos\delta\simeq(-1)^n\left[1-\frac{\delta}{2}\cdots\right] \tag{A.20}$$

Inserting this into A.18 and multiplying through by $(-1)^n$ gives the condition

$$\left|1-\frac{\delta^2}{2}-\varepsilon(-1)^n\sin\theta_1\sin\theta_2\right|>1 \tag{A.21}$$

which is satisfied if

$$|\delta|<(2|\varepsilon\sin\theta_1\sin\theta_2|)^{1/2} \tag{A.22}$$

For real and positive θ_1, θ_2 there always exist ε and δ for which (A.22) is satisfied and which hence cause destabilization.

An important exception to this occurs in the case of rapid switching. In this case we have $t_1 \simeq t_2 \simeq 0$ and hence

$$\theta_1 \simeq \theta_2 \simeq 0 \tag{A.23}$$

We may then show

$$\begin{aligned}|\cos(\theta_1+\theta_2)-\varepsilon\sin\theta_1\sin\theta_2| &\simeq \left|1-\frac{(\theta_1+\theta_2)^2}{2}-\varepsilon\sin\theta_1\sin\theta_2\right| \\ &= \left|1-\left[\frac{\theta_1^2}{2}+\frac{\theta_2^2}{2}+(1+\varepsilon)\theta_1\theta_2\right]\right| \\ &< 1\end{aligned}$$

since $\theta_1, \theta_2, \varepsilon$ are all positive. Thus condition (A.18) is not satisfied and the motion remains stable.

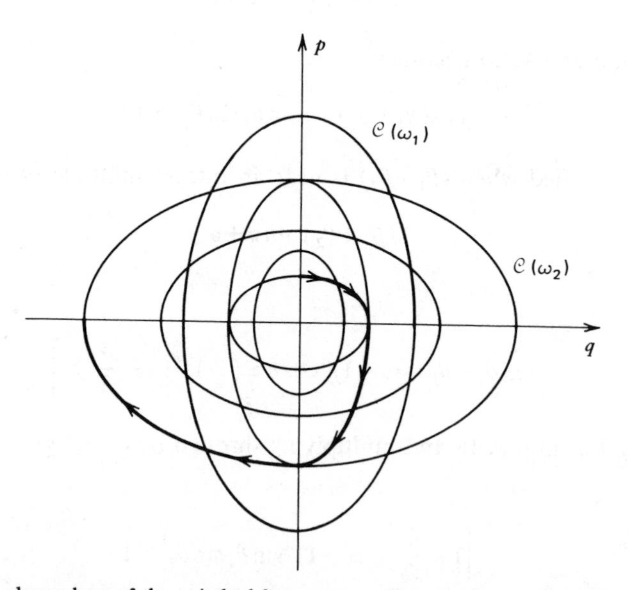

Fig. 27. The phase plane of the switched frequency oscillator is the superposition of the two sets of ellipses, $\mathcal{C}(\omega_1)$ and $\mathcal{C}(\omega_2)$ (corresponding to the two frequencies ω_1 and ω_2). In the optimum pumping case of $\omega_1 t = \omega_2 t = \pi/2$ the particle jumps from one set of ellipses to the other every quarter cycle and rapidly spirals away from the origin.

Optimum pumping occurs when

$$\theta_1 = \theta_2 = \frac{\pi}{2} \tag{A.24}$$

and the swing rapidly spirals away from its equilibrium point. This is easily understood by examining the phase plane of the system shown in Fig. 27

For this system the motion is always locally stable, since at any instant the motion is described by the linear map (A.5), which is always elliptic (stable). However, the overall, global stability of the motion is determined by the product of two elliptic mappings (A.9). As we have seen this product can lead to a hyperbolic (unstable) mapping. This demonstrates how *locally stable motion can be globally unstable.*

Appendix B

Here we fill in some of the details of the application of Chirikov's method to the Walker-Ford Hamiltonian.

The zero-order Hamiltonian is

$$H_0(I_1, I_2) = I_1 + I_2 - I_1^2 - 3I_1 I_2 + I_2^2 \tag{B.1}$$

From Hamilton's equations we can obtain the two frequencies

$$\omega_1 = \frac{\partial H_0}{\partial I_1} = 1 - 2I_1 - 3I_2 \tag{B.2}$$

$$\omega_2 = \frac{\partial H_0}{\partial I_2} = 1 + 2I_2 - 3I_1 \tag{B.3}$$

The matrix $\mathbf{H}^0$ of second derivatives $(\partial^2 H_0 / \partial I_i \partial I_j)$ is easily constructed.

$$\mathbf{H}^0 = \begin{bmatrix} \mathbf{H}_{11}^0 & \mathbf{H}_{12}^0 \\ \mathbf{H}_{21}^0 & \mathbf{H}_{22}^0 \end{bmatrix} = \begin{bmatrix} -2 & -3 \\ -3 & 2 \end{bmatrix} \tag{B.4}$$

We now consider the addition of the 2:2 resonant perturbation, that is,

$$H = H_0(I_1, I_2) + \alpha I_1 I_2 \cos(2\theta_1 - 2\theta_2) \tag{B.5}$$

The phase

$$\psi_1 = 2\theta_1 - 2\theta_2 \tag{B.6}$$

is slow in the vicinity of the resonance

$$\frac{\omega_1(I_1^r, I_2^r)}{\omega_2(I_1^r, I_2^r)} = 1 \tag{B.7}$$

From (B.2) and (B.3) it is easily deduced that this resonance condition is met at all points along the line

$$I_1^r = 5I_2^r \tag{B.8}$$

in I_1, I_2 space.

If we choose $\psi_1 = 2\theta_1 - 2\theta_2$ as the "slow" phase and $\psi_2 = \theta_2$ as the "fast" phase the matrix $\boldsymbol{\mu}$ takes the form

$$\boldsymbol{\mu} = \begin{bmatrix} 2 & -2 \\ 0 & 1 \end{bmatrix} \tag{B.9}$$

From (4.48) the pendulum "mass" M_{11} is given by

$$\begin{aligned}(M_{11})^{-1} &= \mu_{11}^2 \mathbf{H}_{11}^0 + 2\mu_{11}\mu_{12}\mathbf{H}_{12}^0 + \mu_{12}^2 \mathbf{H}_{22}^0 \\ &= 24\end{aligned} \tag{B.10}$$

The resonance width is

$$\Delta J_1^{(2:2)} = 2\sqrt{\alpha V M_{11}} \tag{B.11}$$

where

$$V = I_1^r I_2^r = \frac{(I_1^r)^2}{5} \tag{B.12}$$

from (B.8). Using the Walker-Ford value of $\alpha = 0.02$ (B.11) becomes

$$\Delta J_1^{(2:2)} = 0.02582(I_1^r) \tag{B.13}$$

This equation can then be used to plot the resonant zone in the I_1, I_2 plane, the width components being given by $I_1 = |\mu_{11}\Delta J_1|$ and $I_2 = |\mu_{12}\Delta J_1|$.

The width of the 2:3 perturbation is calculated independently of the 2:2 term, that is, we consider

$$H = H_0(I_1, I_2) + \beta I_1 I_2^{3/2} \cos(2\theta_1 - 3\theta_2) \tag{B.14}$$

The resonance condition is now

$$\frac{\omega_1(I_1^r, I_2^r)}{\omega_2(I_1^r, I_2^r)} = \frac{3}{2} \tag{B.15}$$

which is satisfied along the line

$$I_1^r = \frac{12I_2^r + 1}{5} \tag{B.16}$$

The matrix $\boldsymbol{\mu}$ is now taken to be

$$\boldsymbol{\mu} = \begin{bmatrix} 2 & -3 \\ 0 & 1 \end{bmatrix} \tag{B.17}$$

and hence the "mass" is calculated to be

$$(M_{11})^{-1} = 46 \tag{B.18}$$

Using (B.16) and the value $\beta = \alpha = 0.02$ the width of the $2:3$ zone can be calculated from

$$\begin{aligned} \Delta J_1^{(2:3)} &= 2\sqrt{\beta V M_{11}} \\ &= 0.00647(I_1^r)^{1/2}(5I_1^r - 1)^{3/4} \end{aligned} \tag{B.19}$$

The two zones are plotted in the I_1, I_2 plane and the value of I_1 and I_2 at which they intersect are used to estimate E_{crit} from (B.1).

Acknowledgments

This chapter arose out of a series of talks given at Queen Mary College, London, and I would like to thank Prof. I. C. Percival for providing me with this opportunity. I would also like to thank Profs. J. Ford and S. A. Rice for their encouragement and useful criticism of an early draft of the manuscript. The critical reading of the final version by Drs. M. V. Berry and R. Ramaswamy is greatly appreciated. I would also like to extend my appreciation to T. Bountis, J. M. Greene, M. L. Koszykowski, S. W. McDonald, W. P. Reinhardt, D. L. Rod, and R. E. Wyatt for discussions and/or correspondence. Finally, I would like to thank Prof. R. A. Marcus for support and encouragement. This work was supported in part by the National Science Foundation.

References and Notes

1. J. Lebowitz and O. Penrose, *Phys. Today*, **26** 23 (Feb. 1973); A. S. Wightman, *Statistical Mechanics at the Turn of the Decade*, E. D. G. Cohen Ed., Dekker, New York, 1971.
2. L. J. Laslett, "Some Illustrations of Stochasticity," in *American Institute of Physics Conference Proceedings*, Vol. **46**, AIP, New York, 1978.

3. A. N. Kaufman, "Regular and Stochastic Particle Motion in Plasma Dynamics," to appear in the *Proceedings of the International Workshop on Intrinsic Stochasticity in Plasmas*, Cargese, Corsica, France, June 1979.
4. J. Moser, "Stable and Random Motions in Dynamical Systems," Princeton University Press, 1973.
5. E. Hase, in *Dynamics of Molecular Collisions B*, W. H. Miller, Ed., Plenum, New York, 1976; R. A. Marcus, D. W. Noid, and M. L. Koszykowski, in *Advances in Laser Physics*, A. Zewail, Ed., Springer-Verlag, New York, 1978.
6. S. A. Rice, in *Advances in Laser Chemistry*, A. Zewail, Ed., Springer-Verlag, New York, 1978.
7. I. C. Percival, *J. Phys. B.*, **6**, L229–32 (1973).
8. J. Ford, in *Fundamental Problems in Statistical Mechanics III*, E. D. G. Cohen, Ed., North-Holland, Amsterdam, 1975; *Adv. Chem. Phys.*, **24**, 155 (1973).
9. K. J. Whiteman, *Rep. Prog. Phys.*, **40**, 1033 (1977).
10. M. V. Berry, in *Am. Inst. Phys. Conf. Proc.*, **46**, 16, 1978.
11. Y. M. Treve, *Am. Inst. Phys. Conf. Proc.*, **46**, 147, 1978.
12. P. Brumer, "Intramolecular Energy Transfer: Theories for the Onset of Statistical Behavior," to appear in *Adv. Chem. Phys.*
13. I. C. Percival, *Adv. Chem. Phys.*, **36**, 1 (1977).
14. V. I. Arnol'd and A. Avez, *Ergodic Problems of Classical Mechanics*, Benjamin, New York, 1968.
15. V. I. Arnol'd, *Mathematical Methods of Classical Mechanics*, Graduate Texts in Mathematics, Springer-Verlag, New York, 1978.
16. H. Goldstein, *Classical Mechanics*, Addison-Wesley, Reading, Mass, 1953.
17. M. V. Berry and M. Tabor, *Proc. Roy. Soc.* (*Lond.*), **A349**, 101 (1976).
18. V. I. Arnol'd, *Russ. Math. Surv.*, **18**, 9 (1963).
19. J. Moser, *Nachr. Akad. Wiss. Gottingen, II, Math. Phys. Kl.*, (**1962**) 1.
20. V. I. Arnol'd, *Russ. Math. Surv.*, **18** 85 (1963).
21. M. Henon, *Bull. Astron.* (*Sec* 3) **1**, fasc. 1, 57 and fasc. 2, 49 (1966).
22. M. Henon, *Q. Appl. Math.*, **27** 291 (1969).
23. M. V. Berry, N. L. Balzacs, M. Tabor, and A. Voros, "Quantum Maps," *Ann. Phys. N. Y.* **122**, 26 (1979).
24. J. F. C. Van Velsen, *Phys. Rep.*, **41** 135–190 (1978).
25. M. Henon and C. Heiles, *Astron. J.*, **69**, 73 (1964).
26. G. Benettin, L. Galgani, and J-M. Strelcyn, *Phys. Rev. A.*, **14** 2338 (1976).
27. A. Ichimura and M. Saito, in *Stochastic Behaviour in Classical and Quantum Hamiltonian Systems*, Volta Memorial Conference, Como, 1977, G. Casati and J. Ford, Eds., Vol. 93 of *Lecture notes in Physics*, Springer-Verlag, New York, 1979.
28. J. W. Duff and P. Brumer, *J. Chem. Phys.*, **67**, 4894 (1977); **71**, 2693 (1979).
29. D. W. Noid, M. L. Koszykowski, and R. A. Marcus, *J. Chem. Phys.*, **67**, 404 (1977).
30. R. P. Halmos, *Lectures on Ergodic Theory*, Chelsea Publishing Co. 1956.
31. In practice one computes the spectrum by analyzing the trajectory, integrated over some (long) period $T_{\max}$, with a standard fast Fourier transform program; the resolution is of the order $\omega_{\min} = 2\pi/T_{\max}$. The power spectrum is calculated as the squared moduli of the Fourier coefficients. Some very accurate computations can be found in Ref. 33. For the purposes of comparison with quantal spectra (discussed in Section VII) these classical power spectra can be regarded as representing the intensity of classical dipole radiation.
32. H. L. Swinney, P. R. Fenstermacher, and J. P. Gollub, in *Synergetics*, H. Haken, ed., Springer-Verlag, Berlin, 1977.
33. M. L. Koszykowski, Ph.D. dissertation, University of Illinois, 1978.

34. G. H. Walker and J. Ford, *Phys. Rev.*, **188**, 416 (1969).
35. G. H. Lunsford and J. Ford, *J. Math. Phys.*, **13**, 700 (1972).
36. J. Ford, S. D. Stoddard, and J. S. Turner, *Prog. Theor. Phys.*, **50**, 1547 (1973).
37. M. Henon, *Phys. Rev. B.*, **9**, 1921 (1974).
38. G. Casati and J. Ford, *Phys. Rev.*, **A12**, 1702 (1975).
39. B. Barbanis, *Astron. J.*, **71**, 415 (1966).
40. G. Contopoulos, *Bull. Astron.* (*Ser.* 3), **2**, 223 (1967) and references therein.
41. E. Thiele and D. J. Wilson, *J. Chem. Phys.*, **35**, 1265 (1961).
42. R. H. Tredgold, *Proc. Phys. Soc. Sec. A*, **68**, 920 (1955).
43. Studies of systems of three degrees of freedom can be found in Ref. 35 and a more recent review is found in G. Contopoulos, L. Galgani, and A. Giorgilli, *Phys. Rev. A*, **18**, 1183 (1978). The reader is also referred to an interesting series of studies on the "stochastic" threshold of linear chains by Galgani and co-workers. See L. Galgani and G. Vecchio, *Nuovo Cimento*, **52B** 1 (1979) and references therein.
44. M. Toda, *Phys. Lett.* **48A**, 335 (1974).
45. P. Brumer and J. W. Duff, *J. Chem. Phys.*, **65**, 3566 (1976).
46. C. Cerjan and W. P. Reinhardt, *J. Chem. Phys.*, **71**, 1819 (1979).
47. G. Benettin, R. Brambilla, and L. Galgani, *Physica*, **87A**, 381 (1977).
48. M. C. Gutzwiller, *J. Math. Phys.*, **18** 806 (1977).
49. W. E. Boyce and R. C. DiPrima "Elementary Differential Equations," 2nd ed., Wiley, New York, 1979.
50. B. M. Zaslavskii and B. V. Chirikov, *Sov. Phys. Usp.* (*Engl. Trans.*), **14**, 549 (1972).
51. B. V. Chirikov, "A Universal Instability of Many Dimensional Oscillator Systems," *Phys. Rep.*, **52**, No. 5, 265 (1979).
52. D. W. Noid and J. R. Stine, *Chem. Phys. Lett.*, **65**, 153 (1979).
53. E. F. Jaeger and A. J. Lichtenberg, *Ann. Phys. N. Y.*, **71**, 319 (1972).
54. P. J. Channell, in *Am. Inst. Phys. Conf. Proc. Ser.*, **46**, 248 (1978).
55. A. B. Rechester and T. H. Stix, *Phys. Rev. A*, **19**, 1656 (1979).
55a. In Ref. 51 it is suggested that one chooses all the other rows of $\boldsymbol{\mu}(\mu_{ji}, j \neq k)$ to be orthogonal to $\boldsymbol{\omega}$ in order to eliminate the terms linear in J [see (4.47)]. However, this would render all the other phases, ψ_j, "slow" variables and hence invalidate the averaging procedure. For the (isolated) resonant phase ψ_k the mass M_{kk} depends only on the kth row of $\boldsymbol{\mu}$. Thus the existence of transformations that remove the unwanted terms are, perhaps, more of an aesthetic rather than a practical problem.
56. D. W. Oxtoby and S. A. Rice, *J. Chem. Phys.*, **65**, 1676 (1976).
57. D. W. Noid, Ph.D. Dissertation, University of Illinois, 1976.
58. J. M. Greene, *J. Math. Phys.*, **9** 760 (1968).
59. J. M. Greene, *J. Math. Phys.*, **20** 1183 (1979). I would like to thank the author for sending me an original diagram.
60. R. H. G. Helleman and T. Bountis, in *Stochastic Behaviour in Classical and Quantum Systems*, Volta Memorial Conference, Como, 1977, G. Casati and J. Ford, Eds., Springer-Verlag, Vol. 93 of *Lecture Notes in Physics*, New York, 1979.
61. J. M. Greene, "KAM Surfaces Computed from the Henon-Heiles Hamiltonian," presented at the symposium on Nonlinear Orbit Dynamics of Beam-Beam Interaction, Brookhaven, March 1979.
62. R. C. Churchill, G. Pecelli, and D. L. Rod, in *Stochastic Behaviour in Classical and Quantum Systems*, Volta Memorial Conference Como 1977 G. Casati and J. Ford, Eds, Springer-Verlag, Vol. 93 of *Lecture Notes in Physics*, New York, 1979.
63. J. M. Greene, private communication.
64. T. Bountis, Ph.D. Dissertation, University of Rochester, 1978.
65. C. R. Eminhizer, R. H. G. Helleman, and E. W. Montroll, *J. Math. Phys.*, **17**, 121 (1976).

66. K. C. Mo, *Physica*, **57**, 445 (1972).
67. M. L. Koszykowski, D. W. Noid, M. Tabor and R. A. Marcus "On Correlation Functions and the Onset of Chaotic Motion," to be published in *J. Chem. Phys.*.
68. R. Zwanzig, *Ann. Rev. Phys. Chem.*, **16**, 67 (1965).
69. M. C. Wang and G. E. Uhlenbeck, *Rev. Mod. Phys.*, **17**, 323 (1945).
70. H. Mori, *Prog. Theoret. Phys.* (*Kyoto*), **33**, 423 (1965).
71. B. Berne, in *Physical Chemistry, an Advanced Treatise*, Vol VIIIB *Liquid State*, D. Henderson, Ed., Academic, New York, 1971.
72. R. I. Cukier, K. G. Shuler, and J. D. Weekes, *J. Stat. Phys.*, **5**, 99 (1972).
73. M. Born, *The Mechanics of the Atom*, Ungar, New York, 1960.
74. A. Einstein, *Verh. Dtsch. Phys. Ges.*, **19**, 82 (1917).
75. L. Brillouin, *J. Phys. Radium*, **7**, 353 (1926).
76. J. B. Keller, *Ann. Phys. N. Y.* **4**, 180 (1958).
77. C. Jaffe, Thesis, University of Colorado, Boulder, 1979, unpublished. The algorithm for constructing tori is described in C. Jaffe and W. P. Reinhardt, *J. Chem. Phys.*, **71**, 1862 (1979).
78. This notion is discussed in the thesis of C. Jaffe.[77] One can obtain an idea of what these "pseudo"-tori look like from the comparison of actual surfaces of section of the Henon-Heiles system with those computed by high-order perturbation theory (due to Gustavson) shown in Ref. 9.
79. M. V. Berry, in "Structural Stability in Physics," N. Güttinger and H. Eikemeier, Eds., Springer-Verlag, Berlin, 1979.
80. D. W. Noid, M. L. Koszykowski, M. Tabor, and R. A. Marcus, "Properties of Vibrational Energy Levels in the Quasi-Periodic and Chaotic Regimes," *J. Chem. Phys.*, **72**, 6169 (1980).
81. N. Pomphrey, *J. Phys. B*, **7**, 1909 (1974).
82. S. W. McDonald and A. N. Kaufman, "Ray and Wave Optics of Integrable and Stochastic Systems," presented at the International Workshop on Intrinsic Stochasticity in Plasmas, Cargese, Corsica, France, June 1979. I would like to thank these authors for sending me their original diagrams. (LBL report no. 9465)
83. This type of level crossing has been reported elsewhere, for example, F. T. Hioe, D. MacMillen, and E. W. Montroll, *Phys. Rep.*, **43**, No. 7, 305 (1978).
84. An avoided crossing corresponds semiclassically to the onset of a new resonance. If a state is involved simultaneously in many avoided crossings the wave function takes on a statistical (strongly mixed) character. There may be an analogy between this effect and Chirikov's theory of overlapping resonances (R. Ramaswamy and R. A. Marcus to be published). The effect of an avoided crossing on an eigenstate can be seen in Fig. 7*f* of Ref. 82 (S. W. McDonald, private communication). For an important discussion of avoided crossings see appendix 10 of Arnol'd.[15]
85. M. C. Gutzwiller, *J. Math. Phys.*, **11**, 1791 (1970).
86. R. Balian and C. Bloch, *Ann. Phys. N. Y.*, **69**, 76 (1972).
87. M. C. Gutzwiller, *J. Math. Phys.*, **12**, 353 (1971).
88. M. V. Berry and M. Tabor, *Proc. Roy. Soc.* (*Lond.*) **A356**, 375 (1977).
89. C. F. Porter, Ed., *Statistical Theory of Spectra: Fluctuations*, Academic, New York, 1965.
90. G. M. Zaslavskii, *Sov. Phys. JETP*, **46**, 1094 (1977).
91. S. W. McDonald and A. N. Kaufman, *Phys. Rev. Lett.*, **42**, 1189 (1979).
92. The stadium is an example of a "*K*-system." There are no tori and apart from a special set all trajectories are strongly chaotic. (See, for example, G. Benettin and J.-M Strelcyn, *Phys. Rev. A*, **17** 773 (1978).) This system provides a useful model to test properties of the regular and irregular regimes.

93. J. H. Van Vleck, *Proc. Math. Acad. Sci. US*, **14**, 178 (1928).
94. M. V. Berry and K. E. Mount, *Rep. Prog. Phys.*, **85**, 315 (1972).
95. M. V. Berry, *J. Phys. A.*, **10**, 2083 (1977).
96. M. V. Berry, *Philos. Trans. Roy. Soc. Lond.*, **287**, 237 (1977).
97. K. S. J. Nordholm and S. A. Rice, *J. Chem. Phys.*, **61**, 203 (1974).
98. E. P. Wigner, *Phys. Rev.*, **40**, 749 (1932).
99. See, for example K. Imre, E. Ozizmir, M. Rosenbaum, and P. F. Zweifel, *J. Math. Phys.*, **8**, 1097 (1967).
100. E. J. Heller, *J. Chem. Phys.*, **67**, 3339 (1977).
101. A. Ozorio and J. Hannay, to be published.
102. J. S. Hutchinson and R. E. Wyatt, private communication. I am very grateful to these authors for sending me their results prior to publication.
103. A. Voros, *Ann. Inst. H. Poincare*, **24A**, 31 (1976).
104. S. W. McDonald, private communication.
105. D. W. Noid, M. L. Koszykowski, and R. A. Marcus, "Semiclassical Calculation of Bound States in Multidimensional Systems with Fermi Resonance," *J. Chem. Phys.* **71**, 2864 (1979).
106. P. Pechukas, *J. Chem. Phys.*, **57**, 5577 (1972).
107. R. M. Stratt, N. Handy, and W. H. Miller, "On the Quantum Mechanical Implications of Classical Ergodicity", *J. Chem. Phys.*, **71**, 3311 (1979).
108. C. R. Holt and W. P. Reinhardt, unpublished. W. P. Reinhardt has brought to my attention an amusing and simple example of changes in nodal patterns in R. Courant and D. Hilbert, *Methods of Mathematical Physics*, Vol. 1, Interscience, New York, 1953, p. 456.
109. G. Casati, B. V. Chirikov, F. M. Israelev, and J. Ford, in *Stochastic Behavior in Classical and Quantum Hamiltonian Systems*; G. Casati and J. Ford, Eds., Springer-Verlag, New York, Vol. 93 of *Lecture Notes in Physics*, 1979.
110. G. P. Berman and G. M. Zaslavskii, *Physica*, **91A**, 450 (1978).
111. E. J. Heller, *Chem. Phys. Lett.*, **60**, 338 (1979).
112. E. J. Heller, "Quantum Intramolecular Dynamics: Criteria for Stochastic and Non-Stochastic Flow," J. Chem. Phys., **72**, 1337 (1980)
113. J. Hannay and M. V. Berry, unpublished work.

THERMODYNAMIC ASPECTS OF THE QUANTUM-MECHANICAL MEASURING PROCESS

RONNIE KOSLOFF

The James Franck Institute, The University of Chicago, Chicago, Illinois

CONTENTS

I. INTRODUCTION

Thermodynamics has a special place within the theories of physics. Its main distinction is its empirical character, which is the reason that it survived the revolution of physical theory in the twentieth century. Given the nature of thermodynamic empiricism, it can be an instrument to check the consistency of other theories. In this chapter the relationship between the measurement process and thermodynamic properties is investigated.

Maxwell[1] was the first to point out how a measurement procedure could apparently sort a homogenous system and thereby violate the second law of thermodynamics. To solve the paradox, Szilard[2] suggested that to achieve the measurement operation there is more entropy spent than there is negative entropy obtained from the separated system. This idea has been elaborated by Brillouin,[3] who reformulated the second law of thermodynamics to include information that links the measuring apparatus and the measured object. Brillouin's formulation reads

$$\Delta S - I \geq 0 \tag{1.1}$$

A similar approach was applied by Lindblad[4] to a model of an ideal quantum-mechanical (QM) measurement. He found that the model obeyed the second law of thermodynamics according to Brillouin.

The present investigation applies thermodynamic arguments to the general QM measurement process. The following points are discussed in detail:

1. The use of thermodynamic and information theory functionals, such as entropy and the mutual information functionals, as a means of classifying the various types of measurement processes.
2. Thermodynamic functionals used as a means of comparing the efficiencies and the resolution of different measurement apparatuses.
3. The irreversibility of the common QM measurement, with entropy analysis revealing this fact.

The nature of the quantum-mechanical measurement process is one of the most widely discussed problems in the literature of physics.[5] In fact, the main distinction among the different schools of thought in quantum mechanics originates from the different interpretations of the QM measurement process. The original description of QM measurement can be traced to the Heisenberg gedanken experiment explaining the uncertainty principle,[6] which was later elaborated by Bohr.[7] In those discussions the nature of the QM measurement was related to the properties of the classical apparatus. A different approach was introduced by Von Neumann,[8] who used statistical concepts. In the following years numerous approaches were applied to the analysis of the QM measurement, but at present there is no common agreement on its description or interpretation. The point where the differences among the various schools stands out is in the description of the reduction process. Many discussions use arguments that are outside the usual frame of reference of physics;[9, 10] hence they are confusing to most scientists, who prefer to avoid the measurement problem altogether.

In this chapter the use of thermodynamic and information theory tools leads to the adoption of the statistical interpretation of quantum mechanics. It is in the statistical interpretation of quantum mechanics that there has been progress in formulating and interpreting the measurement process. In the statistical approach QM measurement can be formulated in a complete way that generates its own interpretation. Work along this line has been carried out by Margenau,[11] Park,[12] Moldauer,[13] Peress,[14] Belinfante,[15] Krips,[16] and others. The essence of these descriptions is that it is possible to represent in a consistent way ideal measurements as a mechanical interaction, governed by an interaction Hamiltonian, that links the measured system and the measuring apparatus.

This chapter extends the concepts of the QM ideal measurement model to the more common case of nonideal measurements. A representation and classification is presented of measurements of partial resolution, as well as of destructive measurements. The representations used are borrowed from statistical mechanics, which makes it possible to identify the state of the system with its observables. This formulation presents a clear description of the buildup of measurement correlation among the observables, with constant attention to entropy relations. The main result is an identification of the source of irreversibility in the measurement process as a quantum effect related to the noncommutability of measured operators, and the boundary conditions of separability imposed on the measurement.

The chapter is divided as follows. In Section II a quantitative scheme is developed to classify the different measurement types. The same functionals are utilized to classify different measurement resolutions and the destruction caused by the measurement process. Section III is devoted to the solution of measurement models for a two-level QM system. Section IV summarizes and generalizes the results of Section III.

II. CLASSIFICATION OF THE MEASUREMENT PROCESSES

We start with a rhetorical question: What is a measurement? A measurement is a process in which a connection, or link, is built between the measured system and the measuring apparatus. This link enables the variables of the measured system to be read by scanning the variables of the measuring apparatus. As an illustration, when a thermometer is used to measure the temperature of a sample, the length of the mercury column is linked or is in correspondence with the temperature of the measured sample. There are two main considerations in classifying different types of measurement processes: (*1*) How discriminating of the different values of the parameters of the measured system can be the reading of the apparatus? (How big are the error bars?) This question classifies the measurement according to its resolution. (*2*) How much damage is caused to the measured system by the measuring process? This question classifies the measurement according to its destructive character.

With these two questions in mind, this section presents a framework for dealing with classification of measurements.

All measurement processes have three common features:

1. The measured system (termed S).
2. The measuring apparatus (termed M).
3. A measuring process that builds a correlation between the system S and the apparatus M.

From a mathematical point of view, the measurement process is a mapping from the space defined by the variables of the system to the space defined by the variables of the apparatus. This chapter deals only with mapping processes that are a consequence of physical forces acting between the system and the apparatus. This type of mapping defines the quality of the measurement. A one-to-one correspondence between a variable in S and a variable in M defines a complete resolution measurement. By examining M, one knows completely the value of the measured variable in S. The opposite extreme is no correspondence, a useless measurement.

To include measurements with intermediate resolution, the mapping process is defined as a stochastic process. If a measurement is designed to measure a variable A in S, this variable is mapped into a set of variable values in M. In the classical description of the measurement, this mapping process can be defined by the conditional probability that if the system S's variable A has the value A_i the measuring apparatus possesses the value of the variable B equal to B_j.

$$\{P(B_j|A_i), A_i \in S, B_j \in M\} \tag{2.1}$$

The set of all conditional probabilities defines the mapping completely.

The definition of the mapping in the measurement process is similar to the definition of a noisy communication channel. From this viewpoint the mapping in the measurement process can be considered as a noisy communication link between the systems S and M (this viewpoint is discussed in Appendix B).

In applying the definition (2.1), one encounters the difficulty that quantum systems, unlike classical systems, do not possess definite values of their variables (observables). The value that a variable in a quantum system has is subject to a measurement interaction designed to determine the variable itself. The difficulty can be overcome by defining the mapping process from states of S to states of M, as a generalization of the mapping between variables.

In the framework of quantum mechanics the systems S and M are defined by their abstract Hilbert spaces. The various states the systems S and M can have are defined by the density operators $\tilde{\rho}_S$ and $\tilde{\rho}_M$. The mapping process from states of S to states of M, can be defined analogously to (2.1) by a conditional density operator:

$$\tilde{\rho}_M = \operatorname*{tr}_S (\tilde{\rho}_{M|S} \cdot \tilde{\rho}_S) \tag{2.2}$$

In (2.2) the trace operation replaces the summation in the classical case. A detailed account of the properties of the conditional density operator is presented in Appendix A. Equation 2.2 describes the mapping from states of S to states of M. However, a direct physical interpretation is attached only to the outcomes of variables in S and in M; therefore, a physical interpretation of the measuring process can only be attached to the correspondence between the variables in S to the variables in M. Focusing back on variables, if S is in the eigenstate A_i of the variable A, the probability that the variable B will be found in M, with the value B_j, is given by

$$P(B_j | A_i) = \operatorname*{tr}_{SM} \left(\tilde{P}_{B_j} \cdot \tilde{\rho}_{M|S} \cdot \tilde{P}_{A_i} \right) \tag{2.3}$$

Where $\tilde{P}_x$ is the projection operator x><x.

We now discuss the functional classification of the resolution of the measurement.

In usual cases a good measurement is described by small error bars on the results. Good resolution reduces the error bars on the results; increasing the resolving power reduces the amount of uncertainty about the measured system. For the resolution functionals to have operational meaning, the information obtained by the measurement should be usable. As an illustration, one can examine the situation presented by Maxwells demon.[1] The demon, by performing a measurement, sorts an ensemble of identical systems according to the results of his measurement, thereby using information to achieve operational goals. From a thermodynamic point of view the sorting out of the homogenous ensemble is equivalent to an entropy decrease. This is the basis for the definition of the resolution functional: The resolution of the apparatus is defined as the amount of entropy decreased in the measured system by using the information stored in the apparatus. (A different approach to the resolution problem using the analogy of the measuring process to the noisy communication channel is presented in Appendix B.)

To define the resolution functional, a definition is needed for the entropy and for the conditional entropy, that is the uncertainty that is left after the outcome of the measurement is known. Because the entropy is associated with a physical interpretation, the basic definition is related to the outcome of a measurement of a variable. (In quantum mechanics, because of the uncertainty principle, there is a difference between the entropy of a variable in a definite state and the entropy of the state.[4]) The entropy of a variable A in a state $\tilde{\rho}$ is defined as the mixing entropy of the subensembles that are produced by sorting $\tilde{\rho}$ according to all the possible outcomes of the variable A:

$$S(\tilde{A};\tilde{\rho}) = -\sum_{\alpha} P_\alpha \ln P_\alpha$$

$$P_\alpha = \operatorname{tr}\left(\tilde{P}_{A_\alpha} \cdot \tilde{\rho}\right) \tag{2.4}$$

and $\tilde{P}_{A_k}$ is the projection operator of the outcome k of the variable A. From the information theory point of view, $S(\tilde{A};\tilde{\rho})$ is the uncertainty of a variable A in the state $\tilde{\rho}$.

The entropy of the state is defined as the minimum of mixing entropy of subensembles, produced by sorting $\tilde{\rho}$ to homogenous (pure) subensembles.

$$S(\tilde{\rho}) = \inf_{A \in L}\left[S(\tilde{A};\tilde{\rho})\right] = -\operatorname{tr}(\tilde{\rho}\ln\tilde{\rho}) \tag{2.5}$$

where L is the set of all nondegenerate operators defined on the Hilbert space of S.

As an example of the difference between the two definitions of entropy, consider a state with a definite position. The entropy of the position variable, and also the entropy of the state, is zero, but the momentum of the state is completely undetermined. Therefore, the entropy of the momentum variable is infinite.

The conditional entropy of two variables A and B in a state $\tilde{\rho}$ is defined by the uncertainty left in the variable A after $\tilde{\rho}$ is sorted according to the outcome of B:

$$S(\tilde{A}|\tilde{B};\tilde{\rho}) = \sum_k P_k S\left(\tilde{A};\tilde{P}_{B_k}\cdot\tilde{\rho}\cdot\tilde{P}_{B_k}\right) \tag{2.6}$$

Clearly if $\tilde{A}$ and $\tilde{B}$ do not commute, the conditional entropy can be larger than the entropy of $\tilde{A}$ before $\tilde{B}$ was measured. In the common measurement setup, the variables $\tilde{A}$ and $\tilde{B}$ commute because the measured system and the apparatus belong to different Hilbert spaces. One gets the relations:

$$S(\tilde{A}|\tilde{B};\tilde{\rho}) \leq S(\tilde{A};\tilde{\rho}) \tag{2.7}$$

The uncertainty about the measured object is always decreased by the measurement process. By using the definition of the entropy and the relative entropy (the entropy left after the results of the measurement are known), it is possible to define the functional that classifies the measurement according to its resolving power.

A. The Mutual Information Functional

The resolving power of a measurement is defined by the uncertainty in a variable A in S that has been decreased by knowledge of the variable B in M. This functional is called the mutual information between the variable A and the variable B:

$$\begin{aligned} I(\tilde{A},\tilde{B};\tilde{\rho}) &\equiv S(\tilde{A};\tilde{\rho}) - S(\tilde{A}|\tilde{B};\tilde{\rho}) \\ &\equiv S(\tilde{B};\tilde{\rho}) - S(\tilde{B}|\tilde{A};\tilde{\rho}) \end{aligned} \tag{2.8}$$

where $\tilde{\rho}$ is the joint state of S and M defined on the tensor product space of the Hilbert spaces of S and M. The mutual information is bounded from above by the entropy of the measured system, for which one gets:

$$I(\tilde{A},\tilde{B};\tilde{\rho}) \leq S(\tilde{A};\tilde{\rho}) \tag{2.9}$$

and from below by the useless measurement when no correspondence exists between $\tilde{A}$ and $\tilde{B}$:

$$I(\tilde{A},\tilde{B};\tilde{\rho}) \geq 0 \tag{2.10}$$

In summary, the larger the mutual information, the greater is the resolving power of the measurement.

The mutual information between states can be defined as:

$$I(\tilde{\rho}_s,\tilde{\rho}_M) = \sup_{\substack{A\in L_S \\ B\in L_M}} |I(\tilde{A},\tilde{B};\tilde{\rho}) \tag{2.11}$$

$I(\tilde{\rho}_S,\tilde{\rho}_M)$ is the maximum mutual information between a variable $\tilde{A}$ in S and a variable $\tilde{B}$ in M out of the set of all the nondegenerate variables $\tilde{A}$ in S and $\tilde{B}$ in M.

B. The Relative Mutual Information

The absolute amount of mutual information between A and B does not reveal what percentage of the initial uncertainty the measurement has succeeded in resolving. Therefore, it is useful to define a relative information functional by the ratio of information specified by $\tilde{B}$ in M to the initial uncertainty of $\tilde{A}$ in S:

$$\eta(\tilde{A}|\tilde{B};\tilde{\rho}) \equiv \frac{I(\tilde{A},\tilde{B};\tilde{\rho})}{S(\tilde{A};\tilde{\rho})} = 1 - \frac{S(\tilde{A}|\tilde{B};\tilde{\rho})}{S(\tilde{A};\tilde{\rho})} \tag{2.12}$$

In a measurement that is completely resolving, $\eta = 1$ and $\eta = 0$ when there is no correspondence between the variable A in S and the variable B in M. similarly, the relative mutual information between states can be defined as:

$$\eta(\tilde{\rho}_S|\tilde{\rho}_M) \equiv \frac{I(\tilde{\rho}_S|\tilde{\rho}_M)}{S(\tilde{\rho}_S)} \tag{2.13}$$

C. The Capacity Functional

On comparing different kinds of measuring apparatuses, it is desirable to define a functional that is related only to the resolving power of each apparatus. The capacity functional is designed to achieve this goal by choosing the mutual information that is maximized by all possible states of S and is therefore independent of the measured object:

$$C(\tilde{A},\tilde{B};\tilde{\rho}_{M|S}) \equiv \sup_{\tilde{\rho}_S} \left[I(\tilde{A},\tilde{B};\tilde{\rho}_{M|S}\cdot\tilde{\rho}_S) \right] \tag{2.14}$$

The capacity functional can also be defined by its relation to the mutual information between states:

$$C(\tilde{\rho}_{M|S}) \equiv \sup_{\tilde{\rho}_S} \left[I(\tilde{\rho}_M;\tilde{\rho}_S) \right] \tag{2.15}$$

The capacity functional can be used as an upper bound for the amount of information that can be measured by a particular apparatus:

$$I(\tilde{A},\tilde{B};\tilde{\rho}) \leq C(\tilde{A},\tilde{B};\tilde{\rho}_{M|S}) \leq C(\tilde{\rho}_{M|S}) \tag{2.16}$$

The capacity functional is adequate for the comparison of different measuring apparatuses.

It is clear that if a system has higher initial uncertainty than the capacity of the apparatus that measures it, the system cannot be resolved completely by the apparatus. At this point a more general problem arises: Can an apparatus whose correspondence with the measured object is subject to error completely resolve the state of the system S, even if the capacity of the measuring device is higher than the initial uncertainty of the measured object?

To answer this question, recall the analogy between the mapping in the measuring process and the noisy communication channel. For example, one can imagine transforming a message on a noisy telephone line. A way to overcome the noise is to repeat the message, or to produce redundancy in the message sequence. A similar procedure is possible for the QM measurement. The details of this analogy are found in Appendix B, where the second coding theorem[17, 18] of communication theory is applied to the measurement process. It is found that if the initial uncertainty of the system S is less than the capacity of the measuring apparatus, it is possible by a suitable coding procedure to sort an ensemble of systems S into homogenous (pure) subensembles, with the probability of an error (in the sorting procedure) being zero, when the initial ensemble becomes infinitely large.

A functional related to the capacity is the relative capacity:

$$\eta_C(\tilde{A}, \tilde{B}; \tilde{\rho}_{M|S}) \equiv \frac{C(\tilde{A}, \tilde{B}; \tilde{\rho}_{M|S})}{S(\tilde{A}; \tilde{\rho}_S^C)}$$

$$\eta_C(\tilde{\rho}_{M|S}) \equiv \frac{C(\tilde{\rho}_M, \tilde{\rho}_S^C)}{S(\tilde{\rho}_S^C)} \tag{2.17}$$

when ρ_S^C is the state of S that achieved capacity.

This functional gives the ratio of information transferred at capacity operation to the uncertainty of the initial variable or state of S. The relative capacity is a measure of the amount of noise in the mapping process; $\eta_C = 1$ defines an errorless transmission from S to M or a one-to-one mapping process; and $\eta_C = 0$ defines a useless measuring apparatus since no information can be transferred from S to M. The relative capacity functional is also a measure of the amount of redundency needed in a code to transmit an errorless message.

a. Classifying the measurement processes according to the destruction generated in the measured system. When mechanical forces are responsible for the buildup of correspondence between the apparatus and the measured object, it is possible that the state of the system S is altered considerably in the process. Two extreme situations emerge: (*1*) the harmless measurement (known as a measurement of the first kind), where the measured system remains intact and (*2*) a completely destructive measurement (known as a measurement of the second kind). In this latter case no correlation exists between the states of the measured system before and after the measurement interaction takes place. The whole range of intermediate destructive measurements can be classified by the functionals introduced previously to classify the resolution of the measurement. To achieve this the view is taken that the system S, after the interaction with M takes place, is the measurement apparatus for the system S at previous times. The evolution of the object system S during the measurement process is viewed as a mapping process from the state of the system before the measurement to the state of the system after the measurement:

$$\tilde{\rho}_{\mathrm{S}(t')} = \operatorname*{tr}_{\mathrm{S}(t)} \left[\tilde{\rho}_{\mathrm{S}(t')|\mathrm{S}(t)} \cdot \tilde{\rho}_{\mathrm{S}(t)} \right] \tag{2.18}$$

(The conditional density operator is used to define the mapping process.) This description enables investigation of partially destructive measurements in which the mapping of S into itself is not one to one. The second stage consists of utilization of the fuctionals that classify the resolution to classify the destruction. For example, when the evolution of the system S into itself is one to one, the mutual information between the state of S at time t' and time t is

$$I\left(\tilde{\rho}_{\mathrm{S}(t')}, \tilde{\rho}_{\mathrm{S}(t)}\right) = S\left(\tilde{\rho}_{\mathrm{S}(t')}\right) = S\left(\tilde{\rho}_{\mathrm{S}(t)}\right) \tag{2.19}$$

The mapping of the measured object into itself in the measuring process is strictly positive (see Appendix A), Therefore, it can be proved[19] that the mutual information is bound:

$$I\left(\tilde{\rho}_{\mathrm{S}(t'')}, \tilde{\rho}_{\mathrm{S}(t)}\right) \leq I\left(\tilde{\rho}_{\mathrm{S}(t')}, \tilde{\rho}_{\mathrm{S}(t)}\right) \tag{2.20}$$

where

$$t'' > t' > t$$

Equation (2.20) reveals that any measurement interaction decreases the correlation with the initial state, and that the nondestructive measurement is a limiting case.

It is important to notice that although the formal tools for classifying the resolution and the destruction are similar, there is a difference between the mutual information defined between S and M, and the mutual information defined between the states of S at times t and t'. Only the mutual information between S and M can be used to sort the S ensemble. Information about the past does not have operational use and cannot be used for sorting.

D. The Measurement and the Second Law of Thermodynamics

The use of information gained by the measurement process to sort an ensemble of systems S and reduce the entropy is discussed in the preceding sections. At first glance there is an apparent contradiction to the second law of thermodynamics, which in one of its formulations states that a homogenous system cannot spontaneously sort itself into its components. Actually since we believe the second law, the arguments that extend its validity are based on the idea that to achieve such a reducing entropy operation, the entropy of the measured apparatus is raised at least by an amount that is sufficient to compensate the entropy reduction of the system S. As with a mechanical model of the measurement, the second law of thermodynamics is conserved in a trivial manner, the reason being that the combined motion of the measured object and the measuring apparatus is a one-to-one mapping process described by a unitary transformation that conserves the total entropy[4, 20]:

$$S(\tilde{\rho}_{S\otimes M(t')}) = S(\tilde{\rho}_{S\otimes M(t)}) \tag{2.21}$$

where

$$\tilde{\rho}_{S\otimes M(t')} = \tilde{U}_{t'-t}\tilde{\rho}_{S\otimes M(t)}\tilde{U}^{\dagger}_{t'-t}$$

and $\tilde{U}$ is a unitary operator generated by the combined Hamiltonian:

$$\tilde{U}_{t'-t} = T\exp\left(i\int_t^{t'} \tilde{H}\, dt''\right) \tag{2.22}$$

Instead of the global approach to the entropy changes, one can apply a local approach that accounts for the entropy changes in the individual systems S and M and subtracts the mutual information linking S and M that can be used to reduce the entropy of S. The initial entropy of S and M is the total entropy (2.21), because S and M are uncorrelated. The final entropy is

$$S(\mathrm{S},\mathrm{M};t') = S(\tilde{\rho}_{S(t')}) + S(\tilde{\rho}_{M(t')}) - I(\tilde{\rho}_{S(t')},\tilde{\rho}_{M(t')}) \tag{2.23}$$

An inequality due to Lindblad[4] proves that the summed local entropy of the individual systems, less their useful mutual information, is greater than the initial entropy. A question arises as to the source of this entropy increase, in contradiction to the entropy conservation in the unitary evolution of the combined systems. The solution to the ambiguity is related to the fact that only information that has operational meaning (for a sorting procedure) is counted as decreasing entropy. In the quantum-mechanical measuring process there is the possibility that not all correlations between S and M commute: therefore, not all correlations can be used simultaneously in a sorting procedure and information is lost. A measuring process that posesses such noncommuting correlations is presented in Section III. Noncommuting correlations are also the essence of the Einstein, Podolsky and Rosen (EPR) paradox.[21] In the EPR process the challenge to the completeness of quantum mechanics arises from an attempt to localize the description of the physical variables. In the measurement process a similar situation emerges because the outcome of the process is localized by the definition of the measuring process to the variables of the measured system and the measuring apparatus. When the entropy is accounted for by local variables and when noncommuting correlations exist, information is lost and the entropy increases.

a. The Efficiency of the Measurement Process. It is shown earlier that if the second law of thermodynamics is conserved, every information gain between S and M is accompanied by at least an equal amount of entropy increase. Figure 1 shows various possibilities for information and entropy changes. The efficiency of the measurement process is defined as the ratio

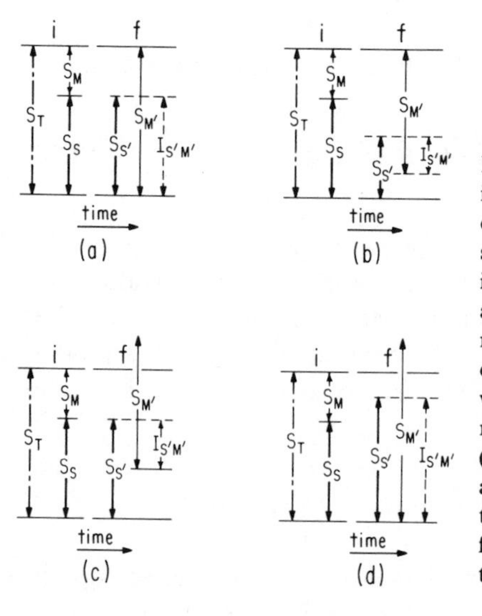

Fig. 1. Various possible entropy and mutual information changes in the measurement process. The left-hand side of each figure represents the initial state. The lack of correlation is indicated by the additivity of the entropy of S and M. (*a* and *c*) Nondestructive measurement in which the entropy of S has not changed during the process. (*a* and *d*) Measurements in which complete resolution is achieved. The mutual information covers the uncertainty in S. (*b* and *c*) Only a partial resolution has been achieved. Irreversible measurements in which the entropy of S and M, less the mutual information between them, is larger than the initial entropy.

of mutual information gained in the measurement process to the entropy increase needed to bring it about:

$$\Theta \equiv \frac{I(\tilde{\rho}_S, \tilde{\rho}_M)}{\Delta S(\tilde{\rho}_S) + \Delta S(\tilde{\rho}_M)} \tag{2.24}$$

where

$$\Delta S(\tilde{\rho}) = S(\tilde{\rho}_{(t')}) - S(\tilde{\rho}_{(t)})$$

For a reversible measurement, the efficiency is 1. The lower the efficiency, the more entropy is wasted to gain the same amount of information (which usually means bigger power supplies).

In summary, thermodynamics and the mathematical theory of communication can be used to define common functionals designed to classify the resolution and the destruction in the measurement process. The classifying functionals can also be used to describe the relation between the measurement and the second law of thermodynamics and to define the efficiency of the measurement process.

III. THE QUANTUM-MECHANICAL MEASUREMENT MODEL

This section presents a solvable QM model that represents all important features of the measurement process in a traceable way. The investigation can be divided into three strongly interlinked levels. The starting point is the construction of a mechanical model of the measurement process. This model has to be solvable and simple enough that the buildup of the correlation in time can be traced. The second level of investigation is concerned with the functionals classifying the measurement. A relation is sought among the mechanical parameters, the functionals classifying the resolution, and the destruction inherent to the measurement. The third level of investigation examines the irreversible character of the measuring process and its relation to the mechanical parameters and the classifying functionals.

A. The Mechanical Model

Any model of the measurement process must provide a description of its three main building blocks. First, the measuring system and the measuring apparatus have to be defined, which implies construction of an appropriate Hilbert space and of density operators to define the states of S and M. Second, we must define an interaction Hamiltonian that will lead to the desired correlation. Finally, we must define an operational procedure in which the interaction is turned on and off in a manner that maximizes the correlation.

In this section a variety of measurement models are presented. Differences among the models are generated by varying the initial states, the interaction Hamiltonian, or the interaction time. A common procedure to solve the equation of motion is used.

The simplest QM model that contains the significant features we require is the two-level system. The two-level system is used as the basis for our model of the measured system.

B. The Definitions of the States of the Systems S and M

A state of a thermodynamic system is defined completely by a finite set of extensive variables. The states of a QM system can be similarly described by ensemble averages of not necessarily commuting variables. Given these definitions, a close resemblance between the state of the system and its observables is maintained during the evolution of the system. The natural observables in the two-level system are the polarization moments in the directions X, Y, and Z. The operators associated with these observables obey the following relations:

$$
\begin{gathered}
\left[\tilde{\sigma}_i, \tilde{\sigma}_j\right] = i\tilde{\varepsilon}_{ijk}\tilde{\sigma}_k \\
\tilde{\sigma}_i^2 = \tfrac{1}{4}\tilde{I}
\end{gathered}
\tag{3.1}
$$

where $\tilde{\varepsilon}$ is the Levi-Civita antisymmetric tensor.

The operators $\tilde{\sigma}_i$ associated with the natural observables are combined in a canonical density operator that defines the state of S:

$$
\tilde{\rho}_S = \exp\left(\alpha_S \tilde{\sigma}_S^X + \beta_S \tilde{\sigma}_S^Y + \gamma_S \tilde{\sigma}_S^Z + \lambda \tilde{I}\right) \tag{3.2}
$$

The canonical density operator is obtained by maximizing the entropy of the system subject to all known constraints.[22] In the present case, the constraints are the polarization moments in the directions X, Y, and Z. In the two-level system the three polarization observables define the state of the system completely.[23] As a consequence a complete description of the state is given by the canonical density operator $\tilde{\rho}_S$ of (3.2).

The concept of the canonical density operator originated in statistical mechanics, where it describes a state in thermal equilibrium. The canonical density operator displays, as well, the most unbiased description of a state subject to all known information for the system.[22] In what follows, the canonical density operator allows a simple display of the dynamical buildup of the correlation between the measured system and the measuring apparatus. These qualities of the canonical density operator therefore make it a natural choice in the description of measurement processes.

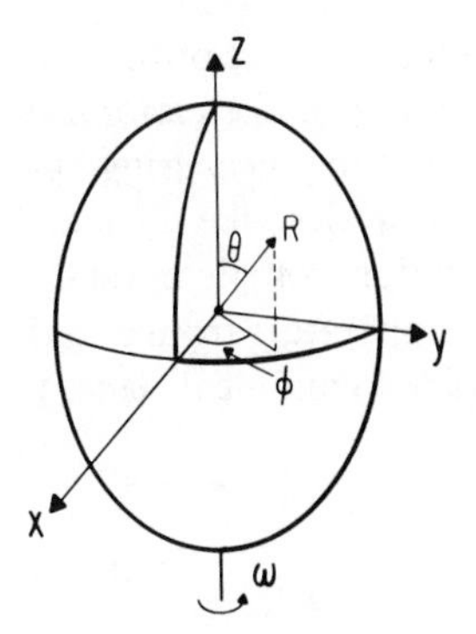

Fig. 2. The polarization of the system as a base for a geometric representation of the state. The surface of the sphere represents pure states. The origin is the completely mixed state.

In the canonical density operator of the two-level system, a pure state is described by the limit where one of the Lagrange parameters becomes infinitely large. A completely mixed state has all the Lagrange parameters equal to zero. If the canonical density operator is given, the angles of polarization θ and φ (relative to the Z-axis) are given by:

$$\varphi = \text{arc tan}\left(\frac{\alpha}{\beta}\right)$$

$$\theta = \text{arc cos}\left(\frac{\gamma}{\Delta}\right) \tag{3.3}$$

where

$$\Delta = (\alpha^2 + \beta^2 + \gamma^2)^{1/2}$$

and the magnitude of the polarization is

$$R = \tanh\left(\frac{\Delta}{2}\right) \tag{3.4}$$

The magnitude and the angle of polarization can be used for a geometric description of the state. Consider the unit sphere: The points on the surface of the sphere are the pure states, while the mixed states are represented by points inside the sphere (Fig. 2).

C. The Measuring Apparatus

The measurement of the polarization along the Z=axis is considered here. A sufficient apparatus for this measurement has to consist of at least two levels. For practical considerations the apparatus can be viewed as an effective two-level system, and projection techniques can be used to reduce the Hamiltonian to an effective Hamiltonian in the Hilbert space of the two levels. This description of the apparatus suggests the use of the canonical density operator of the type (3.2) to describe the state of the measuring apparatus also. A canonical density operator is then used to describe the state of the combined system: $\tilde{\rho}_{S\otimes M}$. To make the description complete, all 16 operators that are formed by the tensor product of the operators from S and from M must be used in the definition of the combined canonical density operator:

$$\tilde{\rho}_{S\otimes M} = \exp\left(\sum_{ij} \alpha^{ij} \tilde{\sigma}_S^i \otimes \tilde{\sigma}_M^j \right) \tag{3.5}$$

D. Dynamical Considerations

The dynamical properties of the canonical density operator and of the observables used to define this operator are closely connected. This fact is the basis for a general procedure for the solution of the equation of motion for canonical density operators developed by Alhasid and Levine[20] and adopted in this work. For the convenience of the reader, a summary of the main results of Alhasid and Levine is presented below.

The equation of motion for the observables is the Heisenberg equation:

$$\dot{\tilde{X}} = \mathcal{L}(\tilde{X}) = i[\tilde{H}, \tilde{X}] \tag{3.6}$$

A solution in closed form to this equation can be found if a set of operators $\tilde{A}$ is closed under the operation of the Liovillian (in this case the commutation relation with the Hamiltonian):

$$\mathcal{L}(\tilde{A}^i) = \sum_j h_i^j \tilde{A}^j \tag{3.7}$$

which reduces (3.6) to a set of coupled linear differential equations. The solution for the set of operators $A(t)$, given the initial set $\tilde{A}(0)$, can be defined by the use of a superevolution operator $\tilde{\tilde{G}}(t-t')$ of the conditional density operator type:

$$\hat{\tilde{A}}(t) = \tilde{\tilde{G}}(t' - t)\hat{\tilde{A}}(t) \tag{3.8}$$

where $\hat{\tilde{A}}(t)$ is the vector of operators $\tilde{A}$.

Turning to the Schroedinger picture, the equation of motion is:

$$\dot{\tilde{\rho}} = \mathcal{L}^*(\tilde{\rho}) = -i[\tilde{H}, \tilde{\rho}] \tag{3.9}$$

A canonical density operator defined by the same set of operators A:

$$\tilde{\rho}_{(t)} = \exp\left(\sum_i \alpha_i(t)\tilde{A}^i\right) \tag{3.10}$$

is used as a trial solution to (3.9). In the Schroedinger picture the operators are constant and the variables are the Lagrange parameters. To get an equation of motion for the Lagrange parameters, notice that because of the unitarity of the evolution any analytic function of $\tilde{\rho}$ is also a solution to (3.9), and so is:

$$\tilde{\varphi} = \sum_i \alpha^i(t)\tilde{A}^i \tag{3.11}$$

By inserting $\tilde{\varphi}$ in (3.9) one finds that it is a solution to the equation of motion provided: (*1*) the Lagrange parameters obey the set of coupled differential equations:

$$\dot{\alpha}^i_{(t)} = \sum \alpha^j(t)h^i_j \tag{3.12}$$

where the h^i_j's are the same coefficients as in equation (3.7); and (*2*) the Lagrange parameters match the Lagrange parameters of the canonical density operator at time $t=0$ [the same result as expressed in (3.12) can be obtained without the use of the unitary evolution requirement by direct calculation using a method developed by Magnus[24]]. The similarity between the equation of motion of the observables (3.7) and of the Lagrange parameters (3.12) is obvious. The Lagrange parameters of (3.12) and the observables of (3.7) evolve with time in an opposite contragradient manner. In the integral form the relation between the solutions is:

$$\tilde{\tilde{G}}_{Lg}(\grave{t}-t) = \tilde{\tilde{G}}_{ob}^{-1}(\grave{t}-t) \tag{3.13}$$

Thus the equations of motion of the observables and the Lagrange parameters representing them in the canonical density operator are closely connected.

E. The Nondestructive Measurement of the Z Component of the Polarization of the System S

The essence of a nondestructive measurement is that it changes only the state of the measuring apparatus. The change in the variables of M as a result of the transformation should be proportional to the measured value in S, when the state of S stays intact. These requirements put a restriction on the Hamiltonian and the states of S and M in the process: (*1*) The evolution of the S system should not depend on the states of the M system, which means that the interaction Hamiltonian has to commute with the state of S. (*2*) If the measurement is designed to determine the polarization in the Z direction of S, the interaction Hamiltonian has to be a function of the operator $\tilde{\sigma}_S^Z$. (*3*) This also means that the state of S has to be a function of $\tilde{\sigma}_S^Z$ alone so that:

$$\tilde{\rho}_{S(0)} = \exp\left(\gamma_S(0)\tilde{\sigma}_S^Z + \lambda_S\right) \tag{3.14}$$

(*4*) The choice of the state of S represented by (3.14) implies a restriction on the Hamiltonian of S in order that the state of S remain constant during the measurement process:

$$\tilde{H}_S = \omega_S \tilde{\sigma}_S^Z \tag{3.15}$$

A measurement scheme for the Z polarization of S that depends on the initial state of the apparatus and that achieves the necessary correlation can be described qualitatively as follows[14]: If the state of S is polarized in the $+Z$ direction the state of M does not change and also points to the $+Z$ direction. If the state of S points to the $-Z$ direction, the state of M rotates by π to the $-Z$ direction. (An alternative scheme can be used when M is initially polarized in the X direction, and the measurement process rotates M clockwise or anticlockwise to the $\pm Z$ directions in correspondence to the state of S.) In such a measurement scheme the eigenstates of the operator $\tilde{\sigma}_M^Z$ are chosen to be the basis for the registration of the results. The simplest case, when the registering states are degenerate, is studied first. The main concern is the influence of the initial state of the apparatus on the outcome of the measurement. First the initial state of M is chosen to be polarized along the Z-axis:

$$\rho_{M(0)} = \exp\left(\gamma_M(0)\tilde{\sigma}_M^Z + \lambda_M\right) \tag{3.16}$$

The interaction Hamiltonian that produces the same motion as that given in the qualitative description of this measurement is:

$$\tilde{H}_I = g(t)\left(\tilde{\sigma}_S^Z - \tfrac{1}{2}\tilde{I}_S\right)\cdot\left(\tilde{\sigma}_M^X - \tfrac{1}{2}\tilde{I}_M\right) \tag{3.17}$$

The stage is now set to follow the buildup of the measurement correlation. The initial state of the combined density operator is:

$$\tilde{\rho}_{S\otimes M(0)} = \exp\left(\gamma_S(0)\tilde{\sigma}_S^Z + \lambda_S\right)\exp\left(\gamma_M(0)\tilde{\sigma}_M^Z + \lambda_M\right) \tag{3.18}$$

The product form of (3.18), as for a joint probability distribution, implies that there is no correlation between S and M. This density operator is defined by the operators $\tilde{\sigma}_S^Z$, $\tilde{\sigma}_M^Z$, and I. A closed set under the commutation relation with the Hamiltonian is now sought. First the commutation relations of the initial set of operators with the Hamiltonian are calculated:

$$\begin{aligned}
&[\tilde{\sigma}_S^Z, \tilde{H}] = 0 \\
&[\tilde{\sigma}_M^Z, \tilde{H}] = ig(t)\left(\tilde{\sigma}_S^Z \cdot \tilde{\sigma}_M^Y - \tfrac{1}{2}\tilde{\sigma}_M^y\right)
\end{aligned} \tag{3.19}$$

One finds that the operators $\tilde{\sigma}_S^Z \tilde{\sigma}_M^Y$ and $\tilde{\sigma}_M^Y$ have to be added to the initial set. The second step is to calculate the commutation relations of the new operators with the Hamiltonian:

$$\begin{aligned}
&[\tilde{\sigma}_S^Z \tilde{\sigma}_M^Y, \tilde{H}] = ig(t)\left(\tfrac{1}{2}\tilde{\sigma}_S^Z \tilde{\sigma}_M^Z - \tfrac{1}{4}\tilde{\sigma}_M^Z\right) \\
&[\tilde{\sigma}_M^Y, \tilde{H}] = ig(t)\left(-\tilde{\sigma}_S^Z \tilde{\sigma}_M^Z + \tfrac{1}{2}\tilde{\sigma}_M^Z\right)
\end{aligned} \tag{3.20}$$

The operator $\tilde{\sigma}_S^Z \tilde{\sigma}_M^Z$ has to be added to the initial set. Repeating the procedure with $\tilde{\sigma}_S^Z \tilde{\sigma}_M^Z$, gives:

$$[\tilde{\sigma}_S^Z \tilde{\sigma}_M^Z, \tilde{H}] = ig(t)\left(\tfrac{1}{4}\tilde{\sigma}_M^Y - \tfrac{1}{2}\tilde{\sigma}_S^Z \tilde{\sigma}_M^Y\right) \tag{3.21}$$

One now finds that the set $\tilde{\sigma}_S^Z$, $\tilde{\sigma}_M^Z$, $\tilde{\sigma}_M^Y$, $\tilde{\sigma}_S^Z \tilde{\sigma}_M^Y$, and $\tilde{\sigma}_S^Z \tilde{\sigma}_M^Z$ is closed, and therefore the density operator at time t that evolves from the initial density operator (3.18) is:

$$\begin{aligned}
\tilde{\rho}_{S\otimes M(t)} = \exp\big(&\gamma_S(0)\tilde{\sigma}_S^Z + \gamma_M(t)\tilde{\sigma}_M^Z + \beta_M(t)\tilde{\sigma}_M^Y + \mu_{SM}^{ZY}(t)\tilde{\sigma}_S^Z \tilde{\sigma}_M^Y \\
&+ \mu_{SM}^{ZZ}(t)\tilde{\sigma}_S^Z \tilde{\sigma}_M^Z + \lambda\big)
\end{aligned} \tag{3.22}$$

The operators in (3.22) that are most significant for the measurement process are $\tilde{\sigma}_S^Z \tilde{\sigma}_M^Y$ and $\tilde{\sigma}_S^Z \tilde{\sigma}_M^Z$. These operators represent correlations between S and M. The growth of the Lagrange parameters associated with these operators describes the buildup of these correlations. The time dependence of the Lagrange parameters still has to be found. The coefficients h_j^i of the equation of motion (3.12) of the Lagrange parameters are recovered from the commutation relations (3.19) to (3.21). The solution becomes simpler by introducing the constants of motion:

$$\begin{aligned}
&\tilde{C}^Z = \tilde{\sigma}_M^Z + 2\tilde{\sigma}_S^Z \tilde{\sigma}_M^Z \\
&\tilde{C}^Y = \tilde{\sigma}_M^Y + 2\tilde{\sigma}_S^Z \tilde{\sigma}_M^Y
\end{aligned} \tag{3.23}$$

and a new closed set of operators

$$\tilde{Z}=\tilde{\sigma}_{\mathrm{M}}^{Z}-2\tilde{\sigma}_{\mathrm{S}}^{Z}\tilde{\sigma}_{\mathrm{M}}^{Z}$$
$$\tilde{Y}=\tilde{\sigma}_{\mathrm{M}}^{Y}-2\tilde{\sigma}_{\mathrm{S}}^{Z}\tilde{\sigma}_{\mathrm{M}}^{Y} \tag{3.24}$$

Transforming back to the initial set of operators, one gets for the time dependence of the Lagrange parameters, with the inital condition defined by (3.18)[25]:

$$\gamma_{\mathrm{M}}(t)=\tfrac{1}{2}\gamma_{\mathrm{M}}(0)\{1+\cos[G(t)]\}$$
$$\beta_{\mathrm{M}}(t)=\tfrac{1}{2}\gamma_{\mathrm{M}}(0)\sin(G(t))$$
$$\mu_{\mathrm{SM}}^{ZY}(t)=\gamma_{\mathrm{M}}(0)\sin G(t)$$
$$\mu_{\mathrm{SM}}^{ZZ}(t)=\gamma_{\mathrm{M}}(0)\{1-\cos[G(t)]\} \tag{3.25}$$

The detailed solution (3.25) opens the door for further examination of the nondestructive measurement. Three mechanical parameters govern the measurement process: (*1*) The Lagrange parameter $\gamma_{\mathrm{S}}(0)$, which describes the state of S throughout the measurement process; (*2*) the Lagrange parameter $\gamma_{\mathrm{M}}(0)$, which descirbes the initial state of M; and (*3*) the effective time duration of the measurement defined by:

$$G(t)=\int_0^t g(t')\,dt' \tag{3.26}$$

The nondestructive character of the measurement is reflected by the fact that the Lagrange parameter $\gamma_{\mathrm{S}}(0)$ that describes the state of S is a constant of the motion.

The quality of the resolution in this measurement depends on the correlation represented by the operator $\tilde{\sigma}_{\mathrm{S}}^{Z}\tilde{\sigma}_{\mathrm{M}}^{Z}$. It is immediately seen that the amount of correlation depends on the Lagrange parameter associated with the correlation operator. Examining (3.25) one finds that the final value of this Lagrange parameter depends on the mechanical parameters associated with the initial state of M and the effective interaction time. The influence of these two parameters can be investigated separately. To examine the influence of the initial state of the apparatus, the effective interaction time is kept at the value that achieves maximum performance: $G(t^*)=\pi$. The combined final density operator then takes the simple form:

$$\tilde{\rho}_{\mathrm{S}\otimes\mathrm{M}(t^*)}=\exp\left(2\gamma_{\mathrm{M}}(0)\tilde{\sigma}_{\mathrm{S}}^{Z}\tilde{\sigma}_{\mathrm{M}}^{Z}+\gamma_{\mathrm{S}}(0)\tilde{\sigma}_{\mathrm{S}}^{Z}+\lambda\right) \tag{3.27}$$

In this case the mapping process is described by a canonical conditional density operator:

$$\tilde{\rho}_{M|S} = \exp(2\gamma_M(0)\tilde{\sigma}_S^Z\tilde{\sigma}_M^Z + \lambda_M) \tag{3.28}$$

Examining (3.28) one can see that the quality of the resolution depends on the magnitude of the Lagrange parameter $\gamma_M(0)$. The measurement process is now examined from a thermodynamic point of view. First we identify the Lagrange parameters with thermodynamic quantities: The Lagrange parameter γ represents a generalized temperature $(1/k_B T)$, since we identify the operator $\tilde{\sigma}^Z$ with energy. The Lagrange parameters μ are similar to the chemical potential since they are associated with operators that represent correlation. The functionals classifying the resolution are functions of these thermodynamic quantities. These functionals become simpler with the introduction of the mixing parameter:

$$U_i = \frac{\exp(\gamma_i/2)}{2\cosh(\gamma_i/2)} \tag{3.29}$$

and

$$V_i = 1 - U_i$$

The mutual information between S and M is now calculated:

$$I(\tilde{\rho}_{S(t^*)}, \tilde{\rho}_{M(t^*)}) = s(U_S U_M + V_S V_M) - s(U_M) \tag{3.30}$$

where

$$s(X) = -X\ln X - (1-X)\ln(1-X)$$

It is found that the mutual information between S and M equals the entropy increase in M (case *b* in Fig. 1). As a consequence, the measurement is reversible because furthermore, the entropy of the state of S does not change during the process. The efficiency coefficient is therefore 1.

The relative information functional is found to be:

$$\eta(\tilde{\rho}_{S(t)} | \tilde{\rho}_{M(t^*)}) = \frac{s(U_S U_M + V_S V_M) - s(U_S)}{s(U_S)} \tag{3.31}$$

Figure 3 presents the dependence of the relative information functional on the initial state of the system S and on the initial state of the apparatus M.

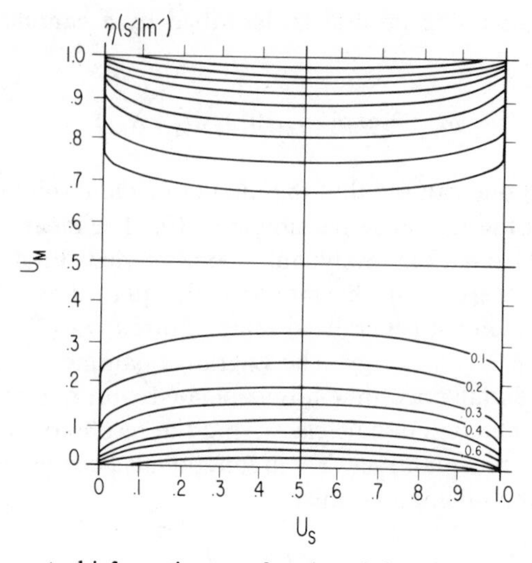

Fig. 3. The relative mutual information as a function of the mixing parameters in S and M. The flatness of the contours shows that the relative mutual information does not depend strongly on the state of S.

The reduction of the quality of the resolution when the apparatus aproaches the completely mixed state (where there is no resolution of S at all) is easily seen. This is an important observation: an apparatus in equilibrium (its initial entropy is maximized) cannot resolve information in a nondestructive measurement because it cannot increase its entropy to compensate for the information gain. When M is initially in a pure state the process is completely resolving. This extreme case is the ideal measurement of Von Nuemann (8) (Fig. 1*a*). The dependence of η on the state of S is weak except when S approaches a pure state. In Fig. 3 one can identify the state of S that achieves capacity $V_S=\frac{1}{2}$. The capacity functional is found to be:

$$C(\tilde{\rho}_{M|S})=\ln 2-s(U_M) \tag{3.32}$$

The influence of the effective interaction time on the outcome of the measurement is described in terms of two related measurement situations: (*1*) when the initial state of the apparatus is polarized out of the *Z*-axis and (*2*) when the mapping of the state of *S* is inbedded into nondegenerate states of *M*. The common feature of these measurement situations is that the resolution is reduced because in the final state the apparatus is not polarized in the *Z*-axis. The simpler case arises when the apparatus is polarized in the *ZY*-plane. The preceding results can be recovered by considering the change in polarization in the direction of the initial state because the interaction Hamiltonian is symmetric with respect to the *ZY*-plane. If the initial polarization is out of the *ZY*-plane the resolution is reduced. The X component of the polarization is a constant of motion so that the influence on the resolution can be viewed as an effective increase in mixing in the Z direction. The significant parameter is $\gamma_{\mathrm{M}}(0)$ and its relation to the angle β between the polarization direction and the ZY-plane is:

$$\gamma_{\mathrm{M}}(0)=\ln\frac{U_{\mathrm{M}}|\chi|^2+V_{\mathrm{M}}|\xi|^2}{V_{\mathrm{M}}|\chi|^2+U_{\mathrm{M}}|\xi|^2} \tag{3.33}$$

where $|\chi|^2=\cos^2\beta$ and $|\xi|^2=\sin^2\beta$. The mutual information functional can now be calculated with the use of (3.29) and (3.30).

The measurement situations when the interaction time is varied and when the degeneracy of the states in M is removed are investigated together. The procedure is intended to calculate the conditional density operator directly, using the fact that $\tilde{\sigma}_{\mathrm{S}}^{Z}$ is a constant of motion. First an effective Hamiltonian for M is defined:

$$\tilde{H}_{ef}=f(t)\left(\tilde{\sigma}_{\mathrm{M}}^{X}-\tfrac{1}{2}\right)+\omega_{\mathrm{M}}\tilde{\sigma}_{\mathrm{M}}^{Z} \tag{3.34}$$

where

$$f(t)=g(t)\left(\tilde{\sigma}_{\mathrm{S}}^{Z}-\tfrac{1}{2}\right)$$

If the term $\omega_{\mathrm{M}}\tilde{\sigma}_{\mathrm{M}}^{Z}$ is added to (3.34), the degeneracy of the registering states in M is reduced. Choosing the initial state of M as (3.16), gives the relevant commutation relations

$$\begin{aligned}
&\left[\tilde{\sigma}_{\mathrm{M}}^{Z},\tilde{H}_{ef}\right]=if(t)\tilde{\sigma}_{\mathrm{M}}^{Y}\\
&\left[\tilde{\sigma}_{\mathrm{M}}^{Y},\tilde{H}_{ef}\right]=if(t)\tilde{\sigma}_{\mathrm{M}}^{Z}-i\omega\tilde{\sigma}_{\mathrm{M}}^{X}\\
&\left[\sigma_{\mathrm{M}}^{X},\tilde{H}_{ef}\right]=i\omega_{\mathrm{M}}\tilde{\sigma}_{\mathrm{M}}^{Y}
\end{aligned} \tag{3.35}$$

and the conditional Lagrange parameters of the three polarization directions are found to be:

$$\begin{aligned}
\gamma_{\mathrm{M}}(t)&=\gamma_{\mathrm{M}}(0)\frac{\omega_{\mathrm{M}}^2+f^2(t)\cos[F(t)]}{f^2(t)+\omega_{\mathrm{M}}^2}\\
\alpha_{\mathrm{M}}(t)&=\gamma_{\mathrm{M}}(0)\frac{\omega_{\mathrm{M}}f(t)}{f^2(t)+\omega_{\mathrm{M}}^2}\{1-\cos[F(t)]\}\\
\beta_{\mathrm{M}}(t)&=\gamma_{\mathrm{M}}(0)\frac{f(t)}{\left[f^2(t)+\omega_{\mathrm{M}}^2\right]^{1/2}}\sin[F(t)]
\end{aligned} \tag{3.36}$$

For the case where $\omega_M \ll f(t)$ we get:

$$F(t)=\int_0^t \left[f^2(t')+\omega_M^2\right]^{1/2} dt'$$

The corresponding angle θ of the final conditional polarization is:

$$\cos^2\theta=\frac{\omega_M^2+f^2(t)\cos(F(t))}{f^2(t)+\omega^2} \tag{3.37}$$

This is always less than $\theta=\pi$ for S polarized in the Z direction corresponding to the completely resolving measurement. With the use of the conditional density operator, the conditional probabilities of M being polarized in the $\pm Z$ direction subject to the state of S being in the $\pm Z$ direction can be calculated:

$$\begin{aligned}
P\left[\tilde{\sigma}_M^Z=\tfrac{1}{2}\,|\,\tilde{\sigma}_S^Z=\tfrac{1}{2}\right]&=U_M\\
P\left[\tilde{\sigma}_M^Z=-\tfrac{1}{2}\,|\,\tilde{\sigma}_S^Z=\tfrac{1}{2}\right]&=V_M\\
P\left[\tilde{\sigma}_M^Z=\tfrac{1}{2}\,|\,\tilde{\sigma}_S^Z=-\tfrac{1}{2}\right]&=V_M|\chi|^2+U_M(1-|\chi|^2)\\
P\left[\tilde{\sigma}_M^Z=-\tfrac{1}{2}\,|\,\tilde{\sigma}_S^2=-\tfrac{1}{2}\right]&=U_M|\chi|^2+V_M(1-|\chi|^2)
\end{aligned} \tag{3.38}$$

where

$$|\chi|^2=\frac{g^2(t)\{1-\cos[G(t)]\}}{2\left[g^2(t)+\omega_M^2\right]}$$

and

$$G(t)=\int_0^t \left[g^2(t)+\omega_M^2\right]^{1/2} dt'$$

The mutual information functional is calculated by using the symmetry of the mutual information to interchange with respect to the S and M indexes:

$$\begin{aligned}
I(\tilde{\rho}_{S(t')},\tilde{\rho}_{M(t')})&=S(\tilde{\rho}_{M(t')})-S(\tilde{\rho}_{M(t')}|\tilde{\rho}_{S(t')})\\
&=s\left(U_SU_M+V_M(|\chi|^2+U_M(1-|\chi|^2))V_S\right)\\
&\quad-U_Ss(U_M)-V_Ss\left(V_M|\chi|^2+U_M(1-|\chi|^2)\right)
\end{aligned} \tag{3.39}$$

When $\omega_M = 0$ and $U_{\hat{M}} = 1$, only the influence of the interaction time is considered. For these conditions the relative capacity is:

$$\eta_C(\tilde{\rho}_{M|S}) = \frac{1}{\ln 2}\left\{s\left[\tfrac{1}{2}(2-|\chi|^2)\right] - \tfrac{1}{2}s(|\chi|^2)\right\} \tag{3.40}$$

where

$$|\chi|^2 = \sin^2\left[\frac{G(t)}{2}\right]$$

Figure 4 displays the dependence of the relative capacity on the interaction time. The point when $G(t) = \pi$ is not stationary, which means that it is difficult to achieve a completely resolving measurement.

The reduction of the resolving power that results from lifting the degeneracy of the registering states can also be studied with the use of (3.40). Even when the interaction time is chosen to maximize the correlation a completely resolving measurement is not possible. For $|\chi|^2$ one gets:

$$|\chi|^2 = \frac{g^2(t)}{g^2(t) + \omega_M^2} \tag{3.41}$$

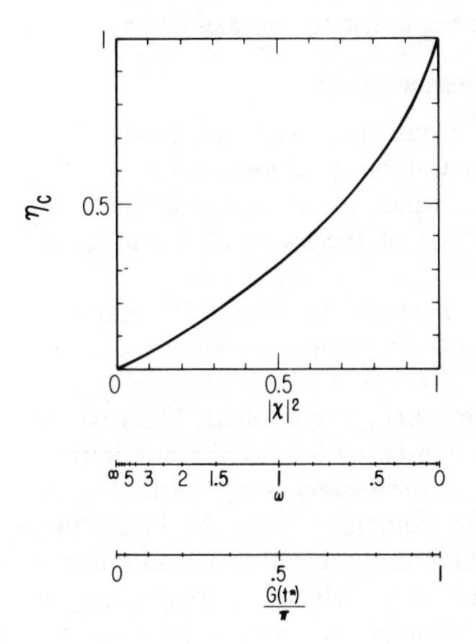

Fig. 4. The dependence of the relative capacity on the angle as well as on the frequency ω_M of the apparatus and the effective time duration of interaction.

and the relative capacity is:

$$\eta_C(\tilde{\rho}_{M|S}) = \frac{1}{\ln 2}\Big\{s\big(\tfrac{1}{2}\big[U_M(2-|\chi|^2)+V_M|\chi|^2\big]\big) - \tfrac{1}{2}\big[s(U_M)+s\big(V_M|\chi|^2+U_M(1-|\chi|^2)\big)\big]\Big\} \quad (3.42)$$

From (3.41) and (3.42) it can be seen that the resolution is reduced when ω_M increases, as also is seen in Fig. 3.

Equation 3.42 implies that a resolving measurement is possible only when the interaction is much stronger than the splitting in the registering states. This result is general and does not depend on the specific model considered here. A visual description will help explain this point. To achieve the correlation when S is pointed to the south pole ($-Z$ direction), M has to turn from the north to the south pole. The lifting of the degeneracy is equivalent to spinning the sphere (describing the state of M) around the Z-axis with an angular velocity of ω_M. The only legitimate path from the north to the south pole is on the surface of the sphere, so that the faster the sphere revolves, the harder it becomes to make the journey. In general, to achieve a QM measurement transition, one always has to pass the equator, and if the states that register are not degenerate, a "gyroscopic effect" hinders the resolution. (This situation is overcome by a macroscopic measuring apparatus that posesses registering states that are almost degenerate.

In this section we discuss nondestructive measurements. The emphasis is on the influence of the apparatuses' performance on the resolution. The same approach is now be used to describe destructive measurements.

F. Destructive Measurements

Even in the two-level system there are numerous ways to achieve a destructive measurement. According to the criteria developed in Section II, a destructive measurement is a process in which the evolution of the measured system is not a one-to-one mapping of the states of S into themselves.

Therefore, almost all measurements in which the interaction Hamiltonian does not commute with the state of the measured system are destructive.

Two examples of destructive measurements are presented. The first example uses the same Hamiltonian that is used earlier for the nondestructive measurement, but the initial state of the system S is polarized in an arbitrary direction so that it no longer commutes with the interaction Hamiltonian (3.17). In the second example the interaction Hamiltonian is changed to represent a measurement model in which the total energy is conserved. One of the outcomes of this example is a process in which the apparatus is in correlation with the state of S in the past and the evolution prepares S in a preselected state.

The main result of this section is the demonstration of the irreversible character of a destructive measurement.

1. Example A

This measurement process is constructed to demonstrate the influence of the initial state of S on the entropy increase. The foundation for the present example has already been set in the nondestructive measurement process, the difference being that now the state of S is no longer constant during motion. The initial state of S is chosen to be polarized on the ZX-plane:

$$\tilde{\rho}_{S(0)} = \exp\left[\gamma_S(0)\tilde{\sigma}_S^Z + \alpha_S(0)\tilde{\sigma}_S^X + \lambda\right] \tag{3.43}$$

This state does not commute with the interaction Hamiltonian because it contains the operator $\tilde{\sigma}_S^X$. The equation of motion of the combined system S⊗M can be solved in a fashion similar to that used for the nondestructive measurement. The set of operators $\tilde{\sigma}_S^X\tilde{\sigma}_M^X$, $\tilde{\sigma}_S^Y\tilde{\sigma}_M^X$, $\tilde{\sigma}_S^X$ and $\tilde{\sigma}_S^Y$ must be added to the initial set. It is found that the new set of operators is not coupled to the previous one. The solution to the equation of motion is:

$$\begin{aligned}\tilde{\rho}_{S\otimes M(t)} = \exp\Big[&\gamma_S(0)\tilde{\sigma}_S^Z + \gamma_M(t)\tilde{\sigma}_M^Z + \alpha_S(t)\tilde{\sigma}_S^Y + \beta_S(t)\tilde{\sigma}_S^x + \beta_M(t)\tilde{\sigma}_M^X \\ &+ \mu_{SM}^{ZZ}(t)\tilde{\sigma}_S^Z\tilde{\sigma}_M^Z + \mu_{SM}^{ZY}(t)\tilde{\sigma}_S^Z\tilde{\sigma}_M^Y + \mu_{SM}^{YX}(t)\tilde{\sigma}_S^Y\tilde{\sigma}_M^X \\ &+ \mu_{SM}^{XX}(t)\tilde{\sigma}_S^X\tilde{\sigma}_M^X + \lambda\Big]\end{aligned} \tag{3.44}$$

where

$$\begin{aligned}\gamma_M(t) &= \tfrac{1}{2}\gamma_M(0)\{1+\cos[G(t)]\} \\ \beta_S(t) &= \tfrac{1}{2}\alpha_S(0)\sin[G(t)] \\ \alpha_S(t) &= \tfrac{1}{2}\alpha_S(0)\{1+\cos[G(t)]\} \\ \beta_M(t) &= \tfrac{1}{2}\gamma_M(0)\sin[G(t)] \\ \mu_{SM}^{ZZ}(t) &= \gamma_M(0)\{1-\cos[G(t)]\} \\ \mu_{SM}^{ZY}(t) &= \gamma_M(0)\sin[G(t)] \\ \mu_{SM}^{YX}(t) &= \alpha_S(0)\sin[G(t)] \\ \mu_{SM}^{XX}(t) &= \alpha_S(0)\{1-\cos[G(t)]\}\end{aligned}$$

Just as for the nondestructive measurement, the correlation is maximized when $G(t^*)=\pi$, in which case the combined density operator has the form:

$$\tilde{\rho}_{S\otimes M(t^*)}=\exp\left[\gamma_S(0)\tilde{\sigma}_S^2+2\gamma_M(0)\tilde{\sigma}_S^Z\tilde{\sigma}_M^Z+2\alpha_S(0)\tilde{\sigma}_S^X\tilde{\sigma}_M^X+\lambda\right]. \quad (3.45)$$

Comparing (3.45) to (3.27), we see that the significant difference between the final density operator of the destructive and nondestructive processes is the new correlation $\tilde{\sigma}_S^X\tilde{\sigma}_M^X$. From the point of view of the apparatus this process is similar to the nondestructive measurement. It can be seen from (3.45) that a one-to-one correspondence between the polarizations in the Z=axis is achieved if $\gamma_M(0)\to\infty$. In this condition the reduced density operators of S and M are:

$$\tilde{\rho}_{S(t^*)}=\operatorname*{tr}_M\left[\tilde{\rho}_{S\otimes M(t^*)}\right]=\exp\left[\gamma_S(0)\tilde{\sigma}_S^Z+\lambda_S\right]$$

$$\rho_{M(t^*)}=\operatorname*{tr}_S\left[\tilde{\rho}_{S\otimes M(t^*)}\right]=\exp\left[\gamma_S(0)\tilde{\sigma}_M^Z+\lambda_S\right] \quad (3.46)$$

If the final state of S is examined, it appears that the X and Y initial polarizations are wiped out by the measurement process (reduction of the wave pocket). Following the definitions of Section II the loss of information on the polarization in the X and Y directions is a consequence of destructive mapping of S into itself. These facts are made clear through the following entropy relations of the individual systems. As before, the entropy increase in the apparatus is exactly compensated by the buildup of the mutual information [for $G(t^*)=\pi$] (case c of Fig. 1):

$$\Delta S_M=I\left(\tilde{\rho}_{S(t^*)},\tilde{\rho}_{M(t^*)}\right) \quad (3.47)$$

Figure 5 shows the dependence of the mutual information on the initial mixing and polarization of S. It is seen that capacity operation is achieved when $\tilde{\gamma}_S=0$. For the S system the final entropy is:

$$S\left(\tilde{\rho}_{S(t^*)}\right)=s\left(U_S|A|^2+V_S(1-|A|^2)\right) \quad (3.48)$$

where

$$|A|^2=\cos^2\frac{\theta}{2}=\frac{\exp\{\gamma_S(0)/2\}}{2\cosh\left(\gamma_S(0)/2\right)}$$

and

$$U_S=\frac{\exp\{\Delta/2\}}{2\cosh\left(\Delta/2\right)}, \qquad V_S=1-U_S$$

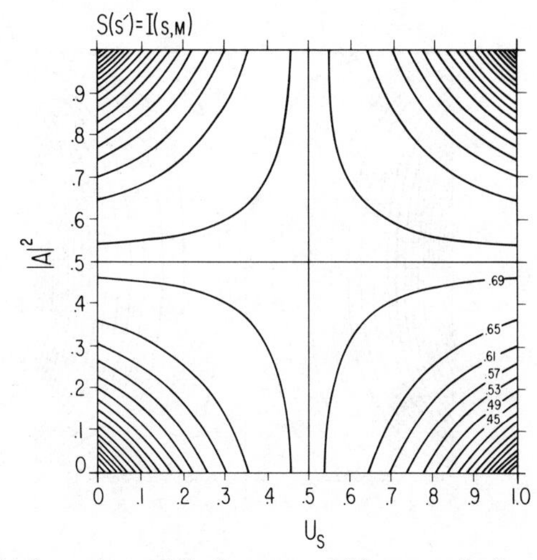

Fig. 5. The mutual information and final entropy of S in a completely resolving measurement as a function of the mixing parameter and the angle of polarization through $|A|^2$.

The entropy increase is therefore:

$$\Delta S = s\big(U_S|A|^2 + V_S(1-|A|^2)\big) - s(U_S) \tag{3.49}$$

Figure 6 represents the entropy increase of the system S during the measurement.

The functional $\eta(\tilde{\rho}_{S(t)}|\tilde{\rho}_{S(t')})$ can be used to demonstrate the destructive character of the measurement to the system S:

$$\eta\big(\tilde{\rho}_{S(t^*)}|\tilde{\rho}_{S(0)}\big) = \frac{s(U_S)}{s\big(U_S|A|^2 + V_S(1-|A|^2)\big)} \leq 1 \tag{3.50}$$

A completely destructive measurement results when the initial state of S is pure and is polarized on the XY-plane, for which the combined density has the form (for S initially polarized on the X-axis):

$$\tilde{\rho}_{S\otimes M(t^*)} = \exp\big[\gamma\big(\tilde{\sigma}_S^Z\tilde{\sigma}_M^Z + \tilde{\sigma}_S^X\tilde{\sigma}_M^X\big) + \lambda\big] \quad \text{and} \quad \gamma\to\infty \tag{3.51}$$

Equation 3.51 represents a double correlation on the X-and Z-axis.

The total entropy change is calculated as follows. The initial entropy is:

$$S_i = S\big(\tilde{\rho}_{S(0)}\big) = s(U_S) \tag{3.52}$$

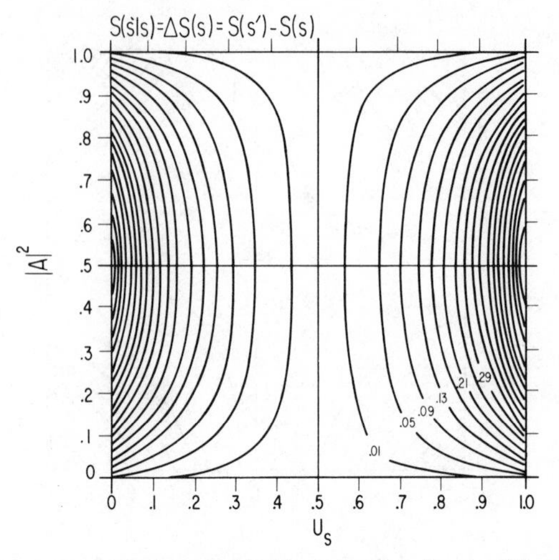

Fig. 6. The entropy increase in the measured object as well as the conditional entropy $S(\tilde{\rho}_{s'}|\tilde{\rho}_s)$ as a function of the mixing parameter and the angle of polarization from the z-axis through $|A|^2$. Maximal entropy increase is when the angle of polarization is $\pi/2$.

The final entropy is (when the mutual information between S and M is subtracted):

$$S_f = S(\tilde{\rho}_{S(t)}) + S(\tilde{\rho}_{M(t)}) - I(\tilde{\rho}_{S(t)}, \tilde{\rho}_{M(t)}) = s(U_S|A|^2 + V_S(1-|A|)^2)) \tag{3.53}$$

The total entropy change is:

$$\Delta S_{i\to f} = s(U_S|A|^2 + V_S(1-|A|^2)) - s(U_S) \geq 0 \tag{3.54}$$

The irreversibility is evident in the entropy increase, which is a consequence of the loss of information on the polarization on the XY-plane.

This information is represented by the operator $\tilde{\sigma}_S^X\tilde{\sigma}_M^X$, which has been left out of the entropy relations, a direct consequence of the definitions introduced in Section II. When accounting for the entropy, only information that has operational meaning for the purpose of defining a sorting procedure is taken into consideration. The correlation $\tilde{\sigma}_S^X\tilde{\sigma}_M^X$ does not commute with the correlation $\tilde{\sigma}_S^Z\tilde{\sigma}_M^Z$ in the Z-axis; therefore, it cannot be used simultaneously with this correlation in a sorting procedure.

The total entropy of the combined system is the minimum amount of mixing entropy between pure subensembles that decompose $\tilde{\rho}_{S\otimes M}$ completely. This procedure is equivalent to finding a representation in which the density operator is diagonal:

$$\tilde{\rho}_{S\otimes M} = \sum_n p_n \phi_n\rangle\langle\phi_n \tag{3.55}$$

and the entropy is:

$$S(\tilde{\rho}_{S\otimes M}) = -\sum_n p_n \ln p_n \tag{3.56}$$

In the measurement process the outcome is localized in the subsystems S and M. This means that the results can be analyzed only by states that are products of local observables. As a consequence, each set of eigenfunctions that can be used to diagonalize the combined density operator is of the form:

$$\phi^{ij}_{S\otimes M} = \varphi^i_S \otimes \varphi^j_M \tag{3.57}$$

Because of this restriction, density operators of the form (3.45) cannot be diagonalized. Thus not all information can be used.

Such a situation is very similar to the process described in the EPR paradox,[21] which is described briefly. A source emits two particles that interacted in the past and therefore are correlated. For example, if the source had an initial angular momentum of zero, the emitted particles have angular momentum that adds up to zero in all directions in space. We now assume that the pair of particles are widely separated and the angular momentum of one of the particles is measured. Because of the conservation law the angular momentum of the other particle is known. The problem that now arises is that the angular momentum operators in different directions do not commute and also that in a local description the particle has a definite direction to its angular momentum. If we measured the X component of one particle, the other particle, without being directly interfered with is found in an eigenstate of the X component. The same happens for any other direction in space. But the particle cannot be in an eigenstate of two different angular momentum directions simultaneously. To Einstein the possibility of an action from a distance is ruled out, so he came to the conclusion that quantum mechanics is not complete since the local description of the particle cannot account for the multiple correlation. As we see earlier in the EPR description, there is an attempt to define a complex system by local variables. In the destructive measurement we demonstrate that this attempt is bound to lose information and therefore is incomplete. Empirical tests of the nonlocal character of the physical world are in accordance with QM predictions.[26]

The same type of irreversibility that is described in this section is found also in scattering situations where the induced separability of the colliding partners imposes irreversibility.[27]

From the preceding analysis, it is clear that the total entropy increase is larger than the information gain. Therefore, the efficiency coefficient is less than 1. The efficiency coefficient is given by:

$$\Theta = \frac{s\left(U_S|A|^2 + V_S(1-|A|^2)\right)}{2s\left(U_S|A|^2 + V_S(1-|A|^2)\right) - s(U_S)} \tag{3.58}$$

which is bounded between 0 and 1.

2. Example B

This model demonstrates a destructive measurement in a frame that is different from that of the ideal measurement. One of the possibilities is a correlation between S and M at different times (this correlation is useless as a basis for a sorting procedure).

In this model the energy is associated with the operator $\tilde{\sigma}^Z$.

$$\begin{aligned} \tilde{H}_S &= \omega \tilde{\sigma}_S^Z \\ \tilde{H}_M &= \omega \tilde{\sigma}_M^Z \end{aligned} \tag{3.59}$$

The interaction Hamiltonian is chosen in such a way that it preserves total energy:

$$H_I = \frac{g(t)\left(\tilde{\sigma}_S^+ \tilde{\sigma}_M^- + \tilde{\sigma}_S^- \tilde{\sigma}_M^+\right)}{\sqrt{2}} \tag{3.60}$$

In the initial stage both S and M are chosen to be polarized in the Z direction (3.18). The combined density operator is solved for any fixed time by a technique similar to that used in the preceding example. (The commutation relations are the same for two interacting harmonic oscillators when $\tilde{\sigma}^{\pm} \to a^{\pm}$. Therefore, the solution that follows is adequate for this case as well.) The results of the calculation are:

$$\tilde{\rho}_{S \otimes M} = \tilde{\rho}_A \otimes \tilde{\rho}_B(t) \tag{3.61}$$

where

$$\tilde{\rho}_A = \exp\left\{\tfrac{1}{2}\left[\gamma_S(0) + \gamma_M(0)\right]\left(\tilde{\sigma}_S^Z + \tilde{\sigma}_M^Z\right) + \lambda\right\}$$

$$\tilde{\rho}_{B(t)} = \exp\left[\gamma_{SM}(t)\tilde{Z} + \beta_{SM}(t)\tilde{Y}\right]$$

and

$$\tilde{Z}=\frac{\sqrt{2}}{2}(\tilde{\sigma}_S^Z-\tilde{\sigma}_M^Z)$$

$$\tilde{Y}=i\frac{\sqrt{2}}{2}(\tilde{\sigma}_S^+\tilde{\sigma}_M^- -\tilde{\sigma}_S^+\tilde{\sigma}_M^-)$$

$$\gamma_{SM}(t)=\frac{\sqrt{2}}{2}[\gamma_S(0)-\gamma_M(0)]\cos[G(t)]$$

$$\beta_{SM}(t)=\frac{\sqrt{2}}{2}[\gamma_S(0)-\gamma_M(0)]\sin[G(t)]$$

Two interaction time durations are considered:

1. $G(t^*)=\pi$, for which the final density operator is:

$$\tilde{\rho}_{S\otimes M(t^*)}=\exp[\gamma_M(0)\tilde{\sigma}_S^Z+\gamma_S(0)\tilde{\sigma}_M^Z+\lambda] \tag{3.62}$$

   By comparing (3.62) to (3.18) it becomes apparent that S and M have changed places, with M in a one-to-one correspondence with the state of S before the interaction took place. On the other hand, S is in correspondence with the initial value of M. In this type of measurement M is registering the state of S in the past. Becuase this state has no relevance to behavior of S in the future no use can be obtained from this information. It is important to notice that when entropy considerations are accounted for, the mutual information between S and M at different times has no operational value. Therefore, it does not reduce the entropy. It can be seen that the measurement is reversible $\Delta S=0$ and also completely destructive $\eta(\tilde{\rho}_{S(t^*)}|\tilde{\rho}_{S(0)})=0$, with S having lost all information on its past.

2. The interaction duration is chosen so that $G(t)=\pi/2$. The resulting density operator is:

$$\tilde{\rho}_{S\otimes M(t)}=\exp\left\{\tfrac{1}{2}[\gamma_S(0)+\gamma_M(0)](\tilde{\sigma}_S^Z+\sigma_M^Z)\right.$$
$$\left.+\frac{\sqrt{2}}{2}[\gamma_S(0)-\gamma_M(0)]\tilde{Y}+\lambda\right\} \tag{3.63}$$

   The operator $\tilde{Y}$ can be written as $\tilde{Y}=\frac{\sqrt{2}}{2}(\tilde{\sigma}_S^X\tilde{\sigma}_M^X-\tilde{\sigma}_S^Y\tilde{\sigma}_M^Y)$. Here the situation has returned to where the correlations exist in two directions,

X and Y, that cannot be followed simultaneously. As a consequence, this measurement is irreversible.

In summary, a measurement process is described in which the measurement system loses all information relevant to its past state. Noncommuting correlations are found, so an irreversible measurement will result.

G. The Measurement Chain and Repeated Measurements

A measurement act usually does not end when the interaction between S and M has been turned off.[8] Usually after the first measurement, a second measuring apparatus M′ is correlated to M and therefore to S. This procedure is repeated until it is eventually terminated at the apparatus M″. As an example, in the Stern Gerlach experiment, the first measurement act correlates the spin to the Z-coordinate. The measurement chain continues when the Z-coordinate is measured by scattered photon that are gathered on a photographic plate and so on. The types of correlations built up this process are examined. In the nondestructive measurement, after the initial stage has been completed the combined density operator is found to have the form:

$$\tilde{\rho}_{\mathrm{S}\otimes\mathrm{M}(t^*)}=\exp\left[2\gamma_{\mathrm{M}}(0)\tilde{\sigma}_{\mathrm{S}}^{Z}\tilde{\sigma}_{\mathrm{M}}^{Z}+\gamma_{\mathrm{S}}(0)\tilde{\sigma}_{\mathrm{S}}^{Z}+\lambda\right] \tag{3.27}$$

A similar apparatus M′ is added, with the same kind of interaction as M:

$$\tilde{H}_{\mathrm{MM}'}=g'(t)\left(\tilde{\sigma}_{\mathrm{M}'}^{X}-\tfrac{1}{2}\right)\left(\tilde{\sigma}_{\mathrm{M}}^{Z}-\tfrac{1}{2}\right) \tag{3.64}$$

and an initial state similar to (3.18) exists. The combined density operator becomes at time t defined by $\int_{t^*}^{t} g'(t'')\,dt''=\pi$

$$\tilde{\rho}_{\mathrm{S}\otimes\mathrm{M}\otimes\mathrm{M}'(t)}=\exp\left[4\gamma_{\mathrm{M}}(0)\gamma_{\mathrm{M}'}(0)\tilde{\sigma}_{\mathrm{S}}^{Z}\tilde{\sigma}_{\mathrm{M}}^{Z}\tilde{\sigma}_{\mathrm{M}'}^{Z}+2\gamma_{\mathrm{M}}(0)\tilde{\sigma}_{\mathrm{M}}^{Z}\tilde{\sigma}_{\mathrm{S}}^{Z}+\gamma_{\mathrm{S}}(0)\tilde{\sigma}_{\mathrm{S}}^{Z}+\lambda\right] \tag{3.65}$$

The new measurement adds the correlation represented by the operator $\tilde{\sigma}_{\mathrm{S}}^{Z}\tilde{\sigma}_{\mathrm{M}}^{Z}\tilde{\sigma}_{\mathrm{M}'}^{Z}$ without disturbing the previous correlations. Adding a sequence of additional apparatuses with similar interactions gives the combined density operator (in the product form):

$$\tilde{\rho}_{\mathrm{S}\otimes\mathrm{M}\ldots\otimes\mathrm{M}^{n}(t)}=\exp\left[\gamma_{\mathrm{S}}(0)\tilde{\sigma}_{\mathrm{S}}^{Z}+\lambda_{\mathrm{S}}\right]\prod_{i=1}^{n}\exp\left[\left(\prod_{k=1}^{n}2\gamma_{\mathrm{M}^{k}}(0)\tilde{\sigma}_{\mathrm{M}^{k}}^{Z}\right)\tilde{\sigma}_{\mathrm{S}}^{Z}+\lambda\right] \tag{3.66}$$

It should be noticed that the correlation is from the highest index to lower ones. The conditional density operator is:

$$\tilde{\rho}_{\mathrm{M}^n|\mathrm{S}\infty\ldots\mathrm{M}^{n-1}}=\exp\left\{\left[\prod_{k=1}^{n} 2\gamma_{\mathrm{M}^k}(0)\tilde{\sigma}^{Z}_{\mathrm{M}^k}\right]\tilde{\sigma}^{Z}+\lambda\right\} \tag{3.67}$$

The fact that the correlation is always down the index ladder makes it possible to terminate the measuring process at any point without changing the conclusions. The Lagrange parameters that represent the magnitude of the downward correlation are:

$$\mu^{Z}_{\mathrm{SM}}\cdots\mu^{Z}_{\mathrm{M}^n}=\prod_{k=1}^{n} 2\gamma_{\mathrm{M}^k}(0) \tag{3.68}$$

It is sufficient that one of the intermediate Lagrange parameters be zero for no correlation to exist between the system S and the last apparatus M^n.

The influence of the measurement chain on the destructive measurement is investigated next. The density operator of the combined system when S was initially polarized in the X direction is:

$$\tilde{\rho}_{\mathrm{S}\otimes\mathrm{M}(t^*)}=\exp\left[\gamma_{\mathrm{M}}(0)\tilde{\sigma}^{Z}_{\mathrm{S}}\tilde{\sigma}^{Z}_{\mathrm{M}}+\alpha_{\mathrm{S}}(0)\tilde{\sigma}^{X}_{\mathrm{S}}\tilde{\sigma}^{X}_{\mathrm{M}}+\lambda\right] \tag{3.69}$$

When the new apparatus M is attached with a similar interaction, one gets for the combined density operator:

$$\begin{aligned}\tilde{\rho}_{\mathrm{S}\otimes\mathrm{M}\otimes\mathrm{M}'(t)}=\exp\big[&4\gamma_{\mathrm{M}}(0)\gamma_{\mathrm{M}'}(0)\tilde{\sigma}^{Z}_{\mathrm{S}}\tilde{\sigma}^{Z}_{\mathrm{M}}\tilde{\sigma}^{Z}_{\mathrm{M}}{}'+4\alpha_{\mathrm{S}}(0)\tilde{\sigma}^{X}_{\mathrm{S}}\tilde{\sigma}^{X}_{\mathrm{M}}\tilde{\sigma}^{X}_{\mathrm{M}'}\\&+2\gamma_{\mathrm{M}}(0)\tilde{\sigma}^{Z}_{\mathrm{S}}\tilde{\sigma}^{Z}_{\mathrm{M}}+\gamma_{\mathrm{S}}(0)\tilde{\sigma}^{Z}_{\mathrm{S}}+\lambda\big]\end{aligned} \tag{3.70}$$

The important point in (3.70) is that the correlation $\tilde{\sigma}^{Z}_{\mathrm{S}}\tilde{\sigma}^{Z}_{\mathrm{M}}$ was not harmed in the process and the correlation $\tilde{\sigma}^{Z}_{\mathrm{S}}\tilde{\sigma}^{Z}_{\mathrm{M}}\tilde{\sigma}^{Z}_{\mathrm{M}}$ enables the measurement of S by M. However, when the correlation $\tilde{\sigma}^{X}_{\mathrm{S}}\tilde{\sigma}^{X}_{\mathrm{M}}$ on the X-axis between S and M disappears a triple correlation $\tilde{\sigma}^{X}_{\mathrm{S}}\tilde{\sigma}^{X}_{\mathrm{M}}\tilde{\sigma}^{X}_{\mathrm{M}}{}'$ emerges instead. When the measurement chain is continued, this correlation is transformed to a multicorrelation operator. In conclusion, the original information on the polarization in the X-axis of S is stored in a multicorrelation operator. To recover this information, a procedure is needed that requires a measurement process operating simultaneously on all the combined systems. It can be said that in a compound measurement this information is lost. The loss of bicorrelation when a third system interacts with a noncommuting interaction is commonly known as the loss of coherence.

In the preceding sections the irreversible character of the measurement is attributed to the failure to utilize all the information contained in the correlation when S and M are separated. When a measurement chain is applied, even the recovery of the lost information (and destruction of the measurement at the same time) becomes more difficult as the chain of measuring apparatuses grows larger.

IV. DISCUSSION

The examples of the QM measurements presented in the preceding sections cover the majority of the problems in the discrete case. The models presented are more general than the two-dimensional Hilbert space may suggest, the reason being that since the dynamics were described by a canonical density operator, they are dependent only on commutation relations, with the Hamiltonian of the operators used to define the state. If a system has commutation relations similar to that of the solved system, it likewise has similar dynamics with no dependence on the dimensionality of the Hilbert space. The difference between the systems is seen only when explicit expectations are calculated.

The two-dimensional model can be presented in a way that makes the generalization easier. The operator that represents the correlation can be written as follows:

$$\tilde{\sigma}_{\mathrm{S}}^{i}\tilde{\sigma}_{\mathrm{M}}^{j} = -\tfrac{1}{2}\left(\tilde{\sigma}_{\mathrm{S}}^{i}-\tilde{\sigma}_{\mathrm{M}}^{j}\right)^{2}+\tfrac{1}{4}I \tag{4.1}$$

In this form the conditional density operator is:

$$\tilde{\rho}_{\mathrm{M|S}} = \exp\left[-\frac{\mu_{\mathrm{SM}}^{ij}}{2}\left(\tilde{\sigma}_{\mathrm{S}}^{i}-\tilde{\sigma}_{\mathrm{M}}^{j}\right)^{2}+\lambda\right] \tag{4.2}$$

Such a description of the mapping emphasizes the overlap in the resolution. In the measurement of continuous variables (which is described in detail elsewhere[28]), even the ideal mapping overlaps. In the canonical form of the conditional density operator, one sees that the quality of the resolution depends on the Lagrange parameter μ. Through the equation of motion it is possible to identify the origin of each Lagrange parameter. In the nondestructive measurement it was found that origin of this Lagrange parameter is the Lagrange parameter that defines the initial state of the apparatus M. For a good resolution, therefore, large initial Lagrange parameters or an initial state with low temperature is required. An apparatus initially in equilibrium is useless as a measuring device.

The states used to register the results are another important consideration of mapping processes. It was found that to achieve a good resolution the mapping must be imbedded into states that are very close to being degenerate. The result is general and is caused by the fact that for the state of M to be correlated to the state of S, a generalized rotation must be carried from the initial state of M to the correlated one. If these states are not degenerate, the rotation from state to state is hindered by the precession motion caused by the energy difference of the registration state and the initial state of M.

In this work great emphasis is placed on tracking down the root of the irreversible character of the QM measurement. It is found that the cause of irreversability stems from the loss of information contained in the combined density operator when restrictions are placed on the representation.[29] This is because by definition the system S and M have to be separated in the final state. In mathematical form, when it is not possible to diagonalize the density operator by a basis set that is a tensor product of the eigenstates from S and M, the measurement is irreversible.[27] An important point in accounting for the entropy is to include information (as negative entropy) only when it has operational consequences. Correlations of the system with the past state and correlations that cannot be used because they do no commute with other correlations are not accounted for in the entropy relations.

Appendix A: The Conditional Density Operator

The conditional density operator is the means to define the most general mapping of states in quantum mechanics. A review of the qualities of the conditional density operator is presented.

1. The conditional density operator maps density operators into density operators:

$$\tilde{\rho}_{\mathrm{B}} = \operatorname*{tr}_{\mathrm{A}}\left(\tilde{\rho}_{\mathrm{B|A}} \cdot \tilde{\rho}_{\mathrm{A}}\right) \tag{A.1}$$

   Because density operators are positive with a unit trace, the conditional density operator is a trace-preserving positive map.
2. If the conditional density operator maps states from system A to B, the mapping is defined on the tensor product space of the Hilbert spaces of A and B. If explicit forms of the conditional density operator are considered, they have a double index system, one from A and one from B. When matrix elements are calculated, they have four indices, that is, they have two from each Hilbert space.
3. The conditional density operator obeys relations similar to the ones of conditional probability:

$$\operatorname*{tr}_{\mathrm{A}}\operatorname*{tr}_{\mathrm{B}}\left(\tilde{\rho}_{\mathrm{A|B}}\right) = 1 \tag{A.2}$$

4. The conditional density operator can be used to define dynamical mapping caused by the evolution of the system. If the motion can be generalized to be produced by a unitary transformation:

$$\tilde{\rho}_{S(t)} = \operatorname*{tr}_{S(0)} \left(\tilde{\rho}_{S(t)|S(0)} \cdot \tilde{\rho}_{S(0)} \right) = \operatorname*{tr}_{R} \left(\tilde{U}_{R \otimes S} \tilde{\rho}_{S(0)} \otimes \tilde{\rho}_{R(0)} \tilde{U}^{\dagger}_{R \otimes S} \right) \quad \text{(A.3)}$$

where R is the index of the reservoir. The conditional density operator is strictly positive as defined by Kraus.[30]

Conditional density operators are found in the literature under different names. The class that describes the mapping of systems into themselves is sometimes called "superoperators." For scattering processes these operators have been used by Snider and Sanctuarry[31] and Kafri and Kosloff.[32] For the description of dynamical mapping caused by the evolution of the system, an operator similar to the relative density operator has been used by Ben Reuven[33] and by Alhasid and Levine.[20] In the C^* literature and conditional density operator is usually referred to as a positive map.

Appendix B: Measurements Working at Capacity

In this appendix the analogy between the measurement process and the noisy communication channel is studied. From this point of view the system S is a source of a message that is transferred through the interaction (the channel) to the apparatus M, which acts as the receiver. Using this analogy we can benefit from the developments of the mathematical theory of communication.

To aid the reader, a brief summary of the main ideas of the theory is given. Suppose our message consists of the composition of an ensemble of N systems S. If the ensemble is pure it is sufficient that one letter describing the state of S is transmitted to M so that the composition of the whole ensemble is known. If the ensemble is completely random, to transmit the message each member needs its own letter describing its composition. In this example we see that the length of the message describing the composition of the ensemble varies from one letter to n letters, depending on the uncertainty of the initial ensemble. The question is how are messages with intermediate uncertainty transferred on the communication line without specifying a letter for each member. The way to compress the length of the message is by using a code. A sequence of ensembles are gathered together. If this sequence is common it receives a short code; infrequent sequences receive longer codes. If we let the length of the sequences grow there are sequences with a probability of occurence that tends to zero; therefore, we do not specify a code for improbable sequences. This description is the essence of the first coding theorem of communication theory,[17, 18] which states that there is a code in the case of an infinitely long sequence for which the average length of the message per member is given by:

$$\bar{n} = \frac{S}{\ln D} \tag{B.1}$$

where $\ln D$ is the size of the alphabet used to construct the code and S is the entropy of the ensemble.

The significance of this theorem from a thermodynamic point of view is that it shows the way to optimize a sorting procedure. The thermodynamic entropy is the average number of operations needed per member of the ensemble to sort it to its components. Up to this point the transmission from S to M is considered errorless. The next step is to consider the more common case when the mapping from S to M is not one to one. To illustrate we can consider a Stern Gerlach device that correlates the spin variable to the coordinate. The device separates an ensemble according to its spin component by use of a slit. Next consider a nonperfect device with errors in the sorting procedure. In the classification of Section II the device has a relative capacity of less than one. Under such circumstances, how is it possible to eliminate the error in the measurement using the device to sort out completely an ensemble into its components? The answer to this question is presented below.

The capacity functional is the maximum amount of mutual information that can be transferred to the apparatus. If in a transmission that is subject to error, the initial uncertainty of the measured object is higher than the capacity, this ensemble cannot be resolved completely. However, the object may be in a state of lower uncertainty than the capacity. How can such a state be resolved? An example may clear the situation. Consider the measurement model of Section III. For the relative capacity to be lower than one of the apparatuses the initial Lagrange parameter has to be finite. For the object an initial state is chosen with uncertainty that is less than the capacity:

$$\tilde{\rho}_{S(0)} = \exp\left[\gamma_S(0)\tilde{\sigma}_S^Z + \nu\right] \tag{B.2}$$

where γ is chosen such that

$$S(\tilde{\rho}_{S(0)}) \leq \ln 2 - S(\tilde{\rho}_{M(0)})$$

At this stage the initial ensemble of S system is divided arbitrarily into S and S′. An interaction is switched on between the two subensembles. The

Hamiltonian

$$\tilde{H}_{S\otimes S'} = g(t)\left(\tilde{\sigma}_S^Z - \tfrac{1}{2}\right)\cdot\left(\tilde{\sigma}_{S'} - \tfrac{1}{2}\right) \tag{B.3}$$

can produce a motion that causes correlations between the subensembles:

$$\tilde{\rho}_{S\otimes S'} = \exp\left[\gamma_S(0)\tilde{\sigma}_S^Z + \gamma_S(0)\tilde{\sigma}_S^Z\tilde{\sigma}_{S'}^Z + \lambda\right] \tag{B.4}$$

The correlations between the subensembles represented by the operator $\tilde{\sigma}_S^Z\tilde{\sigma}_{S'}^Z$ are at the expense of raising the entropy of the subensembles. In the Stern Gerlach setup this is equivalent to two streams of particles with each particle in one stream having its match in the other. These correlations are used to partly eliminate errors in the sorting procedure as follows. Each subensemble is measured separately and the results are matched. In the Stern Gerlach setup the slits are accompanied by a shutter and only when both spins point up (or down) does the shutter open on each slit. If the spins do not match, the result is rejected and remeasured. It is clear that by this method many of the errors are eliminated. Nonetheless, the possibility exists that both devices may give erroneous measurements. To eliminate such a possibility, a more sophisticated coding procedure is needed that correlates a long sequence of subensembles. Such a sophisticated procedure comprises three steps: (*1*) The initial ensemble is divided into N subensembles. (*2*) A unitary transformation through the introduction of an interaction Hamiltonian is applied to the whole ensemble that produces links between the subensembles as a result of their entropy increase. (*3*) Each subensemble is measured by a measuring device. The results are matched in a decoding procedure to eliminate errors. Such a procedure is equivalent to the second coding theorem of communication theory.[18] In this application when the number of subensembles grows to infinity there is a code (a unitary transformation) that will eliminate the errors, with the probability of an error being reduced to zero.

Acknowledgments

I am grateful to Professor R. D. Levine and Professor S. A. Rice for many suggestions and comments and to Dr. F. Novak for carefully reading the manuscript.

References

1. J. C. Maxwell, *Theory of Heat*, New ed., Longmans Green and Co., London, 1897.
2. L. Szilard, *Z. Phys.*, **53** 840 (1929).
3. L. Brillouin, *Science and Information Theory*, Academic, New York, 1956.
4. G. Lindblad, *Commun. Math. Phys.*, **33** 305 (1973); and G. Lindblad, *J. Stat. Phys.*, **11** 231 (1974).

5. See, for example, B. d'Espegnat, *Conceptual Foundations of Quantum Mechanics*, 2nd ed., Addison Wesley, Reading, 1976, or E. Shribe, *The Logical Foundation of Quantum Mechanics*, Pergamon, London 1973, or M. Jammer, *The Philosophy of Quantum Mechanics*, Wiley, New York, 1974.
6. W. Heisenberg, *Z. Phys.*, **43** 1 (1927).
7. N. Bohr, "Discussion with Einstein on Epistemological Problems in "Atomic Phisics," Chapter 7 of *Albert Einstein*; *Philosopher Scientist*, Vol. 7 of the *Library of Living Philosophers*, Evenston, IL, 1949 republished by Open Court, La Salle, Ill., 1970.
8. J. Von Neumann, *Mathematical Foundations of Quantum Mechanics*, Princeton University Press, 1955.
9. E. P. Wigner, *Am. J. Phys.*, **6** 31 (1973); E. P. Wigner, "Remarks on the Mind Body Question," J. J. Good, *The Scientist Speculates*, Basic Books, New York, Chap. 98, E. P. Wigner *Found. Phys.*, **6**, 539, (1976).
10. O. Costa de Beauregard, *Found. Phys.*, **6**, 539 (1976).
11. H. Margenau, *Ann. Phys.*, **23**, 469 (1963).
12. J. Park and H. Margenau, *Int J. Theor. Phys.* **1**, 211 (1968); J. Park, *Philos. Sci.*, **I**, 205; **II** 389 (1968).
13. P. A. Moldauer, *Phys. Rev.*, **D5** 1028 (1972).
14. A. Peress, *Am. J. Phys.* **42** 886 (1974).
15. F. J. Belinfante, *Measurement and Time Reversal in Objective Quantum Theory*, Pergamon, London, 1975.
16. H. Krips, *Found. Phys.*, **4**, 181 (1974); **4**, 381 (1974); **6**, 639 (1976).
17. C. E. Shanon, *Bell Syst. Tech. J.*, **27**, 379 (1948).
18. R. Ash, *Information Theory*, Interscience, New York, 1965.
19. G. Lindblad, *Commun. Math. Phys.*, **40**, 147 (1975).
20. Y. Alhasid and R. D. Levine, *Phys. Rev. A*, **18**, 89 (1978).
21. A. Einstein, B. Podolsky, and N. Rosen, *Phys. Rev.*, **47**, 777 (1935).
22. F. T. Jaynes, *Phys. Rev.*, **106**, 620 (1957); **108**, 171 (1957).
23. W. Band and J. Park, *Found. Phys.*, **1**, 133 (1970).
24. W. Magnus, *Commun. Pure Appl. Math.*, **7** 649 (1954).
25. A. Cohen, *Differential Equations*, 2nd ed., Heath, Boston, 1933.
26. J. F. Clauser and M. A. Horn, *Phys. Rev. D*, **10**, 526 (1974); S. J. Freedman and D. F. Clauser, *Phys. Rev. Lett.*, **23**, 880 (1972); J. F. Clauser, *Phys. Rev. Lett.*, **36**, 1223 (1976).
27. W. Band and J. Park, *Found. Phys.*, **8**, 667 (1978).
28. R. Kosloff, to be published.
29. L. B. Levitin, preprint (1977).
30. K. Kraus, *Ann. Phys.*, **64**, 331 (1971).
31. R. F. Snider and B. C. Sanctuarry, *J. Chem. Phys.*, **55**, 1555 (1971).
32. A. Kafri and R. Kosloff, *Chem. Phys.*, **23**, 257 (1977).
33. A. Ben Reuven, *Phys. Rev. A*, **4**, 2115 (1971).

PASSAGE FROM AN INITIAL UNSTABLE STATE TO A FINAL STABLE STATE

MASUO SUZUKI

Department of Physics, University of Tokyo, Hongo, Bunkyo-ku, Tokyo

CONTENTS

I. INTRODUCTION

Fluctuation and relaxation in far-from-equilibrium systems have been studied extensively by many authors.[1, 2] In particular, the relaxation process from the initial unstable state has been one of challenging problems in nonequilibrium statistical mechanics.[3, 4] Except for the relaxation from the very vicinity of the unstable point, fluctuation near the most probable path $y(t)$ can be treated in the Gaussian approximation.[3–5] That is, the probability distribution function $P(x,t)$ for a stochastic variable x takes the form

$$P(x,t)=\frac{1}{\sqrt{2\pi\varepsilon\sigma(t)}}\exp\left[-\frac{[x-y(t)]^2}{2\varepsilon\sigma(t)}\right] \tag{1.1}$$

for large system size Ω (or small ε; $\varepsilon=\Omega^{-1}$). If $y(t)$ approaches the final equilibrium or stationary value for $t\to\infty$, then the above Gaussian distribution function (1.1) becomes an asymptotically correct one to describe fluctuation near the final stationary state. This situation is discussed again

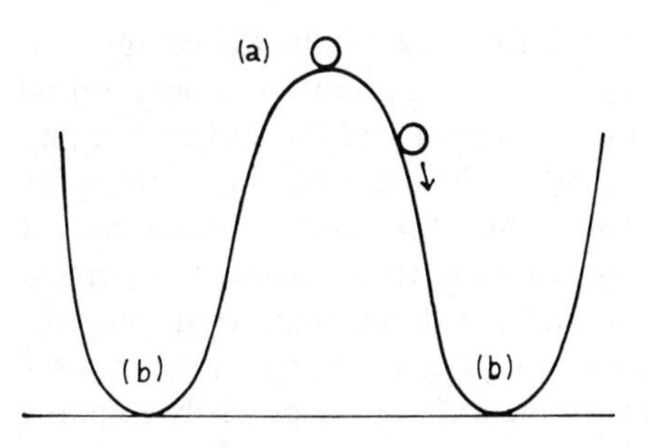

Fig. 1. Schematic explanation of relaxation from (a) the unstable point to (b) the final stable state.

in more detail in Section II. When the initial system is at (or near) the unstable equilibrium state, the above Gaussian treatment cannot be valid and the non-Gaussian property becomes very important. Owing to this nonlinearity, the formation of macroscopic order becomes possible. In fact, the most probable path $y(t)$ does not change in time, if the system starts from the initial unstable equilibrium point. The Gaussian distribution (1.1) with $y(t) \equiv x(0) = y$ (unstable) is a good approximation only in the initial time region, and it cannot describe the relaxation process in an intermediate time region. Thus a nonlinear theory of relaxation from the initial unstable state is necessary and the scaling theory discussed in detail in this chapter is one of such nonlinear theories to describe the transient phenomena near the unstable point, as shown in Fig. 1.

Although several methods have been proposed to treat the above transient nonlinear phenomena, the scaling theory seems to be the simplest and most general approach near the instability point. The essence of the scaling theory is to make an asymptotic evaluation of transient fluctuation, by dividing the time region into three regimes, as shown in Fig. 2, namely, the initial regime in which the Gaussian approximation is valid, the second, nonlinear, scaling regime in which an asymptotic scaling evaluation is substantial, and the remaining regime in which other asymptotic evaluation

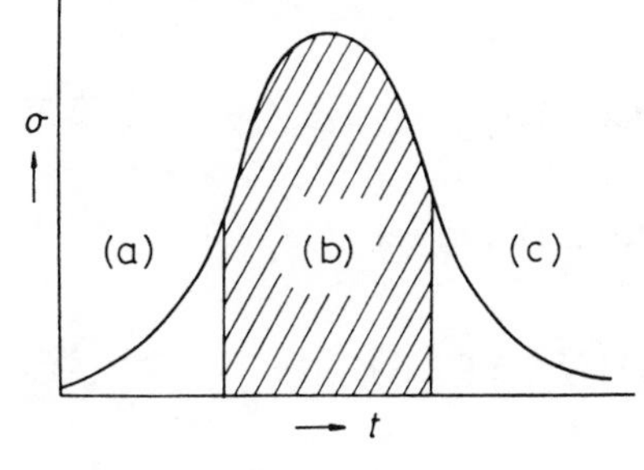

Fig. 2. Division of the whole time region into the three regimes: (a) initial regime, (b) scaling regime, (c) final regime, where σ denotes fluctuation for $\delta \leq \varepsilon^{\mu}$.

methods are required; in particular, in the final regime the Gaussian treatment is also valid. In the initial regime, the random force and initial fluctuation are very important, and the nonlinearity of the system is unimportant. On the other hand, in the second, scaling regime, the nonlinearity of the system plays an important role, while the random force can be asymptotically neglected for a large system size. In this second, nonlinear regime, the macroscopic order sets in, and the onset time lies in this regime, as is discussed in detail in later sections. The division of time into the three regimes is conceptually important and it is not necessarily required technically, as is discussed explicitly later.

II. Ω-EXPANSION AND ANOMALOUS FLUCTUATION THEOREM NEAR THE UNSTABLE POINT

A. Ω-Expansion

As is mentioned in Section I, the Gaussian approximation (1.1) is useful in a normal situation except at the instability point. It is instructive to review here the ordinary Ω-expansion method.[3, 5] Following van Kampen,[5] we separate the stochastic variable into the following two parts:

$$x=y(t)+\sqrt{\varepsilon}\,\xi \tag{2.1}$$

where $y(t)$ denotes the deterministic path and ξ is the remaining fluctuating part. The existence of the smallness parameter $\sqrt{\varepsilon}$ in the second part of (2.1) corresponds to the central limit theorem.

Let us explain the Ω-expansion method for the following Kramers-Moyal expansion

$$\varepsilon\frac{\partial}{\partial t}P(x,t)=\sum_{n=1}^{\infty}\frac{(-\varepsilon)^n}{n!}\frac{\partial^n}{\partial x^n}c_n(x)P(x,t) \tag{2.2}$$

where $c_n(x)$ is the nth moment of the transition probability $w(x,r)$ defined by

$$c_n(x)=\int r^n w(x,r)\,dr \tag{2.3}$$

We transform the variable x according to (2.1) and set

$$\Pi(\xi,t)=P(x,t)=P\big(y(t)+\sqrt{\varepsilon}\,\xi,t\big) \tag{2.4}$$

Then, note that

$$\frac{\partial}{\partial t}\Pi(\xi,t)=\frac{\partial}{\partial t}P\big(y(t)+\sqrt{\varepsilon}\,\xi,t\big)$$
$$=\dot{y}(t)\frac{\partial}{\varepsilon^{1/2}\partial\xi}\Pi(\xi,t)+\frac{\partial}{\partial t}P(x,t) \tag{2.5}$$

Thus from (2.2) we obtain

$$\varepsilon\frac{\partial}{\partial t}\Pi(\xi,t)-\dot{y}(t)\varepsilon^{1/2}\frac{\partial}{\partial\xi}\Pi(\xi,t)$$
$$=-c_1(y)\varepsilon^{1/2}\frac{\partial}{\partial\xi}\Pi-\varepsilon c_1'(y)\frac{\partial}{\partial\xi}(\xi\Pi)+\frac{\varepsilon}{2}c_2(y)\frac{\partial^2\Pi}{\partial\xi^2}+O(\varepsilon^{3/2}) \tag{2.6}$$

By equating like powers of both sides in (2.6), we obtain $\dot{y}(t)=c_1(y(t))$ and

$$\frac{\partial}{\partial t}\Pi(\xi,t)=-c_1'(y(t))\frac{\partial}{\partial\xi}(\xi\Pi)+\tfrac{1}{2}c_2(y(t))\frac{\partial^2}{\partial\xi^2}\Pi \tag{2.7}$$

That is, the probability distribution function Π of the fluctuating variable ξ satisfies the linear Fokker-Planck equation (2.7) with time-dependent coefficients. In other words, the fluctuating part ξ is Gaussian in this asymptotic limit. The solution of (2.7) is given[5] by

$$\Pi(\xi,t)=\frac{1}{\sqrt{2\pi\varepsilon\sigma(t)}}\exp\left[-\frac{\xi^2}{2\sigma(t)}\right] \tag{2.8}$$

for an initial Gaussian distribution with the variance σ_0, where $\sigma(t)$ is the solution of the equation

$$\frac{d}{dt}\sigma(t)=2c_1'(y(t))\sigma(t)+c_2(y(t)) \tag{2.9}$$

Thus the deterministic path $y(t)$ and its variance are determined explicitly in this Ω-expansion approach. The variance $\sigma(t)$ is explicitly given by

$$\sigma(t)=\sigma_0\left[\frac{c_1(y(t))}{c_1(y(0))}\right]^2+c_1^2(y(t))\int_{y(0)}^{y}\frac{c_2(y')}{c_1(y')^3}dy' \tag{2.10}$$

B. Anomalous Fluctuation Theorem and Fluctuation Intensity Relation

It is easily shown[1, 2, 6] from (2.10) that the variance (2.10) takes the following asymptotic form

$$\sigma(t) \simeq (\sigma_0 + \sigma_1) \frac{c_1^2(y(t))}{\gamma^2 \delta^2} \tag{2.11}$$

for large time t of the order

$$t \sim \frac{1}{\gamma} \log\left(\frac{1}{\delta}\right) \tag{2.12}$$

where δ is a deviation of the initial system from the unstable point (i.e., $y_0 = x_0 + \delta$), and $\sigma_1 = c_2(x_0)/(2\gamma)$. Note also that $c_1(y(0)) \cong \gamma\delta$. Here x_0 is an unstable point and $\gamma = c_1'(x_0)$, the growing rate at the unstable point. The maximum value of the variance $\sigma(t)$ is given by

$$\sigma_m = \frac{\sigma_0 + \sigma_1}{\delta^2 \gamma^2} c_1^2(y_m) \sim \frac{1}{\delta^2}; \qquad c_1'(y_m) = 0 \tag{2.13}$$

That is, the enhancement factor R is proportional to δ^{-2}. Namely, the fluctuation is enhanced anomalously around the time (2.12) as the initial system approaches the instability point x_0. This is the so-called *anomalous fluctuation theorem*,[1, 2, 6] and was first discussed by Kubo et al.,[3] who gave simple examples. This result also can be obtained intuitively as follows. First we linearize the equation $\dot{y} = c_1(y)$ around the unstable point x_0 as

$$\frac{d}{dt}(\delta y) = \gamma(\delta y); \qquad \delta y = y(t) - x_0 \tag{2.14}$$

and we simplify (2.9) to

$$\frac{d}{dt}\sigma_l(t) = 2\gamma\sigma_l(t) + c_2(x_0) \tag{2.15}$$

near the unstable point (i.e., $y(t) \simeq x_0$). The solution of (2.14) is given by

$$\delta y = \delta e^{\gamma t}; \qquad \delta \equiv (\delta y)_{t=0} = y_0 - x_0 \tag{2.16}$$

This linear approximation holds only when $|\delta y| \lesssim \Delta$ with a certain constant Δ much less than the distance between the unstable point and the final stable point. In the time region $\delta y \simeq \Delta$, that is, $t \sim t_1 \equiv \gamma^{-1} \log(\Delta/\delta)$, the linear

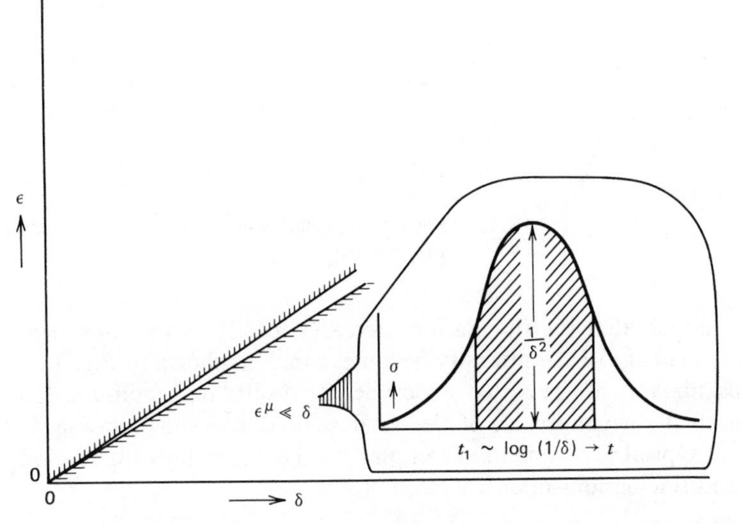

Fig. 3. Qualitative feature of the anomalous fluctuation in the extensive region.

variance becomes anomalously large as follows

$$\sigma_l(t_1)=(\sigma_0+\sigma_1)e^{2\gamma t_1}-\sigma_1 \simeq (\sigma_0+\sigma_1)\frac{\Delta^2}{\delta^2} \sim \frac{1}{\delta^2} \tag{2.17}$$

unless $(\sigma_0+\sigma_1)$ is vanishing. However, the saturation effect cannot be explained in this linear approximation. It comes from the nonlinear effect neglected in the above approximation. The above anomalous fluctuation theorem is shown schematically in Fig. 3.

This anomalous fluctuation theorem can also be expressed more physically. First note that $c_1(y(t))=\dot{y}(t)$ in (2.11), and that $y(0)\simeq\gamma\delta+O(\delta^2)$. For convenience, we call $y(t)$ intensity and write it as $I(t)$. Then the anomalous fluctuation theorem (2.11) can be expressed as the *fluctuation–intensity relation:* The fluctuation $\sigma_I(t)$ of an intensity $I(t)$ is proportional to the square of the change velocity $\dot{I}(t)$ normalized by the initial value $\dot{I}(0)$

$$\sigma_I(t) \propto \left[\frac{\dot{I}(t)}{\dot{I}(0)}\right]^2 \tag{2.18}$$

in the intermediate time region of the order $t \sim \log(1/\delta)$.

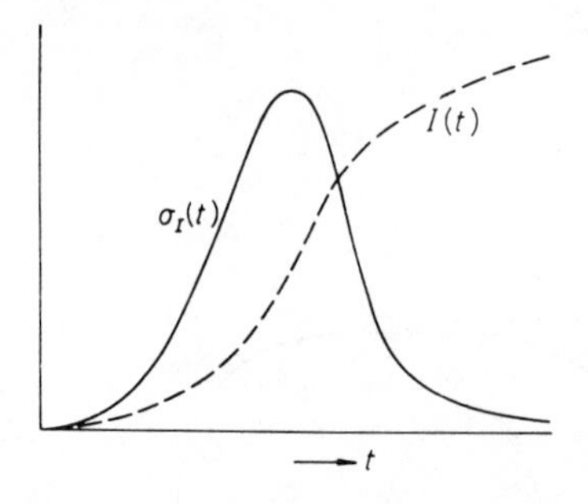

Fig. 4. Explanation of the intensity $I(t)$–fluctuation $\sigma(t)$ relation.

It is natural that the fluctuation increases mostly in the time region at which the change of the intensity becomes rapid, as shown in Fig. 4.

Kabashima et al.[7] reported experimental results on amplitude fluctuations in the transient process of electrical oscillation as shown in Fig. 5. This is one of typical experimental examples for the anomalous fluctuation theorem and fluctuation–intensity relation.

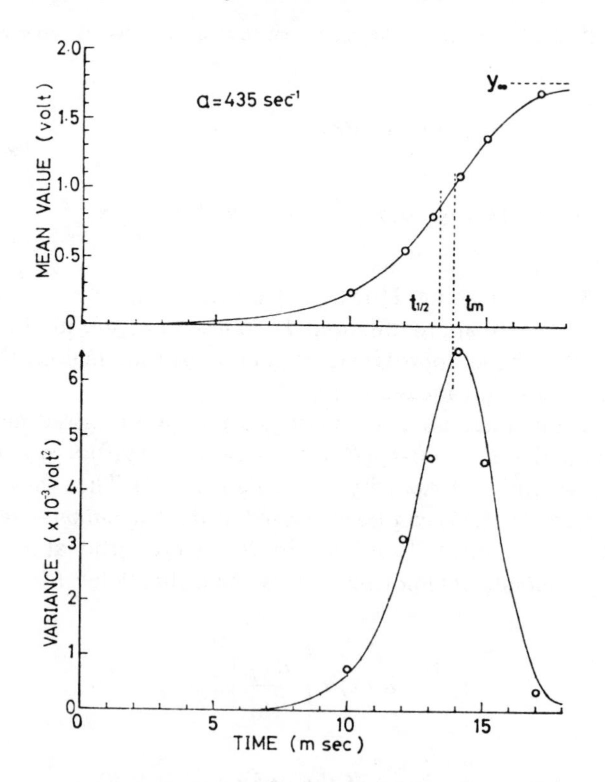

Fig. 5. Evolution of voltage amplitude and time dependence of the variance. (7)

This feature of anomalous fluctuation in the relaxation near the unstable point is expected to be quite general and to appear for many phenomena in various fields associated with instabilities.

III. DYNAMIC MOLECULAR FIELD THEORY IN LANGEVIN'S EQUATION (OR NONLINEAR BROWNIAN MOTION) AND THE FOKKER-PLANCK EQUATION

In this section, we explain the simplest self-consistent treatment of stochastic processes, that is, the self-consistent linearization of nonlinear Langevin's equation and the Fokker-Planck equation. This may be called the dynamic molecular field theory of stochastic process.[2, 8, 9] Even this simplest treatment is helpful in understanding qualitatively the importance of synergism (or cooperative effect) of the initial fluctuation, random force, and nonlinearity of the system for the formation of macroscopic order.

A. Dynamic Molecular Field Theory in Langevin's Equation

To explain the essential point of our method, we discuss here the following typical example of nonlinear Brownian motion

$$\frac{d}{dt}x(t)=\gamma x(t)-gx^3(t)+\eta(t) \tag{3.1}$$

where $\gamma>0$, $g>0$, and $\eta(t)$ is a Gaussian random force satisfying the relation

$$\langle\eta(t)\eta(t')\rangle=2\varepsilon\delta(t-t') \tag{3.2}$$

with the delta function $\delta(t)$. The positivity of γ corresponds to the fact that our system is unstable near $x=0$. Our method can also be applied to other normal situations. Here g is a parameter of nonlinearity and ε denotes a strength of random force, which is usually proportional to the inverse system size Ω^{-1}. The above three parameters, γ, g, and ε, may not be independent, but are related to one another in the equilibrium or stationary state through the fluctuation–dissipation relation.[10] As we are now interested in far-from-equilibrium transient phenomena, we regard them as independent parameters.

The simplest self-consistent linearization of (3.1) may be to replace the nonlinear term $x^3(t)$ in (3.1) by $x(t)\,\langle x^2(t)\rangle$, where the average $\langle x^2(t)\rangle$ is determined self-consistently using the solution of the linearized Langevin equation. The characteristic point of our method is that our simplified equation still contains the random force. Otherwise, our dynamic molecular field theory might not describe the formation process of macroscopic order from the unstable disordered state. Now the equation thus linearized is

$$\frac{d}{dt}x(t)=\gamma(t)x(t)+\eta(t) \tag{3.3}$$

where

$$\gamma(t)=\gamma-g\langle x^2(t)\rangle \tag{3.4}$$

The formal solution of (3.3) is easily given by

$$x(t)=\exp\left[\int_0^t\gamma(t')dt'\right]\cdot\left[\int_0^t\eta(t')\exp\left(-\int_0^{t'}\gamma(s)ds\right)dt'+x(0)\right] \tag{3.5}$$

From (3.5), we have

$$\begin{aligned}\langle x^2(t)\rangle &=\exp\left[2\int_0^t\gamma(t')dt'\right]\left\langle\left[\int_0^t\eta(t')\exp\left(-\int_0^{t'}\gamma(s)ds\right)dt'+x(0)\right]^2\right\rangle\\ &=\exp\left(2\int_0^t\gamma(t')dt'\right)\times\left[\langle x^2(0)\rangle+\int_0^t dt_1\int_0^t dt_2\langle\eta(t_1)\eta(t_2)\rangle\right.\\ &\qquad\left.\times\exp\left(-\int_0^{t_1}\gamma(s)ds-\int_0^{t_2}\gamma(s)ds\right)\right]\\ &=\exp\left[2\int_0^t\gamma(t')dt'\right]\left[\langle x^2(0)\rangle+2\varepsilon\int_0^t dt'\exp\left(-2\int_0^{t'}\gamma(s)ds\right)\right]\end{aligned} \tag{3.6}$$

Here we have used the relation (3.2). Equation (3.6) seems to be a very complicated integral equation. However, this can be easily simplified as follows. First differentiate (3.6) with respect to t. Then, setting $f(t)\equiv\langle x^2(t)\rangle$, we obtain

$$\frac{d}{dt}f(t)=2\gamma(t)f(t)+2\varepsilon \tag{3.7}$$

That is,

$$\frac{d}{dt}f(t)=2(\gamma-gf(t))f(t)+2\varepsilon \tag{3.8}$$

This can be easily solved as follows. First, set

$$(\gamma - gf)f + 2\varepsilon \equiv g(\alpha - f)(f - \beta) \tag{3.9}$$

with $\alpha > \beta$. Thus, integrating (3.8), we obtain

$$\log \frac{f-\beta}{\alpha - f} = 2g(\alpha - \beta)t + \log C \tag{3.10}$$

Here C is a constant. That is,

$$f(t) = \frac{\alpha C \exp[2(\alpha - \beta)gt] + \beta}{1 + C\exp[2(\alpha - \beta)gt]} \tag{3.11}$$

The constant C is determined from the initial condition to be

$$C = \frac{f(0) - \beta}{\alpha - f(0)} \tag{3.12}$$

The roots α and β are given explicitly as

$$\begin{pmatrix} \alpha \\ \beta \end{pmatrix} = \frac{1}{2g}\left[\gamma \pm (\gamma^2 + 4\gamma\varepsilon)^{1/2}\right] \tag{3.13}$$

As we are interested in the asymptotic behavior of $f(t)$ for small ε and small $f(0)$, we make an asymptotic evaluation of the solution (3.10) in the limit of small ε and $f(0)$. Note that $\alpha \simeq \gamma/g$, $\beta \simeq -\varepsilon/g$, and

$$C \simeq \frac{g}{\gamma}\left(\langle x^2(0)\rangle + \frac{\varepsilon}{\gamma}\right) \tag{3.14}$$

Thus we arrive at the final simple form[8, 9]

$$\langle x^2(t)\rangle = \langle x^2\rangle_{\text{st}} \cdot \frac{\tau}{1+\tau}; \qquad \tau = \frac{g}{\gamma}\left(\langle x^2(0)\rangle + \frac{\varepsilon}{\gamma}\right)e^{2\gamma t} \tag{3.15}$$

for small ε and $\langle x^2(0)\rangle$ and for large t, where $\langle x^2\rangle_{\text{st}} = \gamma/g + O(\varepsilon)$. This solution has a very interesting feature, namely, the fluctuation $\langle x^2(t)\rangle$ is expressed only by the so-called scaling variable τ in the intermediate time region.[2, 9, 11]

Before discussing the physical consequences of the above solution, we give here an alternative derivation of the above self-consistent equation (3.8). By multiplying both sides of (3.3) by $2x(t)$ and by taking the average, we obtain

$$\frac{d}{dt}\langle x^2(t)\rangle = 2\gamma(t)\langle x^2(t)\rangle + 2\langle x(t)\eta(t)\rangle \tag{3.16}$$

Now let us evaluate $\langle x(t)\eta(t)\rangle$ in the linear approximation for $x(t)$:

$$\frac{d}{dt}x(t) = \gamma x(t) + \eta(t) \tag{3.17}$$

This yields

$$\begin{aligned}\langle x(t)\eta(t)\rangle &= \left\langle \eta(t)e^{\gamma t}\left[\int_0^t e^{-\gamma t'}\eta(t')dt' + x(0)\right]\right\rangle \\ &= e^{\gamma t}\int_0^t e^{-\gamma t'}\langle\eta(t)\eta(t')\rangle dt' \\ &= 2\varepsilon e^{\gamma t}\int_0^t e^{-\gamma t'}\delta(t-t')dt' = \varepsilon \end{aligned} \tag{3.18}$$

The integration on the delta function in (3.18) is a half-side one, and it gives $\frac{1}{2}$. The above result, (3.18), that is $\langle x(t)\eta(t)\rangle = \varepsilon$, is also easily obtained by using the self-consistent formal solution (3.5). Thus we arrive again at (3.7) or (3.8).

Now we discuss the physical consequence of the solution (3.15). In the usual situation, the initial fluctuation $\langle x^2(0)\rangle$ is of the order ε. Therefore, the scaling variable τ is of the order ε in the initial time region. Thus the average value $\langle x^2(t)\rangle$ changes from the order ε to the order of unity (or $\langle x^2\rangle_{st}$). That is, $\langle x^2(t)\rangle$ expresses simple fluctuation of $x(t)$ for small t, but it shows gradually the characteristic feature of order parameter in the intermediate time region $\tau \sim 1$. This enhancement of fluctuation (of the order $\varepsilon^{-1} = \Omega$) can be regarded as the formation of macroscopic order or structure. The time for this onset of macroscopic order is given by

$$t_0 \simeq \frac{1}{2\gamma}\log\left\{\frac{g}{\gamma}\left[\langle x^2(0)\rangle + \frac{\varepsilon}{\gamma}\right]\right\}^{-1} \tag{3.19}$$

This onset time becomes larger and larger, as the nonlinearity g and the sum of initial fluctuation $\langle x^2(0)\rangle$ and strength of random force (ε/γ) become small. This is called[1, 2, 8, 9] *synergism* of nonlinearity, initial fluctuation, and random force in the formation process of macroscopic order. The functional form of $\langle x^2(t)\rangle$ with respect to the scaling variable τ depends on approximations used to derive it, but the above synergism always appear in any approximation, and the time scale is always given by (3.19). The above self-consistent solution (3.15) gives an asymptotically correct value of $\langle x^2(t)\rangle$ for infinite time. Namely, this describes qualitatively a global feature of the system and can be applied to many other stochastic processes.

B. Self-Consistent Treatment in the Fokker-Planck Equation

It is also instructive to discuss the above self-consistent method in the Fokker-Planck equation, in which not only the moment $\langle x^2(t)\rangle$, but also the distribution function itself, can be obtained self-consistently. The Fokker-Planck equation corresponding to the Langevin equation (3.1) is given by

$$\frac{\partial}{\partial t}P=-\frac{\partial}{\partial x}(\gamma x-gx^3)P+\varepsilon\frac{\partial^2 P}{\partial x^2} \tag{3.20}$$

Here $P=P(x,t)$ denotes the probability of having x at time t. We linearize the drift term of (3.20) and we may rewrite (3.20) as

$$\frac{\partial P}{\partial t}=-\frac{\partial}{\partial x}(\gamma(t)xP)+\varepsilon\frac{\partial^2 P}{\partial x^2} \tag{3.21}$$

in our self-consistent approximation, where

$$\gamma(t)=\gamma-g\langle x^2(t)\rangle \tag{3.22}$$

The average $\langle x^2(t)\rangle$ is defined by

$$\langle x^2(t)\rangle=\int_{-\infty}^{\infty}x^2P(x,t)\,dx \tag{3.23}$$

with the solution $P(x,t)$ of (3.21). Thus (3.21), (3.22), and (3.23) are coupled self-consistent equations. The formal solution of (3.21) with an initial Gaussian distribution is given by

$$P(x,t)=\frac{1}{\sqrt{2\pi f(t)}}\exp\left[-\frac{x^2}{2f(t)}\right] \tag{3.24}$$

and $f(t)$ satisfies the equation

$$\frac{d}{dt}f(t)=2\gamma(t)f(t)+2\varepsilon \tag{3.25}$$

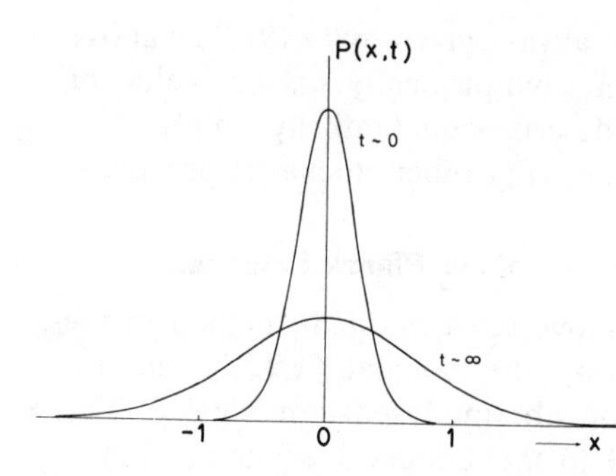

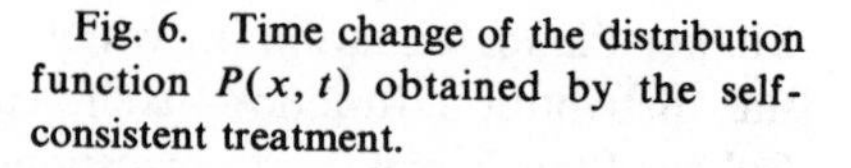
Fig. 6. Time change of the distribution function $P(x, t)$ obtained by the self-consistent treatment.

This is nothing but (3.7). The above treatment is thus completely equivalent to that provided by Langevin's equation, as it should be.

Here it should be noted that the above probability distribution function does not approach the final correct one, because the above approximate $P(x, t)$ remains always Gaussian, as shown in Fig. 6. That is, strictly speaking, this self-consistent linearized approximation cannot describe the onset process of macroscopic order. It gives only qualitative information about fluctuation that becomes anomalously large around the onset time, as is discussed earlier.

The asymptotically exact treatment of transient phenomena near the instability point to describe the formation process of macroscopic order is presented in the succeeding sections.

IV. MOST DOMINANT TERMS IN THE Ω-EXPANSION AND SCALING PROPERTY

As is discussed in the preceding section, the nonlinear terms in ε or g are very important to describe the relaxation from the unstable state. For example, the result (3.14) obtained by the molecular field theory can be expanded to

$$\langle x^2(t)\rangle = \langle x^2\rangle_{st}(\tau - \tau^2 + \tau^3 - \cdots) \tag{4.1}$$

where

$$\tau = \left(\frac{g\varepsilon}{\gamma^2}\right)\exp(2\gamma t) \tag{4.2}$$

Here for simplicity we have assumed that the initial value $\langle x^2(0)\rangle$ is zero. That is, the above result is a partial summation of the higher order terms in a perturbational expansion in ε or g. What kind of partial summation is it?

To answer this question and to explain our scaling idea, we study here perturbational expansions of the solution of the typical nonlinear Langevin equation (3.1):

$$\frac{d}{dt}x(t)=\gamma x(t)-gx^3(t)+\eta(t) \tag{4.3}$$

This is easily transformed into the integral equation

$$x(t)=e^{\gamma t}\int_0^t e^{-\gamma t'}\left[\eta(t')-gx^3(t')\right]dt' \tag{4.4}$$

By iterating this integral equation in g, we obtain

$$\begin{aligned}x(t)=&e^{\gamma t}\int_0^t e^{-\gamma t'}\eta(t')dt'-ge^{\gamma t}\int_0^t e^{-\gamma t'}e^{3\gamma t'}dt'\\&\times\int_0^{t'}dt_1\int_0^{t'}dt_2\int_0^{t'}dt_3e^{-\gamma(t_1+t_2+t_3)}\eta(t_1)\eta(t_2)\eta(t_3)\\&+\cdots\end{aligned} \tag{4.5}$$

Therefore, the fluctuation $\langle x^2(t)\rangle$ is given by

$$\begin{aligned}\langle x^2(t)\rangle=&e^{2\gamma t}\int_0^t dt_1\int_0^t dt_2e^{-\gamma(t_1+t_2)}\langle\eta(t_1)\eta(t_2)\rangle\\&-2ge^{2\gamma t}\int_0^t e^{-\gamma t_4}dt_4\int_0^t e^{2\gamma t'}dt'\int_0^{t'}dt_1\int_0^{t'}dt_2\int_0^{t'}dt_3e^{-\gamma(t_1+t_2+t_3)}\\&\times\langle\eta(t_1)\eta(t_2)\eta(t_3)\eta(t_4)\rangle+\cdots\\=&2\varepsilon e^{2\gamma t}\int_0^t dt' e^{-2\gamma t'}\\&-24\varepsilon^2ge^{2\gamma t}\int_0^t e^{-\gamma t_4}dt_4\int_0^t e^{2\gamma t'}dt'\int_0^{t'}dt_1\int_0^{t'}dt_2\\&\times\int_0^{t'}e^{-\gamma(t_1+t_2+t_3)}\delta(t_1-t_2)\delta(t_3-t_4)\\=&\frac{\varepsilon}{\gamma}(e^{2\gamma t}-1)-3\cdot\frac{g\varepsilon^2}{\gamma^3}(e^{4\gamma t}-4\gamma te^{2\gamma t}-1)+O(\varepsilon^3)\end{aligned} \tag{4.6}$$

The factor 3 in the second term in (4.6) comes from the factorial number in decoupling $\langle\eta(t_1)\eta(t_2)\eta(t_3)\eta(t_4)\rangle$ due to the Wick theorem. Similarly, we can calculate higher order terms in principle, but it is very complicated. The above result (4.6) is rewritten as

$$\langle x^2(t)\rangle = \langle x^2\rangle_{st}\cdot\left[\left(\frac{\varepsilon g}{\gamma^2}\right)(e^{2\gamma t}-1)-\frac{3\varepsilon^2 g^2}{\gamma^4}(e^{4\gamma t}-4\gamma t e^{2\gamma t}-1)+\cdots\right] \tag{4.7}$$

with $\langle x^2\rangle_{st}=\gamma/g$. For small g and large t, the following series becomes most dominant:

$$\langle x^2(t)\rangle \simeq \langle x^2\rangle_{st}\cdot\left[\left(\frac{g\varepsilon}{\gamma^2}e^{2\gamma t}\right)-3\left(\frac{g\varepsilon}{\gamma^2}e^{2\gamma t}\right)^2+\cdots\right] \tag{4.8}$$

If we compare this series with the result (4.1) obtained by the molecular field theory, the numerical factor of the second term in (4.1) already differs from the correct one (4.8). By studying higher order terms perturbationally in a similar way, we may find that the most dominant term in the nth order in $\langle x^2(t)\rangle$ is given by

$$\langle x^2\rangle_{st}\cdot(-1)^{n-1}(2n-1)!!\tau^n; \qquad \tau=\frac{g\varepsilon}{\gamma^2}e^{2\gamma t} \tag{4.9}$$

The systematic derivation of this result is presented in the next section. Thus by summing up the most dominant terms in the perturbational expansion, we obtain

$$\langle x^2(t)\rangle \simeq \langle x^2\rangle_{st}\cdot\sum_{n=1}^{\infty}(-1)^{n-1}(2n-1)!!\tau^n \tag{4.10}$$

This is an asymptotic series, so that we apply the Borel sum in the following. First note that

$$\frac{1}{\sqrt{2\pi}}\int_{-\infty}^{\infty}e^{-\xi^2/2}\xi^{2n}\,d\xi=(2n-1)!! \tag{4.11}$$

Using this formula, we rewrite (4.10) as

$$\langle x^2(t)\rangle/\langle x^2\rangle_{st}\simeq\sum_{n=1}^{\infty}(-1)^{n-1}\frac{1}{\sqrt{2\pi}}\int_{-\infty}^{\infty}e^{-\xi^2/2}\xi^{2n}\tau^n\,d\xi \tag{4.12}$$

The spirit of the Borel sum is to make the summation over n before integration. Thus we obtain

$$\langle x^2(t)\rangle/\langle x^2\rangle_{st} \simeq \frac{1}{\sqrt{2\pi}}\int_{-\infty}^{\infty} e^{-\xi^2/2}\sum_{n=1}^{\infty}(-1)^{n-1}(\xi^2\tau)^n\,d\xi$$

$$= \frac{1}{\sqrt{2\pi}}\int_{-\infty}^{\infty} e^{-\xi^2/2}\frac{\xi^2\tau}{1+\xi^2\tau} \tag{4.13}$$

For more details of the Borel sum, see Appendix A. The above result (4.13) has a very interesting property, that is, the scaling property that the normalized fluctuation $\langle x^2(t)\rangle/\langle x^2\rangle_{st}$ is expressed only by the scaling variable τ. The above integral representation (4.13) is a convenient expression to calculate the fluctuation $\langle x^2(t)\rangle$ explicitly. It should be noted that (4.13) approaches asymptotically a correct stationary value $\langle x^2\rangle_{st}$ for $t\to\infty$ (i.e., $\tau\to\infty$). For $\tau\simeq 1$, we have $\langle x^2(t)\rangle \simeq \frac{1}{2}\langle x^2\rangle_{st}$. Thus the onset time t_0 is again determined by the same equation (3.19) with $\langle x^2(0)\rangle = 0$;

$$t_0 \simeq \frac{1}{2\gamma}\log\left(\frac{g\varepsilon}{\gamma^2}\right)^{-1} \tag{4.14}$$

The above treatment can be easily extended to a more general case $\langle x^2(0)\rangle \neq 0$ and to other general transient phenomena near the instability point. However, perturbational expansions are not practical. A systematic compact method, namely, the scaling theory is much more convenient, as is discussed in later sections.

V. SCALING THEORY OF NONLINEAR BROWNIAN MOTION

A. Scaling Theory—Time-Independent Nonlinear Transformation

Here we present a systematic derivation of the scaling expression by using a nonlinear transformation of a stochastic variable.

Now we consider the following general nonlinear Brownian motion of a single variable x:

$$\frac{d}{dt}x = c_1(x) + \eta(t) \tag{5.1}$$

Let $x=0$ be the instability point of this system; $c_1(0)=0$ and $\gamma \equiv c_1'(0) > 0$. To make an asymptotic evaluation of the solution of (5.1), we make use of the following nonlinear transformation.[2, 8, 9, 12]

$$\xi = F(x) = \exp\left[\int_{a_0}^{x} \frac{\gamma}{c_1(y)}\,dy\right] \tag{5.2}$$

where a_0 is determined so that $F'(0)=1$. Then (5.1) is transformed in the form

$$\frac{d}{dt}\xi = \gamma\xi + [1+f(\xi)]\eta(t) \tag{5.3}$$

where

$$f(\xi) \equiv \frac{\gamma\xi}{c_1(F^{-1}(\xi))} - 1; \qquad f(0)=0 \tag{5.4}$$

This is also transformed into the following integral equation

$$\xi(t) = e^{\gamma t}\left\{\int_0^t e^{-\gamma t'}[1+f(\xi(t'))]\eta(t')\,dt' + \xi(0)\right\} \tag{5.5}$$

It is remarkable that the scaling solution (i.e., asymptotically exact solution in the limit $\varepsilon \to 0$ and $t \to \infty$ for fixed τ) is obtained by neglecting the term $f(\xi(t'))$ in (5.5), that is, it is given by

$$\xi_{\mathrm{sc}}(t) = e^{\gamma t}\int_0^t e^{-\gamma t'}\eta(t')\,dt' + e^{\gamma t}\xi(0) \tag{5.6}$$

The average $\langle \xi_{\mathrm{sc}}^2(t) \rangle$ is easily calculated as

$$\langle \xi_{\mathrm{sc}}^2(t) \rangle = \left[\langle \xi^2(0) \rangle + \frac{\varepsilon}{\gamma}\right] e^{2\gamma t} - \frac{\varepsilon}{\gamma} \tag{5.7}$$

It is shown perturbationally that the contribution of $f(\xi(t'))$ in (5.5) to $\langle \xi^2(t) \rangle$ is of higher order than (5.7). Since $f(0)=0$, the function $f(\xi)$ starts, at least, from the linear term (usually ξ^2 due to the symmetry of the system). The contribution of this term ξ of $f(\xi)$ in the integral (5.5) is

$$\sim e^{\gamma t}\int_0^t dt' e^{-\gamma t'} e^{\gamma t'}\int_0^{t'} ds\, e^{-\gamma s}\eta(t')\eta(s) \sim \varepsilon e^{\gamma t} \sim \sqrt{\varepsilon}\,\tau \sim O(\sqrt{\varepsilon}\,) \tag{5.8}$$

Thus this is of higher order. Similar arguments can be applied to other terms in $f(\xi)$. Consequently, as far as we are concerned with the most dominant terms in ε, the asymptotic solution of (5.1) is given by the following scaling solution

$$x_{sc}(t)=F^{-1}(\xi_{sc}(t)) \tag{5.9}$$

with (5.5). The fluctuation $\langle x^2(t)\rangle$ is expressed by

$$\langle x^2(t)\rangle=\left\langle\left[F^{-1}(\xi_{sc}(t))\right]^2\right\rangle \tag{5.10}$$

in our scaling limit. The average of (5.10) over the random force $\eta(t)$ is easily taken. For example, in the case of a simple Langevin's equation (3.1) discussed in sections III and IV, we have

$$\xi=F(x)=x\left[1-\left(\frac{g}{\gamma}\right)x^2\right]^{-1/2} \tag{5.11}$$

and inversely

$$x=F^{-1}(\xi)=\xi\left[1+\left(\frac{g}{\gamma}\right)\xi^2\right]^{-1/2} \tag{5.12}$$

Therefore, the scaling solution of $\langle x^2(t)\rangle$ is expressed by

$$\begin{aligned}\langle x^2(t)\rangle_{sc}&=\left\langle \xi_{sc}^2(t)\left[1+\left(\frac{g}{\gamma}\right)\xi_{sc}^2(t)\right]^{-1}\right\rangle\\&=\sum_{n=1}^{\infty}\left(\frac{-g}{\gamma}\right)^{n-1}\langle\xi_{sc}^{2n}(t)\rangle\\&=\sum_{n=1}^{\infty}\left(\frac{-g}{\gamma}\right)^{n-1}e^{2n\gamma t}\int_0^t dt_1 e^{-\gamma t_1}\cdots\int_0^t dt_{2n}e^{-\gamma t_{2n}}\langle\eta(t_1)\cdots\eta(t_{2n})\rangle\end{aligned} \tag{5.13}$$

Here, for simplicity, we have set $x(0)=\xi(0)=0$. By use of the Wick theorem, the correlation $\langle\eta(t_1)\ldots\eta(t_{2n})\rangle$ can be calculated easily and we obtain

$$\begin{aligned}\langle x^2(t)\rangle_{sc}=&\sum_{n=1}^{\infty}\left(\frac{-g}{\gamma}\right)^{n-1}(2n-1)!!e^{2n\gamma t}\\&\times\left[\int_0^t dt_1\int_0^t dt_2 e^{-\gamma(t_1+t_2)}\langle\eta(t_1)\eta(t_2)\rangle\right]^n\end{aligned} \tag{5.14}$$

The last factor in (5.14) is now calculated:

$$\int_0^t dt_1 \int_0^t dt_2\, e^{-\gamma(t_1+t_2)} \langle \eta(t_1)\eta(t_2)\rangle = 2\varepsilon \int_0^t e^{-2\gamma t'}\, dt' = \frac{\varepsilon}{\gamma}(1-e^{-2\gamma t}) \tag{5.15}$$

Thus we arrive finally at

$$\begin{aligned}\langle x^2(t)\rangle_{\rm sc} &= \sum_{n=1}^{\infty} \left(\frac{-g}{\gamma}\right)^{n-1} (2n-1)!! \left[\frac{\varepsilon}{\gamma}(e^{2\gamma t}-1)\right]^n \\ &= \langle x^2\rangle_{\rm st} \frac{1}{\sqrt{2\pi}} \int_{-\infty}^{\infty} e^{-\xi^2/2} \frac{\xi^2 \tau}{1+\xi^2\tau}\, d\xi \end{aligned} \tag{5.16}$$

where

$$\tau = \frac{g\varepsilon}{\gamma^2}(e^{2\gamma t}-1) \simeq \frac{g\varepsilon}{\gamma^2} e^{2\gamma t} \tag{5.17}$$

This confirms the previous result (4.11). This calculation can be easily extended to the case $\langle x^2(0)\rangle \neq 0$, and we obtain the same result (5.16) with

$$\tau = \frac{g}{\gamma}\left[\frac{\varepsilon}{\gamma}(e^{2\gamma t}-1) + \langle x^2(0)\rangle e^{2\gamma t}\right] \tag{5.18}$$

In general, the average $\langle Q(x)\rangle$ for an arbitrary function $Q(x)$ of x is calculated in the scaling limit as

$$\begin{aligned}\langle Q(x)\rangle_{\rm sc} &= \langle Q(F^{-1}(\xi_{\rm sc}(t)))\rangle \\ &\equiv \sum_{n=0}^{\infty} a_n \langle \xi_{\rm sc}^n(t)\rangle \\ &= \sum_{n=0}^{\infty} a_{2n}(2n-1)!! \left[\frac{\varepsilon}{\gamma}(e^{2\gamma t}-1) + \langle x^2(0)\rangle\right]^n \\ &= \frac{1}{\sqrt{2\pi}} \sum_{n=0}^{\infty} a_{2n} \int_{-\infty}^{\infty} e^{-\xi^2/2} \left\{\xi^2\left[\frac{\varepsilon}{\gamma}(e^{2\gamma t}-1) + \langle x^2(0)\rangle e^{2\gamma t}\right]\right\}^n d\xi \\ &= \frac{1}{\sqrt{2\pi}} \int_{-\infty}^{\infty} e^{-\xi^2/2} Q\left(F^{-1}\left(\xi\left[\frac{\varepsilon}{\gamma}(e^{2\gamma t}-1) + \langle x^2(0)\rangle e^{2\gamma t}\right]^{1/2}\right)\right) d\xi \end{aligned} \tag{5.19}$$

$$\simeq \frac{1}{\sqrt{2\pi}} \int_{-\infty}^{\infty} e^{-\xi^2/2} Q\left(F^{-1}(\xi\sqrt{\tau}\,)\right) d\xi \tag{5.20}$$

for large t. This is the so-called scaling result for the expectation value $\langle Q(x)\rangle$.

B. Alternative Formulation of the Scaling Theory

An alternative formulation of the above scaling theory is given below.

As we are interested in the scaling limit that $\varepsilon \to 0$ for τ and $\langle x^2(0)\rangle/\varepsilon$ fixed, we write this scaling limit as sc-lim. When

$$\text{sc-lim}\langle[\xi(t)-\xi_{sc}(t)]^2\rangle=0 \tag{5.21}$$

the stochastic variable $\xi(t)$ is called to approach the limiting process in the scaling limit. In the stochastic process (5.1), we can prove (5.21) for $\xi_{sc}(t)$ defined by (5.6). For the proof, see Appendix B.

C. Extended Scaling Theory—Time-Dependent Transformation

Recently de Pasquale and Tombesi[13, 14] applied a time-dependent nonlinear transformation to the nonlinear Langevin equation and reinterpreted the result obtained by the present author[2, 15] in the Fokker-Planck equation. This is a slightly extended interpretation of the above nonlinear transformation (5.2). We introduce here the following time-dependent nonlinear transformation[16]

$$\tilde{\xi}=F^{-1}\left(\left(\frac{\tau_i}{\tau}\right)^{1/2}F(x)\right)=F^{-1}(e^{-\gamma t}F(x)) \tag{5.22}$$

with (5.2) and $\tau_i/\tau=e^{-\gamma t}$. This function (5.22) has been used in Refs. 2 and 16 for deriving the scaling result for the Fokker-Planck equation, as is discussed again later. Clearly, $\tilde{\xi}=\tilde{\xi}(x,t)$ in (5.22) represents the characteristic curve of the stochastic process (5.1), that is, $\tilde{\xi}(x,t)$ is constant along the deterministic path $\dot{x}(t)=c_1(x)$:

$$\begin{aligned}\frac{d}{dt}\tilde{\xi}(x,t)&=\frac{\partial\tilde{\xi}}{\partial x}\frac{dx}{dt}+\frac{\partial\tilde{\xi}}{\partial t}\\ &=[c_1(x)F'(x)-\gamma F(x)]e^{-\gamma t}\left[\frac{d}{dy}F^{-1}(y)\right]_{y=e^{-\gamma t}F(x)}\\ &=0\end{aligned} \tag{5.23}$$

which is easily shown from (5.2).

Now, by the nonlinear transformation (5.22), the stochastic equation (5.1) is changed into

$$\frac{d}{dt}\tilde{\xi}=\frac{\eta(t)e^{-\gamma t}F'\left(F^{-1}\left(e^{\gamma t}F(\tilde{\xi})\right)\right)}{F'(\tilde{\xi})} \tag{5.24}$$

That is, this has no deterministic part, as it should be from the transformation (5.22). If we expand the multiplicative factor of $\eta(t)$ in (5.24) with respect to $\tilde{\xi}$, then we have

$$\frac{e^{-\gamma t}F'\big(F^{-1}\big(e^{\gamma t}F(\tilde{\xi})\big)\big)}{F'(\tilde{\xi})}=e^{-\gamma t}+F_1(t)\tilde{\xi}+\cdots \tag{5.25}$$

It is easily shown perturbationally as before that the first term in (5.25) contributes dominantly in the scaling limit $\varepsilon\to 0$ and τ fixed. Thus (5.24) may be reduced to

$$\frac{d}{dt}\tilde{\xi}=e^{-\gamma t}\eta(t) \tag{5.26}$$

in this limit. Integrating (5.26), we obtain

$$\tilde{\xi}(t)=\int_0^t e^{-\gamma t'}\eta(t')dt'+\xi(0) \tag{5.27}$$

Therefore, we have

$$\tilde{\xi}(t)=e^{-\gamma t}\xi_{sc}(t) \tag{5.28}$$

with (5.6). The original stochastic variable $x(t)$ is given by

$$x(t)=F^{-1}\big(e^{\gamma t}F(\tilde{\xi}(t))\big) \tag{5.29}$$

Since $F^{-1}(y)=y+\cdots$ for small y, the transformation (5.22) is reduced to

$$\tilde{\xi}=e^{-\gamma t}F(x)=e^{-\gamma t}\xi \tag{5.30}$$

for large t. That is, the transformation (5.22) is essentially equivalent to the time-independent one (5.2) for large time t, that is, in the scaling limit. The merit is that it is simple to match the solution with the initial condition because the transformation (5.22) gives $\tilde{\xi}(0)=x(0)$. De Pasquale and Tombesi[13] insisted that the above extended scaling theory can be used even for the critical case $\gamma=0$. In this case, however, to replace the multiplicative factor of $\eta(t)$ in (5.24) by $e^{-\gamma t}$ is not a good approximation, because other neglected terms then are of the same order as the main term, as is easily seen perturbationally. This critical case should be treated separately, as is discussed later.

In general, the average $\langle Q(x)\rangle$ is given in the above extended scaling theory by

$$\langle Q(x)\rangle = \left\langle Q\left(F^{-1}\left(e^{\gamma t}F(\tilde{\xi}(t))\right)\right)\right\rangle \tag{5.31}$$

$$\equiv \sum_{n=0}^{\infty} b_n(t)\langle\tilde{\xi}^n(t)\rangle = \sum_{n=0}^{\infty} b_{2n}(t)(2n-1)!!\left[\frac{\varepsilon}{\gamma}(1-e^{-2\gamma t})+\langle x^2(0)\rangle\right]^n$$

$$= \sum_{n=0}^{\infty} b_{2n}(t)\frac{1}{\sqrt{2\pi}}\int_{-\infty}^{\infty} e^{-\xi^2/2}\left\{\xi^2\left[\frac{\varepsilon}{\gamma}(1-e^{-2\gamma t})+\langle x^2(0)\rangle\right]\right\}^n d\xi$$

$$= \frac{1}{\sqrt{2\pi}}\int_{-\infty}^{\infty} e^{-\xi^2/2}Q\left(F^{-1}\left(e^{\gamma t}F\left(\xi\left[\frac{\varepsilon}{\gamma}(1-e^{-2\gamma t})+\langle x^2(0)\rangle\right]^{1/2}\right)\right)\right)d\xi \tag{5.32}$$

In particular, for the case (4.1), we obtain the fluctuation $\langle x^2(t)\rangle$ in the form

$$\langle x^2(t)\rangle = \langle x^2\rangle_{\text{st}}(1-e^{-2\gamma t})^{-1}\frac{1}{\sqrt{2\pi}}\int_{-\infty}^{\infty} e^{-\xi^2/2}\frac{\xi^2\tilde{\tau}}{1+\xi^2\tilde{\tau}}d\xi \tag{5.33}$$

from the general expression (5.31), where

$$\tilde{\tau} = \frac{g}{\gamma}e^{2\gamma t}\left[\frac{\varepsilon}{\gamma}(1-e^{-2\gamma t})+\langle x^2(0)\rangle\right](1-e^{-2\gamma t}) \tag{5.34}$$

This has been obtained by de Pasquale and Tombesi[13] by using the transformation

$$\tilde{\xi} = xe^{-\gamma t}\left[1-\frac{g}{\gamma}x^2(1-e^{-2\gamma t})\right]^{-1/2} \tag{5.35}$$

which is obtained from (5.29) for $F(x)$ in (5.11).

The time-dependent transformation is equivalent to the time-independent one in the intermediate time region, that is, in the scaling regime.

D. A Simple Scaling Property in the Critical Case

The scaling treatment in Section VA is always valid for $\gamma \neq 0$, even if it is very small. The scaling time region shifts as γ decreases, according to (3.19). Now again, we discuss the example

$$\frac{d}{dt}x = -gx^k + \eta(t) \tag{5.36}$$

where $k=2$ (marginal), 3 (critical),.... As has been discussed by Kubo et al.[3], in the Fokker-Planck equation, we introduce the simple scale transformation

$$x=\left(\frac{\varepsilon}{g}\right)^{1/(k+1)}\xi, \qquad t=\varepsilon^{-(k-1)/(k+1)}g^{-2/(k+1)}t' \tag{5.37}$$

in the Fokker-Planck equation corresponding to (5.36). Then we obtain the following dimensionless Fokker-Planck equation:

$$\frac{\partial P}{\partial t'}=\frac{\partial}{\partial \xi}\xi^k P+\frac{\partial^2}{\partial \xi^2}P \tag{5.38}$$

Therefore, the solution of the original nonlinear Fokker-Planck equation is given by the scaling function

$$P(x,t)=P_{\text{sc}}\left(\left(\frac{g}{\varepsilon}\right)^{1/(k+1)}x,\ \varepsilon^{(k-1)/(k+1)}g^{2/(k+1)}t\right)\left(\frac{g}{\varepsilon}\right)^{1/(k+1)} \tag{5.39}$$

if the initial condition satisfies this property. For example, under the initial condition

$$P(x,0)=\delta\left(\left(\frac{g}{\varepsilon}\right)^{1/(k+1)}x\right)\left(\frac{g}{\varepsilon}\right)^{1/(k+1)} \tag{5.40}$$

the solution has the scaling property (5.39) in the whole region of time. Therefore, the fluctuation $\langle x^2(t)\rangle$ is given by

$$\langle x^2(t)\rangle=\left(\frac{\varepsilon}{g}\right)^{2/(k+1)}G\left(\varepsilon^{(k-1)/(k+1)}g^{2/(k+1)}t\right) \tag{5.41}$$

where

$$G(t)=\int_{-\infty}^{\infty}y^2P_{\text{sc}}(y,t)\,dy \tag{5.42}$$

The characteristic time of this system is of the order

$$t_0\sim g^{-2/(k+1)}\varepsilon^{-(k-1)/(k+1)} \tag{5.43}$$

VI. GENERAL SCALING THEORY

A. General Scaling Ideas

We now present the general scaling theory based on the master equation or Fokker-Planck equation.

The essence of the scaling theory is to make an asymptotic evaluation by relating the system size Ω (or the strength of a random force, ε), nonlinearity g, and time t to each other in evaluating their order, so that we may extract an appropriate time region in which the "macroscopic fluctuation" appears or macroscopic order sets in.

For this purpose, as is mentioned in section I, we divide the time region into three regimes, that is, the initial regime, the second scaling regime, and the remaining regime (or final regime), as shown in Fig. 2.

In the initial regime, the Gaussian approximation or linearization is valid, and the second scaling regime can be treated by introducing an appropriate scaling variable of time τ, as a function of time t, the strength of a random force ε, nonlinearity g, initial fluctuation $\langle x^2(0)\rangle$ and so on

$$\tau = S(t, \varepsilon, g, \langle x^2(0)\rangle, \cdots) \tag{6.1}$$

Such an example is given earlier in sections III to V. This scaling function S can be easily found in specific problems by studying how the initial Gaussian treatment breaks down in the initial regime. To clarify the physics of the transient phenomena, we take the limit $\varepsilon \to 0$ or $g \to 0$ for τ and $\delta\varepsilon^{-\mu}$ fixed in the second scaling regime. Here δ denotes the deviation of the initial system from the instability point, and μ is an appropriate positive exponent, usually $\mu = \frac{1}{2}$.

Before we explain the details of the theory, we explain the main difference between the ordinary Ω-expansion theory and the present scaling theory. The former can be applied to the normal region or extensive region $\varepsilon^{\mu} \ll \delta$ and the latter is powerful in the unstable region $\delta \ll \varepsilon^{\mu}$. Both are schematically shown in Fig. 7.

B. Scaling Theory Based on a Nonlinear Transformation

One of the merits in using the Fokker-Planck equation is that the distribution function itself can be obtained explicitly, and consequently the onset time t_0 can be defined uniquely as is shown later. Now we start from the following Fokker-Planck equation corresponding to the nonlinear Langevin equation (4.3).

$$\frac{\partial}{\partial t} P(x,t) = -\frac{\partial}{\partial x}(c_1(x)P(x,t)) + \varepsilon \frac{\partial^2}{\partial x^2} P(x,t) \tag{6.2}$$

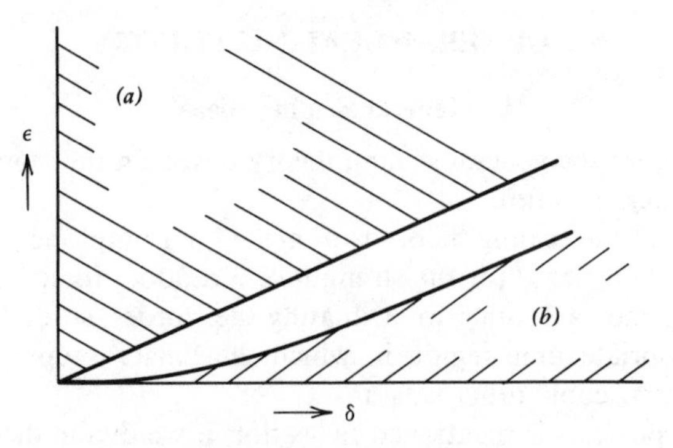

Fig. 7. Division of the ε–δ plane into the two regimes: (a) unstable regime $\delta \lesssim \varepsilon^{\mu}$ and (b) extensive regime $\varepsilon^{\mu} \ll \delta$.

To make an asymptotic evaluation, we use the time-dependent nonlinear transformation (5.22) as before,[16] that is,

$$\tilde{\xi}=\tilde{\xi}(x,t)=F^{-1}(e^{-\gamma t}F(x)) \tag{6.3}$$

Inversely,

$$x=x(\tilde{\xi},t)=F^{-1}\big(e^{\gamma t}F(\tilde{\xi})\big) \tag{6.4}$$

This equation (6.3) is the solution of $\dot{x}=c_1(x)$ with the initial condition $x=\tilde{\xi}$ at $t=0$. This is easily shown below. The solution of $\dot{x}=c_1(x)$ is given by

$$\int_{\tilde{\xi}}^{x}\frac{dy}{c_1(y)}=t \tag{6.5}$$

By using the definition

$$F(x)=\exp\int_{a_0}^{x}\frac{\gamma}{c_1(y)}dy \tag{6.6}$$

(6.5) can be written as

$$e^{\gamma t}=\exp\left(\int_{a_0}^{x}\frac{\gamma}{c_1(y)}dy+\int_{\tilde{\xi}}^{a_0}\frac{\gamma}{c_1(y)}dy\right)=\frac{F(x)}{F(\tilde{\xi})} \tag{6.7}$$

This yields (6.3) and (6.4).

Next we transform the Fokker-Planck equation (6.2) into a new one for $\tilde{P}(\tilde{\xi}, t)$. First note that

$$\tilde{P}(\tilde{\xi}, t)\,d\tilde{\xi} = P(x,t)\,dx \quad \text{and} \quad \frac{d\tilde{\xi}}{dx} = \frac{c_1(\tilde{\xi})}{c_1(x)} \tag{6.8}$$

Using (6.8), we obtain

$$\begin{aligned}
\frac{\partial}{\partial t}\tilde{P} &= \frac{1}{c_1(\tilde{\xi})}\frac{\partial}{\partial t}\left[P\big(x(\tilde{\xi},t),t\big)\cdot c_1\big(x(\tilde{\xi},t)\big)\right] \\
&= \frac{1}{c_1(\tilde{\xi})}\left[\frac{\partial x}{\partial t}\frac{\partial}{\partial x}(Pc_1(x)) + c_1(x)\frac{\partial}{\partial t}P\right] \\
&= \frac{1}{c_1(\tilde{\xi})}\left\{c_1(x)\frac{\partial}{\partial x}(Pc_1(x)) + c_1(x)\left[-\frac{\partial}{\partial x}(Pc_1) + \varepsilon\frac{\partial^2 P}{\partial x^2}\right]\right\} \\
&= \varepsilon\frac{c_1(x)}{c_1(\tilde{\xi})}\frac{\partial^2}{\partial x^2}P = \varepsilon\frac{c_1(x)}{c_1(\tilde{\xi})}\frac{\partial^2}{\partial x^2}\left[\frac{c_1(\tilde{\xi})}{c_1(x)}\tilde{P}(\tilde{\xi},t)\right]
\end{aligned} \tag{6.9}$$

Thus we arrive finally at

$$\frac{\partial}{\partial t}\tilde{P} = \varepsilon\frac{c_1(x)}{c_1(\tilde{\xi})}\frac{\partial^2}{\partial x^2}\left[\frac{c_1(\tilde{\xi})}{c_1(x)}\tilde{P}(\tilde{\xi},t)\right] \tag{6.10}$$

where $x = x(\tilde{\xi}, t)$. This is also rewritten using (6.8) as

$$\frac{\partial \tilde{P}}{\partial t} = \varepsilon\frac{\partial}{\partial \tilde{\xi}}\left\{\frac{c_1(\tilde{\xi})}{c_1(x)}\frac{\partial}{\partial \tilde{\xi}}\left[\frac{c_1(\tilde{\xi})}{c_1(x)}\tilde{P}\right]\right\} \tag{6.11}$$

Thus the drift term in (6.2) has been eliminated formally in (6.11) using the new variable $\tilde{\xi}$. Equation 6.11 is still exact. Then we make an approximation. The guiding principle of our approximation is that we may extract an asymptotically exact solution of (6.2) or (6.11) in the limit of small ε and large t, that is, in the second scaling regime, as in Fig. 2. This is possible by making the approximation

$$\frac{c_1(\tilde{\xi})}{c_1(x)} \simeq \frac{\gamma\tilde{\xi}}{\gamma x} \simeq e^{-\gamma t} \tag{6.12}$$

It is shown perturbationally with respect to ξ in $x=x(\xi, t)$ that this replacement (6.12) yields the dominant contribution up to the scaling regime, that is, in the initial and scaling regimes. Thus (6.11) is reduced to the following simple diffusion equation with a time-dependent coefficient:

$$\frac{\partial}{\partial t}\tilde{P}=\varepsilon e^{-2\gamma t}\frac{\partial^2}{\partial \tilde{\xi}^2}\tilde{P} \tag{6.13}$$

By the transformation

$$s=\left(\frac{-1}{2\gamma}\right)\exp(-2\gamma t) \tag{6.14}$$

(6.13) is reduced to

$$\frac{\partial}{\partial s}\tilde{P}=\varepsilon\frac{\partial^2}{\partial \tilde{\xi}^2}\tilde{P} \tag{6.15}$$

The solution of (6.15) with the initial condition $\tilde{P}(\xi,0)=P(x,0)=P_0(x)$ is given by

$$\tilde{P}(\xi,t)=\frac{1}{[4\pi\varepsilon(s-s_0)]^{1/2}}\int_{-\infty}^{\infty}\exp\left[-\frac{(\xi-x)^2}{4\varepsilon(s-s_0)}\right]P_0(x)dx \tag{6.16}$$

where $s_0=-1/(2\gamma)$. Therefore, the solution of (6.2) is expressed, in our approximation, as

$$P_{sc}(x,t)=\left[\frac{c_1(\tilde{\xi}(x,t))}{c_1(x)}\right]\tilde{P}(\tilde{\xi}(x,t),t) \tag{6.17}$$

with $\tilde{P}(\xi,t)$ in (6.16), and with (6.3) for $\tilde{\xi}(x,t)$. In particular, if the initial distribution function is Gaussian, then we have

$$P_{sc}(x,t)=\frac{1}{\sqrt{2\pi\tau}}F'(x)\left[F'\left(F^{-1}(e^{-\gamma t}F(x))\right)\right]^{-1}$$
$$\times\exp\left\{-\frac{1}{2\tau}\left[e^{\gamma t}F^{-1}(e^{-\gamma t}F(x))\right]^2\right\} \tag{6.18}$$

where

$$\tau = \varepsilon e^{2\gamma t}\left[\sigma_0 + \frac{(1-e^{-2\gamma t})}{\gamma}\right] \tag{6.19}$$

This agrees with the result obtained by the connection procedure in the earlier scaling theory,[2, 11] as is discussed again later. In the second scaling regime, the factor $F^{-1}(e^{-\gamma t}F(x))$ is simplified to

$$F^{-1}(e^{-\gamma t}F(x)) = e^{-\gamma t}F(x) + O(e^{-2\gamma t}) \tag{6.20}$$

for large t. Therefore, (6.18) takes the scaling property

$$P_{sc}(x,t) = \frac{1}{\sqrt{2\pi\tau}} F'(x)\exp\left[\frac{-F^2(x)}{2\tau}\right] \tag{6.21}$$

with the scaling variable τ defined by

$$\tau = \left(\sigma_0 + \frac{1}{\gamma}\right)\varepsilon\exp(2\gamma t) \tag{6.22}$$

This has been derived in Refs. 2 and 11.

It is also possible to confirm the scaling property for a general initial distribution function $P_0(x)$ as follows. In the scaling limit, from (6.16) and (6.17) we have

$$\begin{aligned} P_{sc}(x,t) &= \frac{1}{\sqrt{2\pi\varepsilon/\gamma}} \int_{-\infty}^{\infty} \exp\left\{-\frac{[F(x)e^{-\gamma t} - y]^2}{2\varepsilon/\gamma}\right\} P_0(y)\,dy\, e^{-\gamma t}F'(x) \\ &= \left\{2\pi\left(\frac{\varepsilon}{\gamma}\right)\exp(2\gamma t)\right\}^{-1/2} \int_{-\infty}^{\infty} \exp\left[-\frac{(F(x)-z)^2}{2(\varepsilon/\gamma)\exp(2\gamma t)}\right] P_0(e^{-\gamma t}z) \\ &\quad \times e^{-\gamma t}F'(x) \end{aligned} \tag{6.23}$$

If $P_0(x)$ takes the general form

$$P_0(x) \sim \exp\left[-\frac{1}{\varepsilon}\left(\frac{x^2}{2\sigma_0} + a_3x^3 + \cdots\right)\right] \tag{6.24}$$

then the higher terms $a_3x^3 + $A. in (6.24) give a contribution of higher order in (6.25) and consequently we arrive again at the result (6.21). That is, the non-Gaussian property of the initial distribution function is irrelevant to the transient behavior in the intermediate scaling regime, as far as the Gaussian part in $P_0(x)$ is meaningful ($\sigma_0>0$). We have restricted our arguments to the Gaussian (or delta function type) initial distribution function in our series of papers on transient phenomena. This is justified above.

C. Scaling Theory Based on a Connection Procedure

Here we present the scaling theory in its original formulation.[11] As is mentioned earlier, the transient phenomena near the instability point are very sensitive to the initial condition, namely, the deviation δ of the initial system from the unstable point. They depend also on the smallness parameter ε or inverse system size. Their dependence on these parameters is drastically different on each ε–δ domain in Fig. 7. We are here interested in the unstable region $\delta \lesssim \varepsilon^{\mu}$ (where $\mu=\frac{1}{2}$ in a usual situation).

As we see in the explicit calculations above, existence of the scaling regime for the case of the relaxation from or near the unstable point seems clear. However, it is worthwhile to argue in general the existence of the scaling regime. If the system is located initially just at the unstable point and if $\varepsilon=0$ (i.e., no random force), then the average values of the relevant macrovariable and its fluctuation do not change at all in time, as is easily seen from the analogy to the classical motion in a potential shown in Fig. 1. The smallness parameter is analogous to the Planck constant $\hbar$. All physical quantities such as fluctuation in unstable systems are singular at $\varepsilon=0$ (or $\delta=0$) and $t=0$ (i.e., $z\equiv\exp(-2\gamma t)=0$ for a certain growing constant γ). Therefore, the limiting values of the relevant physical quantities depend on the path, that is, on the ratio $\tau=\varepsilon/z$, as shown in Fig. 8. Thus we conclude that physical quantities depend on the so-called scaling variable τ in the vicinity of the essential singular point $\varepsilon=0$ (or $\delta=0$) and $z=0$ ($t\to\infty$). This yields the existence of the scaling regime for small ε in the intermediate time region.

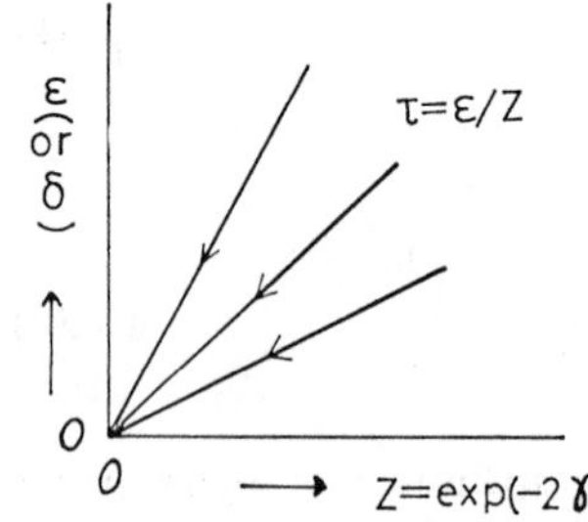

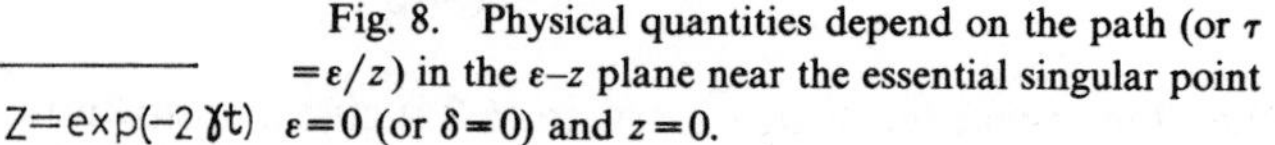

Fig. 8. Physical quantities depend on the path (or $\tau=\varepsilon/z$) in the ε–z plane near the essential singular point $\varepsilon=0$ (or $\delta=0$) and $z=0$.

To explain our idea, we discuss the Kramers-Moyal equation

$$\varepsilon\frac{\partial}{\partial t}P(x,t) = -\mathcal{H}\left(x, \varepsilon\frac{\partial}{\partial x}\right)P(x,t) \tag{6.25}$$

where

$$\mathcal{H}(x,p) = \int(1-e^{-rp})w(x,r) = \sum_{n=1}^{\infty}\frac{(-1)^{n-1}}{n!}p^n c_n(x) \tag{6.26}$$

and

$$c_n(x) = \int r^n w(x,r)\mathrm{d}r \tag{6.27}$$

with the transition probability.

The nth term of the right-hand side of (6.25) is of the order of ε^n in the scaling regime, since all derivatives of $P(x,t)$ with respect to x are of the order of unity because of the scaling property of $P(x,t)$. Thus the first term, that is the drift term, is dominant in the second scaling regime. That is, $P(x,t)$ asymptotically satisfies the drift equation

$$\frac{\partial}{\partial t}P(x,t) + \frac{\partial}{\partial x}c_1(x)P(x,t) = 0 \tag{6.28}$$

The solution of this equation is given by

$$P(x,t) = \left[\frac{c_1(\xi(x,t))}{c_1(x)}\right]P_{\text{ini}}(\xi(x,t),t_1) \tag{6.29}$$

for the initial condition $P(x,t_1) = P_{\text{ini}}(x,t_1)$ at $t=t_1$, and for $\xi(x,t)$ defined by (6.3), namely

$$\xi = \xi(x,t) = F^{-1}(e^{-\gamma(t-t_1)}F(x)) \tag{6.30}$$

Equation (6.29) corresponds to (6.8). This is discussed in detail in Ref. 16.

Now we must find the initial condition near the boundary between the initial and scaling regimes. For this purpose, we solve the problem in the initial regime. In this regime, the linearization of the nonlinear stochastic process is valid, that is, the Gaussian approximation can be used. Thus the solution in the initial regime approximately satisfies the linear Fokker-

Planck equation

$$\frac{\partial P}{\partial t} = -\frac{\partial}{\partial x}\gamma x P + \varepsilon \frac{\partial^2}{\partial x^2} P \tag{6.31}$$

For simplicity we assume that the random force is also Gaussian, so that terms of order higher than the second derivative are neglected in (6.25). Then the solution of (6.31) is given by the Gaussian form

$$P_{\text{ini}}(x,t) = \frac{1}{\sqrt{2\pi\varepsilon\sigma(t)}} \exp\left(-\frac{x^2}{2\varepsilon\sigma(t)}\right) \tag{6.32}$$

with

$$\sigma(t) = (\sigma_0 + \sigma_1)e^{2\gamma t} - \sigma_1 \equiv \sigma e^{2\gamma t} - \sigma_1; \qquad \sigma_1 = \frac{c_2(0)}{\gamma} \tag{6.33}$$

This has the scaling form

$$P_{\text{ini}}^{(\text{sc})}(x,\tau) = \frac{1}{\sqrt{2\pi\tau}} \exp\left(-\frac{x^2}{2\tau}\right) \tag{6.34}$$

for large t, where $\tau = \varepsilon\sigma \exp(2\gamma t)$. It we take this scaling-type solution as the initial condition at $t = t_1$ or at $\tau = \tau_i \equiv \varepsilon\sigma\exp(2\gamma t_1)$, then the solution of $P(x,t)$ in (6.29) is given by

$$P(x,t) = \frac{1}{\sqrt{2\pi\tau}} \frac{F'(x)}{F'(\tilde{\xi}(x,t))} \exp\left\{-\frac{e^{-2\gamma t_1}}{2\varepsilon\sigma}\left[F^{-1}\left(F(x)e^{-\gamma(t-t_1)}\right)\right]^2\right\} \tag{6.35}$$

This equation should be compared with (6.18). It differs from (6.18) in the definition of τ and in that (6.35) contains the extra factor $\exp(-2\gamma t)$. This extra factor is shown, however, to be irrelevant in the second scaling regime. In other words, if we consider the region $t \gg t_1$, that is, $\exp[-\gamma(t-t_1)] \ll 1$, then the factor $F^{-1}(F(x)\exp[-\gamma(t-t_1)])$ is reduced to $\exp[-\gamma(t-t_1)]F(x)$, as before, because of the property that $F(x) = x + \ldots$ or $F^{-1}(\tilde{\xi}) = \tilde{\xi} + \ldots$. Thus we arrive finally at the scaling solution

$$P_{\text{sc}}(x,t) = \frac{1}{\sqrt{2\pi\tau}} F'(x) \exp\left[-\frac{F^2(x)}{2\tau}\right] \tag{6.36}$$

and

$$\tau = \varepsilon\sigma \exp(2\gamma t) \tag{6.37}$$

as is derived in Section VI.B [see (6.21)]. It should be noted also that the connection time t_1 arbitrary and consequently that we may take the limit $t_1 \to 0$ in (6.35). Then we obtain again the same result as (6.18) with the scaling variable τ instead of (6.19). This does not mean that the stochastic process is reduced to the deterministic process. The key point of this trick is that, for the connection procedure, we have used the scaling form (6.32) in the initial regime with $\tau = \sigma\varepsilon \exp(2\gamma t)$ instead of $\tau = (\sigma_0 + \sigma_1)\varepsilon \exp(2\gamma t) - \varepsilon\sigma_1$. This replacement of time by the scaling variable τ results in the inclusion of the effect of the random force [i.e., the addition of σ_1 in τ as $\tau = \varepsilon(\sigma + \sigma_1) \exp(2\gamma t)$]. The above treatments correspond to the procedure in Section VI.B in which the time-dependent transformation is first performed and then the Gaussian random force is taken into account for the new variable obtained by the transformation. This separation of the random force and nonlinearity is one of the essential points in the scaling theory.

D. Operational Formulation and Haake's Theory

Recently Haake[17] applied the following transformation

$$Q(x,t) = \mathfrak{N} \int e^{-(x-y)^2/\eta} P(y,t)\, dy \tag{6.38}$$

to transient phenomena near the instability point. Here $\mathfrak{N}$ denotes the normalization, and η is a parameter to be chosen later. Haake applied (6.38) to the Fokker-Planck equation

$$\frac{\partial}{\partial t} P = -\frac{\partial}{\partial x} x D(x) P + \varepsilon \frac{\partial^2}{\partial x^2} P \tag{6.39}$$

and transformed it into the differential equation

$$\frac{\partial}{\partial t} Q(x,t) = \left\{ -\frac{\partial}{\partial x} x D(x) + \left[\varepsilon - \frac{\eta}{2} D(0) \right] \frac{\partial^2}{\partial x^2} \right\} Q(x,t)$$
$$- \left\{ \frac{\partial}{\partial x} x \left[D\left(x + \frac{\eta}{2} \frac{\partial}{\partial x} \right) - D(x) \right] + \frac{\eta}{2} \frac{\partial^2}{\partial x^2} \left[D\left(x + \frac{\eta}{2} \frac{\partial}{\partial x} \right) - D(0) \right] \right\} Q(x,t)$$
$$\tag{6.40}$$

Haake set $\eta=2\varepsilon/D(0)$ to eliminate the diffusion term in the first term in (6.40) and neglected the second term, which is "effectively" of higher order of η. Thus he obtained asymptotically the drift equation

$$\frac{\partial}{\partial t}Q(x,t)+\frac{\partial}{\partial x}xD(x)Q(x,t)=0 \tag{6.41}$$

The initial condition is changed to

$$Q(x,0)=\mathfrak{N}\int \exp\left[-\frac{(x-y)^2}{\eta}\right]P(y,0)dy \tag{6.42}$$

In particular, for $P(y,0)=\delta(y)$, we have

$$Q(x,0)=\mathfrak{N}\exp\left(\frac{-x^2}{\eta}\right) \tag{6.43}$$

That is, this transformation smears out the initial sharp distribution[17] and consequently makes it easy to handle the instability problem asymptotically or classically. The transformation (6.38) can, in general, not be inverted, but the moments $\langle x^n\rangle_P$ with respect to P can easily be expressed in terms of the moments of Q, that is $\langle x^i\rangle_Q$. Thus the original problem can be solved approximately for small ε.

Clearly, Haake's above theory is essentially equivalent to the scaling theory in Sections VI.B and VI.C. To clarify the relation between the two theories, we reformulate or repeat the above treatment in the following operational form. We introduce the operational transformation[2]:

$$\tilde{P}(x,t)=\exp\left(t_1\varepsilon\frac{\partial^2}{\partial x^2}\right)P(x,t) \tag{6.44}$$

which is equivalent to (6.38) with $\eta=4t_1\varepsilon$. Therefore, the above transformation (6.44) yields (6.40) with $Q=\tilde{P}(x,t)$. Thus $\tilde{P}(x,t)$ satisfies

$$\frac{\partial}{\partial t}\tilde{P}(x,t)=\left[-\frac{\partial}{\partial x}c_1\left(x+2t_1\varepsilon\frac{\partial}{\partial x}\right)+\varepsilon\frac{\partial^2}{\partial x^2}\right]\tilde{P} \tag{6.45}$$

where we have used the formula

$$\exp\left(a\frac{\partial^2}{\partial x^2}\right)f(x)\exp\left(-a\frac{\partial^2}{\partial x^2}\right)=f\left(x+2a\frac{\partial}{\partial x}\right) \tag{6.46}$$

for an arbitrary function $f(x)$. This result (6.45) is easily rewritten as (6.40) with $c_1(x)=xD(x)$. Therefore, for the same reason as before, (6.45) can be simplified to

$$\frac{\partial}{\partial t}\tilde{P}(x,t)+\frac{\partial}{\partial x}c_1(x)\tilde{P}(x,t)=0 \tag{6.47}$$

by setting $t_1=1/2\gamma$ in the scaling regime. The initial condition is given by

$$\tilde{P}(x,0)=\exp\left(t_1\varepsilon\frac{\partial^2}{\partial x^2}\right)P(x,0)=\frac{1}{\sqrt{2\pi\varepsilon\sigma}}\exp\left(-\frac{x^2}{2\varepsilon\sigma}\right) \tag{6.48}$$

with $\sigma=\sigma_0+\sigma_1$ and $\sigma_1=1/\gamma=2t_1$ for an initial Gaussian distribution of the variance σ_0. The solution of (6.47) with (6.48) is given by the right hand side of (6.35). The distribution function $P(x,t)$ is expressed formally as

$$P(x,t)=\exp\left(-t_1\varepsilon\frac{\partial^2}{\partial x^2}\right)\tilde{P}(x,t) \tag{6.49}$$

Here it should be noted that $\tilde{P}(x,t)$ is very flat and of the order of unity in the second scaling regime. Therefore, the operation $\exp[-t_1\varepsilon\frac{\partial^2}{\partial x^2}]$ is replaced by unity in this asymptotic scaling regime. Thus we arrive finally at the earlier result $P(x,t)=\tilde{P}(x,t)$, that is, (6.35). Formally, the scaling solution is also expressed as

$$P_{sc}(x,t)=\exp(t\mathcal{L}_{\text{drift}})\exp(t_1\mathcal{L}_{\text{diffusion}})P(x,0) \tag{6.50}$$

The effect of the random force has been treated by the transformation (6.44), and later the nonlinearity of the system has been taken into account through the drift equation.

It is clear from the above demonstration that the above three formulations in Sections VI.B, VI.C, and VI.D are completely equivalent to each other.

E. Simple Example

Here we apply the above general arguments to the simple Fokker-Planck equation

$$\frac{\partial}{\partial t}P=-\frac{\partial}{\partial x}(\gamma x-gx^3)+\varepsilon\frac{\partial^2}{\partial x^2}P \tag{6.51}$$

which corresponds to (4.3). The solution in the form of (6.18) is given explicitly in this example by

$$P_{sc}(x,t)=\frac{1}{\sqrt{2\pi\tau}}\left[1-\frac{g}{\gamma}x^2(1-e^{-2\gamma t})\right]^{-3/2}$$
$$\times\exp\left\{-\frac{x^2}{2\tau\left[1-(g/\gamma)x^2(1-e^{-2\gamma t})\right]}\right\} \tag{6.52}$$

where τ is given by (6.19). This is a slightly improved solution compared with the previous one[11] without the factor $[1-\exp(-2\gamma t)]$. Of course, the solution[78a] in Ref. 15, is reduced to (6.52) if we set $\tau_i=\varepsilon\sigma$ and use τ defined by (6.19). The merit of this new scaling solution is that it can be used in both the initial and scaling regimes. The time dependence of this distribution function is shown in Fig. 9. The onset time is defined by the time t_0 at which the double peaks just begin to appear, and it is determined[15] by $\tau_0=\gamma/3g$. Consequently, we have

$$t_0=-\frac{1}{2\gamma}\log\left[\frac{3\varepsilon g}{\gamma^2}(1+\gamma\sigma_0)\right] \tag{6.53}$$

which agrees, except for the factor 3, with the result obtained by the self-consistent linearization in Section III. Thus the distribution function de-

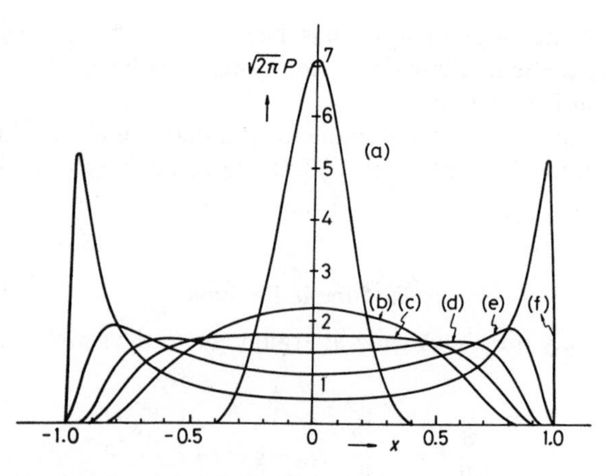

Fig. 9. Distribution function: (a) $\tau=0.02$, (b) $\tau=0.2$ (c) $\tau=\tau_0=\frac{1}{3}$, (d) $\tau=0.5$, (e) $\tau=1$, and (f) $\tau=4$, where $\tau=(\sigma_0+\sigma_1)\,\varepsilon\exp(2\gamma t)$, $g=\gamma$.

termines uniquely the onset time. The fluctuation $\langle x^2(t)\rangle$ can be immediately obtained from (6.52) to give (5.33).

Several other applications are discussed later.

VII. RELATION AMONG THE EXTENSIVITY, Ω-EXPANSION, AND SCALING THEORY

It is instructive to discuss here the relationship among Kubo's extensivity ansatz, van Kampen's Ω-expansion, and the present scaling theory.

A. The Extensivity and Scaling Theory

Kubo (3.18) proposed the following extensivity ansatz: If $P(X, t)$ is the probability function to have a macrovariable X at time t then it takes the asymptotic form

$$P(X,t) \propto \exp[\Omega\varphi(x,t)] \tag{7.1}$$

for large system size Ω and with $x = X/\Omega$, in a Markovian process described by the Kramers-Moyal equation (6.25). This extensivity ansatz has been proved by the present author[19, 20] under quite general conditions.

As is well known,[3] the Gaussian approximation of $\varphi(x, t)$ yields the ordinary Ω-expansion (1.1). This corresponds to the separation of the stochastic variable x, (2.1) into $y(t)$ and ξ. As is easily seen from (6.18) or (6.21), the scaling solution contains a term higher than the extensive part $\Omega\varphi_0(x, t)$. When we expand $P(x, t)$ to

$$P(x,t) \cong C\exp\left[\left(\frac{1}{\varepsilon}\right)\varphi_0(x,t) + \varphi_1(x,t) + \varepsilon\varphi_2(x,t) + \cdots\right] \tag{7.2}$$

the second term $\varphi_1(x, t)$ is of the same order of magnitude as the first extensive part $\Omega\varphi_0(x, t)$ for large time, because the first extensive part, $\varphi_0(x, t)$, becomes exponentially small for large time t when the system starts from the unstable point. Thus we have to study, at least, up to $\varphi_1(x, t)$ in (7.2) in order to discuss the relaxation and fluctuation near the instability point.

Now we derive the temporal evolution equations of $\varphi_0(x, t)$, $\varphi_1(x, t)$ and so on following Kubo et al.[3] For this purpose, we put (7.2) into (6.25) with (6.26) and (6.27), and by equating the like powers in ε, we obtain[16]

$$\frac{\partial\varphi_0}{\partial t} = -\int w(x,r)\left[1 - \exp\left(-r\frac{\partial\varphi_0}{\partial x}\right)\right]dr \tag{7.3}$$

$$\frac{\partial\varphi_1}{\partial t} = \int\left[w(x,r)\left(\frac{r^2}{2}\frac{\partial^2\varphi_0}{\partial x^2} - r\frac{\partial\varphi_1}{\partial x}\right) - r\frac{\partial w}{\partial x}\right]\exp\left(-r\frac{\partial\varphi_0}{\partial x}\right)dr \tag{7.4}$$

For other terms of higher order, $\varphi_2, \varphi_3, \ldots$, see Appendix B in Ref. 16. For the typical Fokker-Planck equation (6.2), (7.3) and (7.4) take the form[3]

$$\frac{\partial \varphi_0}{\partial t} = -c_1(x)\frac{\partial \varphi_0}{\partial x} + \tfrac{1}{2}c_2(x)\left(\frac{\partial \varphi_0}{\partial x}\right)^2 \tag{7.5}$$

$$\frac{\partial \varphi_1}{\partial t} = -c_1(x)\frac{\partial \varphi_1}{\partial x} - c_1'(x) + \tfrac{1}{2}c_2(x)\frac{\partial^2 \varphi_0}{\partial x^2} + c_2'(x)\frac{\partial \varphi_0}{\partial x} + c_2(x)\frac{\partial \varphi_0}{\partial x}\frac{\partial \varphi_1}{\partial x} \tag{7.6}$$

respectively. As is discussed in Ref. 16, the solution of (7.5) with the initial condition $\varphi_0(x) = -x^2/(2\sigma_0)$ can be easily shown to take the asymptotic form $\varphi_0(x,t) \sim \exp(-2\gamma t)\varphi_0(x)$ for large t. Therefore, (7.5) can be reduced asymptotically to the simple linear equation

$$\frac{\partial \varphi_0}{\partial t} = -c_1(x)\frac{\partial \varphi_0}{\partial x} \tag{7.7}$$

because $(\partial\varphi_0/\partial x)^2 \sim \exp(-4\gamma t) \ll \partial\varphi_0/\partial x$. This corresponds to the drift equation for $P_{sc}(x,t)$. Similarly, (7.6) is reduced to

$$\frac{\partial}{\partial t}\varphi_1 = -c_1(x)\frac{\partial \varphi_1}{\partial x} - \frac{d}{dx}c_1(x) \tag{7.8}$$

in the second nonlinear regime (i.e., $\exp(-2\gamma t) \sim \varepsilon \ll 1$). If we solve these equations with the standard initial conditions

$$\varphi_0(x) = \frac{-x^2}{2\sigma_0} \quad \text{and} \quad \varphi_1(x) = -\tfrac{1}{2}\log(2\pi\sigma_0) \tag{7.9}$$

then we obtain the scaling solution (6.18) in the form

$$P_{sc}(x,t) = \exp\left[\frac{1}{\varepsilon}\varphi_0(x,t) + \varphi_1(x,t)\right] \tag{7.10}$$

As is easily seen from the above arguments, the nonlinearity of $\varphi_0(x,t)$ and the x dependence of the second term $\varphi_1(x,t)$ are important in deriving the scaling result.

B. The Ω-Expansion and Scaling Theory

As is clear from the arguments in Section IV, the scaling theory is closely related to the Ω-expansion. The scaling result can be obtained by summing up the most dominant terms in a series of the Ω-expansion. The ordinary Ω-expansion theory retains only the first-order term. It is rather complicated to calculate explicitly terms of higher order in ε, as is seen from the explicit calculation for the example (4.3) in Section 4.

VIII. RELATION BETWEEN LANGEVIN'S EQUATION AND THE FOKKER-PLANCK EQUATION WITH ARBITRARY DIFFUSION

As is well known, the Langevin equation (5.1) is equivalent to the Fokker-Planck equation (6.2). Here we discuss the equivalence of these approaches for general diffusion. Namely, we study the following Langevin equation:

$$\frac{d}{dt}x(t)=c(x)+\beta(x)\eta(t) \tag{8.1}$$

The simplest method to find the corresponding Fokker-Planck equation is to transform (8.1) to the simple Langevin equation (5.1) by the nonlinear transformation[22]

$$y=\int^{x}\frac{dx}{\beta(x)} \tag{8.2}$$

That is, we have

$$\frac{dy}{dt}=\frac{c(x)}{\beta(x)}+\eta(t) \tag{8.3}$$

Here $x=x(y)$, which is the inverse function of (8.2). The corresponding Fokker-Planck equation is given by

$$\frac{\partial}{\partial t}\tilde{P}(y,t)=-\frac{\partial}{\partial y}\frac{c(x)}{\beta(x)}\tilde{P}(y,t)+\varepsilon\frac{\partial^2}{\partial y^2}\tilde{P}(y,t) \tag{8.4}$$

This is transformed back to the Fokker-Planck equation of $P(x,t)=\tilde{P}(y,t)(dy/dx)=\tilde{P}(y,t)/\beta(x)$ as

$$\frac{\partial}{\partial t}P(x,t)=-\frac{dy}{dx}\frac{\partial}{\partial y}c(x(y))P(x(y),t)+\varepsilon\frac{dy}{dx}\frac{\partial}{\partial y}\frac{dx}{dy}\frac{\partial}{\partial x}\beta(x)P(x,t) \tag{8.5}$$

Thus we arrive at

$$\frac{\partial}{\partial t}P(x,t) = -\frac{\partial}{\partial x}c(x)P(x,t) + \varepsilon\frac{\partial}{\partial x}\beta(x)\frac{\partial}{\partial x}\beta(x)P(x,t) \tag{8.6}$$

or equivalently

$$\frac{\partial}{\partial t}P(x,t) = -\frac{\partial}{\partial x}\left[c(x)+\varepsilon\beta(x)\beta'(x)\right]P(x,t) + \varepsilon\frac{\partial^2}{\partial x^2}\beta^2(x)P(x,t) \tag{8.7}$$

The above result can be obtained directly by the definition of the coefficients $c_n(x)$ in (6.27) as follows:[23]

$$\begin{aligned}
c_1(x) &= \int r w(x,r)\,dr = \lim_{\Delta t\to 0}\frac{\langle \Delta x\rangle}{\Delta t} \\
&= c(x) + \lim_{\Delta t\to 0}\frac{1}{\Delta t}\int_t^{t+\Delta t}\langle \beta(x(s))\eta(s)\rangle ds \\
&= c(x) + \lim_{\Delta t\to 0}\frac{1}{\Delta t}\int_t^{t+\Delta t} ds\frac{\partial}{\partial x}\beta(x)\langle \int_t^s \dot{x}(u)du\,\eta(s)\rangle \\
&= c(x) + \lim_{\Delta t\to 0}\frac{1}{\Delta t}\frac{\partial\beta(x)}{\partial x}\cdot\int_t^{t+\Delta t} ds\int_0^s du\beta(x(u))\langle \eta(u)\eta(s)\rangle \\
&= c(x) + \varepsilon\beta(x)\beta'(x)
\end{aligned} \tag{8.8}$$

Similarly, we obtain

$$c_2(x) = \int r^2 w(x,r)\,dr = \lim_{\Delta t\to 0}\frac{\langle (\Delta x)^2\rangle}{\Delta t} = 2\varepsilon\beta^2(x) \tag{8.9}$$

The Fokker-Planck equation (8.7) has been also derived by Kubo[23] using the stochastic Liouville equation.

The extra factor $\varepsilon\beta(x)\beta'(x)$ in the drift term is of higher order in ε and may be omitted for small ε.

All the stochastic differential equations in this chapter are interpreted as being of the Stratonovich type rather than the Itô type.

The above generalized Fokker-Planck equation is used in the next section to present a unified treatment of transient phenomena near the instability point.

IX. UNIFIED TREATMENT OF TRANSIENT PHENOMEMA NEAR THE INSTABILITY POINT

A. General Formulation

The purpose of this section is to present a unified theory to treat transient phenomena near the instability point. The scaling solutions derived in sections V and VI are valid only in the second, nonlinear regime and they shrink to a linear combination of some delta functions in the final time region. This is because the random force is neglected in this time region. Therefore, we must include again the random force in order to obtain the asymptotically correct distribution function in the final regime.

For this purpose it is necessary to take into account[2] the effect of the random force as a Gaussian correction of fluctuation to the scaling result. That is, in the Langevin equation, we express the stochastic variable $x(t)$ as the sum of the following two variables:

$$x(t)=x_{sc}(t)+z(t) \tag{9.1}$$

Here the first term, $x_{sc}(t)$, denotes the scaling solution (5.9) or

$$x_{sc}(t)=F^{-1}\left(e^{\gamma t}F\left(e^{-\gamma t}\xi_{sc}(t)\right)\right) \tag{9.2}$$

with $\xi_{sc}(t)$ defined by (5.6). It should be noted that the first term in (9.1), $x_{sc}(t)$, still contains the random variable $\eta(t)$. The second term in (9.1), z, expresses the remaining fluctuating part. Now let us find the stochastic equation of $z(t)$. For this, we first must find the equation of motion of $x_{sc}(t)$. This can be derived easily through the transformation (5.22) from (5.26) or equivalently from

$$\frac{d}{dt}\xi_{sc}(t)=\gamma\xi_{sc}(t)+\eta(t) \tag{9.3}$$

The result is given by

$$\frac{d}{dt}x_{sc}(t)=c_1(x_{sc})+b_F(t)\eta(t) \tag{9.4}$$

where

$$b_F(t)=\frac{F'\left(e^{-\gamma t}\xi_{sc}\right)}{F'(x_{sc}(t))}=\frac{c_1(x_{sc}(t))}{\gamma F(x_{sc}(t))}F'\left(F^{-1}\left(e^{-\gamma t}F(x_{sc}(t))\right)\right) \tag{9.5}$$

As is shown in Refs. 2 and 9, $b_F(t)$ is reduced to $b_F(t)=c_1(x_{sc})/\gamma F(x_{sc})$ for large t in the second scaling regime. Clearly, $b_F(0)=1$ and $b_F(\infty)=0$. That is, the scaling treatment takes the random force into less and less account, as the time increases. Substituting (9.1) into Langevin's equation (5.1), we obtain

$$\frac{dz}{dt}=c_1(x_{sc}+z)-c_1(x_{sc})+[1-b_F(t)]\eta(t) \tag{9.6}$$

Since $x_{sc}(t)$ is already a good approximation in the intermediate time region, we may assume that the fluctuating part $z(t)$ is small (i.e., of the order of ε, which is true near the stationary state) and consequently we expand (9.6) with respect to z and retain it up to the first order in z. Thus we obtain

$$\frac{dz}{dt}=c_1'(x_{sc}(t))z+[1-b_F(t)]\eta(t) \tag{9.7}$$

The solution of (9.7) is given by

$$z(t)=\left[\exp\int_0^t c_1'(x_{sc}(s))\,ds\right]\int_0^t[1-b_F(s)]\eta(s)\exp\left[-\int_0^s c_1'(x_{sc}(u))\,du\right] \tag{9.8}$$

with the initial condition $z(0)=0$. In principle, any physical quantity $Q(x)$ can be calculated[2] as

$$\langle Q(x)\rangle=\langle Q(x_{sc}(t)+z)\rangle \tag{9.9}$$

in this unified treatment. The probability function $P(x,t)$ is expressed by

$$P(x,t)=\langle\delta(x(t)-x)\rangle=\frac{1}{2\pi i}\int_{c-i\infty}^{c+i\infty}\langle\exp is(x(t)-x)\rangle\,ds$$

$$=\frac{1}{2\pi i}\int_{c-i\infty}^{c+i\infty}ds\,e^{-isx}\sum_{n=0}^{\infty}\frac{(is)^n}{n!}\langle x^n(t)\rangle \tag{9.10}$$

The nth moment $\langle x^n(t)\rangle$ can be, in principle, calculated from (9.9).

B. Relation to Other Methods

To discuss the relationship among the present unified theory, Ω-expansion, and scaling theory, we introduce two parameters, λ and μ, into

(9.4) and (9.7) to obtain

$$\frac{d}{dt}x_{\rm sc}(t)=c_1(x_{\rm sc})+\lambda b_F(t)\eta(t) \tag{9.11}$$

and

$$\frac{dz(t)}{dt}=c_1'(x_{\rm sc})z+\mu[1-\lambda b_F(t)]\eta(t) \tag{9.12}$$

We have the following classification:

1. For $\lambda=\mu=0$, $x_{\rm sc}(t)$ is deterministic.
2. For $\lambda=0$, $\mu=1$, $x_{\rm sc}(t)$ goes to $y(t)$ (the most probable path and we recover the Ω-expansion (2.1). Here we have $z=\sqrt{\varepsilon}\ \xi$ in (2.1); the fluctuation $\langle z^2(t)\rangle\equiv\varepsilon\sigma_z(t)$ calculated from (9.12) is reduced to

$$\frac{d}{dt}\sigma_z(t)=2c_1'(y(t))\sigma_z(t)+2 \tag{9.13}$$

   which agrees with (2.9) for $c_2=2$.
3. For $\lambda=1$, $\mu=0$, the above equations yield the scaling result.
4. The final case $\lambda=1$ and $\mu=1$ describes the general, unified situation.

C. Simple Approximation for the Fluctuating Part $z(t)$

The above unified treatment gives a method to calculate explicitly any kind of fluctuation and even the probability distribution function. However, $x_{\rm sc}(t)$ and z are still stochastic and it is rather complicated to take the average over the random force $\eta(t)$. Here we propose further approximations to calculate $P(x,t)$ explicitly in a compact form.

One of simple approximations may be to replace the coefficients of z and $\eta(t)$ in (9.7) by averages as follows:

$$\frac{dz}{dt}=\langle c_1'(x_{\rm sc}(t))\rangle z+[1-\langle b_F(t)\rangle]\eta(t) \tag{9.14}$$

The probability distribution function $P_z(z,t)$ is easily obtained as

$$P_z(z,t)=\frac{1}{\sqrt{2\pi\varepsilon\sigma_z(t)}}\exp\left[-\frac{z^2}{2\varepsilon\sigma_z(t)}\right] \tag{9.15}$$

where

$$\frac{d}{dt}\sigma_z(t)=2\langle c_1'(x_{\rm sc}(t))\rangle\sigma_z(t)+2[1-\langle b_F(t)\rangle]^2 \tag{9.16}$$

Consequently, the original distribution function $P(x, t)$ for x is given by

$$P(x,t)=\int_{-\infty}^{\infty} P_z(x-y,t)\,P_{sc}(y,t)\,dy \tag{9.17}$$

For $t\to\infty$, we have

$$P_{sc}(x,t)\to\sum_{i=1}^{P} a_i\delta\left(x-x_{st}^{(i)}\right) \tag{9.18}$$

and consequently

$$P(x,t)\to\sum_{i=1}^{P}\frac{a_i}{\sqrt{2\pi\varepsilon\sigma_{st}^{(i)}}}\exp\left[-\frac{\left(x-x_{st}^{(i)}\right)^2}{2\varepsilon\sigma_{st}^{(i)}}\right]\simeq P_{st}(x) \tag{9.19}$$

That is, the above result (9.17) describes correctly the final time region asymptotically for small ε.

D. Time-Dependent Integral Transformation

A result similar to the one above (9.17) can be obtained directly by extending the transformation (6.39) to a time-dependent one. That is, we introduce the transformation

$$\begin{aligned}P(x,t)&=\exp\left[\varepsilon(t)\frac{\partial^2}{\partial x^2}\right]Q(x,t)\\&=\frac{1}{\sqrt{4\pi\varepsilon(t)}}\int_{-\infty}^{\infty}\exp\left[-\frac{(x-y)^2}{4\varepsilon(t)}\right]Q(y,t)\end{aligned} \tag{9.20}$$

The parameter, $\varepsilon(t)$ and $Q(x, t)$ are determined so that we may obtain a good solution in the whole time region. As in Section VI.D we substitute (9.20) into the Fokker-Planck equation (6.2) and obtain

$$\frac{\partial Q}{\partial t}=\left[\varepsilon-\varepsilon'(t)\right]\frac{\partial^2}{\partial x^2}Q-\frac{\partial}{\partial x}c_1\left(x-2\varepsilon(t)\frac{\partial}{\partial x}\right)Q \tag{9.21}$$

We expand the second term of the right hand side of (9.21) to get

$$c_1\left(x-2\varepsilon(t)\frac{\partial}{\partial x}\right)=c_1(x)-2\varepsilon(t)c_1'(x)\frac{\partial}{\partial x}-\varepsilon(t)c_1''(x)+O\left(\varepsilon^2(t)\right) \tag{9.22}$$

This can be easily confirmed by expanding $c_1(y)$ into a power series of y and by expanding again, $[x-2\varepsilon(t)(\partial/\partial x)]^n$

$$\left[x-2\varepsilon(t)\frac{\partial}{\partial x}\right]^n = x^n - 2n\varepsilon(t)x^{n-1}\frac{\partial}{\partial x} - \varepsilon(t)n(n-1)x^{n-2} + O(\varepsilon^2(t)) \tag{9.23}$$

Thus (9.21) is simplified to

$$\begin{aligned}\frac{\partial Q}{\partial t} &= -\frac{\partial}{\partial x}c_1(x)Q + \varepsilon(t)\left[\frac{\partial}{\partial x}(c_1''(x)Q) + 2c_1''(x)\frac{\partial Q}{\partial x}\right] \\ &\quad + \left[\varepsilon + 2\varepsilon(t)c_1'(x) - \varepsilon'(t)\right]\frac{\partial^2 Q}{\partial x^2} + O(\varepsilon^2(t))\end{aligned} \tag{9.24}$$

The second term can be neglected compared with the first term in (9.24), and the third diffusion term may be eliminated effectively by determining $\varepsilon(t)$ as

$$\frac{d}{dt}\varepsilon(t) = 2\langle c_1'(x)\rangle\varepsilon(t) + \varepsilon \tag{9.25}$$

Thus, $Q(x,t)$ satisfies the drift equation

$$\frac{\partial}{\partial t}Q(x,t) + \frac{\partial}{\partial x}c_1(x)Q(x,t) = 0 \tag{9.26}$$

asymptotically for small ε. Thus we may take the scaling solution $P_{sc}(x,t)$ in (6.17) as $Q(x,t)$; that is $Q(x,t) = P_{sc}(x,t)$. Since $P_{sc}(x,0) = P(x,0)$, the initial condition for $\varepsilon(t)$ is given by $\varepsilon(0)=0$. Therefore, the solution for $P(x,t)$ is expressed by the integral (9.20) with $Q(y,t) = P_{sc}(y,t)$. This agrees with (9.17) if we neglect the term $\langle b_F(t)\rangle$ in (9.16) and if we note that $2\varepsilon(t) = \varepsilon\sigma_z(t)$. The average $\langle c_1'(x)\rangle$ in (9.25) may be taken over $P_{sc}(x,t)$. This assures the approach of $P(x,t)$ to the final stationary state in the sense of (9.19).

E. Stochastic Fokker-Planck Equation

To solve the simultaneous stochastic differential equations (9.11) and (9.12) approximately, we propose here a new method, namely, the method of a stochastic Fokker-Planck equation or the separation of the random force into two independent ones. This idea of the stochastic Fokker-Planck equation is analogous to the stochastic Liouville equation proposed by Kubo.[21] In (9.11) and (9.12) for $\lambda = \mu = 1$, the two random forces $b_F(t)\eta(t)$

and $[1-b_F(t)]\eta(t)$ come from the original one $\eta(t)$ applied to the stochastic variable $x(t)$. Although (9.11) and (9.12) can be solved explicitly as is shown earlier, it is rather difficult to evaluate fluctuation explicitly, from the solution thus obtained, using (9.9), because $x_{sc}(t)$ and $z(t)$ in (9.9) contain the same random force $\eta(t)$ in a complicated way and then we must also evaluate the cross effect of $x_{sc}(t)$ and $z(t)$. Note that $\langle x_{sc}(t)z(t)\rangle \neq 0$. If the two random forces appearing in $x_{sc}(t)$ and $z(t)$ are taken to be independent, then the situation is very much simplified as follows. First, we construct the Fokker-Planck equation corresponding to the stochastic process $z(t)$ for given $x_{sc}(t)$, and the solution is denoted by $P_{\text{stoch}}(z, t, x_{sc}(t))$. It should be remarked that this distribution function still contains a different random force. Therefore, the final probability distribution function can be obtained by taking the average over the random force in $x_{sc}(t)$ to be

$$P(x,t)=\langle P_{\text{stoch}}(x-x_{sc}(t),t,x_{sc}(t))\rangle \tag{9.27}$$

The next problem is to find the stochastic Fokker-Planck equation for $P_{\text{stoch}}(z, t, x_{sc}(t))$. For this purpose, we first state the following theorem.

Theorem. The Langevin equation

$$\frac{dx}{dt}=c_1(x)+\alpha(t)\eta(t) \tag{9.28}$$

can be replaced by that with the following two independent random forces:

$$\frac{dx}{dt}=c_1(x)+\alpha_1(t)\eta_1(t)+\alpha_2(t)\eta_2(t) \tag{9.29}$$

where

$$\alpha_1^2(t)+\alpha_2^2(t)=\alpha^2(t) \tag{9.30}$$

and

$$\langle\eta_j(t)\eta_k(t')\rangle=2\varepsilon\delta_{jk}(t-t') \tag{9.31}$$

This is easily proved by noting that

$$\begin{aligned}&\langle[\alpha_1(t)\eta_1(t)+\alpha_2(t)\eta_2(t)][\alpha_1(t')\eta_1(t')+\alpha_2(t')\eta_2(t')]\rangle\\&=2\varepsilon[\alpha_1^2(t)+\alpha_2^2(t)]\delta(t-t')\\&=\alpha^2(t)\langle\eta(t)\eta(t')\rangle\end{aligned} \tag{9.32}$$

By applying this idea to the present problem, we may replace (9.4) and (9.7) respectively by,

$$\frac{dx_{sc}}{dt}=c_1(x_{sc})+b_F(t)\eta_1(t) \tag{9.33}$$

and

$$\frac{dz}{dt}=c_1'(x_{sc}(t))z+\left[1-b_F^2(t)\right]^{1/2}\eta_2(t) \tag{9.34}$$

Here the random force in $x_{sc}(t)$ and $b_F(t)$ is $\eta_1(t)$ and it is independent of $\eta_2(t)$. Thus the probability distribution function $P_{\text{stoch}}(z, t, x_{sc})$ satisfies the stochastic Fokker-Planck equation

$$\frac{\partial}{\partial t}P_{\text{stoch}}=\left[-\frac{\partial}{\partial z}c_1'(x_{sc}(t))z+\varepsilon\frac{\partial^2}{\partial z^2}\left(1-b_F^2(t)\right)\right]P_{\text{stoch}} \tag{9.35}$$

We omit a term of order ε in the first drift term, as is discussed in Section VIII. The solution of (9.35) is given by the Gaussian distribution

$$P_{\text{stoch}}(z,t,x_{sc}(t))=\frac{1}{\sqrt{2\pi\varepsilon\sigma_z(x_{sc},t)}}\exp\left[-\frac{z^2}{2\varepsilon\sigma_z(x_{sc},t)}\right] \tag{9.36}$$

where $\sigma_z(x_{sc}, t)$ is the solution of the equation

$$\frac{d}{dt}\sigma_z(x_{sc},t)=2c_1'(x_{sc}(t))\sigma_z(x_{sc},t)+2\left[1-b_F^2(t)\right] \tag{9.37}$$

This should be compared with (9.14). This variance $\sigma_z(x_{sc}(t), t)$ contains a stochastic variable $x_{sc}(t)$. The solution of (9.37) is easily obtained. Therefore, $P(x, t)$ is calculated from the formula

$$P(x,t)=\left\langle\frac{1}{\sqrt{2\pi\varepsilon\sigma_z(x_{sc}(t),t)}}\exp\left[-\frac{(x-x_{sc}(t))^2}{2\varepsilon\sigma_z(x_{sc}(t),t)}\right]\right\rangle \tag{9.38}$$

To express $P(x, t)$ in an integral form, we make a simple adiabatic approximation in solving (9.14), namely, by fixing the coefficients $2c_1'(x_{sc})$ and $2[1-b_F^2(t)]$, we solve (9.14) to obtain

$$\sigma_z(x_{sc},t)\cong\left[c_1'(x_{sc}(t))\right]^{-1}\left\{\exp\left[2c_1'(x_{sc}(t))t\right]-1\right\}\left\{1-b_F^2(x_{sc}(t),t)\right\} \tag{9.39}$$

Thus we finally arrive at

$$P(x,t) \cong \int_{-\infty}^{\infty} \frac{1}{\sqrt{2\pi\varepsilon\sigma_z(y,t)}} \exp\left[-\frac{(x-y)^2}{2\varepsilon\sigma_z(y,t)} \right] P_{sc}(y,t)\,dy \quad (9.40)$$

We propose several kinds of approximations above. Another simple and useful approximation is to take the average over the random force $\{\eta_1(t)\}$ in (9.34), that is, we have

$$\frac{dz}{dt} = \langle c_1'(x_{sc}(t))\rangle z + \left[1 - \langle b_F^2(t)\rangle\right]^{1/2} \eta_2(t) \quad (9.41)$$

Correspondingly, $P_z(z,t)$ satisfies the following linear equation:

$$\frac{\partial}{\partial t} P_z(z,t) = -\frac{\partial}{\partial z} \langle c_1'(x_{sc}(t))\rangle z P_z + \varepsilon(1 - \langle b_F^2(t)\rangle) \frac{\partial^2}{\partial z^2} P_z \quad (9.42)$$

The solution of this equation takes the Gaussian form (9.36) with the variance $\sigma_z(t)$ satisfying the equation

$$\frac{d}{dt}\sigma_z(t) = 2\langle c_1'(x_{sc}(t))\rangle \sigma_z(t) + 2\left[1 - \langle b_F^2(t)\rangle\right] \quad (9.43)$$

The above result (9.42) can be also obtained more directly as follows. First we assume that

$$P(x,t) = \int P_z(x-y,t) P_{sc}(y,t)\,dy \quad (9.44)$$

where $P(x,t)$ satisfies (6.2) and $P_{sc}(y,t)$ satisfies the Fokker-Planck equation

$$\frac{\partial}{\partial t} P_{sc}(y,t) = -\frac{\partial}{\partial y} c_1(y) P_{sc}(y,t) + \varepsilon \frac{\partial^2}{\partial y^2} b_F^2(t) P_{sc}(y,t) \quad (9.45)$$

Here $\varepsilon b_F(t)\,\partial b_F(t)/\partial y$ is omitted in the drift term because it is of higher order in ε for small ε. Substituting (9.44) into (6.2) and using (9.45), we obtain, by partial integration,

$$\int \frac{\partial P_z(x-y)}{\partial t} P_{sc}(y,t)\,dy - \int \frac{\partial}{\partial x} P_z(x-y,t) c_1(y) P_{sc}(y,t)\,dy + \varepsilon \int \frac{\partial^2}{\partial x^2} P_z(x-y,t) b_F^2(t) P_{sc}(y,t)\,dy = -\int \frac{\partial}{\partial x} P_z(x-y,t) c_1(x) P_{sc}(y,t)\,dy + \varepsilon \int \frac{\partial^2}{\partial x^2} P_z(x-y,t) P_{sc}(y,t)\,dy \quad (9.46)$$

By setting $z=x-y$ and separating integration, this may be approximately reduced to

$$\frac{\partial}{\partial t}P_z(z,t)=-\frac{\partial}{\partial z}[\langle c_1(y+z)\rangle-\langle c_1(y)\rangle]P_z+\varepsilon\frac{\partial^2}{\partial z^2}(1-\langle b_F^2(t)\rangle)P_z \tag{9.47}$$

where

$$\langle Q\rangle=\int QP_{sc}(y,t)dy \tag{9.48}$$

Expanding the drift term in z and keeping the first term in z, we arrive at (9.42). It is well understood from this derivation that the diffusion coefficient $\varepsilon(1-\langle b_F^2(t)\rangle)$ is quite a reasonable approximate expression.

It should be remarked here that all the above approximate solutions approach the correct stationary state asymptotically in the sense of (9.19).

It is useful to derive here a more explicit expression in the case when $P(x,t)$ is given by the convolution of the form (9.17) with the Gaussian distribution (9.15) for $P_z(z,t)$. For convenience, we write $P_z(z,t)$ and $P_{sc}(y,t)$ as

$$P_z(z,t)=\frac{1}{\sqrt{2\pi\varepsilon a(t)}}\exp\left[-\frac{z^2}{2\varepsilon a(t)}\right] \tag{9.49}$$

and

$$P_{sc}(y,t)=\frac{1}{\sqrt{2\pi\varepsilon b(t)}}\exp\left[-\frac{1}{2\varepsilon}\psi_0(y,t)\right]\frac{d\xi}{dy} \tag{9.50}$$

respectively. Here, $a(t)$ denotes $\sigma_z(t)$, which is given, for example, by (9.16) or (9.43), and $b(t)$ denotes τ/ε as given by (6.19). The function $\psi_0(y,t)$ takes the form

$$\psi_0(y,t)=\frac{1}{b(t)}\left[e^{\gamma t}F^{-1}(e^{-\gamma t}F(y))\right]^2 \tag{9.51}$$

which comes from (6.18), and ξ denotes the nonlinear transformation

$$\xi=e^{\gamma t}F^{-1}(e^{-\gamma t}F(y)) \tag{9.52}$$

Thus the distribution function $P(x, t)$ takes the form

$$P(x,t)=\frac{1}{\sqrt{(2\pi\varepsilon)^2 a(t)b(t)}}\int_{-\infty}^{\infty}\exp\left\{-\frac{1}{2\varepsilon}\left[\frac{(x-y)^2}{a(t)}+\psi_0(y,t)\right]\right\}\frac{d\xi}{dy}dy$$

$$=\frac{1}{2\pi\varepsilon\sqrt{a(t)b(t)}}\int_{-\infty}^{\infty}\exp\left[-\frac{1}{\varepsilon}f(\xi,x,t)\right]d\xi \tag{9.53}$$

where

$$f(\xi,x,t)=\frac{[x-y(\xi)]^2}{2a(t)}+\frac{\xi^2}{2b(t)} \tag{9.54}$$

Here $y(\xi)$ is the inverse function of (9.52).

In general, it is very difficult to integrate (9.53) rigorously. Here we are satisfied with the following asymptotic evaluation for small ε. That is, we make use of the saddle point method as follows:

$$P(x,t)\cong\frac{1}{\sqrt{2\pi\varepsilon a(t)b(t)}}\left[\frac{1}{f''(\xi_0(x))}\right]^{1/2}\exp\left[-\frac{1}{\varepsilon}f(\xi_0(x),x,t)\right] \tag{9.55}$$

where the saddle point $\xi_0(x)$ is determined from

$$\frac{\partial f(\xi_0,x,t)}{\partial\xi_0}=0 \tag{9.56}$$

that is,

$$x=y(\xi_0)+\left[\frac{a(t)}{b(t)}\right]\left[\frac{\xi_0}{y'(\xi_0)}\right] \tag{9.57}$$

The second derivative $f''(\xi_0)$ is also given by

$$\frac{\partial^2 f(\xi_0)}{\partial\xi_0^2}=\frac{[y'(\xi_0)]^2}{a(t)}-\frac{y''(\xi_0)(x-y(\xi_0))}{a(t)}+\frac{1}{b(t)} \tag{9.58}$$

These yield $P(x, t)$ in a parameter representation (9.57) in the form of (9.55), or

$$P(x,t) \cong \frac{1}{\sqrt{2\pi\varepsilon a(t)}} \left(1 + \frac{b(t)}{a(t)} \{[y'(\xi_0)]^2 - y''(\xi_0)(x - y(\xi_0))\}\right)$$

$$\times \exp\left[-\frac{1}{2\varepsilon}\left\{\frac{[x - y(\xi_0(x))]^2}{a(t)} + \frac{\xi_0^2(x)}{b(t)}\right\}\right] \tag{9.59}$$

with (9.57). Since $a(t) \to \sigma_{st}$ and $b(t) \to \infty$ for $t \to \infty$, the saddle point ξ_0 goes to infinity as $\xi_0 / y'(\xi_0) \sim b(t) \to \infty$. Therefore, the normalization factor in (9.59) approaches the stationary value $\{2\pi\varepsilon\sigma_{st}\}^{-1/2}$ and $y(\xi_0(x))$ goes to the stationary value x_{st}. Consequently $P(x, t)$ approaches the correct stationary state in the sense of (9.19). It should be remarked that the ε-dependence of the solution is quite the same as that of the scaling solution, namely,

$$P(x,t) \cong \exp\left[\frac{1}{\varepsilon}\varphi_0(x,t) + \varphi_1(x,t)\right] \tag{9.60}$$

in the whole region of time. The first term, $\varepsilon^{-1}\varphi_0(x, t)$, gives Kubo's extensivity property,[3] and the second part, $\varphi_1(x, t)$, gives a correction to it. This x-dependent correction is very essential in describing the fluctuation and relaxation from the instability point, as is shown above.

In particular, for the case of (6.51), we obtain

$$P(x,t) \cong \frac{1}{\sqrt{2\pi\varepsilon a(t)}} \left\{\frac{1 + 4(g/\gamma)\xi_0^2(1 - e^{-2\gamma t})}{1 + (g/\gamma)\xi_0^2(1 - e^{-2\gamma t})} + \frac{b(t)}{a(t)}\left[1 + \xi_0^2 \cdot \frac{g}{\gamma}(1 - e^{-2\gamma t})\right]^{-3}\right\}$$

$$\times \exp\left[-\frac{1}{2\varepsilon}\left(\frac{1}{a(t)}\left\{x - \xi_0\left[1 + \frac{g}{\gamma}\xi_0^2(1 - e^{-2\gamma t})\right]^{-1/2} + \frac{\xi_0^2}{b(t)}\right)\right] \tag{9.61}$$

where the parameter ξ_0 is determined by

$$x = \xi_0\left[1 + \frac{g}{\gamma}\xi_0^2(1 - e^{-2\gamma t})\right]^{-1/2} + \frac{a(t)}{b(t)}\xi_0\left[1 + \frac{g}{\gamma}\xi_0^2(1 - e^{-2\gamma t})\right]^{3/2} \tag{9.62}$$

The function $a(t)$ may be determined from (9.43), namely, in the present case it satisfies the equation

$$\frac{d}{dt}a(t)=2\left[\gamma-3g\langle x_{\rm sc}^2(t)\rangle\right]a(t)+2\left[1-\langle b_F^2(t)\rangle\right] \tag{9.63}$$

with $a(0)=0$, where

$$\begin{aligned}\langle b_F^2(t)\rangle &= \left\langle\left[1-\left(\frac{g}{\gamma}\right)x_{\rm sc}^2(t)(1-e^{-2\gamma t})\right]^3\right\rangle \\ &\simeq\left[1-\left(\frac{g}{\gamma}\right)\langle x_{\rm sc}^2(t)\rangle(1-e^{-2\gamma t})\right]^3\end{aligned} \tag{9.64}$$

The function $b(t)$ is given by $b(t)=\tau/\varepsilon$ from (5.18).

X. PATH INTEGRAL FORMULATION OF THE FOKKER-PLANCK EQUATION

A. Transformation to a Hermitian Form

The original Fokker-Planck operator $\mathcal{H}$ defined by

$$\mathcal{H}P=\left[-\frac{\partial}{\partial x}c_1(x)+\varepsilon\frac{\partial^2}{\partial x^2}\right]P \tag{10.1}$$

is not Hermitian. It is convenient to transform $\mathcal{H}$ into a Hermitian form as follows[24–26]:

$$\tilde{\mathcal{H}}\equiv P_{\rm eq}^{-1/2}\mathcal{H}P_{\rm eq}^{1/2}=\varepsilon\frac{\partial^2}{\partial x^2}-V(x) \tag{10.2}$$

where

$$V(x)=\frac{c_1^2(x)}{4\varepsilon}+\tfrac{1}{2}c_1'(x) \tag{10.3}$$

and $P_{\rm eq}$ denotes the equilibrium probability distribution function

$$P_{\rm eq}\propto\exp\frac{1}{\varepsilon}\int^x c_1(x)\,dx \tag{10.4}$$

Correspondingly, $P(x,t)$ is transformed as

$$P(x,t)=\exp\left(\frac{1}{2\varepsilon}\int^{x}c_1(x)\,dx\right)G(x,t) \tag{10.5}$$

and $G(x,t)$ satisfies

$$\frac{\partial}{\partial t}G(x,t)=\tilde{\mathcal{H}}G(x,t) \tag{10.6}$$

Here $\tilde{\mathcal{H}}$ is Hermitian

B. Path Integral Formulation

The formal solution of (10.6) for the initial condition $G(x,0)=\delta(x)$ is given[22] by the Wiener-Feynman-Kac formula:

$$G(x,t)=\int_{x(0)=0}^{x(t)=x}Dx(t)\left\langle \exp\left[-\int_0^t V(x(s))\,ds\right]\right\rangle \tag{10.7}$$

where $\langle\ldots\rangle$ denotes the expectation on the Wiener process. More explicitly, we have

$$G(x,t)=\lim_{n\to\infty}\int_{-\infty}^{\infty}\cdots\int_{-\infty}^{\infty}\exp\left[-\frac{t}{n}\sum_{k=1}^{n-1}V(x_k)\right]$$
$$\times p(x-x_{n-1},t-t_{n-1})\cdots p(x_1,t_1)\,dx_1\cdots dx_{n-1} \tag{10.8}$$

for $t_k-t_{k-1}=t/n$ $(k=1,\ldots,n;\ t_0=0,\ t_n=t)$, where

$$p(x,t)=\frac{1}{\sqrt{4\pi\varepsilon t}}\exp\left(-\frac{x^2}{4\varepsilon t}\right) \tag{10.9}$$

Here $p(x,t)$ gives the measure of this Wiener integral and the convergence of (10.8) for a continuous positive $V(x)$ has been proved. This is an example of a path integral that can be defined rigorously. In most cases we encounter the problem of measure in the definition of path integral. The expression (10.8) is also rewritten in terms of (10.9) as

$$G(x,t)= \lim_{n\to\infty} \left(\frac{4\pi\varepsilon t}{n}\right)^{-n/2} \int_{-\infty}^{\infty}$$

$$\times \cdots \int_{-\infty}^{\infty} \exp\left\{ -\frac{1}{n}\sum_{k=1}^{n}\left[\frac{1}{4\varepsilon}\left(\frac{x_k - x_{k-1}}{t/n}\right)^2 + V(x_k)\right]\right\} dx_1 \cdots dx_n \tag{10.10a}$$

$$= \int_{x(0)=0}^{x(t)=x} Dx(t) \exp\left\{ -\int_0^t \left[\frac{1}{4\varepsilon}\left(\frac{dx}{ds}\right)^2 + V(x(s))\right] ds\right\} \tag{10.10b}$$

As has been discussed by several authors,[3, 26–29] we must be careful about the difference form in (10.10a). Otherwise, there the problem of non-uniqueness appears. Including the prefactor in (10.5), we obtain the following well-known formula

$$P(x,t)= \int_{x(0)=0}^{x(t)=x} Dx(t) \exp\left(-\int_0^t \left\{ \frac{1}{4\varepsilon}\left[\frac{dx}{ds} - c_1(x(s))\right]^2 + \tfrac{1}{2}c_1'(x(s))\right\} ds\right) \tag{10.11}$$

This formula can be easily extended to the general Fokker-Planck equation

$$\frac{\partial}{\partial t}P(x,t) = -\frac{\partial}{\partial x}c_1(x)P(x,t) + \varepsilon\frac{\partial^2}{\partial x^2}\beta^2(x)P(x,t) \tag{10.12}$$

by applying the transformation (8.2). The transformed probability distribution function $\tilde{P}(y,t)=P(x,t)(dx/dy)=P(x,t)\beta(x)$ satisfies the equation

$$\frac{\partial \tilde{P}}{\partial t} = -\frac{\partial}{\partial y}\tilde{c}(y)\tilde{P} + \varepsilon\frac{\partial^2}{\partial y^2}\tilde{P} \tag{10.13}$$

where

$$\tilde{c}(y) = \frac{c_1(x)}{\beta(x)} - \varepsilon\beta'(x) \tag{10.14}$$

with the relation (8.2) for $x=x(y)$. Therefore, from the (10.11), we obtain

$$\tilde{P}(y,t)= \int_{y(0)=0}^{y(t)=y} Dy(t) \exp\left[-\frac{1}{4\varepsilon}(\dot{y} - \tilde{c}(y))^2 - \tfrac{1}{2}\tilde{c}'(y)\right] \tag{10.15}$$

We now transform this back to $P(x,t)$ and obtain

$$P(x,t)=\tilde{P}(y,t)\frac{dy}{dx}$$

$$=\int_{x(0)=0}^{x(t)=x} Dx(t)\exp\left\{-\frac{1}{4\varepsilon\beta^2(x)}\left[\dot{x}-\beta(x)\tilde{c}(y(x))\right]^2 - \tfrac{1}{2}\beta(x)\frac{d}{dx}\tilde{c}(y(x))\right\} \quad (10.16)$$

This agrees with the results of Horsthemke and Bach,[30] Graham[27] and Mühlschlegel.[22]

The above path integral formulation can be applied to the study of the Ω expansion and scaling theory. For small ε, the term $\frac{1}{2}c_1'(x(s))$ in (10.11) and $\frac{1}{2}\beta(x)(d\tilde{c}/dx)$ in (10.16) can be omitted. The Ω-expansion is reformulated below. The most probable path $y_m(t)$ is determined by maximizing the integrand to yield

$$\dot{y}_m=c_1(y_m(t)) \quad (10.17)$$

for small ε. The fluctuation or variance $\langle[x-y_m(t)]^2\rangle\equiv\varepsilon\sigma_z(t)$ is obtained from the probability distribution for $z=x-y_m(t)$:

$$\hat{P}(z,t)=\int_{z(0)=0}^{z(t)=z} Dz(t)\exp\left\{-\frac{\left(\dot{z}-c_1'(y_m(t))z\right)^2}{4\varepsilon\beta^2[y_m(t)]}\right\} \quad (10.18)$$

which is easily derived from (10.16) for small ε. Thus $\sigma_z(t)$ is easily shown to satisfy the equation

$$\frac{d}{dt}\sigma_z(t)=2c_1'(y_m(t))\sigma_z(t)+2\beta^2(y_m(t)) \quad (10.19)$$

which is equivalent to (2.9) with $c_2(x)=2\beta^2(x)$. The scaling result is derived as follows. First we introduce the nonlinear transformation

$$\xi=\xi(x,t)=e^{\gamma t}F^{-1}(e^{-\gamma t}F(x)) \quad (10.20)$$

which differs from (6.3) by the factor $e^{\gamma t}$. Note that

$$\dot{x}-c_1(x)=\frac{(\dot{\xi}-\gamma\xi)F'(e^{-\gamma t}\xi)}{F'(x)} \quad (10.21)$$

Then we have

$$P(x,t)=\int_{x(0)=0}^{x(t)=x} Dx(t)\exp\left\{-\frac{(\dot{\xi}-\gamma\xi)^2}{4\varepsilon\beta^2(x)}\left[\frac{F'(e^{-\gamma t}\xi)}{F'(x)}\right]^2\right\} \tag{10.22}$$

If we replace in (10.22) $[F'(e^{-\gamma t}\xi)/F'(x)]^2$ by unity and $\beta(x)$ by $\beta(0)$, then we arrive at the scaling solution $P_{sc}(x,t)$ after transforming ξ back to the original variable x in (10.22).

XI. THE WKB METHOD AND SCALING PROPERTY

Quite recently Caroli et al.[31,32] studied the diffusion in a bistable potential corresponding to the model (4.3) using the WKB method. As is discussed in Section X the Fokker-Planck equation is transformed to the following Schrödinger-like equation

$$\frac{\partial}{\partial t}G(x,t)=\left[\varepsilon\frac{\partial^2}{\partial x^2}-V(x)\right]G(x,t) \tag{11.1}$$

Here, $P(x,t)$ is related to $G(x,t)$ through (10.5) and the potential $V(x)$ is given by (10.3). The specific potential corresponding to (5.3) takes the form

$$V(x)=\frac{1}{4\varepsilon}\left[g^2x^6-2g\gamma x^4+(\gamma^2-6\varepsilon g)x^2+2\varepsilon\gamma\right] \tag{11.2}$$

This potential has three minima and two maxima. Consequently, Caroli et al. divided the x-region into seven regimes and studied the following eigenvalue problems:

$$-\varepsilon\frac{d^2}{dx^2}\varphi_n(x)+V(x)\varphi_n(x)=\lambda_n\varphi_n(x) \tag{11.3}$$

The probability function $P(x,t)$ is expressed by the sum of the series

$$P(x,t|x_0)=P_{eq}^{-1/2}(x)P_{eq}^{1/2}(x_0)\sum_{n\geq 0}\varphi_n(x_0)\varphi_n(x)e^{-t\lambda_n/\varepsilon} \tag{11.4}$$

for the initial condition $P_0(x)=\delta(x-x_0)$ at $t=0$. The parameter ε plays a role of $\hbar$ in quantum mechanics and consequently the WKB method is useful for small ε. Caroli et al. distinguished three time regions qualitatively, namely, the initial, intermediate, and Kramers regimes. The final Kramers regime can be described by the knowledge of the first two levels of the potential $V(x)$ in this WKB treatment, and the well-known Kramers relaxation time is obtained.

From our point of view, the treatment of the intermediate regime is quite interesting. The sum of the power series (11.4) has been found to be performed explicitly in this intermediate time regime except for the close vicinity of the final stationary points of x. The result thus obtained agrees exactly with the scaling distribution function $P_{sc}(x,t)$ in (6.52) without the factor $[1-\exp(-2\gamma t)]$. Caroli et al. emphasized that the above confirmation of the scaling result is made only where the WKB approximation is valid, and that the scaling property of $P_{sc}(x,t)$ is violated near the stationary points. Obviously, the scaling property of $P(x,t)$ cannot be valid near $x-x_{st}=O(\sqrt{\varepsilon}\,)$, because the random force has been neglected near $x\approx x_{st}$, as is easily seen from (9.4). However, all the moments $\{\langle x^n(t)\rangle\}$ have the scaling property in the second time region, as we prove in Sections IV and V, and also as is discussed by Caroli et al. The onset time t_0 obtained in Section III.A in the form of (3.19) or (4.14) is the time at which the single peak of $P(x,t)$ begins to decompose into double peaks. Caroli et al. have also obtained the time t_b for obtaining peaks of width comparable with that at equilibrium, namely, the time at which the double peaks have well developed. Their result is

$$t_b=t_0+t_0'; \qquad t_0'=\frac{1}{4\gamma}\log\left(\frac{2\gamma^2}{\varepsilon g}\right)\approx\tfrac{1}{2}t_0 \tag{11.5}$$

Thus t_b is also of the same order as the onset time t_0. Time t_b can be also obtained from the unified theory presented in Section IX, and it is easily confirmed to be of the same order as t_0.

Caroli et al. have also pointed out that the WKB solution, namely, the scaling solution (6.52) takes the following universal Gaussian form:

$$P_{sc}(x,t)dx=\frac{1}{\sqrt{\pi}}e^{-u^2}du \tag{11.6}$$

for the new scaling variable u defined by $u=\xi/\sqrt{2\tau}$, when the initial distribution is a delta function. This new type of scaling property is also confirmed for the Gaussian initial distribution and even for an arbitrary initial distribution, because the general scaling solution (6.23) is reduced to (6.18) with the Gaussian initial distribution.

Caroli et al.[32] extended this treatment to multicomponent systems and obtained results similar to the above one. Quite recently Arimitsu[33] applied this WKB method to the laser problem[34, 35] and confirmed the previous scaling result[35] in the intermediate time region.

XII. EXTENSION OF THE SCALING THEORY TO TIME-DEPENDENT GINZBURG-LANDAU MODEL SYSTEMS

It is quite interesting to extend the present theory to systems with infinite degrees of freedom. In fact, Kawasaki et al.[36] extended the scaling theory to the TDGL model without conservation laws, namely, to the Glauber model in the weak coupling limit. They studied the time evolution of the fluctuations of a system suddenly quenched into a thermodynamically unstable state. The model is described by

$$\frac{\partial}{\partial t}P(\{s\},t)=-\int_{\vec{k}}\frac{\partial}{\partial s_{\vec{k}}}c_{\vec{k}}(\{s\})P(\{s\},t)+L\int_{\vec{k}}\frac{\partial^2}{\partial s_{\vec{k}}\partial s_{-\vec{k}}}P(\{s\},t) \quad (12.1)$$

where

$$c_{\vec{k}}(\{s\})=\gamma_{\vec{k}}s_{\vec{k}}-gL\int_{\vec{k}_1}\int_{\vec{k}_2}\int_{\vec{k}_3}\delta(\vec{k}-\vec{k}_1-\vec{k}_2-\vec{k}_3)s_{\vec{k}_1}s_{\vec{k}_2}s_{\vec{k}_3} \quad (12.2)$$

with $\gamma_{\vec{k}}=\Delta(\kappa^2-k^2)$. The positive sign in front of κ^2 implies that the initial system is unstable. As is shown in Sections V and VI for a single macrovariable, the nonlinearity of the system and the diffusion effect can be treated separately. That is, we first take the diffusion effect into account. This corresponds to replacing the initial probability by the modified one

$$\tilde{P}_0(\{s\})=\text{const}\cdot\exp\left(-\tfrac{1}{2}\int_{\vec{k}}\frac{|s_{\vec{k}}|^2}{\tilde{\chi}_{\vec{k}}}\right) \quad (12.3)$$

where $\tilde{\chi}_{\vec{k}}=\chi_{\vec{k}}+L/\gamma_{\vec{k}}$. The second part $L/\gamma_{\vec{k}}$ in $\chi_{\vec{k}}$ expresses the diffusion effect. The nonlinearity of the system is treated by solving the drift equation

$$\frac{\partial}{\partial t}P(\{s_k\},t)=-\int_{\vec{k}}\frac{\partial}{\partial s_{\vec{k}}}c_{\vec{k}}(\{s_{\vec{k}}\})P(\{s_k\},t) \quad (12.4)$$

with the modified initial condition (12.3). Thus the solution (so-called scaling solution) is formally expressed by

$$\begin{aligned}P(\{s\},t)&=\exp\left[-t\int_{\vec{k}}\frac{\partial}{\partial s_{\vec{k}}}c_{\vec{k}}(\{s\})\tilde{P}_0(\{s\})\right]\\&=N(t)\exp\left\{-\frac{1}{2v_0}\int d\vec{r}\ln\left[1-\alpha s^2(\vec{r})\right]\right\}\\&\quad\times\tilde{P}_0\left(e^{-tL(\kappa^2+\nabla^2)}s(\vec{r})\left[1-\alpha s^2(\vec{r})\right]^{-1}\right)\end{aligned} \quad (12.5)$$

in the second nonlinear time region, where $\alpha = gL/\gamma_0$, $N(t)$ denotes the normalization, and the factor v_0 is the cell volume arising from the division of the system into small cells before taking the continuum limit. Clearly, this is a direct extension of the previous result (6.52) without the extra factor $(1-e^{-2\gamma t})$, which becomes unity for large t. From this microscopic expression for $P_0(\{S\}, t)$, the one-point and two-point correlation functions have been calculated explicitly. One-point function is given by (6.52) obtained by the present author, as it should be. The two-point correlation $C_2(\vec{r}, t)$ is given by a complicated integral.[36] It is expressed[36] as a function of the dimensionless distance x and the dimensionless time $y(t)$ defined by

$$x = \frac{\kappa|\vec{r}|}{(2\gamma_0 t)^{1/2}} \quad \text{and} \quad y(t) = \frac{g\tilde{\chi}_0 \exp(2\gamma_0 t)}{(4\pi)^{d/2}\kappa^2 (2Lt)^{d/2}} \tag{12.6}$$

The $t^{1/2}$-dependence has been confirmed by Kawabata and Kawasaki[37] using the Monte Carlo method.

XIII. COMPUTER SIMULATION OF NONLINEAR BROWNIAN MOTIONS

A. General Formulation

First we explain a general method of the Monte Carlo simulation for stochastic processes described by the Langevin equation

$$\frac{d}{dt}x(t) = c(x) + \beta(x)\eta(t) \tag{13.1}$$

with the Gaussian white random force satisfying (3.2). The time region $0 \le t' \le t$ is divided according to $t_0 = 0$, $t_1 = t/n, \ldots, t_j = jt/n, \ldots t_n = t$. Correspondingly, the random force $\eta(t)$ is also replaced by the following Gaussian white random force η_j defined at discrete times $\{t_j\}$:

$$\langle \eta_i \eta_j \rangle = \varepsilon_p^2 \delta_{ij} \tag{13.2}$$

and

$$P(\eta_j) = \frac{1}{\sqrt{2\pi\varepsilon_p^2}} \exp\left(-\frac{\eta_j^2}{2\varepsilon_p^2}\right) \tag{13.3}$$

Then the Langevin equation (13.1) is rewritten as the difference equation

$$x_{j+1} - x_j = c(x_j)\Delta t + \beta\left(\tfrac{1}{2}(x_j + x_{j+1})\right)\eta_j \Delta t \tag{13.4}$$

Now we obtain the relation between ε and ε_p^2. First note that

$$\int_0^t \eta(t)dt = \lim_{\Delta t \to 0} \sum_j \eta_j \Delta t \tag{13.5}$$

Then, from (3.2), we have

$$\left\langle \int_0^t \eta(t)dt\, \eta(t') \right\rangle = 2\varepsilon \tag{13.6}$$

that is,

$$2\varepsilon = \sum_j \Delta t \langle \eta_j \eta_k \rangle = \sum_j \Delta t\, \varepsilon_p^2 \delta_{j,k} = \varepsilon_p^2 \Delta t \tag{13.7}$$

Therefore, we obtain the relation

$$\varepsilon = \tfrac{1}{2}\varepsilon_p^2 \Delta t \tag{13.8}$$

B. A Simple Example

It is instructive to study the linear Langevin equation

$$\frac{d}{dt}x(t) = -\gamma x(t) + \eta(t) \tag{13.9}$$

in our formulation. From (13.4), we have

$$x_{j+1} = (1-\gamma\Delta t)x_j + \eta_j \Delta t \tag{13.10}$$

By iterating this difference equation, we obtain

$$x_n = \sum_{j=1}^{n} (1-\gamma\Delta t)^{n-j} \eta_{j-1} \Delta t + (1-\gamma\Delta t)^n x_0 \tag{13.11}$$

This is a discrete representation of the solution

$$x(t) = \int_0^t e^{\gamma(t-s)} \eta(s)ds + x_0 e^{\gamma t} \tag{13.12}$$

From (13.11), the expectation value of x_n is given by

$$\langle x_n \rangle = (1-\gamma\Delta t)^n \langle x_0 \rangle \to \langle x_0 \rangle e^{-\gamma t} \qquad \text{for} \qquad \Delta t \to 0 \tag{13.13}$$

with $n\,\Delta t = t$. Similarly, the fluctuation $\langle x_n^2 \rangle$ is expressed as

$$\langle x_n^2 \rangle = (1-\gamma\Delta t)^{2n} \langle x_0^2 \rangle + \sum_{j=1}^{n} (1-\gamma\Delta t)^{2(n-j)} (\Delta t)^2 \langle \eta_{j-1}^2 \rangle$$

$$= (1-\gamma\Delta t)^{2n} \langle x_0^2 \rangle + 2\varepsilon_p^2 (\Delta t)^2 \cdot \left[(1-\gamma\Delta t)^{2n} - 1 \right] \left[1 - (1-\gamma\Delta t)^2 \right]^{-1}$$

$$\rightarrow \langle x_0^2 \rangle e^{-2\gamma t} + \left(\frac{\varepsilon}{\gamma} \right) (1 - e^{-2\gamma t}) \tag{13.14}$$

Here the relation (13.2) is used. All these results are well known and are derived immediately from the integral representation (13.12). However, the above demonstration will be useful in applying the Monte Carlo method to the Langevin equation, on the basis of the difference equation (13.4).

C. Monte Carlo Method for Langevin's Equation

The random force η_j is generated by using the central limit theorem. That is, η_j is generated from the summation

$$\eta_j = \frac{1}{\sqrt{m}} \sum_{k=1}^{m} \alpha_{jk}; \qquad \alpha_{jk} = \pm \varepsilon_p \tag{13.15}$$

for large m. Here, α_{jk} is a stochastic variable that takes two values $\pm \varepsilon_p$ with equal probability. The central limit theorem assures that the above summation approaches the Gaussian distribution (13.3), as m goes to infinity.

We have studied[38] the nonlinear Langevin equation (13.1), for example, with $c(x) = \gamma x - g x^3$ and $\beta(x) = 1$, by applying the Monte Carlo method to the Langevin equation (13.1) on the basis of the above formulation. With the help of (13.15), we generate a series of random forces $\{\eta_j\}$ and correspondingly we obtain a series of $\{x_j\}$. This gives a sample path for this stochastic process. We repeat this procedure many times (say, 100 times) or practically generate many series of $\{\eta_j\}$ and consequently $\{x_j\}$ simultaneously in a high-speed computer, and we take the average over all the samples at each step. A few typical sample paths for $c(x) = \gamma x - g x^3$ are shown in Fig. 10. Since the strength of random force, ε, is very small in this study, as in the scaling theory, the sample paths do not fluctuate very much, except for the initial time region. The initial value $x(0) = 0$ is just an unstable equilibrium point, and consequently it takes much time for the system to deviate from this unstable point. This delay time is stochastic and depends on the strength of random force ε. In the intermediate regime, the nonlinearity of the system is vital as asserted by the scaling theory. The expectation value $\langle x^2(t) \rangle$ taken over 100 samples is shown in Fig. 11 together

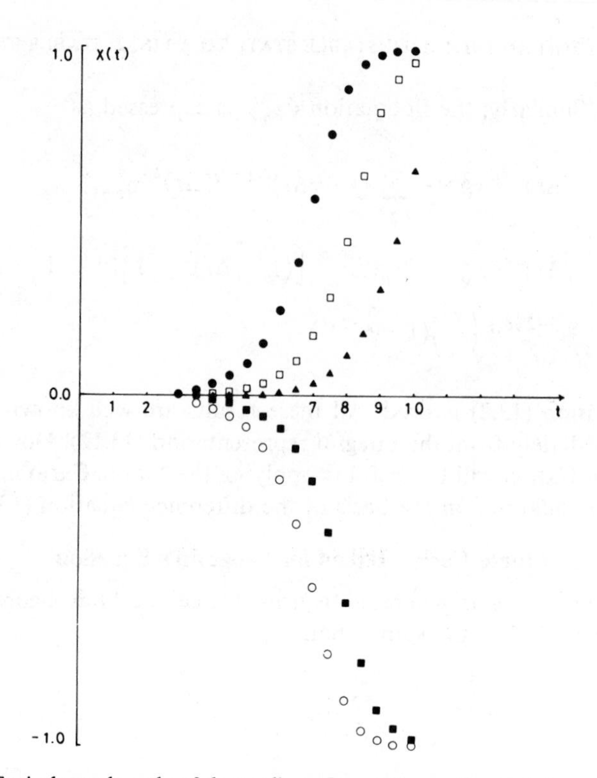

Fig. 10. Typical sample paths of the nonlinear Langevin equation (3.1) for $\gamma = g = 1$, $\varepsilon = 0.5 \times 10^{-6}$, $\varepsilon_p = 0.01$, and $\Delta t = 0.01$.

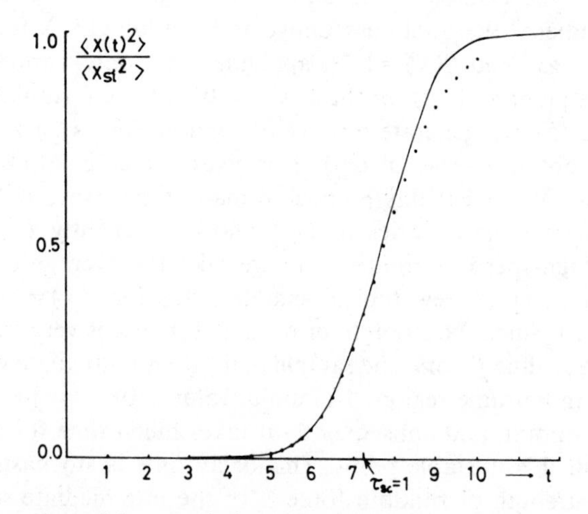

Fig. 11. Comparison of the numerical result of the Monte Carlo simulation with the scaling solution: (——) scaling result and · the numerical result for the values $\gamma = g = 1$, $\varepsilon = 0.5 \times 10^{-6}$, $\varepsilon_p = 0.01$ and $\Delta t = 0.01$.

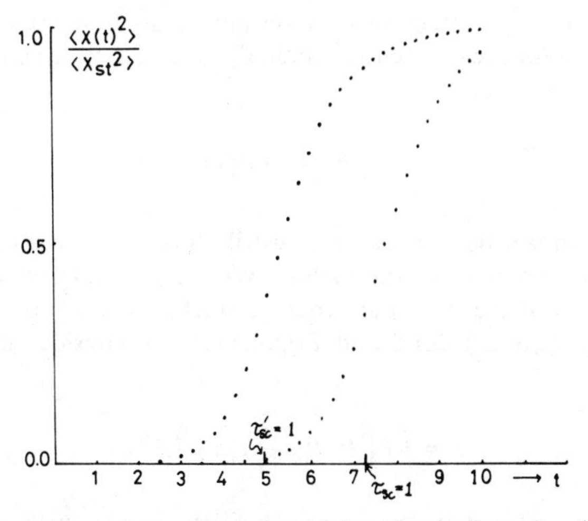

Fig. 12. Two typical numerical results in the real time t for the fluctuation $\langle x^2(t)\rangle$ for typical values of the parameters $\varepsilon_p = 0.1(\varepsilon = 0.5\times 10^{-4}, \Delta t = 0.01)$ and $\varepsilon_p = 0.01(\varepsilon = 0.5\times 10^{-6}, \Delta t = 0.01)$, respectively.

with the scaling result (5.33). The agreement between the two results is remarkable in the initial and second intermediate regimes. If we plot the result obtained by the unified theory given in Section 9, then a much better agreement in the third regime will be made. A few other numerical results are shown in Fig. 12 for different values of parameters. Clearly, all the results are expressed by the universal scaling function of the scaling variable τ. Therefore, the above Monte Carlo simulation of the nonlinear Langevin equation confirms our scaling theory.

The distribution function $P(x, t)$ can also be calculated from the above Monte Carlo simulation, if it is repeated enough times to evaluate the distribution of $x(t)$ at each time. Of course, one can solve the Fokker-Planck equation numerically.[24] However, it takes much time if one starts from the initial delta function $P_0(x) = \delta(x)$ for small diffusion constant ε. Then it is much more convenient to use an initial Gaussian distribution with large variance σ_0.

XIV. VARIATIONAL PRINCIPLES WITH APPLICATIONS TO TRANSIENT PHENOMENA

A. Variational Principles in Nonequilibrium

Onsager and Machlup[39] and Glansdorff and Prigogine[40] proposed variational principles to treat nonequilibrium systems.

Here we focus our attention only on applications of Prigogine's variational principle (or local potential method) to stochastic systems described by the master equation

$$\frac{\partial}{\partial t}P(t)=\Gamma P(t) \tag{14.1}$$

where $P(t)$ denotes the microscopic distribution function and Γ the temporal evolution operator of the system. We are now interested in the formation process of macroscopic order in stochastic systems described by (14.1). Following Glansdorff and Prigogine,[40] we consider the following Lagrangian

$$L=\int F\left(P,P_0,\frac{\partial}{\partial t}P_0;\Gamma\right)d^Nx \tag{14.2}$$

where P is a variational distribution function, P_0 is an auxiliary function that belongs to the same functional space as P, and $\int\dots d^Nx$ denotes the integral or trace over all the stochastic variables. The variation δL should be taken for P_0 fixed, and we set

$$\delta L=0 \tag{14.3}$$

Then $P(t)$ is determined as a functional of $P_0(t)$. This can be regarded as a transformation from $P_0(t)$ to $P(t)$:

$$P(t)=\mathfrak{T}P_0(t) \tag{14.4}$$

Since $P_0(t)$ is arbitrary in a certain functional space, $P(t)$ changes correspondingly and it is not determined uniquely without fixing the function $P_0(t)$. Our purpose here is to find the variational Lagrangian so that the variational solution may give the solution of the original master equation (14.1). Then we construct the Lagrangian (14.2) so that the fixed point function of the transformation

$$P^*(t)=\mathfrak{T}P^*(t) \tag{14.5}$$

may become the solution of (14.1). First we study the property of the Lagrangian (14.2). If Γ is a linear operator as it is in our problem, then the functional F in (14.2) should satisfy the homogeneity

$$F\left(\lambda P,\lambda P_0,\lambda\frac{\partial}{\partial t}P;\Gamma\right)=\lambda^n F\left(P,P_0,\frac{\partial}{\partial t}P;\Gamma\right) \tag{14.6}$$

because if P is the solution of (14.1), then λP is also the solution of it. From the viewpoint of variation, it is desirable that F be bilinear in P. Physically we hope L is related to the entropy production rate (40, 41), and, consequently, we expect $n=1$. The simplest form of F that satisfies these conditions may be

$$L_1=\int PP_0^{-1}\frac{\partial}{\partial t}P_0\,d^Nx-\int P_0^{-1}P\Gamma P\,d^Nx \tag{14.7}$$

For this Lagrangian, the variation L_1 is given by

$$\delta L_1=\int\left[P_0^{-1}\frac{\partial}{\partial t}P_0-P_0^{-1}\Gamma P-\tilde{\Gamma}(P_0^{-1}P)\right]\delta P\,d^Nx \tag{14.8}$$

where $\tilde{\Gamma}$ is an adjoint operator of Γ and is assumed to have the property $\tilde{\Gamma}1=0$. Setting $\delta L_1=0$, we obtain

$$\frac{\partial P_0}{\partial t}=\Gamma P+P_0\tilde{\Gamma}(P_0^{-1}P) \tag{14.9}$$

This gives a transformation from $P_0(t)$ to $P(t)$ or from $P(t)$ to $P_0(t)$. The fixed point function $P^*=P_0=P$ of this transformation satisfies the equation

$$\frac{\partial}{\partial t}P^*(t)=\Gamma P^*(t) \tag{14.10}$$

Therefore, the fixed point function thus obtained is the solution desired.

It is, however, difficult to find whether the solution $P^*(t)$ gives a maximum or minimum function to the Lagrangian (14.7):

$$\delta^{(2)}L_1=-\tfrac{1}{2}\int P_0^{-1}\delta P\Gamma\delta P\,d^Nx \tag{14.11}$$

On the other hand, the simple functional F for $n=2$, namely, L_2 is given by

$$L_2=\int\left[P\frac{\partial}{\partial t}P_0-(P\Gamma P-P\tilde{\Gamma}P_0)\right]d^Nx \tag{14.12}$$

Clearly, $\delta L_2=0$ yields

$$\frac{\partial P_0}{\partial t}=\Gamma P+\tilde{\Gamma}(P-P_0) \tag{14.13}$$

Then the fixed point function $P^*=P=P_0$ satisfies (14.10). Furthermore, the second variation is

$$\delta^{(2)}L_2=-\tfrac{1}{2}\int(\delta P)\Gamma(\delta P)d^Nx\geq 0 \tag{14.14}$$

because Γ is negative definite in ordinary stochastic processes. Therefore, $P(t)$ is minimum for P_0 fixed. However, L_1 may be more physical, because the first term in L_1 corresponds to the entropy production rate and the second term in L_1 expresses a dissipative effect in a wide sense. For more detailed arguments in the case of the Fokker-Planck equation, see a paper by Hasegawa.[41]

B. Application of the Variational Principle to Relaxation Processes in Stochastic Systems

Explicitly we consider here the kinetic Ising model[42–44] which has no energy conservation. The initial distribution is assumed to be given by the equilibrium distribution function

$$P_i=\frac{e^{-\beta_i\mathcal{H}}}{\mathrm{Tr}\,e^{-\beta_i\mathcal{H}}} \tag{14.15}$$

for $\beta_i=1/k_BT_i$ and $T_i>T_c$. To make our arguments clear, we consider the explicit situation where the temperature is changed suddenly from T_i to T_f. For infinite time $t\to\infty$, the distribution function becomes an equilibrium one

$$P_{eq}=\frac{\exp(-\beta_f\mathcal{H})}{\mathrm{Tr}\exp(-\beta_f\mathcal{H})}; \qquad \beta_f=\frac{1}{k_BT_f}. \tag{14.16}$$

Our purpose here is to study how the initial distribution P_i changes into the final one P_f and to evaluate the onset time t_0 at which the macroscopic order begins to appear in the case $\beta_i<\beta_c\leq\beta_f$. For this purpose, we have to take into account enough fluctuation to assure symmetry breaking in the course of time for the thermodynamic limit. This may be accomplished approximately by considering a variational microscopic distribution function, for example,

$$P(t)=\exp[F(t)-\beta(t)\mathcal{H}+\cdots] \tag{14.17}$$

where $F(t)$ is determined by the normalization $\mathrm{Tr}\,P(t)=1$. If we include the complete set of the Hilbert space, then (14.17) becomes exact. This is, however, practically impossible. Thus we are satisfied here with some approximate solutions that contain the first few terms in (14.17).

We define here the time-dependent long-range order (order parameter) as follows:

$$M_s^2(t) = \lim_{H\to+0} \lim_{|i-j|\to\infty} \langle \sigma_i \sigma_j \rangle_t; \qquad \sigma_i = \pm 1 \tag{14.18}$$

for a magnetic field H, where we have used Ising spins $\{\sigma_i = \pm 1\}$ for simplicity, and $\langle \ldots \rangle_t$ denotes the average over $P(t)$. This is a natural extension of spontaneous magnetization in equilibrium.[45] If the system starts from an equilibrium state at T_i above T_c and is quenched in a new heat bath with $T_f < T_c$, then $M_s(0) = 0$ and $M_s(\infty) \neq 0$. Therefore, the long-range order or macroscopic order should appear at a certain time, so-called onset time t_0. This is defined by

$$M_s(t) = 0 \quad \text{for} \quad t \leq t_0 \quad \text{and} \quad M_s(t) \neq 0 \quad \text{for} \quad t > t_0 \tag{14.19}$$

To study this process, we apply the above variational method. As the first approximation, we take the variational function

$$P(t) = Z^{-1}(t)\exp[-\beta(t)\mathcal{H}]; \qquad Z(t) = \operatorname{Tr}\exp[-\beta(t)\mathcal{H}] \tag{14.20}$$

Since we are interested in the fixed point function $P^*(t) = P_0(t) = P(t)$, we take $P_0(t)$ of the same form as (14.21)

$$P_0(t) = Z_0^{-1}(t)\exp[-\beta_0(t)\mathcal{H}]; \qquad Z_0(t) = \operatorname{Tr}\exp(-\beta_0(t)\mathcal{H}) \tag{14.21}$$

Noting that

$$\frac{\partial}{\partial t} P_0(t) = -\beta_0'(t)\mathcal{H} P_0(t) - Z_0'(t) Z_0^{-1}(t) P_0(t) \tag{14.22}$$

we obtain

$$L_1 = Z^{-1}(t) \int \Big\{ \Big[-\beta_0'(t)\mathcal{H} - \frac{d}{dt}\log Z(t) \Big] \exp[-\beta(t)\mathcal{H}] - Z_0(t) Z^{-1}(t) \exp[\beta_0(t)\mathcal{H} - \beta(t)\mathcal{H}] \Gamma \exp[-\beta(t)\mathcal{H}] \Big\} d^N x \tag{14.23}$$

Here the temporal evolution operator Γ is a function of T_f. Now we make L_1 extremum (or $\delta L_1 = 0$). Then we obtain

$$\beta'(t)\mathrm{Tr}(\mathcal{H}-\langle\mathcal{H}_0\rangle)\mathcal{H}P(t)=Z_0(t)Z^{-2}(t)\mathrm{Tr}\,\mathcal{H}e^{(\beta_0-\beta)\mathcal{H}}\Gamma e^{-\beta\mathcal{H}} \tag{14.24}$$

where we omit terms that vanish for $\beta_0=\beta=\beta^*$. Here we set $\beta_0=\beta$ (fixed point). Then we arrive at

$$\frac{d}{dt}\beta(t)=\frac{-Z^{-1}(t)\mathrm{Tr}(\mathcal{H}-\langle\mathcal{H}\rangle_t)\Gamma\exp[-\beta(t)\mathcal{H}]}{\langle(\mathcal{H}-\langle\mathcal{H}\rangle_t)^2\rangle_t} \tag{14.25}$$

This determines the variational parameter $\beta(t)$ as a function of time t. Setting the denominator of (14.25) as $\hat{C}_v(\beta(t))$, and the numerator as $F(\beta(t),\beta_f)$, we rewrite (14.25) as

$$\int_{\beta_i}^{\beta}\frac{\hat{C}_v(x)}{F(x,\beta_f)}dx=t \tag{14.26}$$

Clearly, $F(\beta_f,\beta_f)=0$, and consequently $\beta(t)$ approaches β_f as t goes to infinity. The onset time t_0 is expressed by

$$t_0=\int_{\beta_i}^{\beta_c}\left[\frac{\hat{C}_v(x)}{F(x,\beta_f)}\right]dx \tag{14.27}$$

It should be remarked that this onset time t_0 is finite for $\beta_f>\beta_c$ even in the thermodynamic limit. This is in great contrast to the treatment given up to Section IX, in which t_0 is proportional to the logarithm of the system size Ω_0. In the present microscopic model, fluctuations remain finite even for $\Omega\to\infty$. This makes t_0 finite.

The behavior of $\beta(t)$ can be studied analytically by using the Onsager solution[46] and it is shown explicitly[47] in Fig. 13 for typical values of T_i and T_f. It is found[47] that the onset time t_0 decreases as the difference (T_c-T_f) increases. This may be intuitively interpreted as indicating that as the difference (T_c-T_f) increases, flipping of each process becomes faster.

C. Rate Equation Method

We propose here a simple method to derive the above result (14.25). This is essentially a moment method analogous to Langer's method.[48] We study the temporal evolution of the energy $\langle\mathcal{H}\rangle_t$ in two ways and identify the two

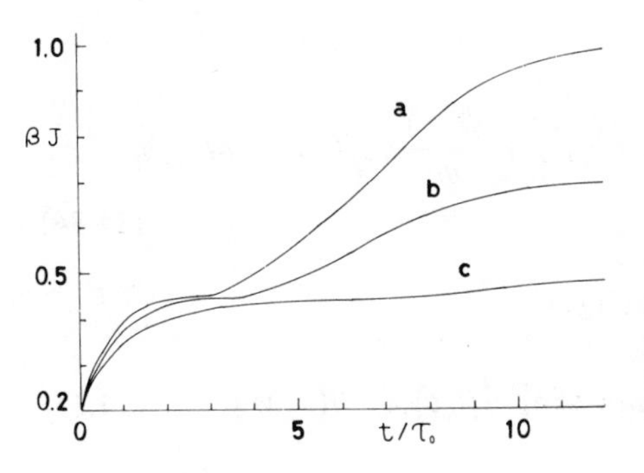

Fig. 13. Time change of the *effective* inverse temperature $\beta(t)$. Typical values of the parameters (a) $\beta_f J=1.0$, (b) $\beta_f J=0.7$, and (c) $\beta_f J=0.5$.

results. First we have

$$\frac{d}{dt}\langle \mathcal{H} \rangle_t = \mathrm{Tr}\, \mathcal{H}\frac{\partial}{\partial t}P(t) = \mathrm{Tr}\, \mathcal{H}\Gamma P(t) \tag{14.28}$$

On the other hand, $\langle \mathcal{H} \rangle_t$ is expressed as

$$\langle \mathcal{H} \rangle_t = Z^{-1}(t)\mathrm{Tr}\, \mathcal{H}\exp(-\beta(t)\mathcal{H}) \tag{14.29}$$

in our approximation (14.20). The rate of variation of the energy is then obtained as

$$\frac{d}{dt}\langle \mathcal{H} \rangle_t = -\frac{d\beta(t)}{dt}\left(\langle \mathcal{H}^2 \rangle_t - \langle \mathcal{H} \rangle_t^2\right) \tag{14.30}$$

by differentiating (14.29). Equating (14.28) and (14.30), we arrive again at (14.25).

This method can be applied also to the case in which a magnetic field changes. The first-order variational distribution function for this case is

$$P(t) = Z^{-1}(t)\exp[-\beta(t)\mathcal{H} + h(t)M] \tag{14.31}$$

where

$$Z(t) = \mathrm{Tr}\exp[-\beta(t)\mathcal{H} + h(t)M] \tag{14.32}$$

Now we study the change rate of energy and magnetization. For the energy, we obtain, as before,

$$\frac{d}{dt}\langle \mathcal{H} \rangle_t = Z^{-1}(t)\mathrm{Tr}\, \mathcal{H}\Gamma\exp[-\beta(t)\mathcal{H} + h(t)M] \tag{14.33}$$

and

$$\frac{d}{dt}\langle\mathcal{H}\rangle_t = -\frac{d\beta(t)}{dt}\langle(\mathcal{H}-\langle\mathcal{H}\rangle_t)^2\rangle + \frac{dh(t)}{dt}\langle\mathcal{H}(M-\langle M\rangle_t)\rangle_t \tag{14.34}$$

Similarly, for the magnetization, we get

$$\frac{d}{dt}\langle M\rangle_t = Z^{-1}(t)\mathrm{Tr}\, M\Gamma\exp[-\beta(t)\mathcal{H}+h(t)M] \tag{14.35}$$

and

$$\frac{d}{dt}\langle M\rangle_t = \frac{dh(t)}{dt}\langle(M-\langle M\rangle_t)^2\rangle - \frac{d\beta(t)}{dt}\langle M(\mathcal{H}-\langle\mathcal{H}\rangle_t)\rangle_t \tag{14.36}$$

Equating (14.33) and (14.34), and also (14.35) and (14.36), we derive the following coupled equations:

$$\begin{cases} \hat{C}_v\dfrac{d\beta(t)}{dt} - \hat{C}\dfrac{dh(t)}{dt} = -\Delta E(t) \\ \hat{C}\dfrac{d\beta(t)}{dt} - \hat{\chi}\dfrac{dh(t)}{dt} = -\Delta M(t) \end{cases} \tag{14.37}$$

Here

$$\begin{aligned} \hat{C}_v &= \langle(\mathcal{H}-\langle\mathcal{H}\rangle_t)^2\rangle_t \\ \hat{C} &= \langle\mathcal{H}(M-\langle M\rangle_t)\rangle_t = \langle M(\mathcal{H}-\langle\mathcal{H}\rangle_t)\rangle_t \\ \hat{\chi} &= \langle(M-\langle M\rangle_t)^2\rangle \\ \Delta E(t) &= Z^{-1}(t)\,\mathrm{Tr}\,\mathcal{H}\Gamma\exp[-\beta(t)\mathcal{H}+h(t)M] \\ \Delta M(t) &= Z^{-1}(t)\,\mathrm{Tr}\, M\Gamma\exp[-\beta(t)\mathcal{H}+h(t)M] \end{aligned} \tag{14.38}$$

By solving (14.37), we obtain

$$\frac{d\beta(t)}{dt} = \frac{\hat{C}\,\Delta M(t) - \hat{\chi}\,\Delta E(t)}{\Delta(t)}, \qquad \frac{dh(t)}{dt} = \frac{\hat{C}_v\Delta M(t) - \hat{C}\,\Delta E(t)}{\Delta(t)} \tag{14.39}$$

and

$$\Delta(t) = \hat{C}_v\hat{\chi} - \hat{C}^2$$

In general, these are coupled nonlinear differential equations, and consequently it is difficult to solve them analytically. Thus we must solve them numerically in order to study the time-dependent behavior of $\beta(t)$ and $h(t)$. These solutions are omitted here.

The above rate equation method is expressed in the compact form

$$\int\left(\frac{\partial}{\partial t}P(t)-\Gamma P(t)\right)\delta\log P(t)d^N x=0 \tag{14.40}$$

for an arbitrary trial function $P(t)$. Consequently, this method is generally equivalent to the variational method by Gransdorff and Prigogine. More general arguments on the equivalence of the two methods will be published elsewhere.[78]

D. Validity of the Present Approximation

As is discussed by Mori[49] and Zubarev,[50] the local equilibrium approximation does not describe dissipative phenomena or transport phenomena for systems described by the Liouville equation. However, our starting point is a stochastic process described by (14.1) and consequently the local equilibrium distribution is expected to describe relaxation phenomena approximately.

To see the feature of this approximations, we study the linear response in (14.1) on the basis of the above formulation and compare this result with Kubo's formula[51] based on time correlation functions. For example, the relaxation time τ_A of a physical quantity A is given by[52]

$$\tau_A=\int_0^\infty\frac{\langle\delta A\,\delta A(t)\rangle}{\langle(\delta A)^2\rangle}dt=-\langle\delta A\frac{1}{\Gamma}\delta A\rangle\langle(\delta A)^2\rangle^{-1} \tag{14.41}$$

in the linear response theory, where $\delta A=A-\langle A\rangle$. On the other hand, the present treatment yields

$$\tau_A'=\frac{-\langle(\delta A)^2\rangle}{\langle\delta A\Gamma\delta A\rangle} \tag{14.42}$$

in the linear approximation. Thus the present method corresponds to the approximation

$$\langle\delta A\frac{1}{\Gamma}\delta A\rangle\simeq\frac{\langle(\delta A)^2\rangle^2}{\langle\delta A\Gamma\delta A\rangle} \tag{14.43}$$

in the linear regime. That is, the long time correlation is approximated by the initial short time correlation in our method.

XV. NONEQUILIBRIUM PHASE TRANSITION AND CRITICAL SLOWING DOWN

Here we study the critical slowing down near the nonequilibrium phase transition described by Langevin's equation or a stochastic differential of the form

$$\frac{d}{dt}x(t)=c(x)+\beta(x)\eta(t) \tag{15.1}$$

with the Gaussian random force $\eta(t)$ satisfying the relation

$$\langle\eta(t)\eta(t')\rangle=2\varepsilon\delta(t-t') \tag{15.2}$$

The probability distribution function $P(x, t)$ satisfies the F-P equation (8.6). Consequently, the stationary solution $P_{st}(x)$ is given by

$$P_{st}(x)\sim\exp\left[\frac{1}{\varepsilon}\varphi(x,\lambda)\right] \tag{15.3}$$

where

$$\varphi(x,\lambda)=\int^{x}\frac{c(x)}{\beta^2(x)}dx-\varepsilon\log\beta(x) \tag{15.4}$$

Here λ denotes a certain parameter describing phase transition. The critical point λ_c is determined by the condition that a certain most probable point x_0 begins to change to an unstable state, namely,

$$\varphi'(x_0,\lambda_c)=[c(x_0)-\varepsilon\beta(x_0)\beta'(x_0)]\beta^{-2}(x_0) \tag{15.5}$$

and

$$\varphi''(x_0,\lambda_c)=\Big[c'(x_0)-\varepsilon\{[\beta'(x_0)]^2+\beta(x_0)\beta''(x_0)\}\Big]\beta^{-2}(x_0) \tag{15.6}$$

For example, in the case of $\beta(x)=1$, we obtain

$$\varphi'(x_0,\lambda_c)=c(x_0)=0 \qquad \text{and} \qquad \varphi''(x_0,\lambda_c)=c'(x_0)=0 \tag{15.7}$$

If we set $x(t)=x_0+\xi(t)$, then in the limit of small ε we have

$$\frac{d}{dt}\xi(t)=c'(x_0)\xi(t)+\eta(t)+O(\xi^2) \tag{15.8}$$

Thus the time correlation function takes the form

$$\langle \xi(0)\xi(t)\rangle = \langle \xi^2(0)\rangle e^{\lambda t} \tag{15.9}$$

where $\lambda = c'(x_0)$. Consequently, $\lambda_c = 0$ and the relaxation time τ is proportional to

$$\tau \sim \frac{1}{|\lambda|} \sim \frac{1}{|\lambda - \lambda_c|} \tag{15.10}$$

in the limit of small ε. This is the so-called critical slowing down. Alternatively, this critical slowing down is also explained in the following time change of fluctuation:

$$\langle \xi^2(t)\rangle = \frac{\varepsilon}{\lambda} + \left(\langle \xi^2(0)\rangle - \frac{\varepsilon}{\lambda}\right)\exp(2\lambda t) \tag{15.11}$$

In this simple example, the critical slowing down always appears just at the phase transitions point $\lambda = \lambda_c$ in the limit of small ε. For the arguments on finite ε, see Refs. 25 and 57. However, the general situation is not so simple, as is shown in the following multiplicative stochastic process[53, 54]:

$$\frac{d}{dt}x = \gamma x - x^n + x\eta(t) \tag{15.12}$$

The stationary distribution function is given by

$$P_{st}(x) \simeq C x^{(\gamma-\varepsilon)/\varepsilon} \exp\left[\frac{-x^{n-1}}{\varepsilon(n-1)}\right] \tag{15.13}$$

The phase transition occurs at $\gamma = \varepsilon$, as is easily seen for any value of n larger than unity. Schenzle and Brand[54] solved the Fokker-Planck equation corresponding to (15.12); they obtained all eigenvalues of the Fokker-Planck operator with the result

$$\lambda_m = -m(n-1)[\gamma - m(n-1)\varepsilon] \tag{15.14}$$

for $m = 0, 1, 2, \ldots, [\gamma/(n-1)\varepsilon]$. The first negative eigenvalue λ_1 takes the form

$$\lambda_1 = -(n-1)[\gamma - (n-1)\varepsilon] \tag{15.15}$$

for $\gamma > (n-1)\varepsilon$. The critical slowing down may be governed by λ_1, which goes to zero proportionally to $\gamma - (n-1)\varepsilon$. The relaxation time τ is proportional to $[\gamma-(n-1)\varepsilon]^{-1}$. On the other hand, the phase transition occurs at $\gamma=\varepsilon$. Therefore, the so-called critical slowing down appears at the transition point only for $n=2$, except for $\varepsilon=0$.

This situation is easily understood from the following formal compact solution of the original Stratonovich stochastic differential equation (15.12):

$$x(t)=\exp\left[\gamma t+\int_0^t \eta(t')dt'\right]$$
$$\times\left[(n-1)\int_0^t \exp\left[(n-1)\left(\gamma t'+\int_0^{t'}\eta(s)ds\right)\right]dt'+x_0^{1-n}\right]^{-1/(n-1)} \tag{15.16}$$

This formal solution is easily obtained by transforming the variable $x(t)$ into a new variable $y(t)$ as $y(t)=x(t)^{1-n}$ in (15.12). Then we arrive at the linearized equation

$$\frac{d}{dt}y(t)=-(n-1)(\gamma+\eta(t))y(t)+(n-1) \tag{15.17}$$

The average $\langle y(t)\rangle=\langle x(t)^{1-n}\rangle$ is easily calculated from (15.16) or (15.17) as

$$\langle y(t)\rangle=\langle y_0\rangle e^{-(n-1)\gamma t}\left\langle \exp\left[-(n-1)\int_0^t\eta(t')dt'\right]\right\rangle$$
$$+(n-1)\int_0^t\left\langle \exp\left[(n-1)\gamma t'-(n-1)\int_{t'}^t\eta(s)ds\right]\right\rangle dt'$$
$$=\langle y_0\rangle e^{-(n-1)\gamma t}\exp\left[\tfrac{1}{2}(n-1)^2\int_0^t dt_1\int_0^t dt_2\langle\eta(t_1)\eta(t_2)\rangle\right]$$
$$+(n-1)\int_0^t \exp[(n-1)\gamma t']\exp\left[\tfrac{1}{2}(n-1)^2\int_{t'}^t dt_1\int_{t'}^t dt_2\langle\eta(t_1)\eta(t_2)\rangle\right]$$
$$=\left(\langle y_0\rangle-\frac{1}{\gamma-(n-1)\varepsilon}\right)e^{-(n-1)[\gamma-(n-1)\varepsilon]t}+\frac{1}{\gamma-(n-1)\varepsilon} \tag{15.18}$$

That is, the relaxation time of $\langle y(t)\rangle$ is given by $\tau=|\lambda_1|^{-1}$. This corresponds to the above argument based on the eigen spectrum of Schenzle and Brand.[54]

A more intuitive explanation of this result $\tau=|\lambda_1|^{-1}$ is as follows. From the stationary solution (15.13), $\langle y\rangle_{st}=\langle x^{1-n}\rangle_{st}$ diverges at $\gamma=(n-1)\varepsilon$, that is at $\lambda_1=0$. Consequently, the relaxation time becomes infinite at this point.

This is a nice example of the general criterion for the appearance of slowing down in stochastic processes, that one physical mode Q, at least, should exist such as $\langle Q \rangle_{st} = \infty$ at some point for the appearnace of slowing down. Recently Shiino[55] studied the same problem for $n=2$, in the limit $\varepsilon \to 0$ and $t \to 0$ for εt fixed, using the perturbational method up to the first order of ε, and he obtained

$$\langle \xi(t)\xi(0) \rangle \sim \frac{\gamma}{2} \exp[-(\gamma - \varepsilon)t] \tag{15.19}$$

for $x(t) = \gamma + (2\varepsilon)^{1/2}\xi(t)$ in the above limit. Kitahara[56] has also discussed the critical slowing down in a two-level noise system different from (15.1).

Details of the present arguments concerning the critical slowing down in nonequilibrium systems will be published elsewhere.[57] See also an Extension of the Scaling Treatment to Multiplicative Stochastic Processes (Section XVI.C).

XVII. SOME APPLICATIONS

Several applications of the present scaling treatment of transient phenomena have been previously reported.[1, 2, 9, 13, 14, 31, 32, 35, 36, 58, 59, 60] Here we discuss briefly some of these applications.

A. Superradiance

According to Agarwal[61] and Bonifacio et al.,[62] the quantal master equation for atoms relevant to superradiance is given by

$$\frac{\partial}{\partial t}\rho = \tfrac{1}{2} I_1 \{ [S^-, \rho S^+] + [S^- \rho, S^+] \}; \qquad I_1 = \frac{2g^2}{\kappa} \tag{16.1}$$

where ρ denotes the density matrix of the system, $\{S\}$ describes relevant atom states, $S^{\pm}$ '$S^x \pm iS^y$, g denotes the coupling constant between the atoms and radiation, and κ is the damping constant of superradiance field. All the relaxation and fluctuation of superradiance field ($b^\dagger$ and b) are calculated through the relation[61–64]

$$\langle (b^\dagger)^n b^n \rangle_t \infty \langle (S^+)^n (S^-)^n \rangle_t \tag{16.2}$$

This is also expressed by $\{\langle (S^z)^m \rangle_t\}$. Thus it is sufficient to study the generating function $\Psi(\lambda, t)$ defined[58] by

$$\Psi(\lambda, t) = \langle \exp(\lambda s_z) \rangle_t; \qquad s_z = \frac{S^z}{S} \tag{16.3}$$

As before, we scale the time t as $SI_1 t \to t$ (new time variable). Then $\Psi(\lambda, t)$ is shown[65] to satisfy the equation

$$\frac{\partial \Psi}{\partial t} = (e^{-\varepsilon\lambda} - 1)\left[(S+1) + \frac{\partial}{\partial \lambda} - S\frac{\partial^2}{\partial \lambda^2}\right]\Psi; \qquad \varepsilon = S^{-1} \tag{16.4}$$

This is transformed[58] into the following Kramers-Moyal equation:

$$\frac{\partial P}{\partial t} = \left[\frac{1-s^2+\varepsilon}{\varepsilon}(e^{\varepsilon(\partial/\partial s)} - 1) - s(e^{\varepsilon(\partial/\partial s)} + 1)\right]P(s,t) \tag{16.5}$$

for the distribution function

$$P(s,t) = \frac{1}{2\pi i}\int_{c-i\infty}^{c+i\infty} e^{-\lambda s}\Psi(\lambda, t)\,d\lambda = \langle \delta(s_z - s)\rangle \tag{16.6}$$

which is defined in the range $-1 \le s \le 1$.

According to the general procedure (5.22), we introduce the time-dependent transformation

$$\tilde{\xi} = F^{-1}(e^{-\gamma t}F(s)); \qquad F(x) = \frac{1-x}{1+x} \tag{16.7}$$

that is,

$$\tilde{\xi} = \frac{s + \tanh\left(\frac{1}{2}t\right)}{1 + s\tanh\left(\frac{1}{2}t\right)} \tag{16.8}$$

Then the scaling solution is given by

$$P_{\mathrm{sc}}(s,t) \propto \frac{1}{\varepsilon\sigma(t)}\exp\left(-\frac{1-\tilde{\xi}}{\varepsilon\sigma(t)}\right) \tag{16.9}$$

where

$$\sigma(t) = -\left(\log\left\{1 + \left[\exp\left(\frac{-1}{\sigma_0}\right) - 1\right]e^{-2t}\right\}\right)^{-1} \tag{16.10}$$

This result can be valid in the initial and second scaling regimes. The intensity $I(t)$ of superradiance and fluctuation $\sigma_f(t)$ can be easily calculated from this result as

$$I(t) \propto S^2 M_1 \qquad \text{and} \qquad \sigma_f(t) = M_2 - M_1^2 \tag{16.11}$$

with $M_n = \langle (1-s_z^2)^n \rangle_t$. These results agree with those obtained by Glauber and Haake,[64] who used the method of the directed angular momentum representation.

B. Nucleon Transport in Nonlinear Systems

Quite recently R. Schmidt and G. Wolschin[60] applied the present scaling idea and connection procedure to the nucleon transport phenomena in low-energy heavy-ion collisions (particularly ^{238}U(7.42 MeV/nucleon) + ^{238}U). Because of the effect of nuclear shell structure, the driving potential for transport becomes nonlinear, and Schmidt and Wolschin assumed for simplicity that the driving potential $U(x,T)$ is

$$U(x,T) = k(T)x^2\left(1-\tfrac{1}{2}x^2\right) + \text{const.} \tag{16.12}$$

Here x denotes the mass asymmetry variable, and it is chosen to be ± 1 at the shell minima and zero at mass symmetry. The corresponding Fokker-Planck equation takes the form of (6.51) and consequently all the previous arguments can be applied to this problem. In particular, Schmidt and Wolschin effectively used the connection procedure presented in Section VI.C. That is, they connected the initial Gaussian solution to a nonlinear solution at $t=t_1$, when the variance σ becomes $\sigma_1 = \frac{1}{3}$, and they also connected it to the final time region at $t=t_2$, when $\sigma=1$. Thus they found physical solutions of the transport equation that change from the initial Gaussian solution to the final double-peak solution. For more details, see their original paper.

C. Extension of the Scaling Treatment to Multiplicative Stochastic Processes

Several authors[53–55, 66] have discussed multiplicative stochastic processes described by the Langevin equation

$$\frac{d}{dt}x(t) = \gamma x - gx^n + x\eta(t) \tag{16.13}$$

Here we generalize this process as

$$\frac{d}{dt}x(t) = c(x) + x\eta(t) \tag{16.14}$$

with $\gamma = c'(0) > 0$. That is, the point $x=0$ is an unstable one. We assume the existence of stable point x_0. The purpose of this section is to extend the scaling theory to the transient phenomena of this multiplicative stochastic process near the instability point. If we apply the time-dependent nonlinear transformation (5.22), that is,

$$\tilde{\xi}=F^{-1}\left(e^{-\gamma t}F(x)\right) \tag{16.15}$$

with $F(x)$ defined by

$$F(x)=\exp\int_{a_0}^{x}\frac{\gamma}{c(y)}\,dy \tag{16.16}$$

then the stochastic process (16.14) is transformed into

$$\frac{d}{dt}\tilde{\xi}=\frac{\eta(t)e^{-\gamma t}F^{-1}\left(e^{\gamma t}F(\tilde{\xi})\right)F'\left(F^{-1}\left(e^{\gamma t}F(\tilde{\xi})\right)\right)}{F'(\tilde{\xi})} \tag{16.17}$$

As in Section V, this equation can be approximated by

$$\frac{d}{dt}\tilde{\xi}=\eta(t)\tilde{\xi} \tag{16.18}$$

in the intermediate nonlinear time region. This corresponds to (5.26). It should be noted that the scaling solution now satisfies the pure multiplicative stochastic process (16.18) without a drift term. The solution of (16.18) is given by

$$\tilde{\xi}(t)=x(0)\exp\int_0^t\eta(t')\,dt' \tag{16.19}$$

Therefore, the scaling solution for this pure multiplicative process (16.2) is expressed as

$$x_{\text{sc}}(t)=F^{-1}\left(e^{\gamma t}F\left(x(0)\exp\int_0^t\eta(t')\,dt'\right)\right) \tag{16.20}$$

With the use of this scaling solution, we can discuss the fluctuation and formation process of macroscopic order. Here we compare this scaling solution with the exact solution (15.16) for the simple example (16.13). The corresponding scaling solution takes the form

$$x_{\text{sc}}(t)=\tilde{\xi}(t)e^{\gamma t}\left[1+g\gamma^{-1}\left(e^{(n-1)\gamma t}-1\right)\tilde{\xi}^{n-1}(t)\right]^{-1/(n-1)} \tag{16.21}$$

with $\tilde{\xi}(t)$ defined by (16.19), because the function $F(x)$ in (16.16) for this system is given by

$$F(x)=x(1-g\gamma^{-1}x^{n-1})^{-1/(n-1)} \tag{16.22}$$

Thus we find that our scaling solution is very close to the exact one (15.16) in the above nonlinear system, and that the above scaling treatment is equivalent to the following approximation in the exact solution (15.16):

$$\int_0^t \exp\left[(n-1)\left(\gamma t' + \int_0^{t'} \eta(s)\,ds\right)\right] dt'$$

$$\simeq \int_0^t e^{(n-1)\gamma t'} dt' \cdot \exp(n-1)\int_0^t \eta(s)\,ds = \frac{e^{(n-1)\gamma t}-1}{\gamma}\exp(n-1)\int_0^t \eta(s)\,ds \tag{16.23}$$

Furthermore, the critical slowing down appears in the average $\langle x_{sc}^{-1}(t)\rangle$ near the threshold $\gamma=\varepsilon$ as

$$\langle x_{sc}^{-1}(t)\rangle = \langle x^{-1}(0)\rangle e^{-(\gamma-\varepsilon)t} + g\gamma^{-1}(1-e^{-\gamma t}) \tag{16.24}$$

for $n=2$. The average $\langle x_{sc}(t)\rangle$ for $n=2$ is also given by

$$\langle x_{sc}(t)\rangle = \frac{\gamma}{g(1-e^{-\gamma t})} - \frac{\gamma^2}{g^2x(0)(1-e^{-\gamma t})}e^{-(\gamma-\varepsilon)t} + o(e^{-(\gamma-\varepsilon)t}) \tag{16.25}$$

for large t and for $\gamma>2\varepsilon$. Thus for $\langle x(t)\rangle$, there is seemingly a slowing down up to the region $\gamma=2\varepsilon$, but there occurs[57] no critical slowing down at $\gamma=\varepsilon$. This will be discussed in more detail elsewhere.[57]

XVII. CONCLUSION AND DISCUSSION

In this Chapter, we present the general scheme of treating transient phenomena near the instability point and discuss various aspects of instability phenomena. Particularly, we clarify the mechanism of formation of macroscopic order from disordered states and emphasize the importance of the synergism of the initial fluctuation, random force, and nonlinearity of the system. This synergism is closely related to Haken's synergetics[34] and may be one of the central concepts in synergetics.

Some sections in this chapter are not satisfactory representations of completed theories, but they provide sprouts of possible new fields in the future. In particular, the relation between phase transition in nonlinear open systems and critical slowing down near the generalized phase transition point is a very interesting problem, as is discussed in Section XV and will be discussed again in more detail in the near future.

The extension of the present scaling theory to nonuniform systems also will be of great interest for the study of spinodal decomposition at a later stage.

It may be possible[1, 2] to apply the present treatment to chemical reaction described, for example, by Prigogine, Lefever, and Nicolis,[68] and by Schlögl and Pasquale et al.,[66, 69] and to biological problems as well.[70–72]

Appendix A: Borel summation

Consider the asymptotic series of the form

$$f(x)=f(0)+\sum_{n=0}^{\infty} a_n x^{n+1} \tag{A.1}$$

If a_n is confirmed to be of the order $n!$ for large n, then we consider the integral

$$f(x)=f(0)+\int_0^{\infty} e^{-\xi/x} g(\xi)\,d\xi \tag{A.2}$$

where

$$g(\xi)=\sum_{n=0}^{\infty} \frac{a_n}{n!}\xi^n \tag{A.3}$$

If $g(\xi)$ is convergent in a certain range of ξ and the integral

$$\int_0^{\infty} \exp\left(\frac{-\xi}{x}\right) g(\xi)\,d\xi \tag{A.4}$$

exists, then the integral (A.2) is called the Borel sum of the series (A.1).

This is easily extended to the general case, where a_n is of the order $(n!)^m$ for large n in (A.1), where $m=1,2,\ldots$. In this general case, we consider the multiple integral

$$f(x)=f(0)+\int_0^{\infty} d\xi_1 \int_0^{\infty} d\xi_2 \cdots \int_0^{\infty} d\xi_m \exp\left(-x^{-1/m}\sum_{j=1}^{m}\xi_j\right) \times \sum_{n=0}^{\infty} \frac{a_n}{(n!)^m}(\xi_1\cdots\xi_m)^n \tag{A.5}$$

If the summation in (A.5) converges in a certain range of variables and the multiple integral (A.5) exists, then we may call it a generalized Borel sum of (A.1) for $a_n \sim (n!)^m$. These formulas are useful in nonlinear many-body problems.

Appendix B: Proof of (5.21)

The proof of (5.18) is as follows:

$$
\begin{aligned}
&\mathrm{sc}-\lim\langle(\xi(t)-\xi_{\mathrm{sc}}(t))^2\rangle \\
&\quad = \mathrm{sc}-\lim e^{2\gamma t}\int_0^t dt_1\int_0^t dt_2\langle f(\xi(t_1))f(\xi(t_2)\eta(t_1)\eta(t_2))\rangle e^{-\gamma(t_1+t_2)} \\
&\quad = \mathrm{sc}-\lim e^{2\gamma t}\int_0^t dt_1\int_0^t dt_2\langle f(\xi(t_1))f(\xi(t_2))\rangle\langle\eta(t_1)\eta(t_2)\rangle e^{-\gamma(t_1+t_2)}
\end{aligned}
\tag{B.1}
$$

In the last equality of (B.1), we may have many other decoupling routes, because $f(\xi(t))$ contains the random force through $\xi(t)$. However, such other decoupling can be shown to be of the same or of higher order in ε than the above decoupling term (B.1). Even the above term in (B.1) is shown to vanish, as is discussed below, and consequently all other decoupling terms go to zero. Now we show that the last expression in (B.1) vanishes. Since $f(0)=0$, $f(\xi)$ can be written as $f(\xi)=\xi g(\xi)$ with $g(0)<\infty$. Then $\langle f^2(\xi)\rangle$ can be expressed as

$$
\langle f^2(\xi)\rangle = \langle \xi^2 g^2(\xi)\rangle = \varepsilon e^{2\gamma t}h(t,\xi); \qquad h(t,0)<\infty \tag{B.2}
$$

Using (B.2), we obtain

$$
\begin{aligned}
&\mathrm{sc}-\lim\langle(\xi(t)-\xi_{\mathrm{sc}}(t))^2\rangle \\
&\qquad = \mathrm{sc}-\lim 2\varepsilon\int_0^t e^{2\gamma(t-t')}\langle f^2(\xi(t'))\rangle dt' \\
&\qquad = \mathrm{sc}-\lim(2\varepsilon e^{2\gamma t})\cdot \mathrm{sc}-\lim(\varepsilon t)\cdot \mathrm{sc}-\lim\frac{1}{t}\int_0^t h(t',\varepsilon)dt' \\
&\qquad = 0
\end{aligned}
\tag{B.3}
$$

where we use the mean value theorem of integral calculus, and assume the existence of the scaling limit $\mathrm{sc}-\lim\langle f^2(\xi)\rangle$, and consequently of the limit $\mathrm{sc}-\lim h(t,\varepsilon)$.

Acknowledgments

The author would like to thank Professor R. Kubo for his continual encouragement and fruitful discussions, and also T. Arimitsu, S. Miyashita, F. Sasagawa, and K. Kaneko for their cooperation on the present work. This study was partially financed by the Mitsubishi Foundation.

References

1. *Synergetics; Far from Equilibrium—Instabilities and fluctuations*, A. Pacault and C. Vidal, Eds., Springer-Verlag, 1979. In particular, see a paper by the present author, p. 94.
2. M. Suzuki, *Proceedings on the XVIIth Solvay Conference on Physics*, 1978, Brussels, John Wiley & Sons, Inc. 1980.
3. R. Kubo, K. Matsuo, and K. Kitahara, *J. Stat. Phys.*, **9**, 51 (1973).
4. N. G. van Kampen, *Adv. Chem. Phys.*, **34**, 245 (1976).
5. N. G. van Kampen, *Can. J. Phys.*, **39**, 551 (1961).
6. M. Suzuki, *Prog. Theor. Phys.*, **56**, 477 (1976).
7. S. Kabashima, M. Itsumi, T. Kawakubo, and T. Nagashima, *J. Phys. Soc. Jap.*, **39**, 1183 (1975).
8. M. Suzuki, *Phys. Lett.*, **67A**, 339 (1978).
9. M. Suzuki, *Prog. Theor. Phys. Suppl.*, **64**, 402 (1978).
10. R. Kubo, *Rep Prog. Phys.*, **XXIV**, Part 1, 255 (1966).
11. M. Suzuki, *Prog. Theor. Phys.*, **56**, 77 (1976), and *J. Stat. Phys.* **16**, 11 (1977).
12. M. Suzuki, *Prog. Theor. Phys.*, **57**, 380 (1977).
13. F. de Pasquale and P. Tombesi, *Phys. Lett.*, **72A**, 45 (1979).
14. F. de Pasquale, P. Tartaglia, and P. Tombesi, *Physica*, **99A**, 587 (1979).
15. M. Suzuki, *J. Stat. Phys.*, **16**, 11 (1977).
16. M. Suzuki, *J. Stat. Phys.*, **16**, 477 (1977).
17. F. Haake, *Phys. Rev. Lett.*, **41**, 1685 (1978).
18. R. Kubo, in *Synergetics*, H. Haken and B. G. Teubner Eds., Stuttgart, 1973.
19. M. Suzuki, *Prog. Theor. Phys.*, **53**, 1657 (1975).
20. M. Suzuki, *J. Stat. Phys.*, **20**, 163 (1979).
21. R. Kubo, *J. Math. Phys.*, **4**, 174 (1963).
22. B. Mühlschlegel, *Path Integrals and Their Applications in Quantum Statistical and Solid State Physics*, G. J. Papadopoulos and J. T. Devreese, Eds., Plenum Press, 1978.
23. R. Kubo, *Statistical Mechanics*, Vol. 6 of *Foundation of Modern Physics*, Iwanami, 1972 in Japanese.
24. H. Tomita, A. Ito, and K. Kidachi, *Prog. Theor. Phys.*, **56**, 786 (1976).
25. H. Dekker and N. G. van Kampen, *Phys. Lett.*, **73A**, 374 (1979).
26. F. Langouche, D. Roekaerts, and E. Tirapegui, J. Phys. **A13**, 449 (1980).
27. R. Graham, *Z. Phys.*, **B26**, 281 (1977).
28. H. Dekker, preprint referred to in Ref. 26.
29. C. P. Enz, *Statphys*, **13**, Haifa, (September 1977).
30. W. Horsthemke and A. Bach, *Z. Phys.* **B22**, 189 (1975).
31. B. Caroli, C. Caroli, and B. Roulet, J. Stat. Phys. **21**, 415 (1979).
32. B. Caroli, C. Caroli, and B. Roulet, Physica **101A**, 581 (1980).
33. T. Arimitsu, to be published.
34. H. Haken, *Synergetics*, Springer-Verlag, Berlin 1977, and references cited therein.
35. T. Arimitsu and M. Suzuki, *Physica*, **93A**, 574 (1978).
36. K. Kawasaki, M. C. Yalabik, and J. D. Gunton, *Phys. Rev.*, **17**, 455 (1978).

37. C. Kawabata and K. Kawasaki, *Phys. Lett.*, **65A**, 137 (1978).
38. M. Suzuki and S. Miyashita, to be published.
39. L. Onsager and S. Machlup, *Phys. Rev.*, **91**, 1505 (1953).
40. P. Glansdorff and I. Prigogine, *Thermodynamic Theory of Structure, Stability and Fluctuations*, Wiley, New York, 1971.
41. H. Hasegawa, *Prog. Theor. Phys.*, **58**, 128 (1977).
42. R. J. Glauber, *J. Math. Phys.*, **4**, 294 (1963).
43. M. Suzuki and R. Kubo, *J. Phys. Soc. Jap.*, **24**, 51 (1968).
44. K. Kawasaki, *Phase Transition and Critical Phenomena*, Vol. 2, C. Domb and M. S. Green Eds., Academic, New York, 443.
45. T. D. Shultz, D. C. Mattis, and E. H. Lieb, *Rev. Mod. Phys.*, **36**, 856 (1964).
46. L. Onsager, *Phys. Rev.*, **65**, 117 (1944), B. Kaufman and L. Onsager, *Phys. Rev.*, **76**, 1244 (1949).
47. M. Suzuki and F. Sasagawa, to be published in Prog. Theor. Phys.
48. J. S. Langer, M. Bar-on, and H. D. Miller, *Phys. Rev.*, **A11**, 1417 (1975).
49. H. Mori, *Phys. Rev.*, **112**, 1829 (1958).
50. D. H. Zubarev, *Nonequilibrium Statistical Thermodynamics*, Consultants Bureau, New York, 1974.
51. R. Kubo, *J. Phys. Soc. Jap.*, **12**, 570 (1957).
52. M. Suzuki, *Int. J. Magn.*, **1**, 123 (1971).
53. W. Horsthemke and M. Malek-Mansour, *Z. Phys.*, **B24**, 307 (1976).
54. A. Schenzle and H. Brand, *Phys. Lett.*, **69A**, 313 (1979).
55. M. Shiino, reported at the Meeting of the Physical Society of Japan in Matsuyama, October 1979.
56. K. Kitahara, reported at the Meeting of Physical Society of Japan in Matsuyama, October 1979.
57. M. Suzuki, K. Kaneko, and F. Sasagawa, submitted to Prog. Theor. Phys..
58. M. Suzuki, *Physica*, **86A**, 622 (1977).
59. T. Arimitsu and M. Suzuki, *Physica*, **90A**, 303 (1978).
60. R. Schmidt and G. Wolschin, Z. Physik A—Atoms and Nuclei **29b**, 215 (1980).
61. G. S. Agarwal, *Phys. Rev.*, **A2**, 2038 (1970).
62. R. Bonifacio, P. Schwendimann, and F. Haake, *Phys. Rev.*, **A4**, 302 and 854 (1971).
63. F. Haake and R. J. Glauber, *Phys. Rev.*, **A5**, 1457 (1972).
64. R. J. Glauber and F. Haake, *Phys. Rev.*, **A13**, 357 (1976).
65. M. Suzuki, *Physica*, **84A**, 48 (1976).
66. F. de Pasquale, P. Tartaglia, and P. Tombesi, preprint.
67. L. M. Narducci and V. Bluemel, *Phys. Rev.*, **A11**, 1354 (1975).
68. I. Prigogine and R. Lefever, *J. Chem. Phys.*, **48**, 1695 (1968); R. Lefever, *J. Chem. Phys.*, **48**, 4977 (1968); R. Lefever and G. Nicolis, *J. Theor. Biol.*, **30**, 267 (1971).
69. F. Schlögl, *Z. Phys.*, **147**, 253 (1972).
70. K. Kometani and S. Shimizu, *Z. Stat. Phys.*, **13**, 473 (1975).
71. R. C. Desai and R. Zwanzig, *J. Stat. Phys.*, **19**, 1 (1978).
72. I. Prigogine, *Advan. Biol. Med. Phys.*, **16**, 241 (1977).

References added in proof

73. H. Horner and K. Jüngling, Z. Physik **B36**, 97 (1979).
74. H. Lemarchand, Physica **101A**, 518 (1980).
75. Y. Hamada, Ph.D. Thesis (Tokyo Institute of Technology, 1980).

76. T. Arimitsu, Ph.D. Thesis (University of Tokyo, 1980).
77. M. Suzuki, Proceedings of the VI Sitges Conference on Statistical Mechanics, to be published by Springer-Verlag as "Lecture Notes in Physics", in wich the renormalized perturbation expansion scheme is explained in detail.
78. M. Suzuki, F. Sasagawa, K. Kaneko and Fang Fu-Kang, to be published in Physica A, in which an application of the variational method to the nonlinear Fokker-Planck equation is also discussed to study the onset time.
79. For related papers, see also
T. Shimizu, Physica **91A**, 534 (1978).
K. Matsuo, J. Stat. Phys. **16**, 169 (1977).
G. Nicolis and J. W. Turner, Physica **89A**, 326 (1977).
M. Moreau, Physica **90A**, 410 (1978).
Y. Aizawa, Prog. Theor. Phys. **59**, 1399 (1978).
C. Murakami and H. Tomita, Prog. Theor. Phys. **60**, 683 (1978).
R. Kubo, The Boltzmann award lecture, 1977, August 24, Haifa.
K. Tanabe, Prog. Theor. Phys. **61**, 354L (1979).

Note added in proof

Even if we use the time-independent nonlinear transformation (5.2), we obtain the solution which can be used both in the initial and scaling regimes, namely

$$P(x,t)=\int_{-\infty}^{\infty} dy \frac{1}{\sqrt{4\pi s\varepsilon}} e^{-\gamma t}F'(x)\exp\left[-\frac{\left(e^{-\gamma t}F(x)-y\right)^2}{4s\varepsilon}\right]$$

$$\times \frac{1}{\sqrt{2\pi\varepsilon\sigma_0}}\left[\frac{d}{dy}F^{-1}(y)\right]\exp\left[-\frac{\left(F^{-1}(y)\right)^2}{2\varepsilon\sigma_0}\right]$$

with $s=\left[1-e^{-2\gamma t}\right]\beta^2(0)/(2\gamma)$.

THEORY OF THE LIQUID–VAPOR INTERFACE

MYUNG S. JHON[a] AND JOHN S. DAHLER

Departments of Chemical Engineering and Chemistry, University of Minnesota, Minneapolis, Minnesota
and

RASHMI C. DESAI

Department of Physics, University of Toronto, Toronto, Ontario, Canada

CONTENTS

[a] Permanent address: Department of Chemical Engineering, Carnegie-Mellon University, Pittsburgh, PA, 15213

I. INTRODUCTION

We recently have attempted to develop a novel theoretical approach to the analysis of inhomogeneous, two-phase systems. Our objective here is not to construct a balanced review of theoretical research on the liquid–vapor interface, but rather to present these ideas of our own in the context of earlier theories, to draw connections where connections can be found, and to interpret our results, whenever possible, in terms of older, more familiar points of view. The pervasive theme of our theory is identifying and then separating for examination a thin transition zone or interfacial region of the two-phase, many-body system that is coupled dynamically to the much more massive bulk phases. The static and dynamic characteristics of the liquid and vapor bulk phases are assumed to be well understood. The interfacial region has intrinsic properties, such as surface tension, and a dynamics of its own, but interactions with the surrounding bulk phases usually dominate its irreversible behavior.

The dynamics of this two-phase, many-body system is treated by means of linear response theory and, in particular, extensive use is made of the correlation function and projection operator formalisms due principally to Zwanzig and Mori. Two very different types of projections are involved. In hydrodynamic considerations use is made of projections onto collective variables (conserved and symmetry-restoring) with long dynamic lifetimes. To deal with the interface itself we introduce projections in physical space that separate the transition zone from the surrounding bulk phases.

This chapter is divided into three major parts: (*1*) first, a consideration of liquid–vapor equilibrium with special attention to surface tension; (*2*) then, the hydrodynamic or continuum mechanical theory of interfacial dynamics; and (*3*) finally, a kinetic theory of interfacial dynamics. The last two of these differ with regard to the amounts of interfacial detail involved. The hydrodynamic characterization of the interfacial region is confined to a few quasimacroscopic variables, such as density and momentum, whereas the kinetic theory deals with the much richer description provided by the phase space distribution function.

A number of mathematical items concerning correlation functions and the solutions of kinetic equations have been relegated to appendices so they do not disrupt the intended logical flow of the presentation. Some of the material included in these appendices is new and some is not. However, we think that it may prove useful to have all of this material at one's fingertips while reading this chapter.

II. EQUILIBRIUM

A. Thermodynamics

Because the thermodynamics of heterogeneous, multiphase systems is a well-established and prolifically documented subject,[1–3] we consider here only those topics that bear directly on the microscopic theory that follows. For simplicity we restrict our attention to an equilibrium system consisting of one vapor and one liquid phase separated by a single, connected interfacial region. Furthermore, unless otherwise stated, this interface is taken to be planar. An interface with this geometry is, of course, stable only in the presence of an external body-force of fixed direction, for example, gravity. It is this external field that determines the spatial location of the interface: without it one of the fluid phases would form into one or more spherical regions imbedded within the second phase. However, it is convenient to suppose that even in the absence of gravity the liquid forms a single spherical phase so large that its surface may be treated as if it were effectively flat.

With this understanding and with reference to Fig. 1*a*, we see that the heterogeneous system can be separated into three parts: (*1*) the vapor phase, $x_3 > \varepsilon$, (*2*) the liquid phase, $x_3 < -\varepsilon$, and (*3*) the transition zone, $|x_3| < \varepsilon$. Here x_3 is a coordinate measured in the 3 direction, normal to the phase boundary. The scale parameter ε characterizes the thickness of the interfacial, transition zone and may, conveniently, be defined in terms of the fluid (molar or molecular number) density $n_0(x_3)$. Thus $\nabla_3 n_0(x_3) \neq 0$ for $|x_3| < \varepsilon$ and $\nabla_3 n_0(x_3) = 0$ for $|x_3| > \varepsilon$. Strictly speaking, the gradient $\nabla_3 n_0(x_3)$ never vanishes, but the assumption that it equals zero outside of a narrow transition zone is a convenient fiction and an excellent approximation. Thus, in a typical van der Waals fluid (far from the critical point), $n_0(\varepsilon) = 1.01\ n_v$ and $n_0(-\varepsilon) = 0.99 n_l$ for $\varepsilon = 13.5$ Å.[4] Because this transition zone is so very thin, it is usual in thermodynamics (and in other branches of continuum mechanics as well) to treat the interface as if it were a mathematical surface of separation between two uniform bulk (B) phases. The density profile corresponding to this model is given by

$$n_0^B(x_3) = n_l \theta(-x_3) + n_v \theta(x_3) \tag{1}$$

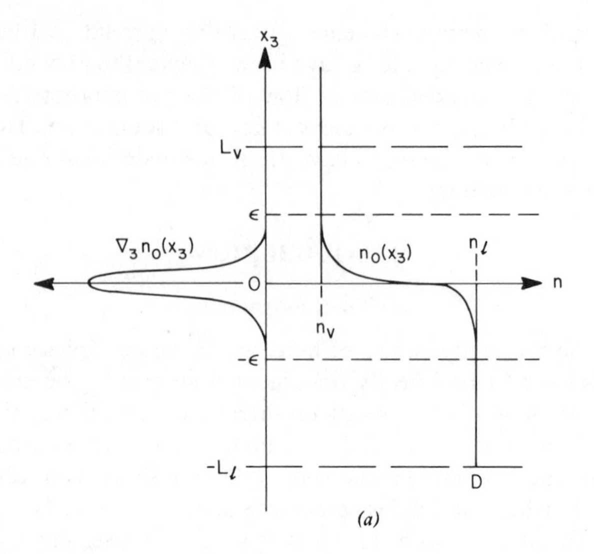

Fig. 1. (a) The density, $n_0(x_3)$, and density gradient, $\nabla_3 n_0(x_3)$, in the vicinity of a liquid–vapor interface. Three regions are depicted: (1) the *gas phase*, $x_3 \geq \epsilon$; (2) the *liquid phase*, $x_3 \leq -\epsilon$; and (3) the *transition region*, $|x_3| < \epsilon$. Here ϵ is defined by $\nabla_3 n_0(x_3) \neq 0$ for $|x_3| < \epsilon$ and $\nabla_3 n_0(x_3) = 0$ for $|x_3| \geq \epsilon$; n_v and n_l are the densities of the bulk vapor and liquid phases, respectively, and L_v and L_l are macroscopic lengths defined in the text.

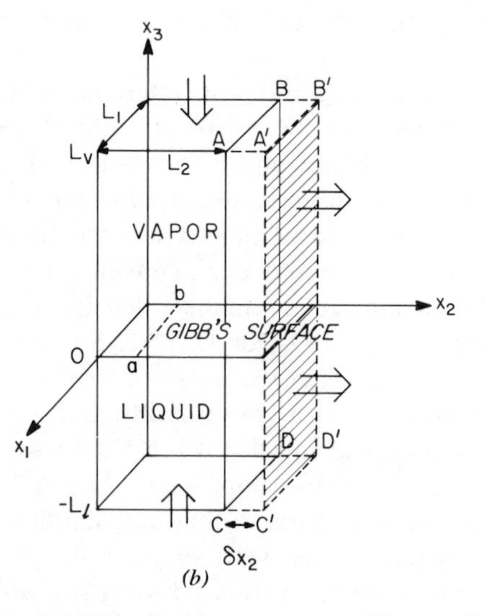

Fig. (b) Schematic representation of the liquid–vapor system. The meanings of the various symbols appearing here are given in the text.

with $\theta(x)$ denoting the unit step function. Furthermore, it is customary to assign the location of this dividing surface (i.e., the definition of $x_3=0$) according to the Gibbs construction[5] $\int_{-L_l}^{L_v}[n_0(x_3)-n_0^B(x_3)]dx_3=0$ or, equivalently,

$$\int_{-L_l}^{0}\left[n_l-n_0(x_3)\right]dx_3=\int_0^{L_v}\left[n_0(x_3)-n_v\right]dx_3 \tag{2}$$

Here L_l and L_v are lengths of macroscopic scale so that the fluid properties at the two points $x_3=-L_l$ and $x_3=L_v$ are indistinguishable from those of the bulk liquid and vapor phases; specifically, L_l and L_v are both much greater than ε. Figure 1*a* includes schematic representations of the actual and hypothetical density profiles n_0 and n_0^B and of the Gibbs dividing surface, $x_3=0$. Also included (at the left) is a graph of the density gradient, $\nabla_3 n_0(x_3)$.

We already have alluded indirectly to the well-known fact that small liquid drops in air and small gas bubbles in water are very nearly spherical. Thus it appears that the natural tendency is to minimize the interfacial area. This, as well as a multitude of related observations, supports the hypothesis that the boundary between two phases possesses a special form of energy in an amount proportional to the interfacial area. The proportionality coefficient is called the surface tension, σ.

One of the best known relationships involving σ is the Young-Laplace equation,

$$P_{\text{ex}}=\sigma\left(\frac{1}{R_1}+\frac{1}{R_2}\right) \tag{3}$$

It relates the principal radii of curvature, R_1 and R_2, at a point on a curved interface to the value of the "excess pressure", P_{ex}, that is needed to maintain local mechanical equilibrium. This excess is the pressure difference between the fluids on the concave and convex sides of the interface.

Related to these energetic considerations is the interpretation of σ as a force or stress. To illustrate this let us examine the two-phase system confined within the container of volume $V=L_1L_2L_3$ shown in Fig. 1*b*. Here $L_3=L_v+L_l$ is the sum of the vertical extensions of the vapor and liquid phases. We consider an isothermal, constant volume process that increases the interfacial area from $A=L_1L_2$ to $A+\delta A$, $\delta A=L_1\delta x_2$. This change of state is accomplished by shifting the vertical wall *ABCD* to the right by an amount δx_2 (a step that increases the volume of the system by $\delta V=L_1L_3\,\delta x_2\equiv L_3\,\delta A$) and then pushing the top and bottom toward one another just enough to reduce the volume of the system to its original value. The amounts of reversible work that the system must perform on its surroundings to cause these two deformations are $L_1 \delta x_2 \int_{-L_l}^{L_v} P_T(x_3)\, dx_3$ and $-P_N \delta V = -L_1 \delta x_2 \int_{-L_l}^{L_v} P_N dx_3$, respectively. Here $P_T(x_3)$ is the force per unit area acting in the direction 2, tangential to the interface. P_N is the normal component of the force per unit area. If we ignore minor, barometric variations due to gravity, P_N is equal to the value of the pressure common to both of the equilibrium phases. Therefore, the net work that must be done on the system to effect the desired isothermal, isochoric change of interfacial area is $\delta W = \sigma\, \delta A$ with

$$\sigma = \int_{-L_l}^{L_v} [P_N - P_T(x_3)]\, dx_3 \tag{4}$$

Far from the interface $P_T(x_3)$ becomes equal to P_N and so the limits of integration in this expression are immaterial provided both are greater than the interfacial thickness ε.

This formula identifies the surface tension σ with the sum of the stress (force per unit area) acting in the direction tangential to the interface and the uniform normal pressure P_N. As a further interpretive device, consider an imaginary vertical surface that intersects the Gibbs surface along a line such as $a-b$ shown in Fig. 1*b*. The fluid on one side of this line *pulls* that on the other side with a tension whose magnitude is σ. It is this mechanical interpretation of σ that is used to formulate the statics and dynamics of continuum systems involving phase boundaries.

There is an equivalent thermodynamic identification of σ with the surface excess free energy per unit area. This equivalence is a direct consequence of our previous observation that $\sigma\, dA$ equals the amount of reversible work associated with an isothermal, isochoric process that increases the interfacial area by an amount dA. The changes of internal energy and entropy that accompany this isochoric change of state are related to the reversible work by $dU = dQ_{\text{rev}} + dW_{\text{rev}} = T\,dS + \sigma\, dA$. Consequently, the relationship between $\sigma\, dA$ and the Helmholtz free energy, $F = U - TS$, is given by $dF = -S\,dT + \sigma\, dA$, from which follows

$$\sigma = \left(\frac{\partial F}{\partial A}\right)_{T,V} \tag{5}$$

It is a remarkable characteristic of the surface tension that its value is independent of the interfacial area A. This behavior is totally unlike that of an elastic film, whose tension increases roughly in proportion to the amount by which it is stretched. The mechanistic explanation of this hinges on the realization that the interfacial zone does not consist of a fixed amount of material. The quantity needed to form an additional piece of phase boundary is drawn from the bulk phases. Thus the surface tension is an intrinsic interfacial property that is unrelated to the value of A but dependent on temperature and, in a multicomponent system, on composition as well. Because of this the integral of (5) can be written in the form

$$F(T,V,A)=\sigma(T)A+F_B(T,V) \tag{6}$$

where F_B denotes the free energy of the bulk phases. It is this relationship that permits us to identify σ with the surface excess free energy per unit area.

The surface tension of unassociated liquids closely conforms to the empirical relationship,

$$\sigma(T)=\sigma_0(1-T/T_c)^{\mu} \tag{7}$$

over a wide range of temperatures.[6] Thus the value of σ decreases with rising temperature and vanishes at the critical point. From extensive studies near T_c the value of the critical coefficient μ has been found to be 1.28 ± 0.06.[7] This is to be compared with the value of 1.27 ± 0.02[6] obtained from data collected over a much broader range of temperatures and from experiments on a large number of liquids. Classical mean field theory predicts that μ should be 1.5.[8]

The behavior of σ near T_c is reviewed in Ref. 7. Also, an interesting historical account of early experiments and of the classical theory of surface tension is given in Ref. 9. Our considerations here have been restricted to a single-component system. A recent survey of the more general, multicomponent theory has been presented by Harasima.[3] His paper contains many useful thermodynamic results, including a Gibbs-Duhem relationship for the interface from which an expression is obtained for the interfacial energy density.

B. Molecular Theories of Surface Tension

The theoretical considerations that follow are restricted to the planar liquid–vapor interface of a single-component system composed of structureless molecules. The Hamiltonian of this system can be written as

$$H=\sum_{\alpha}\frac{\mathbf{p}_{\alpha}^{2}}{2m}+\tfrac{1}{2}\sum_{\alpha\neq\beta}\mathcal{V}(\mathbf{r}^{\alpha},\mathbf{r}^{\beta}) \tag{8}$$

where $\mathbf{r}^{\alpha}$ and $\mathbf{p}^{\alpha}$ denote the position and momentum of molecule α. $\mathcal{V}(\mathbf{r}^{\alpha},\mathbf{r}^{\beta})\equiv V(|\mathbf{r}^{\alpha}-\mathbf{r}^{\beta}|)$ is the energy of interaction between molecules α and β.

Among the variables that concern us most are the number density, $n(\mathbf{r}, t)=\Sigma_\alpha \delta(\mathbf{r}-\mathbf{r}^\alpha(t))$ and the Cartesian components of the momentum density, $g_i(\mathbf{r}, t)=\Sigma_\alpha p_i^\alpha \delta(\mathbf{r}-\mathbf{r}^\alpha(t))$. These many-body, collective variables are the four lowest order momentum moments of the single-particle phase space density, $f(\mathbf{r},\mathbf{p}, t)=\Sigma_\alpha \delta(\mathbf{r}-\mathbf{r}^\alpha(t))\delta(\mathbf{p}-\mathbf{p}^\alpha(t))$, that is,

$$n(\mathbf{r}, t)=\int d^3p\, f(\mathbf{r},\mathbf{p}, t)=\sum_\alpha \delta(\mathbf{r}-\mathbf{r}^\alpha(t)) \tag{9a}$$

$$g_i(\mathbf{r}, t)=\int d^3p\, p_i f(\mathbf{r},\mathbf{p}, t)=\sum_\alpha p_i^\alpha \delta(\mathbf{r}-\mathbf{r}^\alpha(t)) \tag{9b}$$

As shown in Fig. 1*a*, the equilibrium number density, $n_0(x_3)=\langle \Sigma_\alpha \delta(\mathbf{r}-\mathbf{r}^\alpha)\rangle$, assumes the limiting values of n_v and n_l as x_3 tends to $+\infty$ and to $-\infty$, respectively. Later we see that $\nabla_3 n_0(x_3)$, shown schematically on the left of Fig. 1*a*, provides a much more useful characterization of the interfacial region than does $n_0(x_3)$ itself. This macroscopic density gradient is related to the equilibrium pair distribution function, $n_2(\mathbf{r},\mathbf{r}')=\langle \Sigma_{\alpha\neq\beta}\delta(\mathbf{r}-\mathbf{r}^\alpha)\delta(\mathbf{r}'-\mathbf{r}^\beta)\rangle$, through the relationship[10]

$$\nabla_i n_0(x_3)=\delta_{i,3}\nabla_3 n_0(x_3)=-\beta\int d^3r'\, n_2(\mathbf{r},\mathbf{r}')\nabla_i \mathcal{V}(\mathbf{r},\mathbf{r}') \tag{10}$$

wherein $\beta=1/k_BT$. The angular brackets used above in defining $n_0(x_3)$ and $n_2(\mathbf{r},\mathbf{r}')$ denote the grand canonical ensemble average for the inhomogeneous system. Although it is not explicit in our notation [e.g., in our choice of the Hamiltonian given by (8)], this inhomogeneity can be formally incorporated in the homogeneous system ensemble by means of a weak external potential.[10]

Now that these matters of notation and definition have been attended to, we can proceed to the central task of this section, that of deriving from statistical mechanics, formulas for the surface tension σ. Most of these formulas can be assigned to one of two categories: (*1*) *potential form* (σ_P) or (*2*) *density functional form* (σ_D), each of which is characterized by the mathematical form of the results and by the theoretical constructs employed in the derivation. In both cases the surface tension is identified with the work (or excess free energy) that must be done to cause a unit increase of interfacial area. A third category of surface tension theories, (*3*) *correlation function form*, has recently emerged from our own investigations[11–13] of interfacial mechanics. This theory not only generates novel formulas for σ, but also provides valuable insights into the relationship between the other two, more traditional theories.

1. Potential Form

Prime examples of theories belonging to this category are those of Kirkwood and Buff,[14, 15] who based their development on the relationship (4) that exists between surface tension and the fluid stress tensor. However, instead of reproducing their derivation we derive the Kirkwood-Buff formula by using an infinitesimal scaling transformation.[15–17]

Consider again the system depicted in Fig. 1*b* and the isothermal, isochoric process described in the preceding section. For simplicity assume that $L_1 = L_2 = L_3 = L$. The container wall *ABCD*, whose normal lies in the direction 2, is shifted outward through a distance $\delta x_2 = \varepsilon L$, and the top of the container simultaneously is depressed by an equal amount. These displacements are performed isothermally and reversibly. Therefore, it follows that the changes of (free) energy and interfacial area are connected by $\langle \delta H \rangle = \sigma\, dA$.[17] The surface tension, consequently, can be identified with the differential quotient

$$\sigma = \lim_{\varepsilon \to 0} \frac{\langle \delta H \rangle}{\delta A} = \lim_{\varepsilon \to 0} \frac{\langle H' - H \rangle}{\varepsilon A}$$

where the difference between H and H' is due to the different ranges of coordinate values that are accessible to the particles before and after the deformations have occurred. Let us now assume that these distortions are distributed uniformly throughout the system and that every particle within the container obeys the same infinitesimal scaling transformation, namely,

$$\begin{aligned} &X_1^\alpha = x_1^\alpha, \qquad X_2^\alpha = x_2^\alpha + \varepsilon x_2^\alpha, \qquad X_3^\alpha = x_3^\alpha - \varepsilon x_3^\alpha \\ &P_1^\alpha = p_1^\alpha, \qquad P_2^\alpha = p_2^\alpha - \varepsilon p_2^\alpha, \qquad P_3^\alpha = p_3^\alpha + \varepsilon p_3^\alpha \end{aligned} \tag{11}$$

The generating function for this infinitesimal contact transformation, defined by[18]

$$p_i^\alpha = P_i^\alpha + \varepsilon\left(\frac{\partial G}{\partial x_i^\alpha}\right); \qquad X_i^\alpha = x_i^\alpha + \varepsilon\left(\frac{\partial G}{\partial p_i^\alpha}\right)$$

is easily shown to be[19]

$$G(\mathbf{r}, \mathbf{P}) = \text{const.} + \sum_\alpha \left(x_2^\alpha P_2^\alpha - x_3^\alpha P_3^\alpha\right) \tag{12}$$

From this we immediately are able to calculate the change of the energy function due to the distortion,

$$\delta H = \varepsilon[H, G]_{P.B} = \varepsilon\left\{-\sum_\alpha \frac{1}{m}\left(p_2^{\alpha^2} - p_3^{\alpha^2}\right) + \sum_{\alpha \neq \beta}\left(x_2^\alpha \nabla_2^\alpha - x_3^\alpha \nabla_3^\alpha\right) V(|\mathbf{r}^\alpha - \mathbf{r}^\beta|)\right\}$$

as well as its ensemble average,

$$\langle \delta H\rangle = -2\varepsilon\left\langle \sum_{\alpha} \frac{1}{2m}\left(p_2^{\alpha^2}-p_3^{\alpha^2}\right)\right\rangle$$

$$+\frac{\varepsilon}{2}\left\langle \sum_{\alpha\neq\beta} \frac{(x_2^\alpha - x_2^\beta)^2-(x_3^\alpha-x_3^\beta)^2}{|\mathbf{r}^\alpha-\mathbf{r}^\beta|}\frac{dV(|\mathbf{r}^\alpha-\mathbf{r}^\beta|)}{d|\mathbf{r}^\alpha-\mathbf{r}^\beta|}\right\rangle$$

$$=\frac{\varepsilon}{2}\int d^3r\int d^3r'\frac{(x_2-x_2')^2-(x_3-x_3')^2}{|\mathbf{r}-\mathbf{r}'|}V'\ (|\mathbf{r}-\mathbf{r}'|)$$

$$\times\left\langle \sum_{\alpha\neq\beta}\delta(\mathbf{r}-\mathbf{r}^\alpha)\delta(\mathbf{r}'-\mathbf{r}^\beta)\right\rangle$$

$$=\tfrac{1}{2}\varepsilon A\int dx_3\int d^3r'\frac{(x_2-x_2')^2-(x_3-x_3')^2}{|\mathbf{r}-\mathbf{r}'|}V'(|\mathbf{r}-\mathbf{r}'|)n_2(\mathbf{r},\mathbf{r}')$$

It then follows from this and our previous identification of $\langle\delta H\rangle$ with $\delta W_{\text{rev}}=\sigma dA$ that the surface tension is given by the Kirkwood-Buff formula,

$$\sigma_P=\tfrac{1}{2}\int dx_3\int d^3(r-r')\frac{(x_2-x_2')^2-(x_3-x_3')^2}{|\mathbf{r}-\mathbf{r}'|}V'(|\mathbf{r}-\mathbf{r}'|)n_2(\mathbf{r},\mathbf{r}')$$

$$=\tfrac{1}{4}\int dx_3\int d^3(r-r')\frac{(\mathbf{r}-\mathbf{r}')^2-3(x_3-x_3')^2}{|\mathbf{r}-\mathbf{r}'|}V'(|\mathbf{r}-\mathbf{r}'|)n_2(\mathbf{r},\mathbf{r}') \quad (13)$$

The quantal counterpart of this formula is obtained by utilizing the unitary operator $\hat{U}=\exp(i\varepsilon\hat{G}/\hbar)$ with $\hat{G}=\frac{1}{2}\Sigma_\alpha(x_2^\alpha P_2^\alpha + P_2^\alpha x_2^\alpha - x_3^\alpha P_3^\alpha - P_3^\alpha x_3^\alpha)$. Thus it is readily verified that the operator transformation $\mathbf{R}^\alpha = \hat{U}\mathbf{r}^\alpha\hat{U}^{-1}$, $\mathbf{P}^\alpha=\hat{U}\mathbf{p}^\alpha\hat{U}^{-1}$ is formally identical to the contact transformation of (11). The construction of δH proceeds as before, but now the ensemble average $\langle A\rangle$ is to be identified with $\{\mathrm{Tr}(A\rho)/\mathrm{Tr}\rho\}$ where Tr denotes the trace and ρ is the equilibrium density matrix. For the surface tension we obtain the formula,

$$\sigma=\frac{1}{A}\frac{2\mathrm{Tr}\left[-\dfrac{\hbar^2}{2m}\displaystyle\sum_\alpha\left(\partial^2/\partial x_3^{\alpha 2}-\partial^2/\partial x_2^{\alpha 2}\right)\rho(\mathbf{r};\mathbf{r}')\right]}{\mathrm{Tr}\rho(\mathbf{r};\mathbf{r}')}$$

$$+\tfrac{1}{2}\int d^3r\int d^3r'\frac{(x_2-x_2')^2-(x_3-x_3')^2}{|\mathbf{r}-\mathbf{r}'|}V'(|\mathbf{r}-\mathbf{r}'|)n_2^Q(\mathbf{r},\mathbf{r}') \quad (14)$$

which first was derived by Toda.[20] Here $\rho(\mathbf{r};\mathbf{r}')$ is the position representative of the density operator for the inhomogeneous fluid and $n_2^Q(\mathbf{r},\mathbf{r}')$ is the quantum-mechanical pair correlation function.

Partly because they lacked detailed information about the functional forms of $n_0(x_3)$ and $n_2(\mathbf{r},\mathbf{r}')$ and partly because of mathematical difficulties, Fowler,[21] as an initial approximation, and Kirkwood and Buff,[14] to facilitate numerical calculations, both resorted to the expedient of shrinking the transition zone to a density discontinuity located at the Gibbs surface. The singlet and pair correlation functions appropriate to this model are

$$n_0^F(x_3) = n_v\theta(x_3) + n_l\theta(-x_3) \tag{15a}$$

and

$$\begin{aligned} n_2^F(\mathbf{r},\mathbf{r}') &= n_v^2 g_v(|\mathbf{r}-\mathbf{r}'|)\theta(x_3)\theta(x_3') + n_l^2 g_l(|\mathbf{r}-\mathbf{r}'|)\theta(-x_3)\theta(-x_3') \\ &\quad + n_v n_l[\theta(x_3)\theta(-x_3') + \theta(-x_3)\theta(x_3')] \end{aligned} \tag{15b}$$

respectively. Here the subscripts v and l refer to the vapor and liquid phases, the superscript F refers to Fowler, and $g_\gamma(r)$ is the equilibrium radial distribution function of the bulk phase γ.

The expression obtained by substituting this approximation into the Kirkwood-Buff formula is[22]

$$\sigma_P \doteq \frac{\pi}{8}\int_0^\infty dr\, r^4 V'(r)[n_l^2 g_l(r) - n_v^2 g_v(r)] \tag{16}$$

By neglecting the second of the integrand factors in comparison to the first we obtain from (16) the Fowler formula,

$$\sigma_F = \frac{\pi}{8} n_l^2 \int_0^\infty dr\, r^4 V'(r) g_l(r) \tag{17}$$

What makes this approximation so attractive is that it provides a simple way to construct estimates of surface tension from well-known properties of the bulk liquid phase. Considering the severity of the approximation on which the formula is based, its numerical predictions are surprisingly accurate. This success is especially remarkable in the light of detailed studies by Harasima in which the Fowler approximation was applied separately to the normal and tangential stresses, P_N and P_T. Thus, Harasima[3] found for one typical and representative case that the normal stress predicted by this theory was equal to 200 atm at the Gibbs surface, attained a maximum of 660 atm at $x_3 = -2.4$ Å, and then fell nearly to zero at about $x_3 = -10$ Å. Since hydrodynamic stability requires that P_N be constant throughout the two-phase system, it would appear that the numerical success of the Fowler formula depends on a fortuitous cancellation of errors between the values it predicts for P_N and P_T [cf. (4)]. Pastor and Goodisman[23] have examined this approximate theory in great detail and suggested means for correcting its most serious deficiencies.

In contrast to the generally satisfactory estimates of σ that it produces for simple fluids, the Fowler formula generates totally unrealistic results ($\sigma \ll 0$!) for molten salts. Pastor and Goodisman[23] have traced this behavior to the Coulomb interactions among the components and the failure of the model to impose the necessary microscopic electroneutrality. They then proposed a simple correction for this deficiency that produces noticeably improved predictions of σ.

There have been several more systematic and physically palatable investigations of the Kirkwood-Buff formula that combine in a self-consistent formalism some functional ansatz for $n_2(\mathbf{r},\mathbf{r}')$ (or the related direct correlation function) with the criterion (10) of hydrostatic stability. These studies are discussed in a recent survey by Evans.[24]

2. *Density Functional Formulas for Surface Tension*

A formula for the surface tension that involves both the density gradient and a direct correlation function of the inhomogeneous, two-phase fluid was first presented by Yvon,[25] whose method of derivation has been described by Lovett et al.[26] as "a leading virial coefficient type of argument." More than two decades then elapsed before systematic derivations of this formula were independently formulated by Triezenberg and Zwanzig,[27] who related the excess surface energy to the fluctuations of density that occur near the interface, and by Lovett et al.,[26] who utilized the Young-Laplace formula (3). Here we present (a special case of) the Triezenberg and Zwanzig derivation in a way intended to illustrate how closely it is related to the derivation of Lovett et al.

Figure 2 shows the physical origin of this similarity. Lovett et al. consider an initially planar interface ($x_3 = 0$), which is then slightly bent, as in Fig. 2*a*, by imposing an external field. We assume that this isochoric deformation of the fluid involves the *vertical* displacement of each molecule according to the transformation $X_1^\alpha = x_1^\alpha$, $X_2^\alpha = x_2^\alpha$, $X_3^\alpha = x_3^\alpha + \zeta(x_1, x_2)$. The variation of density associated with the deformation then can be identified with

$$\delta n = \langle n \rangle_{\text{bent}} - \langle n \rangle_{\text{flat}} = \Big\langle \sum_\alpha \left[\delta(x_3 - x_3^\alpha - \zeta(x_1, x_2)) - \delta(x_3 - x_3^\alpha) \right] \Big\rangle$$

$$= n_0(x_3 - \zeta) - n_0(x_3) \doteq -\zeta(x_1, x_2) \nabla_3 n_0(x_3)$$

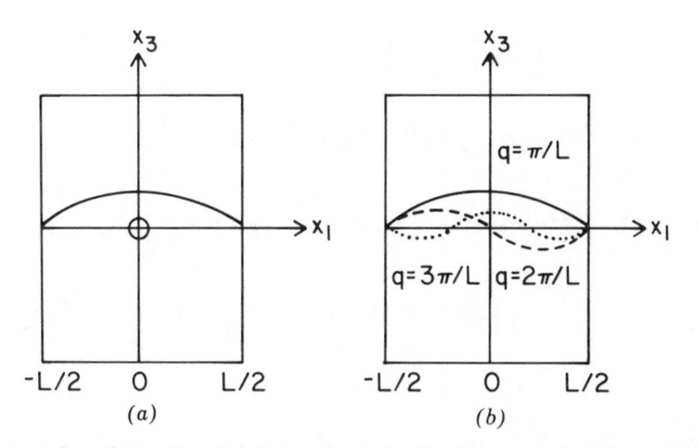

Fig. 2. Long-wavelength modes of deformation of a liquid–vapor interface. Here q is a wave number and L is the linear extension of the interface.

where $x_3 = \zeta(x_1, x_2)$ is the equation of the deformed, nonplanar interface.

The variations of density involved in the theory of Triezenberg and Zwanzig are the spontaneous fluctuations, $\delta n = n - \langle n \rangle$, that occur near the interface of a heterogeneous system that is in a state of macroscopic equilibrium. As a consequence of these fluctuations the local and instantaneous position of the Gibbs surface can be represented by a superposition of the various modes depicted in Fig. 2*b*. From these we now select for consideration the surface mode with the longest possible wavelength, $2L$, and hence with the smallest allowable wave number, $q_{\min} = 2\pi/2L = \pi/L$. For this special case the Triezenberg-Zwanzig and Lovett et al. interpretations of δn are identical.[28]

The vertical shift associated with this mode can be modeled by the formula $\zeta(x_1) = h \cos \pi x_1/L$; the corresponding change of interfacial area is

$$\Delta A = \tfrac{1}{2} \int_{-L/2}^{L/2} dx_1 \int_{-L/2}^{L/2} dx_2 |\nabla_1 \zeta(x_1)|^2 = \tfrac{1}{4}\pi^2 h^2$$

The critical assumption that we now invoke is that the change of (Helmholtz) free energy caused by this deformation of the liquid–vapor interface is a functional of δn. Consequently, the value of this change can be calculated (to order δn^2) from the well-known formula

$$\Delta F = \tfrac{1}{2}\beta^{-1} \int d^3r \int d^3r' \, \delta n(\mathbf{r}) K(\mathbf{r};\mathbf{r}') \delta n(\mathbf{r}') \tag{18}$$

where $K(\mathbf{r};\mathbf{r}')$ denotes the inverse of the density autocorrelation function, $\langle \delta n(\mathbf{r}) \delta n(\mathbf{r}') \rangle$. With $\delta n(r) \doteq -\zeta(x_1)\nabla_3 n_0(x_3)$ and $\zeta(x_1) = h \cos \pi x_1/L$ the change of free energy then becomes

$$\Delta F=\tfrac{1}{2}\beta^{-1}h^2\int dx_3\int dx_3'[\nabla_3 n_0(x_3)][\nabla_3' n_0(x_3')]$$

$$\times\int d^2\rho\int d^2\rho'\cos\left(\frac{\pi x_1}{L}\right)\cos\left(\frac{\pi x_1'}{L}\right)K(\boldsymbol{\rho}-\boldsymbol{\rho}';x_3,x_3')$$

$$=\tfrac{1}{4}\beta^{-1}L^2h^2\int dx_3\int dx_3'[\nabla_3 n_0(x_3)]K\left(\frac{\pi}{L};x_3,x_3'\right)[\nabla_3' n_0(x_3')] \quad (19)$$

with

$$K\left(\frac{\pi}{L};x_3,x_3'\right)\equiv\int d^2(\rho-\rho')\cos\frac{\pi}{L}(x_1-x_1')K(\boldsymbol{\rho}-\boldsymbol{\rho}';x_3,x_3')$$

$$=K_0(x_3,x_3')+\left(\frac{\pi}{L}\right)^2K_2(x_3,x_3')+O(L^{-4}) \quad (20)$$

Here the invariance of the two-phase system with respect to translations parallel to the plane of $\boldsymbol{\rho}=(x_1,x_2)$ has been used to reexpress $K(\mathbf{r};\mathbf{r}')$ as $K(\boldsymbol{\rho}-\boldsymbol{\rho}';x_3,x_3')$.

In the following section the important identity

$$\int dx_3\int dx_3'[\nabla_3 n_0(x_3)]K_0(x_3,x_3')[\nabla_3' n_0(x_3')]=0 \quad (21)$$

is proved. Here we use this result, together with (19) and (20), to conclude that for $L\gg 1$, ΔF is equal to $\sigma_D\Delta A$ with σ_D given by the Triezenberg-Zwanzig formula,[27]

$$\sigma_D=\beta^{-1}\int dx_3\int dx_3'[\nabla_3 n_0(x_3)]K_2(x_3,x_3')[\nabla_3' n_0(x_3')] \quad (22)$$

The condition $L\gg 1$ corresponds to the small wave-number limit of $q_{\min}=\pi/L\to 0$. Furthermore, according to the model adopted here the principal radii of curvature of the distorted liquid–vapor interface are $R_1=L^2/2h$ and $R_2\to\infty$. The spirit of the present derivation is, therefore, much the same as that of Lovett et al., who used the Young-Laplace relationship, (3), to relate surface tension to the radii of curvature of a very nearly planar interface.

The Triezenberg-Zwanzig formula for surface tension can be rewritten in terms of the direct correlation function,

$$C(\mathbf{r};\mathbf{r}')=\frac{\delta(\mathbf{r}-\mathbf{r}')}{n_0(x_3)}-K(\mathbf{r};\mathbf{r}') \quad (23)$$

the (partial) Fourier transform of which may be expanded in the manner

$$C(q;x_3,x_3')\equiv\int d^2(\rho-\rho')e^{-i\mathbf{q}\cdot(\rho-\rho')}C(\boldsymbol{\rho}-\boldsymbol{\rho}';x_3,x_3')$$

$$=C_0(x_3,x_3')+q^2C_2(x_3,x_3')+\ldots \tag{24}$$

Thus, by equating terms in the wave-number expansions of K and C, it follows that

$$K_0(x_3,x_3')=\frac{\delta(x_3-x_3')}{n_0(x_3)}-C_0(x_3,x_3') \tag{25a}$$

and

$$K_2(x_3,x_3')=-C_2(x_3,x_3') \tag{25b}$$

From these relationships we see that (21) can be rewritten in the form

$$\int dx_3\int dx_3'[\nabla_3 n_0(x_3)]C_0(x_3,x_3')[\nabla_3' n_0(x_3')]$$

$$=\int dx_3[\nabla_3 n_0(x_3)]\nabla_3\ln n_0(x_3)$$

or, equivalently, as

$$\int dx_3'\int d^2(\rho-\rho')C(\mathbf{r};\mathbf{r}')[\nabla_3' n_0(x_3')]=\nabla_3\ln n_0(x_3) \tag{26}$$

Therefore, the Triezenberg-Zwanzig formula can be cast into the alternative form

$$\sigma_D=\beta^{-1}\int dx_3\int dx_3'[\nabla_3 n_0(x_3)][-C_2(x_3,x_3')][\nabla_3' n_0(x_3')]$$

$$=\tfrac{1}{4}\beta^{-1}\int dx_3\int dx_3'\int d^2(\rho-\rho')(\boldsymbol{\rho}-\boldsymbol{\rho}')^2C(\mathbf{r};\mathbf{r}')[\nabla_3 n_0(x_3)][\nabla_3' n_0(x_3')] \tag{27}$$

which is identical to the result obtained by Lovett et al.[26]

A distinguishing characteristic of the direct correlation function is its extremely short range. Because of this, it seems reasonable to replace the integral appearing in (27) with the approximation

$$\int dx_3 |\nabla_3 n_0(x_3)|^2 \int d^3(r-r')(\rho-\rho')^2 C[n_0(x_3); |\mathbf{r}-\mathbf{r}'|]$$

$$= \tfrac{2}{3} \int dx_3 |\nabla_3 n_0(x_3)|^2 \int d^3 r' r'^2 C[n_0(x_3); r']$$

This results in the estimate of σ_D,

$$\sigma_D \doteq \sigma_{\text{YFG}} \equiv 2 \int dx_3 f_2[n_0(x_3)] |\nabla_3 n_0(x_3)|^2 \tag{28a}$$

with

$$f_2[n] = \frac{1}{12\beta} \int d^3 r\, r^2 C(n; r) \tag{28b}$$

which has been derived previously by Yang et al.[29] If one makes the further approximation of ignoring the density dependence of f_2, one obtains in place of σ_{YFG} the formula

$$\sigma_{\text{vdW}} = \frac{l^2}{6\beta} \int dx_3 |\nabla_3 n_0(x_3)|^2 \tag{29}$$

of van der Waals,[30] wherein l is the characteristic length defined by $l^2 = \int d^3 r\, r^2 C(\bar{n}; r)$. This same result is obtained by replacing $C(\mathbf{r}; \mathbf{r}')$ in (27) with the local Orstein-Zernike approximation,

$$C(q; x_3, x_3') \doteq C_0(x_3)\delta(x_3 - x_3') - \frac{l^2}{6}(q^2 - \nabla_3^2)\delta(x_3 - x_3')$$

The surface tension expressions σ_{YFG} and σ_{vdW} are especially interesting because they are identical in form to results obtained from the van der Waals theory of interfacial tension.[29-33] The basic assumption of this theory (and of its generalizations) is that the free energy of the inhomogeneous two-phase system can be written as the volume integral of a free energy density,

$$f(r) = f^\dagger[n(x_3)] + \tfrac{1}{2} A[n(x_3)] |\nabla_3 n(x_3)|^2$$

Here $f^\dagger(n)$ denotes the free energy density of a homogeneous (single-phase) system with density n, and $A[n]$ is an undetermined, positive-valued function of n. By requiring that the density profile minimize the free energy [subject to the constraint that $\int d^3r\, n(x_3)=N=\text{constant}$] one readily obtains for $n(x_3)$ the differential equation

$$A(n)\nabla_3^2 n+\tfrac{1}{2}A'(n)|\nabla_3 n|^2=\mu^\dagger(n)-\mu$$

Here $\mu^\dagger(n)\equiv df^\dagger(n)/dn$ is the chemical potential of the hypothetical, homogeneous fluid and μ is that of the stable, two-phase system. By multiplying this differential equation by $\nabla_3 n_0(x_3)$, we find that it can be rewritten in the form $\frac{1}{2}A(n)|\nabla_3 n|^2=\omega(n)-\omega(n_l)$, where $\omega(n)\equiv f^\dagger(n)-n\mu$. Finally, the Helmholtz free energy is given by $F\equiv\int d^3r\, f(\mathbf{r})=N\mu-PV+\sigma A$ with $\sigma=\int dx_3 A[n(x_3)]|\nabla_3 n(x_3)|^2$.

An excellent account of the classical van der Waals theory (in which the function A is independent of n) and of its application to critical phenomena has been given by Widom.[34] Yang et al.[29] have extended this theory to density-dependent A functions and, in addition, derived (28a) and (28b). A different modification of the van der Waals theory has been proposed by Bongiorno and Davis.[33]

One obvious extension of the van der Waals theory is to perform a "partial summation" of the gradient expansion and adopt for F the ansatz [see (18) for comparison]

$$F[n]=\int d^3r\, f^\dagger[n(\mathbf{r})]+\tfrac{1}{2}\beta^{-1}\int d^3r\int d^3r'\, K[\bar{n};|\mathbf{r}-\mathbf{r}'|][n(\mathbf{r})-n(\mathbf{r}')]^2$$

Here K is the correlation function for the homogeneous fluid and $\bar{n}$ is some average of the local densities, such as $\bar{n}=\frac{1}{2}[n(\mathbf{r})+n(\mathbf{r}')]$. This is the scheme used by Ebner et al.[35] and, in a somewhat different context, by Carey et al.[36]

3. *Correlation Function Form of the Surface Tension*

Although it is generally believed that the surface tension formulas for σ_P and σ_D [given by (13) and (22), respectively] are exact and complementary to one another, the two are strikingly different in appearance. This difference becomes even more dramatic if one replaces σ_P and σ_D with the corresponding approximations σ_F and σ_{vdW}, respectively. Thus the integrand of the Fowler formula, (17), depends exclusively on properties of the bulk liquid phase, whereas that of the van der Waals formula, (29), is different from zero only within the narrow transition zone, $|x_3|<\varepsilon$, that separates the two coexisting bulk phases. The connection between these two, seemingly very different theories of surface tension is not immediately obvious.

The numerical equivalence of σ_P and σ_D has been established by Leng et al.,[37] but only in the mean field approximation and for the special model of penetrable spheres. As we soon see, there is a way of proving that σ_P and σ_D are equal in the vicinity of the critical point, where fluctuations of density are the overwhelmingly dominant contributors to the surface excess of the free energy. However, the very same method that permits us to draw this conclusion casts some doubt on the contention that σ_P and σ_D are precisely equivalent. These connections between σ_P and σ_D to which we refer are drawn from a theory that expresses the surface tension in terms of correlation functions. It is this theory that we now consider.

It was mentioned earlier that the equilibrium two-phase system is invariant with respect to translations parallel to the plane of x_1 and x_2 (or $\boldsymbol{\rho}$) and that, because of this, it often is useful to introduce partially Fourier-transformed equilibrium correlation functions such as

$$\begin{aligned}\langle a(\mathbf{r})b(\mathbf{r}')\rangle(\mathbf{q},\mathbf{q}') &\equiv \int d^2\rho \int d^2\rho' e^{-i[\mathbf{q}\cdot\boldsymbol{\rho}+\mathbf{q}'\cdot\boldsymbol{\rho}']}\langle a(\boldsymbol{\rho}, x_3)b(\boldsymbol{\rho}', x_3')\rangle \\ &= (2\pi)^2\delta(\mathbf{q}+\mathbf{q}')\int d^2(\rho-\rho')e^{-i\mathbf{q}\cdot(\boldsymbol{\rho}-\boldsymbol{\rho}')}\langle a(\boldsymbol{\rho}-\boldsymbol{\rho}', x_3)b(\mathbf{0}, x_3')\rangle \\ &\equiv (2\pi)^2\delta(\mathbf{q}+\mathbf{q}')\langle a(\mathbf{r})b(\mathbf{r}')\rangle(\mathbf{q})\end{aligned}$$

Furthermore, because the relevant scale of length in the $\boldsymbol{\rho}$-plane is large compared with the average intermolecular spacing, our interest generally is limited to very small values of the wave number q.

In Section II.A we saw that the surface tension is closely related to the stress tensor of the inhomogeneous system. Therefore, it should come as no surprise to learn that it is the autocorrelation function of the force component normal to the interface that plays the central role in the correlation function theory of surface tension. Specifically, the object on which our attention is focused is

$$G_n(q) = \int dx_3 \int dx_3' \langle \dot{g}_3(\mathbf{r},0)\dot{g}_3(\mathbf{r}',0)\rangle(\mathbf{q}) \tag{30}$$

Here $g_3(\mathbf{r},0)$ is the normal (n) component of the momentum density defined by (9b). The autocorrelation function of $\dot{g}_3$ is given by the sum rule[11]

$$\begin{aligned}\langle \dot{g}_3(\mathbf{r},0)\dot{g}_3(\mathbf{r}',0)\rangle(\mathbf{q}) &\equiv \int d^2(\rho-\rho')e^{-i\mathbf{q}\cdot(\boldsymbol{\rho}-\boldsymbol{\rho}')}\langle \dot{g}_3(\mathbf{r},0)\dot{g}_3(\mathbf{r}',0)\rangle \\ &= \beta^{-2}(q^2+3\nabla_3\nabla_3')(n_0(x_3)\delta(x_3-x_3')) \\ &\quad -\beta^{-1}\left\{\delta(x_3-x_3')\int d\bar{x}_3\tau_{33}(\mathbf{q};x_3,\bar{x}_3)\right. \\ &\quad \left. -\tau_{33}(\mathbf{q};x_3,x_3')\right\}\end{aligned} \tag{31a}$$

wherein

$$\tau_{33}(\mathbf{q}; x_3, x_3') = \int d^2(\rho-\rho')\exp[-i\mathbf{q}\cdot(\boldsymbol{\rho}-\boldsymbol{\rho}')]n_2(\mathbf{r};\mathbf{r}')\nabla_3^2 V(|\mathbf{r}-\mathbf{r}'|) \tag{31b}$$

The result of substituting (31a) into the definition of $G_n(q)$ is the formula

$$G_n(q) = \beta^{-2}q^2\int dx_3 n_0(x_3) + \beta^{-1}\int dx_3\int dx_3'\int d^2(\rho-\rho')$$
$$\times\{1-\exp[-i\mathbf{q}\cdot(\boldsymbol{\rho}-\boldsymbol{\rho}')]\}[\nabla_3 n_2(\mathbf{r};\mathbf{r}')][\nabla_3 V(|\mathbf{r}-\mathbf{r}'|)] \tag{32}$$

We simplify the task of evaluating this expression by adopting for $n_0(x_3)$ and $n_2(\mathbf{r};\mathbf{r}')$ the same approximations, $-n_l\theta(-x_3)$ and $n_0(x_3)n_0(x_3')$ $g_l(|\mathbf{r}-\mathbf{r}'|)$, respectively, that are used in Section II.A to obtain the Fowler approximation, σ_F, from σ_P. To this approximation the integrand factor $\nabla_3 n_2(\mathbf{r};\mathbf{r}')$ appearing in (32) becomes

$$\nabla_3 n_2(\mathbf{r};\mathbf{r}') = -n_l\delta(x_3)n_0(x_3')g_l(|\mathbf{r}-\mathbf{r}'|) + n_0(x_3)n(x_3')\nabla_3 g_l(|\mathbf{r}-\mathbf{r}'|)$$

Thus, $G_n(q)$ itself is the sum, $G_n^S(q)+G_n^B(q)$, of a "surface" term

$$G_n^S(q) = \beta^{-1}n_l^2\int d^3r\{1-\exp[-i\mathbf{q}\cdot\boldsymbol{\rho}]\}\theta(-x_3)g(r)\nabla_3 V(r)$$
$$= q^2\beta^{-1}\sigma_F + O(q^4) \tag{33}$$

with σ_F given by (17), and a "bulk" contribution

$$G_n^B(q) = \beta^{-1}q^2\int_{-\infty}^{+\infty} dx_3 n_0(x_3) + \beta^{-1}\int d^3(r-r')\int_{-\infty}^{+\infty} dx_3' n_0(x_3)n_0(x_3')$$
$$\times\{1-\exp[-i\mathbf{q}\cdot(\boldsymbol{\rho}-\boldsymbol{\rho}')]\}\nabla_3 g_l(|\mathbf{r}-\mathbf{r}'|)\nabla_3 V(|\mathbf{r}-\mathbf{r}'|)$$
$$= \lim_{L_3\to\infty}[L_3\beta^{-1}n_l q^2\tilde{\gamma}(q)] \tag{34a}$$

wherein

$$\tilde{\gamma}(q) = q^{-2}\left\{\beta^{-1}q^2 + n_l\int d^3r(1-\exp[-i\mathbf{q}\cdot\boldsymbol{\rho}])\theta(-x_3)[\nabla_3 g_l(r)][\nabla_3 V(r)]\right\}$$
$$= \beta^{-1} + \frac{2\pi}{15}n_l\int_0^\infty dr\, r^2 g_l'(r)V'(r) + O(q^2) \tag{34b}$$

The feature that distinguishes $G_n^B(q)$ from $G_n^S(q)$ is, of course, the proportionality of the former to L_3, the linear extension of the bulk phases in the direction normal to the phase boundary.

To isolate the part of $G_n(q)$ that is intrinsic to the interface, the contributions of the bulk phases must be subtracted from the whole. This we do by introducing the surface projection operator δ_S, whose action generates the surface excess or difference,

$$\delta_S\langle\ldots\rangle=\langle\ldots\rangle-\langle\ldots\rangle_B \tag{35}$$

between the average computed for the true, two-phase system and that computed for a hypothetical system composed of two undistorted, bulk (B) phases butted together at the Gibbs surface $x_3=0$. The singlet and pair distribution functions of this hypothetical, reference system are (partially) defined by

$$n_0^B(x_3)=\begin{cases} n_v; & x_3>0 \\ n_l; & x_3<0 \end{cases} \tag{36a}$$

$$n_2^B(\mathbf{r};\mathbf{r}')=\begin{cases} n_{2v}(|\mathbf{r}-\mathbf{r}'|); & x_3>0, \quad x_3'>0 \\ n_{2l}(|\mathbf{r}-\mathbf{r}|); & x_3<0, \quad x_3'<0 \\ n_l n_v;\ x_3>0, & x_3'<0 \quad \text{or} \quad x_3<0, \quad x_3'>0 \end{cases} \tag{36b}$$

with $n_{2k}(r)\equiv n_k^2 g_k(r)$ for $k=v$ or l. The purpose of introducing this hypothetical reference system is to separate intrinsic surface properties from those of the bulk phases. This separation should be done in such a way that the discontinuities of n_0^B and n_2^B do not give rise to unphysical singularities. To guard against this contingency we require that

$$\nabla_3 n_0^B(x_3)=0, \qquad \forall x_3 \tag{37a}$$

and

$$\nabla_3 n_2^B(\mathbf{r};\mathbf{r}')=\begin{cases} \nabla_3 n_{2v}; & x_3>0, \quad x_3'>0 \\ \nabla_3 n_{2l}; & x_3<0, \quad x_3'<0 \\ 0; & \text{otherwise} \end{cases} \tag{37b}$$

The operator δ_S acts as the identity upon G_n^S but it annihilates the quantity G_n^B, defined by (34a). Therefore, *subject to the approximations that we have adopted for n_0 and n_2*, it follows that

$$\delta_S G_n(q)=q^2\beta^{-1}\sigma_F+O(q^4) \tag{38a}$$

with

$$\sigma_{\mathrm{F}} = \lim_{q\to 0} \beta q^{-2} \int dx_3 \int dx_3' \, \delta_s \langle \dot{g}_3 \dot{g}_3 \rangle(\mathbf{q}) \tag{38b}$$

The conditions for this derivation of a correlation-function expression for the surface tension are, to be sure, far too restrictive. However, the computational simplifications that flow from the Fowler approximation of a vanishingly thin transition zone permit us to present the theory in a form that is uncluttered by annoying details. Indeed, in this special case we could have extracted the surface contribution to G_n simply by subtracting G_n^B. The true utility of the surface projection operator δ_S becomes apparent only when one considers less restrictive models of the transition zone, in which cases the bulk phases are distorted by the presence of the interface. Because of this there is a nonvanishing "distorted bulk" contribution to $\delta_S G_n^B$ that is found to be of order ε^3. This is both good and bad. On the one hand, it means that the distorted bulk term is negligible provided that the real interfacial zone is very thin, that is, provided that the state of the system is remote from its critical point. Consequently, when this is the case we can use the same procedure as before to prove that for the real (but thin) transition zone

$$\delta_S G_n(q) = q^2 \beta^{-1} \sigma_P + O(q^4) \tag{39}$$

where σ_P denotes the "exact," Kirkwood-Buff surface tension defined by (13). Because the algebraic details of this generalization are rather tedious, we do not present them here but refer the reader to Ref. 38.

Thus, we have proved that when the interfacial region is thin (far from the critical point) the quantity defined by

$$\sigma_{\infty} = \lim_{q\to 0} q^{-2} \beta \int dx_3 \int dx_3' \, \delta_S \langle \dot{g}_3 \dot{g}_3 \rangle(\mathbf{q}; x_3, x_3') \tag{40}$$

is equal to the exact, potential form for the surface tension, derived by Kirkwood and Buff. This new formula, (40), relates the surface tension to a sum rule for the inhomogeneous, two-phase system and it has the form of a correlation function. It is the latter of these two qualities that permits us to establish a relationship between σ_∞ (or, equivalently, σ_P) and the quantity σ_D defined by the density functional formula, (27).

The trick is to use a complete set of dynamical variables, $\{A_i\}$, to generate the relationship

$$\sigma_\infty = \lim_{q\to 0} q^{-2}\beta \int dx_3 \int dx_3' \delta_S \left\{ \sum_{i,j} \langle \dot{g}_3 | A_i \rangle \langle A_i | A_j \rangle^{-1} \langle A_j | \dot{g}_3 \rangle \right\}(\mathbf{q}) \geq \sigma(\{B_\alpha\}) \tag{41}$$

between σ_∞ and the function

$$\sigma(\{B_\alpha\}) \equiv \lim_{q\to 0} q^{-2}\beta \int dx_3 \int dx_3' \delta_S \left\{ \sum_{\alpha,\beta} \langle \dot{g}_3 | B_\alpha \rangle \langle B_\alpha | B_\beta \rangle^{-1} \langle B_\beta | \dot{g}_3 \rangle \right\}(\mathbf{q}) \tag{42}$$

defined in terms of the *in*complete set $\{B_\alpha\} \subset \{A_i\}$

For example, let us choose $\{B_\alpha\}$ to be the fluctuation of number density, $\delta n(\mathbf{r},0) = \Sigma_\alpha \delta(\mathbf{r}^\alpha - \mathbf{r}) - n_0(x_3)$. In this case the discrete index is replaced with the continuous variable $\mathbf{r}$ and (42) becomes

$$\sigma(\{\delta n\}) = \lim_{q\to 0} q^{-2}\beta \int dx_3 \int dx_3' \delta_S$$

$$\times \left\{ \int d^3\bar{r} \int d^3\bar{r}' \langle \dot{g}_3(\mathbf{r}) \delta n(\bar{\mathbf{r}}) \rangle \langle \delta n(\bar{\mathbf{r}}) \delta n(\bar{\mathbf{r}}') \rangle^{-1} \langle \delta n(\bar{\mathbf{r}}') \dot{g}_3(\mathbf{r}') \rangle \right\}(\mathbf{q})$$

$$= \lim_{q\to 0} q^{-2}\beta \int dx_3 \int dx_3' [\nabla_3 n_0(x_3)] K(q; x_3, x_3') [\nabla_3' n_0(x_3')] \tag{43}$$

with $K^{-1}(\mathbf{q}; x_3, x_3') = \langle \delta n(\mathbf{r},0) \delta n(\mathbf{r}',0) \rangle(\mathbf{q})$. To obtain the second of these expressions from the first use has been made of the sum rule,[12]

$$\langle \dot{g}_i(\mathbf{r},0) \delta n(\mathbf{r}',0) \rangle = \beta^{-1} \nabla_i [n_0(x_3) \delta(\mathbf{r} - \mathbf{r}')] \tag{44}$$

and of the Fourier-transform relationship, $\langle a(\mathbf{r}) b(\mathbf{r}') \rangle^{-1}(\mathbf{q}) = \{\langle a(\mathbf{r}) b(\mathbf{r}') \rangle(\mathbf{q})\}^{-1}$, which is consequence of the system's translational invariance in the interfacial plane of $\boldsymbol{\rho}$.

For the limit appearing in (43) to exist the first term of the series $K(q; x_3, x_3') = K_0(x_3, x_3') + q^2 K_2(x_3, x_3') + \dots$ must satisfy (21). Lovett et al.[39] used functional analysis (based on a density functional form for the free energy) to derive this important identity. It also can be obtained by applying the Schwartz inequality to the autocorrelation function of $\dot{g}_3$.[11]

Thus from (31a) it follows that

$$\lim_{q\to 0}\int dx_3 \int dx_3' \langle \dot{g}_3 \dot{g}_3 \rangle(\mathbf{q}; x_3, x_3') = 0 \tag{45}$$

Therefore, by integrating the inequality $\langle \dot{g}_3(\mathbf{r})\dot{g}_3(\mathbf{r}')\rangle \geq \langle g_3(\mathbf{r})n(\bar{\mathbf{r}})\rangle K(\bar{\mathbf{r}};\bar{\mathbf{r}}') \langle n(\bar{\mathbf{r}}')g_3(\mathbf{r}')\rangle$[40] over all values of x_3 and x_3' and then passing to the limit of $q\to 0$, we find that

$$0 \geq \int dx_3 \int dx_3' [\nabla_3 n_0(x_3)] K_0(x_3, x_3') [\nabla_3' n_0(x_3')] \geq 0$$

and so conclude that the integral must vanish.

The application of this result to (43) establishes that $\sigma(\{\delta n\})$ is equal to the Triezenberg and Zwanzig surface tension σ_D, defined by (22). From (41) it then follows that $\sigma_\infty \geq \sigma([\delta n\}) = \sigma_D$. Therefore, we are led to the conclusion that *for a thin interfacial zone* $\sigma_P = \sigma_\infty \geq \sigma(\{\delta n\}) = \sigma_D$ or, finally,

$$\sigma_P \geq \sigma_D \text{ (for a thin transition zone)} \tag{46}$$

In an effort to obtain a more restrictive result than $\sigma_\infty \geq \sigma(\{\delta n\}) \equiv \sigma_D$, we expanded the incomplete set $\{B_\alpha\}$ to include fluctuations of all the "single-particle field variables," $\Sigma_\alpha \psi(\mathbf{r}^\alpha, \mathbf{p}^\alpha)\delta(\mathbf{r}^\alpha - \mathbf{r})$.[38] This was done by identifying $\{B_\alpha\}$ with the fluctuations of the single-particle density $f(\mathbf{r},\mathbf{p}) \equiv \Sigma_\alpha \delta(\mathbf{p}^\alpha - \mathbf{p})\delta(\mathbf{r}^\alpha - \mathbf{r})$. What we found was the sensible but somewhat unexpected result that $\sigma(\{\delta f\}) = \sigma(\{\delta n\})$: of all the single-particle field variables only δn contributes to σ! The conclusions that can be drawn from this observation and the relationship (46) are that, for conditions remote from the critical point, σ_P provides an upper bound on σ_D and that the (possibly zero) difference between the two is due exclusively to fluctuations of multiparticle fields, such as particle interaction energy.

We have demonstrated above that, in the limit of a thin transition zone, there is a simple relationship between σ_P and the autocorrelation function $G_n(q)$. However, as the critical point is approached, the width of the transition zone increases greatly and so it no longer is possible to neglect the contributions to $\delta_S G_n(q)$ from the distorted bulk phases. Our way of contending with this difficulty has been to invent a correlation function expression for surface tension that is precisely equivalent to the Kirkwood-Buff formula, regardless of the thickness of the interface. This expression,

$$\sigma = \lim_{q\to 0} q^{-2}\beta\,\delta_S[G_n(q) - G_t(q)] \tag{47}$$

contains, in addition to the normal force autocorrelation function, $G_n(q)$, the analogous tangential (t) force correlation function

$$G_t(q)=\int dx_3\int dx_3'\int d^2(\rho-\rho')e^{-iq(x_3-x_3')}\langle \dot{g}_t(\mathbf{r},0)\dot{g}_t(\mathbf{r}',0)\rangle \tag{48a}$$

with

$$\langle \dot{g}_t\dot{g}_t\rangle=\tfrac{1}{2}\{\langle \dot{g}_1\dot{g}_1\rangle+\langle \dot{g}_2\dot{g}_2\rangle\} \tag{48b}$$

In the vicinity of the critical point the behavior of the fluid is dominated by fluctuations of density. Therefore, it should be an excellent approximation to replace the correlation functions $\langle g_i(\mathbf{r})g_i(\mathbf{r}')\rangle$ with $\langle g_i(\mathbf{r})n(\bar{\mathbf{r}})\rangle K(\bar{\mathbf{r}};\bar{\mathbf{r}}')\langle n(\bar{\mathbf{r}}')g_i(\mathbf{r}')\rangle$.[40] To this approximation $\delta_S G_t(q)$ is zero, and so

$$\sigma=\sigma_P\doteq\sigma_D \text{ (near the critical point)} \tag{49}$$

4. *Concluding Remarks*

All the results presented in this section are specific to the planar interface separating the liquid and vapor phases of a conventional three-dimensional system. However, computer dynamics experiments and some model calculations as well are more readily (and cheaply) performed for hypothetical, two-dimensional bulk phases. The corresponding analogue of the planar interface is a straight-line boundary. All three of the derivations that we perform above for three-dimensional bulk phases can be extended to this case and, indeed, with only a slightly greater expenditure of effort they can be extended to the general case of two k-dimensional bulk phases separated by a planar interface. Because the methods of derivation are precisely analogous to those for the three-dimensional case, we report the results without proofs.

The generalization of the Kirkwood-Buff formula can be derived by considering an isochoric deformation of a k-dimensional hypercube with edge length L. The surface tension is given by

$$\sigma_p^{(k)}=\lim_{\varepsilon\to 0}\frac{\langle\delta H\rangle}{\varepsilon L^{k-1}}$$

$$=\frac{1}{2(k-1)}\int dx_n\int d^k(r-r')\frac{|\mathbf{r}-\mathbf{r}'|^2-k|x_n-x_n'|^2}{|\mathbf{r}-\mathbf{r}'|}V'(|\mathbf{r}-\mathbf{r}'|)n_2(\mathbf{r};\mathbf{r}') \tag{50}$$

Here $\mathbf{r}$ is the position vector of a point in this space and x_n is its component in the direction normal to the flat interface (straight-line interface if $k=2$).

To obtain the k-dimensional analogue of the Triezenberg and Zwanzig formula we can confine our attention to fluctuations of the surface mode with the longest possible wavelength, namely, $2L$. For a sinusoidal fluctuation with this wavelength and a maximum amplitude of h, the attendant change of surface area is $\delta A^{(k)} = \frac{1}{4}\pi^2 h^2 L^{k-3}$. The change of free energy associated with this fluctuation is equal to $\delta F^{(k)} = \sigma_D^{(k)} \delta A^{(k)}$, with

$$\sigma_D^{(k)} = \beta^{-1} \int dx_n \int dx'_n [\nabla_n n_0(x_n)] K_2^{(k)}(x_n, x'_n) [\nabla'_n n_0(x'_n)] \quad (51)$$

Here $K_2^{(k)}(x_n, x'_n)$ is the coefficient of q^2 in the series expansion of the correlation function $K^{(k)}(q; x_n, x'_n) = \int d^{k-1}(\rho - \rho') e^{-i\mathbf{q}\cdot(\boldsymbol{\rho}-\boldsymbol{\rho}')} K^{(k)}(\boldsymbol{\rho} - \boldsymbol{\rho}'; x_n, x'_n)$wherein $\boldsymbol{\rho}$ is the projection of the k-dimensional position vector $\mathbf{r}$ on the hyperplane perpendicular to the direction of the surface normal. To derive (51) use has been made of the generalization,

$$\int dx_n \int dx'_n [\nabla_n n_0(x_n)] K_0^{(k)}(x_n, x'_n) [\nabla'_n n_0(x'_n)] = 0$$

of the fundamental identity (21).

The correlation function theory of Section II.B.3 also can be extended to k-dimensional systems. The expressions appearing in this general formulation differ from those of the three-dimensional theory only by the replacement of x_3 with x_n and of two- and three-dimensional integrations with the corresponding $k-1$ and k-dimensional integrals.

Although the only systems of physical interest are those composed of two-and three-dimensional bulk phases, the general expressions (50) and (51) are presented here to expose more clearly the effect of dimensionality on the potential and density functional formulas for surface tension. [Because we have chosen to replace the integrand factor $(x_i - x'_i)^2 - (x_n - x'_n)^2$ with the equivalent and more symmetric expression $(k-1)^{-1}[(\mathbf{r}-\mathbf{r}')^2 - k(x_n - x'_n)^2]$, formula (50) for $\sigma_P^{(k)}$ appears to have an explicit dependence on the dimension k, which is absent from $\sigma_D^{(k)}$, given by (51).] One could, of course, use the general formulas to generate two-dimensional analogues of all the approximate surface tension expressions we already have discussed. However, what we really learn from these general formulas and their manners of derivation is that the concept and theory of surface tension do not vary significantly with the dimensionality of the bulk phases and that the surface tension formulas themselves depend on this dimensionality only in the most obvious ways.

III. DYNAMICS OF THE LIQUID–VAPOR INTERFACE

In the preceding section our objectives were limited to the characterization of the equilibrium states and thermomechanical properties of heterogeneous, two-phase systems. Here the focus is on the *dynamics* of the phase boundary. The bases of the traditional, continuum mechanical theory of interfacial dynamics are the continuity and Navier-Stokes equations of fluid mechanics and the macroscopic boundary condition provided by the Young-Laplace relationship, (3). We briefly review this theory to set the stage for our subsequent presentation of a many-body, collective coordinate theory of the liquid–vapor interface. We seek with this generalization of the hydrodynamic theory a more fundamental description of interfacial dynamics that avoids the imposition of a macroscopic boundary condition but at the same time provides insights into its microscale interpretation. To achieve this objective and, in particular, to elucidate the microscopic origin of surface ripples we invoke the broken symmetry concept and make use of the techniques of generalized hydrodynamics, due to Zwanzig and Mori. The theory obtained in this manner is intermediate in its degree of dynamical detail to conventional hydrodynamics and the kinetic theory of Section IV.

A. Hydrodynamic Theory of Capillary Waves

In continuum fluid mechanics, interfaces are treated as sharp boundaries between bulk phases. The forces acting at these interfaces provide boundary conditions for the hydrodynamic equations. Free fluid surfaces tend to assume equilibrium shapes, spherical in the absence of all external forces and planar in the presence of the force of gravity. Perturbations of the equilibrium shapes then give rise to propagating surface waves. For a planar surface subject both to gravity and to surface tension forces, the hydrodynamic description of the propagating surface waves is well established.[41, 42] In the simplest analysis the perturbation is assumed to be so small that the wave amplitude is negligible in comparison with the wavelength. This enables us to discard the inertial term $\mathbf{v}\cdot\text{grad }\mathbf{v}$ from the Navier-Stokes equation. Furthermore, the wavelength of the disturbance is assumed to be so large that we can treat the fluid as if it were inviscid. With these approximations the Navier-Stokes equation reduces to $\partial\mathbf{v}/\partial t = -\nabla(p\rho^{-1} + gx_3)$. From this it follows that the flow is irrotational and so is characterized by a velocity potential $\vartheta(\mathbf{v} = \text{grad }\vartheta)$ that satisfies the Laplace equation, $\nabla^2\vartheta = 0$. Here g denotes the gravitational constant, ρ is the liquid density, and x_3 is, as before, the coordinate measured normal to the equilibrium liquid–vapor interface.

By applying this equation at the free surface, whose equation is given by $x_3 = \zeta(x_1, x_2, t)$, we obtain the boundary condition, $p = -\rho(g\zeta + \partial\vartheta/\partial t)$.

The constant ambient pressure is easily eliminated by a simple change of gauge so that p, the fluid pressure immediately adjacent to the free surface, may be identified with the excess pressure due to curvature of the liquid–vapor interface. This, in turn, is given by the Young-Laplace relationship, (3), which for small ζ can be written as $P_{\mathrm{ex}} \doteq -\sigma(\partial^2\zeta/\partial x_1^2 + \partial^2\zeta/\partial x_2^2)$. The Young-Laplace equation then becomes

$$\rho g\zeta + \rho\frac{\partial\vartheta}{\partial t} - \sigma\left(\frac{\partial^2\zeta}{\partial x_1^2} + \frac{\partial^2\zeta}{\partial x_2^2}\right) = 0 \tag{52}$$

The normal component of the interfacial velocity is related to the velocity potential according to $v_3 = \partial\vartheta/\partial x_3$ and can, for small amplitude motions, be equated to the time derivative of ζ, that is $v_3 \doteq \partial\zeta/\partial t$ and

$$\zeta(x_1, x_2; t) = \int^t \left[\frac{\partial\vartheta(x_1, x_2, x_3; s)}{\partial x_3}\right]_{x_3=0} ds \tag{53}$$

Therefore, by differentiating (52) with respect to t we obtain for ϑ the boundary condition,

$$\rho g\frac{\partial\vartheta}{\partial x_3} + \rho\frac{\partial^2\vartheta}{\partial t^2} - \sigma\frac{\partial}{\partial x_3}\left(\frac{\partial^2\vartheta}{\partial x_1^2} + \frac{\partial^2\vartheta}{\partial x_2^2}\right) = 0 \tag{54}$$

The function

$$\vartheta = \vartheta_0 e^{qx_3}\cos(qx_1 - \omega t) \tag{55}$$

is a solution of the Laplace equation, $\nabla^2\vartheta = 0$, that satisfies the boundary condition $v_3 \to 0$ as $x_3 \to -\infty$ and is descriptive of a surface wave propagating in the direction x_1. This solution also satisfies the Young-Laplace boundary condition, (54), provided that the frequency and wave number conform to the dispersion relation

$$\omega^2 = gq + \left(\frac{\sigma}{\rho}\right)q^3 \tag{56}$$

The first term on the right-hand side of this equation is associated with the [large-wavelength or small wave-number, $q \ll (g\rho/\sigma)^{1/2}$] dispersion relation, $\omega^2 = gq$, characteristic of deep-water gravity waves. At the opposite extreme of very short wavelengths, (56) reduces to the dispersion relation, $\omega^2 = (\sigma/\rho)q^3$, for capillary waves or ripples.

The group velocity of gravity waves in only one-half the phase velocity, whereas the group velocity for ripples is $\frac{3}{2}$ times the corresponding phase velocity. This has a number of interesting consequences, all of which are subject to immediate visual confirmation from commonplace observations of flowing water.[43] Consider, for example, the waves produced by a moving source or, equivalently, by a stationary source imbedded in a moving fluid. The fastest waves generated by this source are little (small wavelength) ripples that run out ahead, whereas the long wavelength gravity waves follow after.

The dispersion relation for surface waves depends on the depth of the channel, that is, on the thickness of the bulk liquid phase. Thus the dispersion relation appropriate to a channel of finite depth d is given by

$$\omega^2 = \left[gq + \left(\frac{\sigma}{\rho} \right) q^3 \right] \tanh(qd) \tag{57}$$

instead of (56). The two are, of course, identical for $qd \gg 1$.

In the absence of gravity the equilibrium configuration of a liquid is spherical. Except for obvious geometric differences the theory of capillary waves on the surface of a sphere is much the same as that for a planar surface. The characteristic modes of oscillation (solutions of the Laplace equation for the velocity potential) are of the form $(ar^l + br^{-l-1})Y_l^m(\hat{r})$, where a and b are constants and Y_l^m is a spherical harmonic. If the unperturbed liquid fills an entire sphere of radius R, then the constant b must vanish. In this case the dispersion relation for capillary waves assumes the form[41, 42]

$$\omega^2 = \left(\frac{\sigma}{\rho R^3} \right) l(l-1)(l+2) \tag{58}$$

Suppose, however, that the sphere of radius R is a liquid shell of thickness d, covering a rigid core of radius $R-d$. The dispersion relation is then

$$\omega^2 = \left[\left(\frac{\sigma}{\rho R^3} \right) l(l-1)(l+2) \right] f_l\left(\frac{d}{R} \right) \tag{59a}$$

with

$$f_l(x) = \frac{1-(1-x)^{2l+2}}{1+[l/(l+1)](1-x)^{2l+2}} \tag{59b}$$

Consider now the limit in which both l and R become unbounded while the value of their quotient $l/R \equiv q$ remains fixed. In this limit the curvature of the spherical surface tends to zero,

$$\lim_{l\to\infty,\, R\to\infty,\, l/R=q} f_l\left(\frac{d}{R}\right) = \tanh(qd)$$

and the dispersion relation, for a spherical shell of thickness d (59.1), reduces to $\omega^2 = (\sigma/\rho)q^3 \tanh(qd)$. Apart from the additive contribution due to gravity, this is identical to the dispersion relation for waves on the *planar* surface of a liquid of finite depth (57)!

The value of $l=2$ corresponds to the minimum frequency,

$$\omega_{\text{min}} = \left(\frac{8\sigma}{\rho R^3}\right)^{1/2} \tag{60}$$

with which the surface of a spherical drop can perform natural oscillations. The radial velocity of the mode with $l=0$ vanishes identically. Thus this is not a spherical "breathing mode" as stated in Refs. 41 and 42. The amplitudes of the zero-frequency modes associated with $l=1$ must be set equal to zero: because the directions of the associated velocities are unchanging in time they describe deformations that grow uniformly with time and so are incompatible with our initial assumption of small-amplitude motions.

The hydrodynamic theory can be extended to include the attenuation of surface waves due to viscous coupling between the interface and the bulk phases, especially the liquid.[42] In the case of a low-viscosity liquid the capillary wave motion is the same as for an inviscid fluid. However, the wave amplitude is damped by a factor of $\exp(-2\nu q^2 t)$ where $\nu \equiv \eta/\rho$ denotes the kinematic viscosity of the liquid. Huang and Webb[44] discuss the theory of the dispersion and attenuation of capillary waves in the vicinity of the critical point. Valuable experimental information about these dynamical characteristics of capillary waves has been extracted from inelastic light scattering. The interested reader should consult Refs. 7 and 44 to 47. In addition to these there is a recent review by Edwards and Saam[48] devoted to phenomena associated with the free surface of liquid helium.

As we have seen above, the principal result of the hydrodynamic theory of surface waves is the "boundary condition," (54). This relationship was obtained as the result of a straightforward but somewhat involved fluid-mechanical argument. Therefore, it may come as something of a surprise to learn that it also can be derived directly (and rather convincingly) from our correlation function expression for the surface tension, σ_∞.[11] To establish this connection between the correlation function theory and the hydrodynamic boundary condition we begin by rewriting (40) in the form,

$$-\rho\int dx_3\int dx_3'\,\delta_S\langle\dot{g}_3\dot{g}_3\rangle=\rho\int dx_3\int dx_3'\,\delta_S\langle\ddot{g}_3g_3\rangle=-\rho\beta^{-1}q^2\sigma_\infty \quad (61)$$

which is, of course, valid only for very small values of q. Here, $\rho\equiv\rho_l-\rho_v=m(n_l-n_v)$, with m denoting the molecular weight.

Next, with the help of the sum rule[11]

$$\langle g_3(\mathbf{r},0)g_3(\mathbf{r}',0)\rangle(\mathbf{q})=mn_0(x_3)\beta^{-1}\delta(x_3-x_3') \quad (62)$$

we transform $\rho\beta^{-1}q^2\sigma_\infty$ according to

$$\begin{aligned}\rho\beta^{-1}q^2\sigma_\infty&=q^2\sigma_\infty\beta^{-1}m(n_l-n_v)\\&=-mq^2\sigma_\infty\beta^{-1}\int dx_3\int dx_3'[\nabla_3n_0(x_3)]\delta(x_3-x_3')\\&=-q^2\sigma_\infty\int dx_3\int dx_3'\nabla_3\{\delta_S\langle g_3g_3\rangle\}(\mathbf{q};x_3,x_3')\end{aligned} \quad (63)$$

From (61) and (63) it then follows that

$$\int dx_3\int dx_3'\{\rho\delta_S\langle\ddot{g}_3g_3\rangle-\sigma_\infty\sigma^2\delta_S\langle(\nabla_3g_3)g_3\rangle\}=0 \quad (64)$$

proving that the integrand factor $\delta_S\langle[\rho\ddot{g}_3-\sigma_\infty q^2\nabla_3g_3]g_3\rangle$ vanishes identically in the long-wavelength limit. Consequently, near the Gibbs surface $(x_3=0)$

$$\rho\ddot{g}_3(\mathbf{q};x_3,0)-\sigma_\infty q^2\nabla_3g_3(\mathbf{q};x_3,0)=h(\mathbf{q};x_3,0) \quad (65)$$

where the fluctuating generalized force h is "orthogonal" to $g_3(\mathbf{q};x_3,0)$ in the sense that $\delta_S\langle h(\mathbf{q};x_3,0)g_3(\mathbf{q};x_3',0)\rangle=0$.

Let us now adopt the mean field approximation, thereby discarding the fluctuating term from (65) and assuming that the resulting equation is valid for finite times. Furthermore, we assume the flow to be irrotational so that g_3 is proportional to the gradient of a velocity potential ϑ. These approximations lead from (65) to the (physical space) relationship,

$$\rho\frac{\partial^2\vartheta(\mathbf{r},t)}{\partial t^2}+\sigma_\infty\nabla_3(\nabla_1^2+\nabla_2^2)\vartheta(\mathbf{r},t)=0 \qquad \text{(at the Gibbs surface, } x_3=0) \quad (66)$$

which, aside from a term involving the gravitational force, is identical in form to the hydrodynamic boundary condition, (54). It is appropriate that σ_∞ should appear in this equation because the derivation of its hydrodynamic analogue, (54), was applicable only to a thin interfacial zone.

Our derivation based on the correlation function theory provides a new interpretation for the hydrodynamic boundary conditions. Specifically, it can be identified as the high-frequency (short-time) limit of an equation of change that is closely analogous to the Vlasov (or mean field) equation for a homogeneous system.[49]

To derive the Young-Laplace boundary condition from the correlation function theory we found it necessary to restrict our considerations to interfacial zones so thin that $\sigma_P = \sigma_\infty$. When the zone grows beyond this limit one no longer can neglect the distortions of the bulk phases represented by $\delta_S G_n^B(q)$ or, equivalently, the contribution to the surface tension associated with the tangential force autocorrelation function $\delta_S G_t(q)$. We suspect that this signals a failure of linear response theory to properly account for the dynamics of thick, noncritical interfacial zones. To illustrate the meaning of this remark let us briefly consider a conceptually related aspect of the dynamics of homogeneous fluids. Molecular dynamics experiments[50] have provided dramatic evidence that the motion of a "tagged" particle distorts the transverse and longitudinal components of the surrounding fluid in such a way as to generate a vortex ring of circulating particles; this is the microscale analogue of a well-known hydrodynamic result. Linear response theory would treat the longitudinal and transverse velocity fields as uncoupled and so be unable to account for this vortex flow. To cope with it one must involve nonlinear response theory or introduce mode couplings.[51, 52]

By analogy, let us identify an element of the interfacial zone with the tagged or test particle and the adjacent surface and bulk phases (especially the liquid) with its surroundings. Consider then a local, vertical displacement $\zeta(\boldsymbol{\rho})$ of the interfacial zone, associated with a density fluctuation δn. Because of the virtual incompressibility of the surrounding fluid, this displacement and the concomitant vertically directed flow of mass are parts of a highly collective event, which necessarily involves flow of an equal amount of mass in directions tangential to the liquid–vapor interface. Thus there is a coupling of normal and tangential flows. For a vertical displacement of given amplitude the magnitude of this mass transfer depends on the thickness of the interfacial zone. When the zone is thin, the mass transfer is correspondingly small and so the tangential flow is only weakly excited. However, a tangential flow of greater magnitude is needed to support the vertical displacment of a thicker transition zone. Thus, according to this simple picture, an ascending surface element is "supported" by a local vortex ring, the strength of which is greater for thick than for thin interfacial zones. From this we conclude that a nonlinear or bilinear, mode–mode coupling sort of theory is needed to properly take account of the large-amplitude tangential flows associated with fluctuations occurring in thick interfacial zones.

B. Generalized Hydrodynamics of a Two-Phase System

1. Broken Symmetry and Lifetimes of Dynamic Variables

The many-body system of interest to us here possesses an unimaginably large number of degrees of freedom. It is clear, then, that if the dynamical analysis of this system is to be reduced to manageable proportions, some means must be found for effecting a simplification of comparable enormity. This could be done if we succeeded in discovering collective fluid variables with properties that closely mimicked those of the normal coordinates of a coupled-oscillator system. However, the little that is known about the microscale dynamics of fluids destroys all hope of inventing a complete set of fluid-collective variables whose equations of motion would be even remotely as simple as those of a crystalline solid. Discouraging as this may be, it does not tell the whole story, for the other and more basic characteristic of the crystalline normal modes is their persistence in time. It is this lifetime concept that can be just as usefully applied to a (partially) disordered system, such as a fluid, as it can to a crystalline solid. Thus most of the infinitely many variables of an (anharmonically) interacting many-body system quickly relax toward their equilibrium values and so contribute little to its behavior beyond providing a rapidly fluctuating "background noise," which can be modeled by a suitable stochastic ansatz. Because of this, one is able to describe the relevant dynamics of the many-body system by considering in detail the behavior of only a few long-lived, "relevant" variables. The most familiar of these are the microscale analogues of fields studied in continuum mechanics, namely, the densities of mass, momentum and energy. More generally, the coordinates of a many-body system can be separated into two categories, the irrelevant variables, $\{A^I(\mathbf{r}, t)\}$, whose values rapidly die away, and the relevant variables, $\{A^R(\mathbf{r}, t)\}$, whose values are more persistent.

Although there are other standards for dividing the dynamical variables into complementary relevant and irrelevant subsets[13, 53] (see Section IV), this lifetime criterion is perfectly suited to the long-wavelength, low-frequency hydrodynamic limit on which our attention is focused.

To acquaint ourselves with the long-lived, relevant variables and identify their characteristic properties and the physical phenomena for which they provide descriptions, we begin with an examination of a homogeneous, single-component fluid. Then, in the following section, it is shown how concepts and mathematical constructs specific to the homogeneous fluid can be extended to inhomogeneous systems. And finally, we study in considerable detail the many-body, collective variable theory of surface waves.

Throughout these considerations our attention is confined to systems displaced only slightly from equilibrium. This actually is not as restrictive as one might suppose, for it leaves open to investigation thermodynamics and transport processes, as well as the physical and chemical phenomena that are probed by ultrasonics, lasers, neutron scattering, and molecular dynamics computer "experiments." The simplification achieved by dealing with a system near equilibrium is that its observables are the averages

$$G_{ij}(\mathbf{r}, t; \mathbf{r}', t') = \langle A_i(\mathbf{r}, t) A_j(\mathbf{r}', t') \rangle \tag{67}$$

over a equilibrium ensemble, of correlations among the dynamical variables $A_i(\mathbf{r}, t)$. Because the ensemble representative of this equilibrium, homogeneous fluid is stationary and spatially isotropic, $G_{ij}(\mathbf{r}, t; \mathbf{r}', t') = G_{ij}(\mathbf{r}-\mathbf{r}'; t-t') = \langle A_i(\mathbf{r}-\mathbf{r}', t-t') A_j(\mathbf{0}, 0) \rangle$.

In most cases it actually is not the functions $G_{ij}(\mathbf{r}; t)$, but their Fourier transforms

$$G_{ij}(\mathbf{k}, \omega) \equiv \int_{-\infty}^{+\infty} dt \int d^3r \, G_{ij}(\mathbf{r}; t) \exp[-i(\mathbf{k}\cdot\mathbf{r} - \omega t)] \tag{68}$$

that are the experimental observables. However, from a theoretical standpoint it is more convenient to deal with the complex fluctuation functions (Fourier-Laplace transforms)

$$\begin{aligned} G_{ij}(\mathbf{k}, z) &\equiv \langle A_i(\mathbf{r}, t) A_j(\mathbf{0}, 0) \rangle (\mathbf{k}, z) \\ &\equiv i \int_0^{\infty} dt \int d^3r \, G_{ij}(\mathbf{r}; t) \exp[-i(\mathbf{k}\cdot\mathbf{r} - zt)] \end{aligned} \tag{69}$$

which are related to the corresponding Fourier transforms by the formulas

$$G_{ij}(\mathbf{k}, z) = \int_{-\infty}^{+\infty} \frac{d\omega}{2\pi} \frac{G_{ij}(\mathbf{k}, \omega)}{\omega - z} \tag{70a}$$

To obtain the inverse relationship one uses the identity $(x \pm i\varepsilon)^{-1} = \mathcal{P}(x^{-1}) \mp i\pi\delta(x)$, where $\mathcal{P}$ denotes the principal value, and finds that

$$G_{ij}(\mathbf{k}, \omega) = 2[\operatorname{Im} G_{ij}(\mathbf{k}, z)]_{z=\omega+i\varepsilon} \tag{70b}$$

For the purposes of this section it is sufficient to consider a single collective variable $A(\mathbf{r}, t)$. To develop an understanding of the associated autocorrelation function $G_{AA}(\mathbf{k}, z)$ we use the projection operator formalism. Thus, by introducing the projection operator $P_0 \equiv 1 - Q_0 \equiv |\mathrm{A}>\chi_{\mathrm{AA}}^{-1}<\mathrm{A}|$, with $\chi_{AA}(\mathbf{k}) = \langle A|A\rangle(\mathbf{k}) \equiv \langle A^* A\rangle(\mathbf{k})$, it is possible [see Appendix A; specifically, (A.14), (A.15), and (A.16) with $\{A_i\}$ taken equal to the single element A] to obtain for G_{AA} the formula

$$G_{AA}(\mathbf{k}, z) = -\frac{\chi_{AA}(\mathbf{k})}{z - \Omega_{AA}(\mathbf{k}) - \Sigma_{AA}^{(c)}(\mathbf{k}, z)}. \tag{71}$$

Because only a single dynamical variable is involved, the frequency matrix $\Omega_{AA}(\mathbf{k}) \equiv i\langle \dot{A}|A\rangle(k)\chi_{AA}^{-1}$ vanishes by time-reversal invariance. All the complex dynamical information aspects of the many-body system are then compacted into the memory function

$$\Sigma_{AA}^{(c)}(\mathbf{k}, z) = \langle \dot{A}|Q_0(z - Q_0 \mathfrak{L} Q_0)^{-1} Q_0|\dot{A}\rangle(\mathbf{k})\chi_{AA}^{-1}(\mathbf{k}) \tag{72}$$

wherein $\mathfrak{L}$ denotes the Liouville operator, which is so defined that $\dot{A} = i\mathfrak{L}A$.

Our immediate objective is to identify those characteristics of a collective variable that ensure its persistence in time. Because we are seeking the analogue in many-body physics of the foregoing fluid mechanical theory, our search for this information is confined to correlations with characteristic lengths that are long compared to the intermolecular spacing. In this long-wavelength, low-frequency, "hydrodynamic limit" the memory function reduces to the sum $\Sigma'_{AA} - i\Sigma''_{AA}$ of a reactive part

$$\Sigma'_{AA} = \lim_{\substack{k \to 0 \\ \omega \to 0}} \langle \dot{A}|Q_0 \mathcal{P}(\omega - Q_0 \mathfrak{L} Q_0)^{-1} Q_0|\dot{A}\rangle(\mathbf{k})\chi_{AA}^{-1}(\mathbf{k})$$

and a part

$$\Sigma''_{AA} = \lim_{\substack{k \to 0 \\ \omega \to 0}} \pi\langle \dot{A}|Q_0\, \delta(\omega - Q_0 \mathfrak{L} Q_0) Q_0|\dot{A}\rangle(\mathbf{k})\chi_{AA}^{-1}(\mathbf{k})$$

which accounts for the spatial damping of the correlation function. It then follows from (70b), (71), and (72) that the spectral function $G_{AA}(\mathbf{k}, \omega)$ can be written in the form,

$$G_{AA}(\mathbf{k}, \omega) = \frac{2\Sigma''_{AA}\chi_{AA}}{(\omega - \Sigma'_{AA})^2 + (\Sigma''_{AA})^2} \tag{73}$$

from which we are able to identify Σ'_{AA} as a frequency shift and Σ''_{AA} as a line width, inversely proportional to the lifetime of the dynamical variable A.[54] Therefore, in the hydrodynamic limit the lifetime of the collective variable A is given by the limit, $\tau_A = \lim_{\mathbf{k}\to 0} \tau_A(\mathbf{k})$, of the positive definite function

$$\tau_A(\mathbf{k}) \propto \lim_{\omega\to 0} \{\Sigma''_{AA}(\mathbf{k}, \omega+i\varepsilon)\}^{-1} = \frac{\chi_{AA}(\mathbf{k})}{\pi\langle \dot{A}|Q_0\,\delta(Q_0\mathfrak{L}Q_0)Q_0|\dot{A}\rangle(\mathbf{k})} \tag{74}$$

Equation 74 is the key to our search for relevant dynamical variables. According to the criterion we adopted earlier, a collective variable is judged to be relevant only if its lifetime is very large compared to a characteristic microscale unit of time, such as the inverse of the collision frequency. There are three distinct ways by which a variable can satisfy this condition for relevancy: (*a*) the denominator of (74) may vanish in the limit $k\to 0$; (*b*) the numerator $\chi_{AA}(\mathbf{k})$ may become unbounded in this limit; and (*c*) either or both of the preceding two conditions may be satisfied by nonlinear combinations of variables that individually satisfy *a* or *b*. We now examine these three cases individually.

a. All Conserved Variables are Relevant.[55] A collective coordinate $A_i^{(1)}(\mathbf{r}, t)$ that satisfies the global balance equation $(d/dt)\int d^3r A_i^{(1)}(\mathbf{r}, t) = 0$ is called (the density of) a conserved variable. The differential form of this balance equation is $\dot{A}_i^{(1)}(\mathbf{r}, t) + \nabla_l j_i^l(\mathbf{r}, t) = 0$ where $\mathbf{j}_i$ denotes the flux of the variable $A_i^{(1)}$. The Fourier transform of this equation is $\dot{A}_i^{(1)}(\mathbf{k}, t) + ik_l j_i^l(\mathbf{k}, t) = 0$. Consequently, for a conserved variable the denominator of (74) becomes

$$\langle \dot{A}_i^{(1)}|Q_0\,\delta(Q_0\mathfrak{L}Q_0)Q_0|\dot{A}_i^{(1)}\rangle(\mathbf{k}) = k^2\langle j_i^k|Q_0\delta(Q_0\mathfrak{L}Q_0)Q_0|j_i^k\rangle$$

where j_i^k is the component of $\mathbf{j}_i$ parallel to the wave vector $\mathbf{k}$.

By substituting this result into (74) we find that $\tau_{A_i}^{(1)}(\mathbf{k}) \propto k^{-2}$ and so conclude that the lifetime of the conserved variable $A_i^{(1)}$ is infinite in the hydrodynamic limit. Therefore, according to our criterion, *conserved variables are relevant*. Because particle number, momentum, and energy are conserved variables, this result provides a partial explanation and a useful interpretation of the choices that usually are made for the ensembles of equilibrium statistical mechanics. It also explains why one must include all conserved variables in the relevant set to obtain the correct hydrodynamics. It was by doing just this that Kadanoff and Martin[56] succeeded in formulating the exact hydrodynamic limit of the many-body problem. Furthermore, as these authors showed, the global nature of the conservation laws gives rise to propagating modes, namely, *sound waves*.

b. Symmetry-Restoring Variables are Relevant.[57] In addition to the conserved variables, $\{A_i^{(1)}(\mathbf{r}, t)\}$, there is a second category of long-lived, cooperative hydrodynamic processes that result not from microscopic conservation laws but from broken symmetries. A system that can experience a "spontaneous breakdown of symmetry" is one whose Hamiltonian possesses a symmetry that is not shared by all its equilibrium states. Thus a ferromagnetic state of a magnet (in the absence of an external field) includes a specification of the direction of spin orientation and so has a lower symmetry than its spherically symmetric Hamiltonian. This is analogous to one of the inhomogeneous, two-phase states of a fluid (in the absence of gravity), which includes a specified direction for the interfacial normal and so exhibits a lower symmetry than does the associated, translationally (and spherically) symmetric Hamiltonian. In both of these examples and in the more general case, as well, the thermodynamic state breaks the symmetry associated with one or more conserved variables, that is, $[\rho_{\text{eq}}, \mathcal{Q}_i^{(1)}] \neq 0$ for $\mathcal{Q}_i^{(1)} \equiv \int d^3r A_i^{(1)}(\mathbf{r}, t)$. Here ρ_{eq} is the classical distribution function or quantal density matrix for the equilibrium state of lowered symmetry and the symbol $[a, b]$ denotes either the classical Poisson bracket of the two functions a and b or the commutator of the two corresponding operators.

Let us (assume that it is possible to) associate with the breakable symmetry $A_i^{(1)}$ a "symmetry-restoring" variable $B_\alpha^{(1)}$, which is so defined that the ensemble average of $[A_i^{(1)}, B_\alpha^{(1)}]$ is nonvanishing. Then, from the properties of the trace operator and the conventional definition of the ensemble average it follows that

$$\operatorname{Tr}\left(\left[\rho_{\text{eq}}, A_i^{(1)}\right] B_\alpha^{(1)}\right) = \operatorname{Tr}\left(\rho_{\text{eq}}\left[A_i^{(1)}, B_\alpha^{(1)}\right]\right) \equiv \langle\left[A_i^{(1)}, B_\alpha^{(1)}\right]\rangle \equiv \langle C_{i\alpha}^{(1,1)}\rangle \neq 0 \tag{75}$$

In the ground state of an isotropic ferromagnet all of the spins are up. Thus, with the common direction of these spins selected to lie parallel to the x_3-axis, the value of the magnetization $\langle M_3 \rangle$ differs from zero. It is the existence of this preferred direction that constitutes the breakdown of symmetry characteristic of the ferromagnetic state. In this situation $A_i^{(1)}$ and $B_\alpha^{(1)}$ can be identified with the components M_1 and M_2 of the total magnetization so that $[A_i^{(1)}, B_\alpha^{(1)}] \rightarrow [M_1, M_2] = i\hbar M_3$ and $\langle C_{i\alpha}^{(1,1)} \rangle \rightarrow i\hbar \langle M_3 \rangle \neq 0$.[58] More generally, $A_i^{(1)}$, $B_\alpha^{(1)}$ and the associated symmetry-breaking variable $C_{i\alpha}^{(1,1)}$ need not be components of the same vector-valued variable. Thus, in the case of a nematic liquid crystal,[57] $A_i^{(1)}$ is the total angular momentum while $B_\alpha^{(1)}$ and $C_{i\alpha}^{(1,1)}$ are proportional to components of the molecular mass quadrupole moment or "director."[59]

The importance of the relationship (75) is that it permits us to establish that the static correlation function of the symmetry-restoring variable

$B_\alpha^{(1)}(\mathbf{r}, t)$ exhibits long-ranged order, that is, $\langle B_\alpha^{(1)}(\mathbf{r},0)B_\alpha^{(1)}(\mathbf{r}',0)\rangle \propto |\mathbf{r}-\mathbf{r}'|^{-1}$ or $\langle B_\alpha^{(1)}(\mathbf{r},0)B_\alpha^{(1)}(\mathbf{r}',0)\rangle(\mathbf{k}) \propto k^{-2}$. Thus, by using the Schwartz inequality,[60]

$$\langle B_\alpha^{(1)} | B_\alpha^{(1)}\rangle\langle \dot{A}_i^{(1)} | \dot{A}_i^{(1)}\rangle \geq |\langle \dot{A}_i^{(1)} | B_\alpha^{(1)}\rangle|^2 \tag{76}$$

and the well-known relationship, $\langle \dot{A}B\rangle = \beta^{-1}\langle[A, B]\rangle$,[61] we conclude that

$$\chi_{B_\alpha^{(1)} B_\alpha^{(1)}} \equiv \langle B_\alpha^{(1)} | B_\alpha^{(1)}\rangle \geq \frac{|\langle \dot{A}_i^{(1)} | B_\alpha^{(1)}\rangle|^2}{\langle \dot{A}_i^{(1)} | \dot{A}_i^{(1)}\rangle} = \beta^{-2}\frac{|\langle C_{i\alpha}^{(1,1)}\rangle|^2}{k^2\langle j_i^k | j_i^k\rangle} \tag{77}$$

Therefore, in the long-wavelength limit

$$\chi_{B_\alpha^{(1)} B_\alpha^{(1)}}(\mathbf{k}) \to (\beta\rho_s k^2)^{-1} \tag{78a}$$

with

$$\rho_s \leq \frac{\beta}{|\langle C_{i\alpha}^{(1,1)}\rangle|^2} \lim_{k\to 0} \langle j_i^k | j_i^k\rangle \tag{78b}$$

In each specific case a definite physical meaning can be attached to the upper bound on ρ_s given by the right-hand side of this equation. For example, in superfluidity ρ_s can be identified with the density of the superfluid component and in the case of a liquid–vapor interface ρ_s is, as we see in Section III.B.2, closely related to the surface tension.

For the ferromagnet we conclude from (78a) that $\chi_{M_1 M_1}(\mathbf{k}) = \chi_{M_2 M_2}(\mathbf{k}) \propto k^{-2}$, which demonstrates the existence of variables (transverse magnetization) with static correlation functions that exhibit long-ranged order. Above the transition temperature the spins are randomly oriented and the average magnetization is equal to zero. In this paramagnetic state there are no variables with which one can associate long-ranged order.

In the hydrodynamic limit the sole criterion for judging a collective variable to be relevant is that its lifetime be infinite. Now from (74) and (77) we see that even if $\lim_{k\to 0} \langle \dot{B}_\alpha^{(1)} | Q_0 \delta(Q_0 \mathcal{L} Q_0) Q_0 | \dot{B}_\alpha^{(1)}\rangle(\mathbf{k})$ is finite, $\tau_{B_\alpha^{(1)}}(\mathbf{k})$ still is proportional to k^{-2} and so becomes infinite in the long-wavelength limit. Therefore, the *symmetry-restoring variable* $B_\alpha^{(1)}(\mathbf{r}, t)$ *is relevant*. Associated with this relevant variable is a new propagating mode that first was encountered in high-energy physics (quantum field theory) and called a *Goldstone boson*.[62] The nonrelativistic counterpart of this theory[63] has been extensively studied in connection with magnetic materials,[53] liquid crystals,[59] and many other systems.[57] It has proved to be extremely useful in providing a qualitative understanding of homogeneous many-body systems. In Section III B.2 we establish the existence of hydrodynamic Goldstone modes in *in*homogeneous systems,[12] illustrate their physical significance, and demonstrate their usefulness.

c. Composite (Nonlinear) Variables Associated with Classes a and b are Relevant. The arguments presented above can be extended to include non linear combinations of conserved and symmetry-restoring variables, such as $A_{ij}^{(2)}(\mathbf{k};\mathbf{k}_1)=A_i^{(1)}(\mathbf{k}-\mathbf{k}_1)A_j^{(1)}(\mathbf{k}_1)$, $AB_{i\alpha}^{(2)}(\mathbf{k};\mathbf{k}_1)=A_i^{(1)}(\mathbf{k}-\mathbf{k}_1)B_\alpha^{(1)}(\mathbf{k}_1)$, and $B_{\alpha\beta}^{(2)}(\mathbf{k};\mathbf{k}_1)=B_\alpha^{(1)}(\mathbf{k}-\mathbf{k}_1)B_\beta^{(1)}(\mathbf{k}_1)$. Thus the lifetimes of these composite variables are infinite in the hydrodynamic limit, provided that the intermediate wave numbers $\mathbf{k}_1$ are small. This means that the set of relevant variables can be extended to $\{A_i^{(1)}, A_{ij}^{(2)},\ldots; B_\alpha^{(1)}, B_{\alpha\beta}^{(2)}\ldots\}$. The introduction of these additional, composite variables modifies the damping (transport coefficients) of the propagating modes associated with the conserved variables, $\{A_i^{(1)}\}$, but does not alter their global nature. Among the observable consequences of the higher order terms, and especially the mode–mode coupling terms $A_{ij}^{(2)}$, are critical slowing down,[64] long-time tails of nonlinear hydrodynamics,[65] and some aspects of diffusion-controlled chemical reactions in dense fluids.[52]

In the linearized hydrodynamics to which we restrict our attention in the remainder of this chapter, all composite variables are ignored. This leaves the qualitative characteristics of the propagating modes intact but diminishes the accuracy with which damping is described.

What we have found is that the class of long-lived, relevant variables consists of the conserved variables, $\{A_i^{(1)}\}$, the symmetry-restoring variables, $\{B_\alpha^{(1)}\}$, and a vast array of associated composite variables. However, the much smaller set, $\{A^R\}\equiv\{A_i\}\equiv\{A_i^{(1)}, B_\alpha^{(1)}\}$ is sufficient for linearized hydrodynamics. Once the members of this relevant set have been identified, one then can make use of the generalized Langevin theory. Specifically, we can use for the complex fluctuation functions $G_{ij}(\mathbf{k},z)$ the equations of motion (see Appendix A)

$$\left[z\,\delta_{il}-\Omega_{il}(\mathbf{k})-\Sigma_{il}^{(c)}(\mathbf{k},z)\right]G_{lj}(\mathbf{k},z)=-\chi_{ij}(\mathbf{k}) \tag{79}$$

where

$$\Omega_{ij}(\mathbf{k})=i\langle \dot{A}_i|A_l\rangle(\mathbf{k})\chi_{lj}^{-1}(\mathbf{k}) \tag{80a}$$

$$\Sigma_{ij}^{(c)}(\mathbf{k})=\langle \dot{A}_i|Q(z-Q\mathfrak{L}Q)^{-1}Q|\dot{A}_l\rangle(\mathbf{k})\chi_{lj}^{-1}(\mathbf{k}) \tag{80b}$$

and $\chi_{ij}(\mathbf{k})=\langle A_i|A_j\rangle(\mathbf{k})$. The operator Q is the complement of the projection operator $P=|A_i\rangle\chi_{ij}^{-1}\langle A_j|$. Here and henceforth we adopt the summation convention for the suffixes labeling the different collective variables.

2. *Surface waves*

Our objective here is to illustrate how the broken symmetry concept developed in Section III.B.1 can be adapted to the analysis of an inhomogeneous, two-phase fluid. From our previous examination of the traditional fluid dynamical theory of interfacial dynamics it can be anticipated that the symmetry-breaking collective coordinate will be closely related to the continuum mechanical variable $\zeta(\boldsymbol{\rho})$, which characterizes the displacement of the interface from the equilibrium (Gibbs) surface of separation. The long-ranged correlated motion (Goldstone mode) associated with this collective coordinate then would be a capillary wave or ripple. Verification of these speculations depends on whether we are able to invent suitable many-body analogues for the continuum mechanical coordinates and then succeed in proving that the associated propagating mode is, indeed, the many-body counterpart of a capillary wave. However, before getting too deeply involved in all this, let us briefly consider a simple, nonrigorous demonstration that $\zeta(\boldsymbol{\rho})$ does, in fact, exhibit long-ranged correlations.

We begin by recalling that the inverse of the correlation function

$$K^{-1}(\mathbf{q}; x_3, x_3') \equiv \langle \delta n(\mathbf{r},0)\delta n(\mathbf{r}',0)\rangle(\mathbf{q}) \tag{81}$$

can be expanded in the power series

$$K(\mathbf{q}; x_3, x_3') = K_0(x_3, x_3') + q^2 K_2(x_3, x_3') + O(q^4)$$

the first two coefficients of which satisfy the relationships (see Section II)

$$\int dx_3 \int dx_3' \,\nabla_3 n_0(x_3) K_0(x_3, x_3') \nabla_3' n_0(x_3') = 0 \tag{22}$$

$$\int dx_3 \int dx_3' \,\nabla_3 n_0(x_3) K_2(x_3, x_3') \nabla_3' n_0(x_3') = \beta\sigma_D \tag{23}$$

From these it follows that

$$\int dx_3 \int dx_3' \,\nabla_3 n_0(x_3) K(\mathbf{q}; x_3, x_3') \nabla_3' n_0(x_3') = 0 + q^2\beta\sigma_D + O(q^4) \tag{82}$$

We now adopt from Section II.B.2 [see (18)] the quasicontinuum mechanical relationship, $\delta n(\mathbf{r}) \doteq -\zeta(\boldsymbol{\rho})\nabla_3 n_0(x_3)$, between the fluctuation δn, the gradient $\nabla_3 n_0(x_3)$, and the interfacial displacement $\zeta(\boldsymbol{\rho})$. The result of substituting this approximation into the definition (81) is the formula

$$K^{-1}(\mathbf{q}; x_3, x_3') \doteq [\nabla_3 n_0(x_3)] \langle \zeta(\boldsymbol{\rho})\zeta(\boldsymbol{\rho}')\rangle(\mathbf{q}) [\nabla_3' n_0(x_3')].$$

Next we insert this into the identity $\int dx_3' K(\mathbf{q}; x_3, x_3')K^{-1}(\mathbf{q}; x_3', x_3'') = \delta(x_3 - x_3'')$, multiply the result by $\nabla_3 n_0(x_3)$, and integrate with respect to x_3. The result is the expression

$$\int dx_3 \int dx_3' [\nabla_3 n_0(x_3)] K(\mathbf{q}; x_3, x_3') [\nabla_3' n_0(x_3')] \langle \zeta\zeta \rangle(\mathbf{q}) [\nabla_3'' n_0(x_3'')] = \nabla_3'' n_0(x_3'')$$

which, because of (82), leads us to conclude that

$$\langle \zeta\zeta \rangle(\mathbf{q}) = \frac{1}{0 + \beta\sigma_D q^2 + O(q^4)} \underset{q \to 0}{\to} (\beta\sigma_D q^2)^{-1} \tag{83}$$

This establishes what we set out to prove. Moreover, it shows that for the two-phase system, the parameter ρ_s, defined by (78), is equal to the surface tension σ_D. An equally important aspect of this proof is the significance it attaches to the basic relationship (22): were it not that $\nabla_3 n_0(x_3)$ is an eigenfunction of $K_0(x_3, x_3')$ with eigenvalue zero, the surface displacement $\zeta(\boldsymbol{\rho})$ would not exhibit long-ranged correlations and there would be no associated propagating disturbance (Goldstone boson).

Although intuitively appealing, the foregoing demonstration really amounts to only half a proof because of its reliance on the quasicontinuum mechanical relationship $\delta n \doteq -\zeta \nabla_3 n_0$ to identify the symmetry-restoring variable ζ. However, despite its flaws, the success of this simple theory indicates that we are on the right track and furthermore suggests that a more satisfactory, less heuristic theory can be obtained by careful refinement of the concepts we consider above. The remainder of this section is devoted to the development of just such a theory.

We begin with the two sum rules

$$\begin{aligned} \langle \dot{n}(\mathbf{r},0) g_3(\mathbf{r}',0) \rangle(\mathbf{q}) &= -\langle n(\mathbf{r},0) \dot{g}_3(\mathbf{r}',0) \rangle(\mathbf{q}) \\ &= \beta^{-1} \langle [n(\mathbf{r},0), g_3(\mathbf{r}',0)] \rangle(\mathbf{q}) \\ &= -\beta^{-1} \nabla_3 [n_0(x_3) \delta(x_3 - x_3')] \end{aligned} \tag{84a}$$

and

$$\begin{aligned} \langle \dot{g}_3(\mathbf{r},0) n(\mathbf{r}',0) \rangle(\mathbf{q}) &= -\beta^{-1} \nabla_3' [n_0(x_3) \delta(x_3 - x_3')] \\ &= \langle \dot{n}(\mathbf{r},0) g_3(\mathbf{r}',0) \rangle(\mathbf{q}) + \beta^{-1} \delta(x_3 - x_3') \nabla_3 n_0(x_3) \end{aligned} \tag{84b}$$

which are easily proved with the aid of the relationship $\langle \dot{A}B \rangle = \beta^{-1} \langle [A, B] \rangle$.[61] From the first of these it then follows that

$$\int_{-\varepsilon^+}^{\varepsilon^+} dx_3' \langle [n_0(\mathbf{r},0), g_3(\mathbf{r}',0)] \rangle(\mathbf{q}) = -\nabla_3 n_0(x_3); \qquad -\varepsilon \le x_3 \le \varepsilon \quad (85a)$$

and

$$\int_{-\varepsilon^+}^{\varepsilon^+} dx_3 \langle [n_0(\mathbf{r},0), g_3(\mathbf{r}';0)] \rangle(\mathbf{q}) = 0; \qquad -\varepsilon \le x_3' \le \varepsilon \qquad (85b)$$

wherein $\varepsilon^+ = \varepsilon + 0$. The integral appearing on the left-hand side of (85a) differs from zero only if $|x_3\rangle < \varepsilon$. This is closely analogous to (75), the relationship that is so intimately associated with the breakdown of symmetry. The sharp distinction between (85a) and (85b) provides a dramatic illustration of how sensitively the results depend on the ordering of the spatial arguments x_3 and x_3'. This spatial order is dealt with in greater detail in Ref. 12.

The inhomogeneous states of the fluid exhibit a broken symmetry that is not shared by the homogeneous states. For homogeneous states the gradient $\nabla_3 n_0(x_3)$ is everywhere equal to zero. These homogeneous states are the analogues of the paramagnetic states of the spin system. On the other hand, $\nabla_3 n_0(x_3) = 0$ for the inhomogeneous, two-phase states of the fluid, and these, in turn, are analogous to the ferromagnetic states of the spin system. It is this breakdown of translational symmetry that is the inhomogeneous-fluid counterpart of the loss of rotational symmetry that accompanies the establishment of a ferromagnetic state of the spin system. The propagating modes connected with these inhomogeneous states are surface (capillary) waves; the associated quasiparticles are called ripplons. They are analogous to the characteristic spin waves of the ferromagnet.

It is only within the interface itself that the gradient $\nabla_3 n_0(x_3)$ differs from zero. This immediately suggests that within this transition region there exists a long-ranged variable [$\zeta(\boldsymbol{\rho}, t)$ defined by (87a)] and that the associated order parameter is $\nabla_3 n_0(x_3)$. An alternative, but essentially equivalent order parameter is the difference

$$\delta_S n_0(x_3) = n_0(x_3) - n_0^B(x_3) = \left\{ \begin{array}{ll} 0; & x_3 > \varepsilon \\ \left\{ \begin{array}{ll} n_0(x_3) - n_v; & \varepsilon > x_3 > 0 \\ n_0(x_3) - n_l; & 0 > x_3 > -\varepsilon \end{array} \right\} & \\ 0; & -\varepsilon > x_3 \end{array} \right\} \qquad (86)$$

between $n_0(x_3)$ and the hypothetical reference profile, $n_0^B(x_3)$, shown in Fig. 1 and defined by (36a). This function shares with $\nabla_3 n_0(x_3)$ the property of vanishing everywhere except within the transition zone, $|x_3|<\varepsilon$.

This idea of introducing an order parameter that distinguishes the liquid–vapor interfacial zone from the bulk phases is, perhaps, sufficiently novel to warrant further discussion. Thus, to guard against possible confusion and, at the same time, to introduce another analogy between ferromagnets and two-phase fluids, let us compare a two-phase lattice fluid[66, 67] to the Bloch-wall problem.[68] Pictorial representations of these systems are shown in Fig. 3. We assume that the cells of the lattice fluid are so small that multiple occupancy is impossible; a cell is therefore either empty (indicated by the symbol ○ in Fig. 3) or singly occupied (indicated by ●). Accordingly, the state of a cell, say the ith, is fully characterized by the value of a variable, e_i, that equals zero when the cell is empty and unity when it is occupied.

The reader will recall that the variable S_i of the Ising model is similarly defined: $S_i = 1$ when the spin on lattice site i is "up" and $S_i = -1$ when this spin is "down." Therefore, we can connect the cell occupancy variables of the lattice fluid to the spin variables of the Ising model by the relationship $e_i = \frac{1}{2}(S_i + 1)$ or $S_i = 2e_i - 1$. This establishes a one-to-one mapping between the characteristics of the lattice fluid and those of the Ising model. Thus, with the "local densities" of the "liquid" and "vapor" phases of the inhomogeneous lattice fluid, we associate the values of $\langle e_i \rangle = 1$ and $\langle e_i \rangle = 0$, respectively. These, of course, correspond to the two macroscopic regions of the spin system throughout which the local magnetization $\mu_3(x_3) \equiv \mu\langle S_i \rangle(x_3)$ has the values $\mu_3 = \mu$ and $\mu_3 = -\mu$, respectively. The liquid–

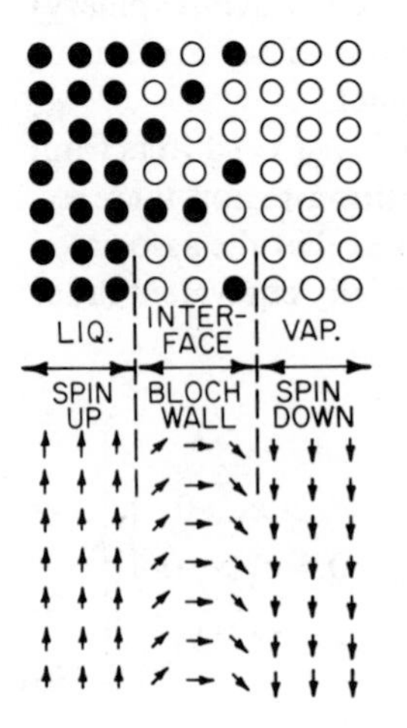

Fig. 3. Schematic representations of the comparable lattice-gas, liquid–vapor interface (● occupied cells and ○ unoccupied cells) and the Bloch wall separating two ferromagnetic domains with oppositely directed magnetizations.

vapor interfacial zone is the region in which the value of $n_0(x_3) \equiv \langle e_i \rangle (x_3)$ drops from unity to zero. It is the analogue of the Bloch wall, across which the value of the magnetization varies from μ to $-\mu$. Order parameters for these two (equivalent) systems, differing from zero only within the interfacial zone or Bloch wall, are given by

$$\delta_S n_0(x_3) \equiv n_0(x_3) - n_0^B(x_3) = \langle e_i \rangle (x_3) - \theta(-x_3)$$

and

$$\delta_S \mu_3(x_3) \equiv \mu_3(x_3) - \mu_3^B(x_3) = \mu[\langle S_i \rangle (x_3) + \theta(x_3) - \theta(-x_3)]$$

respectively.

To construct an explicit mathematical representation of the long-ranged correlation function associated with the order parameter $\nabla_3 n_0(x_3)$, we introduce the two collective variables $\zeta(\boldsymbol{\rho}, t)$ and $\phi(\boldsymbol{\rho}, t)$ defined by

$$\zeta(\boldsymbol{\rho}, t) = \frac{1}{\Delta n} \int_{-\infty}^{+\infty} dx_3 \, \delta_S[\delta n(\mathbf{r}, t)] \tag{87a}$$

and

$$\phi(\boldsymbol{\rho}, t) = \int_{-\infty}^{+\infty} dx_3 \, \delta_S[g_3(\mathbf{r}, t)] \tag{87b}$$

Here, $\delta n = n - n_0$ and $\Delta n = n_l - n_v$. The variable $\zeta(\boldsymbol{\rho}, t)$ can be interpreted as the normal displacement of the surface and $\phi(\boldsymbol{\rho}, t)$ as the normal component of the interfacial momentum density.

Our objective is to establish that the static correlation function $\langle \zeta(\boldsymbol{\rho}, 0) \zeta(\boldsymbol{\rho}', 0) \rangle$ exhibits long-ranged order. This we do by employing the Schwartz inequality,[69]

$$\langle \zeta(\boldsymbol{\rho}, 0) \zeta(\boldsymbol{\rho}', 0) \rangle (\mathbf{q}) \geq \frac{|\langle \dot{\phi}(\boldsymbol{\rho}, 0) \zeta(\boldsymbol{\rho}', 0) \rangle (\mathbf{q})|^2}{\langle \dot{\phi}(\boldsymbol{\rho}, 0) \dot{\phi}(\boldsymbol{\rho}', 0) \rangle (\mathbf{q})} \tag{88}$$

thereby transforming our problem into that of evaluating the two correlation integrals $\langle \dot{\phi} \dot{\phi} \rangle$ and $\langle \dot{\phi} \zeta \rangle$. For the first of these we have from (38) the relationship

$$\langle \dot{\phi} \dot{\phi} \rangle (\mathbf{q}) = \int dx_3 \int dx_3' \, \delta_S \langle \dot{g}_3 \dot{g}_3 \rangle (\mathbf{q}; x_3, x_3') = \beta^{-1} \sigma_P q^2 + O(q^4) \tag{89}$$

which is valid for a *thin* interfacial zone. To calculate $\langle \dot{\phi} \zeta \rangle$ in this same limit of a thin transition zone we replace ζ and ϕ defined by (87a) and (87b) with the approximations

$$\zeta(\boldsymbol{\rho}, t) \doteq \frac{1}{\Delta n} \int_{-\varepsilon}^{\varepsilon} dx_3 \, \delta n(\mathbf{r}, t) \tag{90a}$$

$$\phi(\boldsymbol{\rho}, t) \doteq \int_{-\varepsilon}^{\varepsilon} dx_3 \, g_3(\mathbf{r}, t) \tag{90b}$$

We then introduce the operator $I(\varepsilon)$, which is so defined that these two expressions can be written in the forms $(\Delta n)\zeta(\boldsymbol{\rho}, t) = I(\varepsilon)[\delta n(\mathbf{r}, t)]$ and $\phi(\boldsymbol{\rho}, t) = I(\varepsilon)[g_3(\mathbf{r}, t)]$, respectively. [The operator I can be interpreted as multiplication by a bra vector $\langle I(\varepsilon)|$, whose x_3 representative is $\langle I(\varepsilon)|x_3\rangle = \int_{-\varepsilon}^{\varepsilon} dx_3$. This, in turn, is closely related to the approximate surface projection operator $P_S \equiv |I(\varepsilon)\rangle\langle I(\varepsilon)|$ with $\langle x_3|I(\varepsilon)\rangle = \nabla_3 n_0(x_3)/\Delta n$, which is considered at greater length in the following section.]

Next, to avoid ambiguity in the evaluation of integrals (so-called space ordering), we adopt for $\dot{\zeta}$ and $\dot{\phi}$ the definitions $\dot{\zeta} = (\Delta n)^{-1} I(\varepsilon^+)[\delta \dot{n}]$ and $\dot{\phi} = I(\varepsilon^+)[\dot{g}_3]$ with $\varepsilon^+ = \varepsilon + 0$. Then from (84b) it follows that

$$I(\varepsilon^+)I'(\varepsilon)\{\langle \dot{n}(\mathbf{r},0) g_3(\mathbf{r}',0)\rangle(\mathbf{q}) - \langle \dot{g}_3(\mathbf{r},0) n(\mathbf{r}',0)\rangle(\mathbf{q})\}$$

$$= I(\varepsilon^+)I'(\varepsilon)\{-\beta^{-1}\delta(x_3 - x_3')\nabla_3 n_0(x_3)\} = \beta^{-1}\Delta n$$

or

$$\langle \dot{\zeta}(\boldsymbol{\rho},0)\phi(\boldsymbol{\rho}',0)\rangle(\mathbf{q}) - \langle \dot{\phi}(\boldsymbol{\rho},0)\zeta(\boldsymbol{\rho}',0)\rangle(\mathbf{q}) = \beta^{-1} \tag{91}$$

The final relationship that we require is

$$\langle \dot{\zeta}\phi\rangle(\mathbf{q}) \doteq \frac{q}{m\,\Delta n}\langle \phi\phi\rangle(\mathbf{q}) \tag{92}$$

for from this and (91) it follows that

$$\langle \dot{\phi}\zeta\rangle(\mathbf{q}) = -\beta^{-1} + \frac{q}{m\,\Delta n}\langle \phi\phi\rangle(\mathbf{q}) \tag{93}$$

Then, by combining (88), (89), and (93) we find that in the long-wavelength limit ($q \to 0$)

$$\langle \zeta(\boldsymbol{\rho},0)\zeta(\boldsymbol{\rho}',0)\rangle(\mathbf{q}) = (\beta \rho_s q^2)^{-1} \tag{94a}$$

with

$$\rho_s \leq \sigma_P \tag{94b}$$

This means that as a consequence of translational symmetry breaking, the static correlation function of the collective variable $\zeta(\boldsymbol{\rho}, t)$, defined by (90a), exhibits long-ranged order.

The one part of the argument that remains unproved is the relationship given by (92). In Ref. 12 a quasihydrodynamic proof is presented. However, in Ref. 13 and in the following section, as well [see (142)] we use a more direct argument that relies almost exclusively on sum rules.

It has been mentioned previously that long-lived variables may give rise to propagating modes. In a homogeneous fluid the only long-lived variables are conserved variables and the associated propagating modes are sound waves. However, if the fluid has additional structure connected with long-ranged order there then could be other propagating modes. We have discovered collective variables $\zeta(\boldsymbol{\rho}, t)$ and $\phi(\boldsymbol{\rho}, t)$ associated with the breakdown of translational symmetry that accompanies the establishment of an inhomogeneous, two-phase state of a single-component fluid. Using these variables we now construct the long-wavelength limit of the dispersion relation for the propagating modes associated with spatial symmetry breaking.

In what follows use is made of the condensed notation, $\langle ab\rangle = \langle a(\boldsymbol{\rho},0)b(\boldsymbol{\rho}',0)\rangle(\mathbf{q})$ and $\langle \dot{a}|\mathfrak{R}_1|\dot{b}\rangle = \langle \dot{a}(\boldsymbol{\rho},0)|\mathfrak{R}_1(z)|\dot{b}(\boldsymbol{\rho}',0)\rangle(\mathbf{q})$ wherein $\mathfrak{R}_1(z) \equiv Q_1(z - Q_1\mathfrak{L}Q_1)^{-1}Q_1$ and Q_1 is the complement of the projection operator $P_1 = |\zeta\rangle\langle\zeta\zeta\rangle^{-1}\langle\zeta| + |\phi\rangle\langle\phi\phi\rangle^{-1}\langle\phi|$. Then, because of time-reversal invariance it follows [from (79) and (80)] that the frequency matrix, $\Omega_1(\mathbf{q})$, and the memory function, $\Sigma_1^{(c)}(z,\mathbf{q})$, are given by

$$\Omega_1 = i\begin{bmatrix} \langle\dot{\zeta}\zeta\rangle & \langle\dot{\zeta}\phi\rangle \\ \langle\dot{\phi}\zeta\rangle & \langle\dot{\phi}\phi\rangle \end{bmatrix}\begin{bmatrix} \langle\zeta\zeta\rangle & \langle\zeta\phi\rangle \\ \langle\phi\zeta\rangle & \langle\phi\phi\rangle \end{bmatrix}^{-1}$$

$$= i\begin{bmatrix} 0 & \langle\dot{\zeta}\phi\rangle \\ \langle\dot{\phi}\zeta\rangle & 0 \end{bmatrix}\begin{bmatrix} \langle\zeta\zeta\rangle & 0 \\ 0 & \langle\phi\phi\rangle \end{bmatrix}^{-1}$$

$$= \begin{bmatrix} 0 & i\langle\dot{\zeta}\phi\rangle\langle\phi\phi\rangle^{-1} \\ i\langle\dot{\phi}\zeta\rangle\langle\zeta\zeta\rangle^{-1} & 0 \end{bmatrix}$$

and

$$\Sigma_1^{(c)} = \begin{bmatrix} \langle\dot{\zeta}|\mathfrak{R}_1|\dot{\zeta}\rangle & \langle\dot{\zeta}|\mathfrak{R}_1|\dot{\phi}\rangle \\ \langle\dot{\phi}|\mathfrak{R}_1|\dot{\zeta}\rangle & \langle\dot{\phi}|\mathfrak{R}_1|\dot{\phi}\rangle \end{bmatrix} \begin{bmatrix} \langle\zeta\zeta\rangle & \langle\zeta\phi\rangle \\ \langle\phi\zeta\rangle & \langle\phi\phi\rangle \end{bmatrix}^{-1}$$

$$= \begin{bmatrix} \langle\dot{\zeta}|\mathfrak{R}_1|\dot{\zeta}\rangle & 0 \\ 0 & \langle\dot{\phi}|\mathfrak{R}_1|\dot{\phi}\rangle \end{bmatrix} \begin{bmatrix} \langle\zeta\zeta\rangle & 0 \\ 0 & \langle\phi\phi\rangle \end{bmatrix}^{-1}$$

$$= \begin{bmatrix} \langle\dot{\zeta}|\mathfrak{R}_1|\dot{\zeta}\rangle\langle\zeta\zeta\rangle^{-1} & 0 \\ 0 & \langle\dot{\phi}|\mathfrak{R}_1|\dot{\phi}\rangle\langle\phi\phi\rangle^{-1} \end{bmatrix}$$

respectively.

The task before us now is that of evaluating the four components of these two matrices. We begin with the component, $i\langle\dot{\phi}\zeta\rangle\langle\zeta\zeta\rangle^{-1}$, which according to (93) and (94), equals $-i\rho_s q^2 \equiv -i\hat{\sigma}_s q^2$. Next, it follows from (92) that, in the hydrodynamic limit, the value of the matrix element $i\langle\dot{\zeta}\phi\rangle\langle\phi\phi\rangle^{-1}$ is $i(q/m\,\Delta n)\langle\phi\phi\rangle^{-1}\langle\phi\phi\rangle = iq/m\,\Delta n$.

Let us now examine the elements of the memory function matrix, $\Sigma_1^{(c)}(z,\mathbf{q})$. Because ϕ is a linear functional of the conserved variable g_3, we easily extract the hydrodynamic limit of $\langle\dot{\phi}|\mathfrak{R}_1|\dot{\phi}\rangle\langle\phi\phi\rangle^{-1}$ and define

$$\hat{\sigma}_1 = \lim_{q\to 0}\lim_{\omega\to 0}\left\{\langle\dot{\phi}(\boldsymbol{\rho},0)|\mathfrak{R}_1(\omega+i0)|\dot{\phi}(\boldsymbol{\rho}',0)\rangle(\mathbf{q})\chi_{\phi\phi}^{-1}(\mathbf{q})q^{-2}\right\}$$

where $\chi_{\phi\phi} = \langle\phi\phi\rangle(\mathbf{q})$. Then since

$$\dot{\zeta}(\mathbf{q},z) = \frac{1}{\Delta n}\int_{-\varepsilon}^{\varepsilon} dx_3\, F.T.\{\dot{n}(\mathbf{r},t)\} = -\frac{1}{m\,\Delta n}\int_{-\varepsilon}^{\varepsilon} dx_3\, \text{F.T.}\{\nabla_j g_j(\mathbf{r},t)\}$$

$$= -\frac{1}{m\,\Delta n}\left[\{g_3(\varepsilon,\mathbf{q},t) - g_3(-\varepsilon,q,t)\} - iq\psi(\mathbf{q},t)\right]$$

with $\psi(\mathbf{q},t) \equiv \int_{-\varepsilon}^{\varepsilon} dx_3\, \hat{q}\cdot\mathbf{g}(x_3,\mathbf{q},t)$, it follows that $Q_1\dot{\zeta}(\mathbf{q},t) = iq\psi(\mathbf{q},t)/m\,\Delta n$. We then can define $\hat{\sigma}_2$ by

$$\beta\rho_s \lim_{q\to 0}\lim_{\omega\to 0}\left\{\langle\dot{\zeta}(\boldsymbol{\rho},0)|\mathfrak{R}_1(\omega+i0)|\dot{\zeta}(\boldsymbol{\rho}',0)\rangle(q)q^{-2}\right\}$$

and so conclude that

$$z-\Omega_1(\mathbf{q})-\Sigma_1^{(c)}(z,\mathbf{q})=\begin{bmatrix} z-q^4\hat{\sigma}_2 & \dfrac{-iq}{\rho_m} \\ iq^2\hat{\sigma}_s & z-q^2\hat{\sigma}_1 \end{bmatrix}$$

with $\rho_m = m\,\Delta n$.

The coupled equations of motion for the matrix of correlation functions $G=\{G_{ij}(\mathbf{q}, z)\}$ are $(z-\Omega_1-\Sigma_1^{(c)})G=-\chi$ and the corresponding dispersion relation is

$$\det\left(z-\Omega_1-\Sigma_1^{(c)}\right)=\left(z-q^2\hat{\sigma}_1\right)\left(z-q^4\hat{\sigma}_2\right)-q^3\left(\frac{\hat{\sigma}_s}{\rho_m}\right)=0 \tag{95}$$

From this we deduce that

$$z=\pm\left(\frac{\hat{\sigma}_s}{\rho_m}\right)^{1/2} q^{3/2}+\left(\frac{\hat{\sigma}_1}{2}\right)q^2+O(q^{5/2}) \tag{96}$$

By resolving the complex number $\hat{\sigma}_1=\hat{\sigma}_1'-i\hat{\sigma}_1''$ into its real (reactive) part, $\hat{\sigma}_1'$, and its imaginary part, $\hat{\sigma}_1''$, defined by

$$\hat{\sigma}_1''=\pi\lim_{q\to 0}\left[\langle\dot{\phi}(\boldsymbol{\rho},0)|Q_1\delta(Q_1\mathfrak{L}Q_1)Q_1|\dot{\phi}(\boldsymbol{\rho}',0)\rangle(q)\chi_{\phi\phi}^{-1}(\mathbf{q})q^{-2}\right]\geq 0$$

we see that these propagating modes are, indeed, damped. The wave-number dependencies of the reactive and dissipative parts of this dispersion relation provide a qualitative description of the surface waves that agrees with the predictions of the phenomenological theory.[46, 47]

The set of relevant variables $\{\zeta(\boldsymbol{\rho}, t), \phi(\boldsymbol{\rho}, t)\}$ can be expanded to include other conserved variables, such as the tangential component of momentum ψ. This results in a dispersion relation whose qualitative features are the same as those of (95). The values of the damping coefficients are somewhat altered, but the factor $(\hat{\sigma}_s/\rho_m)^{1/2}q^{3/2}=(\rho_s/\rho_m)^{1/2}q^{3/2}$, which determines the propagation velocity of the capillary waves, is unchanged. Thus, the surface tension that enters into this linearized dynamical theory of a thin interface is $\rho_s \leq \sigma_P$. We suspect, but have not yet proved, that the introduction of mode–mode coupling alters the surface tension that occurs in the dynamical theory.

IV. KINETIC THEORY OF INHOMOGENEOUS FLUIDS

We turn now to the task of formulating a truly microscopic, kinetic theory of two-phase systems. There are two elements in the theory we propose: (*1*) a kinetic equation for the singlet distribution function of the inhomogeneous, two-phase system and (*2*) a means for extracting from this equation information specific to the interface. In the linear response regime to which our present considerations are confined, the generalized Langevin (kinetic) equation is formally exact and so provides the perfect (formal) solution to the first of our problems. Furthermore, the surface-excess projection operator defined in Section II is a device that meets the second of our requirements. Accordingly, we first present (Section IV.A) a formal demonstration of how these two elements of the microscale theory can be combined to produce a kinetic theory of the interfacial zone.

To proceed much beyond this point, approximations (models) are needed, both for the kinetic equation and for the surface projection operator. In Section IV.B a number of approximate surface projection operators are examined. Then, in Section IV.C attention is directed to the problem of generating approximate kinetic equations for the two-phase system. In Section IV.D we first derive kinetic sum rules that involve the inhomogeneity of the system in an essential way. These sum rules are the tools that then are used to construct mean field (collisionless) and generalized Fokker-Planck approximations to the exact kinetic equation.

A. Formal Structure of the Theory

Here, as in Section III, our concern is with the spontaneous fluctuations that occur in an inhomogeneous, two-phase system. A conventional and convenient way of studying the microscale behavior of these fluctuations is to use a kinetic equation, that is, an equation which governs the time evolution of the phase space density fluctuation

$$\delta f(1,t)=f(1,t)-\langle f(1,t)\rangle \tag{97a}$$

Here

$$f(1,t)=\sum_{\alpha}\delta(\mathbf{r}_1-\mathbf{r}^{\alpha}(t))\delta(\mathbf{p}_1-\mathbf{p}^{\alpha}(t)) \tag{97b}$$

is the particle density and the symbol 1 denotes the phase space coordinates $(\mathbf{r}_1,\mathbf{p}_1)$. The sum in (97b) extends over all particles belonging to the system. As is shown in Appendix A and in Section IV.C the equation of motion for δf is the generalized Langevin-equation,

$$\partial_t[\delta f(1,t)]+\int_0^t d\bar{t}\,\Sigma(1,\bar{1};t-\bar{t})\delta f(\bar{1},\bar{t})+\hat{F}(1,t)=0 \tag{98}$$

with ∂_t denoting the partial derivative with respect to time. Here we continue to use the "summation convention" introduced previously (see Refs. 13 and 40), according to which the occurrence of a repeated, overscored coordinate such as $\bar{1}$ or $\bar{x}_3$ implies integration over the domain of the variable.

The symbol $\hat{F}$ represents a random fluctuating force that is orthogonal to the initial fluctuation of density, that is,

$$\langle \hat{F}(t)\delta f(0)\rangle = \langle \delta f(0)\hat{F}(t)\rangle = 0$$

The nonlocal and non-Markovian memory function Σ is descriptive of the molecular dynamics and of the static fluid structure as well. Because of the restricted translational invariance of the two-phase system, $\Sigma(1,\bar{1};t-\bar{t})$ depends separately on x_3 and $\bar{x}_3$ but only on the differences, $x_1-\bar{x}_1$ and $x_2-\bar{x}_2$, of the particle coordinates measured parallel to the planar interface. This memory function can be resolved into the sum

$$\Sigma(1,1';t-t') = \hat{\Omega}(1,1')\delta(t-t') - \Sigma^{(c)}(1,1';t-t') \tag{99}$$

of instantaneous and delayed responses. The mean field term, $\hat{\Omega}$, contains information about the static structure of the fluid, whereas $\Sigma^{(c)}$ encompasses the non-Markovian characteristics stemming from the intricate microscopic dynamics of the many-body system. $\Sigma^{(c)}$ is related to the autocorrelation function of the fluctuating force $\hat{F}$.

In Section IV.C the two functions $\hat{\Omega}$ and $\Sigma^{(c)}$ are examined in greater detail. It is, of course, utterly impossible to evaluate $\Sigma^{(c)}$ exactly and, in the absence of an identifiable small parameter, there is no way of developing a systematic perturbation theory. Furthermore, even for homogeneous fluids it is only for a few very special cases, such as the Vlasov[70, 71], Fokker-Planck,[71–74] and modified BGK[75] models, that one knows how to construct analytic solutions of the resulting kinetic equations. Despite these limitations, there is at present rather widespread agreement that the memory function and fluctuations of simple, homogeneous fluids are fairly well understood.[76] It is this understanding of homogeneous fluids that we intend to use and build on to elucidate the microscale dynamics of inhomogeneous two-phase fluids.

The other essential component of our theory is the surface projection operator δ_S defined in Section II [see (35)]. The virtue of this operator is its ability to separate the intrinsic properties of the interface from those of the much more massive bulk phases in which the surface is imbedded. In practice (see Section IV.B) it often proves necessary, and sometimes even desirable, to replace the exact surface-projection operator with simpler, more tractable operators that share its most important properties.

These practical, computational difficulties connected with the memory function and projection operator do not prevent us from examining the formal aspects of the kinetic theory of interfacial dynamics. We begin by replacing the Langevin equation, (98), with its Laplace transform (defined here by $\mathfrak{L}[F(t)] \equiv F(z) \equiv i \int_0^\infty dt \exp(izt) F(t)$, $\operatorname{Im} z > 0$)

$$[z1 - \Sigma(z)]\,\delta f(z) = -\delta f(t=0) - i\hat{F}(z) \tag{100}$$

Acting on this equation with the surface projection operator, $P_S = \delta_S$, and its complement, the bulk-phase projection operator, $P_B = 1 - P_S$, we then obtain the coupled equations

$$(zP_s - \Sigma_{SS})\,\delta f_S(z) - \Sigma_{SB}\,\delta f_B(z) = \Lambda_S \tag{101a}$$

$$(zP_B - \Sigma_{BB})\,\delta f_B(z) - \Sigma_{BS}\,\delta f_S(z) = \Lambda_B \tag{101b}$$

for the two components $\delta f_X(z) \equiv P_X\,\delta f(z)$, $X = S, B$, of the density fluctuation. Here, $\Sigma_{XY} = P_X \Sigma P_Y$ and $\Lambda_X = P_X\{-\delta f(t=0) - i\hat{F}(z)\}$. The result of substituting the solution to the second of these into the first is the closed equation

$$[zP_S - \Sigma_S]\,\delta f_S(z) = -\Pi_S[\delta f(t=0) + i\hat{F}(z)] \tag{102}$$

wherein

$$\Sigma_S \equiv \Sigma_{SS} + \Sigma_{SB}(zP_B - \Sigma_{BB})^{-1}\Sigma_{BS} \tag{103a}$$

and

$$\Pi_S = P_S + \Sigma_{SB}(zP_B - \Sigma_{BB})^{-1} P_B \tag{103b}$$

The first contributor to the effective surface memory function Σ_S appearing in (103a) is the "bare" memory, Σ_{SS}, specific to the transition zone itself. The second term of (103a) contains in addition to the factors Σ_{BS} and Σ_{SB}, which couple the surface to the bulk phases, the bulk-phase propagator $(zP_B - \Sigma_{BB})^{-1}$, which, according to our earlier remarks, may be considered as a "known" and well-understood object.

The equation of motion, (102), illustrates how the dynamics of the interface can be regarded formally as an intrinsic, transition-zone dynamics (associated with Σ_{SS}) "perturbed" by couplings to the surrounding bulk phases. In this theory the counterparts to the macroscopic Young-Laplace and dissipative (viscous) boundary conditions are buried in the surface projection operator and the bulk-surface couplings.[77] Thus, for example, we show in Section IV.D that the dispersion relation $\omega^2=(\sigma/\rho)q^3$ for small amplitude surface waves is the result of neglecting all bulk–surface couplings and retaining only the nondissipative, mean-field portion $P_S\hat{\Omega}P_S$ of Σ_{SS}. The function $P_S\Sigma^{(c)}P_S$ presumably provides a dissipative contribution to the surface dispersion relation specific to the transition zone itself (surface viscosity), while the functions Σ_{BS} and Σ_{SB} provide means by which the properties (e.g., viscosity) of the bulk phases can contribute to the damping of surface disturbances. In highly viscous fluids the last of these effects may be so great that the propagating modes associated with the dispersion relation $\omega^2=(\sigma/\rho)q^3$ are replaced by strongly damped motions attributable to diffusive poles of $(zP_B-\Sigma_{BB})^{-1}$. This sort of behavior can, of course, be anticipated from the phenomenological theory of surface waves.[69]

Finally, we call attention to the close analogy that can be drawn between this description of interfacial dynamics and that of a polymeric fluid.[78] Thus the bulk-surface coupling is analogous to the "hydrodynamic interactions"[79] between polymer and solvent molecules, and the dissipative modes attributable to $P_S\Sigma^{(c)}P_S$ appear to be conceptually related to the polymeric "internal viscosity."[80]

B. Approximate Projection Operators and Models for the Density Profile

The surface-excess projection operator defined in Section II [see (35)] is excellently suited to the task of separating the attributes of the transition zone, be these static or dynamic, from those of surrounding bulk phases. This projection operator is especially important not only because it so closely mimics the surface excess definitions of thermodynamics, but also because it plays such a critical role in the connection we have established in Section II.B.3 between force autocorrelation functions and the Kirkwood-Buff formula for the surface tension. But despite the special and almost unique qualities of this particular operator, it is by no means the only one that qualifies as a projection onto the interfacial zone. Indeed, as a glance at Fig. 1 will verify, the density gradient $\nabla_3 n_0(x_3)$ is ready-made for the job—it is finite within the transition zone and (almost) equal to the zero outside of it. Thus the function $\nabla_3 n_0(x_3)$ provides a natural basis for the construction of surface projection operators. The simplest of these is defined by

$$P_S A(x_3)=\frac{-1}{n_l-n_v}\nabla_3 n_0(x_3)\int_{-\varepsilon}^{\varepsilon} dx_3 A(x_3) \tag{104}$$

This operator clearly succeeds in extracting from the position-dependent function $A(x_3)$ its projection onto the interfacial region and it also exhibits the idempotent property $P_S^2 = P_S$ essential to a projection operator. It sometimes is useful to write this operator in the form $P_S = 1 - Q_S = |S\rangle\langle S|$, where $Q_S \equiv P_B$, the complement of P_S, projects onto the bulk phases and where the x_3 representatives of the bra and ket are defined by

$$\langle S|x_3\rangle = \int_{-\varepsilon}^{\varepsilon} dx_3 \tag{105a}$$

and

$$\langle x_3|S\rangle = -\frac{1}{n_l - n_v}\nabla_3 n_0(x_3) \tag{105b}$$

respectively.

The action of P_S can be illustrated by adopting some model for the density profile $n_0(x_3)$. Although this profile has been the subject of extensive experimental, theoretical, and molecular dynamics studies,[4, 81] we are more concerned with its general characteristics than with the intricacies of real interfaces. Therefore, as a first example let us choose for $n_0(x_3)$ the simple model

$$n_0(x_3) = \begin{cases} n_l; & x_3 < -\varepsilon \\ n_l - \dfrac{n_l - n_v}{2\varepsilon}(x_3 + \varepsilon); & -\varepsilon < x_3 < \varepsilon \\ n_v; & \varepsilon < x_3 \end{cases} \tag{106a}$$

for which

$$\nabla_3 n_0(x_3) = -\frac{n_l - n_v}{2\varepsilon}\left[\theta(x_3 + \varepsilon) - \theta(x_3 - \varepsilon)\right] \tag{106b}$$

and

$$P_S A(x_3) = \left[\theta(x_3 + \varepsilon) - \theta(x_3 - \varepsilon)\right] A^{(S)} \tag{106c}$$

Here $A^{(S)} \equiv (2\varepsilon)^{-1}\int_{-\varepsilon}^{\varepsilon} dx_3 A(x_3)$ is the unweighted average of A across the interfacial zone. In the (hypothetical) limit of an infinitely thin transition zone (106b) and (106c) reduce to $\nabla_3 n_0(x_3) \to -(n_l - n_v)\delta(x_3)$ and $P_S A(x_3) \to 2\varepsilon\delta(x_3)A(0) = 0$, respectively.

A second, somewhat more realistic approximation is to assume that the shape of the density profile is that of an error function. The density gradient is then the Gaussian,

$$\nabla_3 n_0(x_3) = \frac{n_l - n_v}{(2\pi\mu_2)^{1/2}} \exp\left(\frac{-x_3^2}{2\mu_2}\right) \tag{107}$$

whose second moment is equal to μ_2. As the value of μ_2 falls to zero the function $\nabla_3 n_0(x_3)$ defined by (107) tends to the same limit, $-(n_l - n_v)\delta(x_3)$, that was encountered in our earlier example. Indeed, this behavior is common to all models in the limit of a thin interfacial zone.

Finally, we consider the density profile,

$$n_0(x_3) = \tfrac{1}{2}\left[(n_l + n_v) - (n_l - n_v)\tanh\left(\frac{x_3}{a}\right)\right] \tag{108a}$$

associated with the planar interface of a Maxwell-van der Waals fluid.[31, 82] By neglecting the vapor density, n_v, we obtain in place of (108a) the expression,

$$n_0(x_3) = n_l\left[1 + \exp\left(\frac{2x_3}{a}\right)\right]^{-1} \tag{108b}$$

which is analogous in form to the Fermi-Dirac energy distribution for a system of noninteracting fermions. [In Ref. 12 this analogy is examined in considerable detail.] According to (108b) the particle density tends to zero in the limit $x_3 \to +\infty$: as $x_3 \to -\infty$ the value of $n_0(x_3)$ approaches the liquid density n_l. The gradient $\nabla_3 n_0(x_3)$ associated with this model has the following properties: (*1*) It is negative semidefinite, (*2*) it is an even function of x_3, and (*3*) it is bounded from above by the value zero and exhibits a minimum of $-n_l/2a$ at the location, $x_3 = 0$, of the "Gibbs dividing surface." These general properties are shared by many other models of the density profile, including the two considered earlier.

The most commendable feature of the surface projection operator defined by (104) is, perhaps, its simplicity. More refined operators would incorporate other measures of the system's inhomogeneity in addition to the density gradient $\nabla_3 n_0(x_3)$ and more subtle aspects of the function A than its unweighted integral across the interface. These refinements could involve an analysis in terms of spatial moments comparable in scope to the analysis in terms of the momentum basis functions, $\psi^{(n)}_{i_1 i_2 \ldots i_n}(\mathbf{p})$, defined in Appendix B. Thus, for example, one can associate with the Gaussian model, (107), the set of basis functions

$$\chi_p(x_3) \equiv \frac{(-\mu_2^{1/2})^p}{\nabla_3 n_0(x_3)} \nabla_3^{p+1} n_0(x_3) \tag{109}$$

the first three of which are $\chi_0(x_3)=1$, $\chi_1(x_3)=x_3/\mu_2^{1/2}$, and $\chi_2(x_3)=(x_3/\mu_2^{1/2})^2-1$. It can be arranged so that the members of this basis set are orthogonal. Thus, in terms of the scalar product defined by

$$\langle\psi(x_3)|R(x_3,x_3')|\phi(x_3')\rangle\equiv\int dx_3\int dx_3'\psi(x_3)[R(x_3,x_3')\nabla_3'n_0(x_3')]\phi(x_3') \tag{110a}$$

it follows that

$$\langle\chi_n(x_3)|\delta(x_3-x_3')|\chi_m(x_3')\rangle\equiv\langle\chi_n|\chi_m\rangle=-(n_l-n_v)n!\delta_{nm} \tag{110b}$$

This more elaborate analysis almost certainly is unnecessary for thin transition zones, but we anticipate that it will be very useful in future studies of extended interfaces.

C. Approximate Kinetic Equations

The task before us here is that of constructing approximations to the formally exact Langevin kinetic equation, (98), or more specifically, to the memory function Σ, whose mean-field and non-Markovian parts are, according to (A.14) and (A.18) of Appendix A,

$$\begin{aligned}\hat{\Omega}(1,1')=i\Omega(1,1')&=-\langle\dot{f}(1)|\delta f(\bar{1})\rangle\langle\delta f(\bar{1})|\delta f(1')\rangle^{-1}\\&=-S_1(1,\bar{1})S_0^{-1}(\bar{1},1)\end{aligned} \tag{111}$$

and

$$\Sigma^{(c)}(1,1';t-t')=-\langle\dot{f}|Qe^{-iQ\mathfrak{L}Q(t-t')}Q|\dot{f}\rangle(1,\bar{1})S_0^{-1}(\bar{1},1') \tag{112}$$

respectively. Here Q is the complement of the projection operator $P\equiv|\delta f\rangle S_0^{-1}\langle\delta f|$, $\mathfrak{L}$ is the Liouville operator of the system, and S_0 and S_1 are, respectively, the first and second members of the set of *kinetic sum rules*

$$S_n(1,1')\equiv\left\langle[\partial_t^n\delta f(1,t)]_{t=0}\delta f(1',0)\right\rangle \tag{113}$$

In the remainder of this chapter S_2 is the only additional sum rule that is needed. The evaluation of these three functions is greatly facilitated by the use of the relationship $\langle\dot{A}B\rangle=\beta^{-1}\langle[A,B]\rangle$, and the resulting formulas are somewhat less cumbersome when the dimensionless variable

$$\xi=\frac{\mathbf{p}}{mv_0},\qquad v_0^2=(m\beta)^{-1}$$

is used in place of the momentum $\mathbf{p}$. Specifically, the two-dimensional Fourier transforms of S_0, S_1, and S_2 are given by the formulas,

$$[S_0(1,1')](\mathbf{q})=[\delta(1-1')+n_0(x_3')h(q;x_3,x_3')\Phi(\xi')]n_0(x_3)\Phi(\xi) \tag{114a}$$

$$\begin{aligned}[S_1(1,1')](\mathbf{q})=&-v_0[i\boldsymbol{\xi}_\perp\cdot\mathbf{q}+\xi_3\nabla_3][n_0(x_3)\Phi(\xi)\delta(1-1')]\\ &-v_0[\nabla_3 n_0(x_3)]\partial_3[\Phi(\xi)\delta(1-1')]\end{aligned} \tag{114b}$$

and

$$\begin{aligned}[S_2(1,1')](\mathbf{q})=&-v_0^2[i\boldsymbol{\xi}_\perp\cdot\mathbf{q}+\xi_3\nabla_3][-i\boldsymbol{\xi}'_\perp\cdot\mathbf{q}+\xi_3'\nabla_3][n_0(x_3)\Phi(\xi)\delta(1-1')]\\ &-v_0^2\{[\nabla_3' n_0(x_3')]\partial_3'[i\boldsymbol{\xi}_\perp\cdot\mathbf{q}+\xi_3\nabla_3]\\ &\quad+[\nabla_3 n_0(x_3)]\partial_3[-i\boldsymbol{\xi}'_\perp\cdot\mathbf{q}+\xi_3'\nabla_3']\}\Phi(\xi)\delta(1-1')\\ &-v_0^2[\nabla_3^2 n_0(x_3)]\partial_3\partial_3'[\Phi(\xi)\delta(1-1')]\\ &+\frac{1}{m}\Big\{\tau_{ij}(\mathbf{q};x_3,x_3')\partial_j\partial_i'[\Phi(\xi)\Phi(\xi')]\\ &\quad-\int d\bar{x}_3\tau_{ij}(\mathbf{q}=0;x_3,\bar{x}_3)\partial_j\partial_i'[\Phi(\xi)\delta(1-1')]\Big\}\end{aligned} \tag{114c}$$

with $\delta(1-1')\equiv\delta(x_3-x_3')\delta(\boldsymbol{\xi}-\boldsymbol{\xi}')$, $\partial_i\equiv\partial/\partial\xi_i$, $\nabla_i\equiv\partial/\partial x_i$ and where $\boldsymbol{\xi}_\perp$ is the component of $\boldsymbol{\xi}$ perpendicular to the direction of the density gradient, that is, the projection of $\boldsymbol{\xi}$ in the interfacial plane. Other quantities appearing in these formulas are the Maxwell-Boltzmann momentum distribution $\Phi(\xi)=(2\pi)^{-3/2}\exp(-\frac{1}{2}\xi^2)$ and

$$\tau_{ij}(\mathbf{q};x_3,x_3')=\int d^2\rho\, e^{i\mathbf{q}\cdot\boldsymbol{\rho}}n_2(\boldsymbol{\rho};x_3,x_3')\nabla_i\nabla_j V(\boldsymbol{\rho},x_3-x_3') \tag{115}$$

the second rank tensor of which τ_{33} defined by (31b) is an element. Finally, $h(\mathbf{q},x_3,x_3')$ is the Fourier transform of the spatial correlation function $h(\mathbf{r},\mathbf{r}')$, which is related to the equilibrium pair distribution function $n_2(\mathbf{r},\mathbf{r}')$ by

$$n_2(\mathbf{r},\mathbf{r}')=\Big\langle\sum_\alpha\sum_{\beta\neq\alpha}\delta(\mathbf{r}-\mathbf{r}^\alpha)\delta(\mathbf{r}'-\mathbf{r}^\beta)\Big\rangle=n_0(x_3)n_0(x_3')[1+h(\mathbf{r},\mathbf{r}')] \tag{116a}$$

The function $h(\mathbf{q};x_3,x_3')$ also is connected to the direct correlation function of Ornstein and Zernike[76, 83] through the integral relationship

$$h(\mathbf{q}; x_3, x_3') = C(\mathbf{q}; x_3, x_3') + C(\mathbf{q}; x_3, \bar{x}_3) n_0(\bar{x}_3) h(\mathbf{q}; \bar{x}_3, x_3') \quad (116b)$$

Then since

$$K^{-1}(\mathbf{q}; x_3, x_3') \equiv \langle \delta n(\mathbf{r}) \delta n(\mathbf{r}') \rangle(\mathbf{q}) = n_0(x_3) n_0(x_3') h(\mathbf{q}; x_3, x_3') + n_0(x_3) \delta(x_3 - x_3') \quad (116c)$$

we obtain the Fourier transform,

$$K(\mathbf{q}; x_3, x_3') = \frac{\delta(x_3 - x_3')}{n_0(x_3)} - C(\mathbf{q}; x_3, x_3'), \quad (116d)$$

of the relationship (23), between K and C.

The last of these is closely related to the (easily verified) formula

$$[S_0^{-1}(1,1')](\mathbf{q}) = \frac{\delta(1-1')}{n_0(x_3)\Phi(\xi)} - C(\mathbf{q}; x_3, x_3') \quad (117)$$

for the functional inverse of the sum rule S_0.

1. *Mean-Field Approximation (Collisionless Kinetic Equation)*[13]

The simplest nontrivial approximation to the Langevin equation is the collisionless equation

$$\partial_t[\delta f(1,t)] + \hat{\Omega}(1,\bar{1}) \delta f(\bar{1},t) = 0 \quad (118)$$

obtained by totally ignoring the fluctuating force $\hat{F}$ and the associated non-Markovian memory function $\Sigma^{(c)}$. From the form of this equation one sees that the "frequency function" $\hat{\Omega}(1,1')$ plays a role analogous to the Hartree-Fock energy of many-body physics and to the Vlasov field of plasma physics.

From (114a) and (117) and the defining relationship, (111), we find that this frequency function is the sum of a contribution

$$[\hat{\Omega}^B(1,1')](\mathbf{q}) = v_0[i\xi_\perp \cdot \mathbf{q} + \xi_3 \nabla_3][\delta(1-1') - n_0(x_3)\Phi(\xi) C(\mathbf{q}; x_3, x_3')] \quad (119a)$$

which is completely analogous to the operator that occurs in the collisionless (Vlasov) equation for a homogeneous fluid, and an additional, surface-related contribution

$$[\hat{\Omega}^S(1,1')](\mathbf{q}) = v_0[\nabla_0 \ln n_0(x_3)]\partial_3\delta(1-1') \\ + v_0\xi_3\Phi(\xi)[\nabla_3 n_0(x_3)]C(\mathbf{q}; x_3, x_3') \tag{119b}$$

Consequently, the collisionless kinetic equation, (118), can be written in the form

$$\{\partial_t + v_0(i\boldsymbol{\xi}_\perp\cdot\mathbf{q} + \xi_3\nabla_3) + v_0[\nabla_3 \ln n_0(x_3)]\partial_3\}\delta f(\mathbf{q}; x_3, \boldsymbol{\xi}, t) \\ = \Phi(\xi)[v_0(i\boldsymbol{\xi}_\perp\cdot\mathbf{q} + \xi_3\nabla_3)]\left[n_0(x_3)C(\mathbf{q}; x_3, \bar{x}_3)\delta f(\mathbf{q}; \bar{x}_3, \bar{\boldsymbol{\xi}}, t)\right] \\ - v_0\xi_3\Phi(\xi)[\nabla_3 n_0(x_3)]C(\mathbf{q}; x_3, \bar{x}_3)\delta f(\mathbf{q}; \bar{x}_3, \bar{\boldsymbol{\xi}}, t) \tag{120}$$

The first and second terms on the left-hand side of this equation form the usual streaming operator. The third can be written as the product $F(x_3)(\partial/\partial p_3)$, where $F(x_3) \equiv \beta^{-1}\nabla_3 \ln n_0(x_3) \leq 0$ is an effective force that acts in the $-x_3$ direction and is due solely to the presence of the interface. The first collection of terms on the right-hand side is precisely analogous to the usual mean-field contribution to the kinetic equation for a homogeneous fluid. The last term on the right-hand side is a mean-field contribution specific to the inhomogeneous, two-phase system.

It is well known that the collisionless kinetic equation for a homogeneous fluid is descriptive of a collective mode with the dispersion relation characteristic of high-frequency sound. In Section IV.D we demonstrate that the surface-specific parts of the mean-field frequency function appearing in (120) lead to the correct, nondissipative dispersion relation for capillary waves or surface ripples.

2. *Fokker-Planck Approximation*[71–74]

We next examine the adaptation to two-phase systems of Akcasu and Duderstadt's[84] theory of homogeneous fluids. In this theory the function $\Sigma^{(c)}$ is replaced with the approximation

$$\Sigma^{(c)}(1,1';t) \doteq e^{-t/\tau}\Sigma^{(c)}(1,1';0) \tag{121a}$$

wherein

$$\Sigma^{(c)}(1,1';0) \equiv -\Sigma_\infty(1,1') = -\langle \dot{f}|Q|\dot{f}\rangle(1,\bar{1})S_0^{-1}(\bar{1},1') \tag{121b}$$

This simple ansatz for the non-Markovian memory function is more sophisticated than it might at first appear. It can be viewed as a formula for interpolating between the low-frequency hydrodynamic regime, characterized by the relaxation time τ, and the high-frequency information represented by the sum rules that determine Σ_∞. The effect of this ansatz for $\Sigma^{(c)}$ is to represent with a single relaxation time the decays of all those processes that are not exclusively associated with the phase density $f(\mathbf{r}, \boldsymbol{\xi}, t)$. The mathematical approximation corresponding to this statement is our replacement of the many-body operator $Q\exp[-iQ\mathfrak{L}Q(t-t')]Q$ occurring in (112) with the projection $Q=1-P=1-|\delta f\rangle\langle\delta f|\delta f\rangle^{-1}\langle\delta f|$. Furthermore, by stipulating that τ be real valued, we chose to ignore possible resonance structures that could produce a frequency shift. Appropriate values for τ can be extracted from low-frequency, hydrodynamic information. It is sensible to choose for this relaxation time a density-dependent function $\tau=\tau(n_0(x_3))$ such as $\tau=\tau_l\theta(-x_3)+\tau_v\theta(x_3)$, where τ_l and τ_v are relaxation times characteristic of the coexisting liquid and vapor phases.

In contrast to the somewhat empirical nature of the relaxation time τ, the function Σ_∞ is fully determined by the three sum rules S_0, S_1, and S_2. Thus, since

$$\begin{aligned}\langle\dot{f}|Q|\dot{f}\rangle(1,\bar{1})&=\langle\dot{f}|\dot{f}\rangle(1,\bar{1})-\langle\dot{f}|P|\dot{f}\rangle(1,\bar{1})\\&=-\langle\ddot{f}|\delta f\rangle(1,\bar{1})-\langle\dot{f}|\delta f\rangle(1,\bar{\bar{1}})\langle\delta f|\delta f\rangle^{-1}(\bar{\bar{1}},\bar{\bar{1}}')\langle\delta f|\dot{f}\rangle(\bar{\bar{1}}',\bar{1})\\&=-\Big[S_2(1,\bar{1})+S_1(1,\bar{\bar{1}})S_0^{-1}(\bar{\bar{1}},\bar{\bar{1}}')S_1(\bar{\bar{1}}',\bar{1})\Big]\end{aligned}$$

it follows from (112) that

$$\Sigma_\infty(1,1')=-\Big[S_2(1,\bar{1})+S_1(1,\bar{\bar{1}})S_0^{-1}(\bar{\bar{1}},\bar{\bar{1}}')S_1(\bar{\bar{1}}',\bar{1})\Big]S_0^{-1}(\bar{1},1') \tag{122a}$$

Then, by using (114) and (117) we find that

$$\begin{aligned}\Sigma_\infty(1,1')=&-v_0^2\big[\nabla_3^2\ln n_0(x_3)\big]\partial_3(\partial_3+\xi_3)\delta(1-1')\\&-\int d\bar{x}_3\,\hat{\tau}_{ij}(\mathbf{q}=0;x_3,\bar{x}_3)\partial_j(\partial_i+\xi_i)\delta(1-1')-\hat{\tau}_{ij}(\mathbf{q};x_3,x_3')\xi_j\xi_i'\Phi(\xi)\\&+v_0^2n_0(x_3)\Phi(\xi)\big[i\xi_\perp\cdot\mathbf{q}+\xi_3\nabla_3\big]\big[-i\xi_\perp'\cdot\mathbf{q}+\xi_3'\nabla_3'\big]C(\mathbf{q};x_3,x_3')\end{aligned} \tag{122b}$$

wherein

$$\hat{\tau}_{ij}(\mathbf{q};x_3,x_3')=\frac{1}{mn_0(x_3')}\tau_{ij}(\mathbf{q};x_3,x_3') \tag{123}$$

The contributions arising from $\Sigma^{(c)}$ may, of course, be added to those from the mean-field term $\hat{\Omega}$ to obtain for δf a less restrictive but more complicated equation of motion than (120). In terms of the Laplace transform

$$\hat{G}_{ff}(\mathbf{q};1,1';z)\equiv\langle\delta f^*(\mathbf{q};1,t)\delta f(\mathbf{q};\bar{1},0)\rangle(z)S_0^{-1}(\bar{1},1') \tag{124}$$

of the phase space autocorrelation function this equation is of the form

$$\left\{z\,\delta(1-\bar{1})-\Sigma_1(\mathbf{q};1,\bar{1};z)-\Sigma_2(\mathbf{q};1,\bar{1};z)\right\}\hat{G}_{ff}(\mathbf{q};\bar{1},1';z) = -n_0(x_3)\Phi(\xi)\delta(1-1') \tag{125}$$

with

$$\Sigma_1(\mathbf{q};1,1';z) = \left\{ \begin{array}{c} v_0(\boldsymbol{\xi}_\perp\cdot\mathbf{q}-i\xi_3\nabla_3)-iv_0(\nabla_3\ln n_0(x_3))\partial_3 \\ -\dfrac{1}{z+i\tau^{-1}}\left[\begin{array}{c} v_0^2(\nabla_3^2\ln n_0(x_3))\partial_3(\partial_3+\xi_3) \\ +\int d\bar{x}_3\hat{\tau}_{ij}(\mathbf{q}=0;x_3,\bar{x}_3)\partial_j(\partial_i+\xi_i)\end{array}\right]\end{array}\right\}\delta(1-1') \tag{126a}$$

and

$$\begin{aligned}\Sigma_2(\mathbf{q};1,1';z) = {} & v_0[-\boldsymbol{\xi}_\perp\cdot\mathbf{q}+i\xi_3\nabla_3]\{n_0(x_3)C(\mathbf{q};x_3,x_3')\}\Phi(\xi) \\ & -iv_0\xi_3[\nabla_3\ln n_0(x_3)]C(\mathbf{q};x_3,x_3')\Phi(\xi) \\ & -\frac{1}{z+i\tau^{-1}}\{\hat{\tau}_{ij}(\mathbf{q};x_3,x_3')\xi_i'\xi_j-v_0^2n_0(x_3)[i\boldsymbol{\xi}_\perp\cdot\mathbf{q}+\xi_3\nabla_3] \\ & \times[-i\boldsymbol{\xi}_\perp'\cdot\mathbf{q}+\xi_3'\nabla_3']C(\mathbf{q};x_3,x_3')\}\Phi(\xi)\end{aligned} \tag{126b}$$

The first and second terms of Σ_1 and Σ_2, respectively, are the streaming and mean field contributions discussed in the preceding section. The remaining terms are concerned with collisional contributions to the self and distinct motions. Specifically, the sum of the last two terms of Σ_1 may be written in the form

$$\left\{\alpha_\perp\boldsymbol{\partial}_\perp\cdot(\boldsymbol{\partial}_\perp+\boldsymbol{\xi}_\perp)+\alpha_\parallel\partial_3(\partial_3+\xi_3)\right\}\delta(1-1') \tag{127a}$$

with

$$\alpha_{\perp} = -\frac{1}{z+i\tau^{-1}} \int d\bar{x}_3 \hat{\tau}_{\perp}(\mathbf{q}=0; x_3, \bar{x}_3) \tag{127b}$$

$$\alpha_{\parallel} = -\frac{1}{z+i\tau^{-1}} \left\{ \begin{aligned} &\int d\bar{x}_3 \hat{\tau}_{33}(\mathbf{q}=0; x_3, \bar{x}_3) \\ &+ v_0^2 [\nabla_3^2 \ln n_0(x_3)] \end{aligned} \right\} \tag{127c}$$

and where, because of the two-dimensional isotropy of the liquid–vapor system, $\hat{\tau}_{\perp} = \hat{\tau}_{11} = \hat{\tau}_{22}$. Thus these terms comprise an inhomogeneous Fokker-Planck operator with components that depend explicitly on the density gradient. The anisotropy of this operator implies that the motion of a test particle lying near the interface is correspondingly biased.

The factor $\nabla_3^2 \ln n_0(x_3)$ that contributes to $\alpha_{\parallel}$ is negative valued: for the van der Waals model of (108b) it equals $-(2/a)^2 \exp(2x_3/a)[1+\exp(2x_3/a)]^{-2}$.

3. *Modified BGK Model*

As a third and final example we replace the non-Markovian memory function with the simple approximation

$$\Sigma^{(c)}(1,1';t) \doteq \frac{1}{\tau} \delta(t) Q_K \delta(1-1') \tag{128}$$

where the positive-valued parameter τ is to be interpreted as a relaxation time similar to that in the Fokker-Planck model. The symbol Q_K denotes the complement of the projection operator

$$P_K = 1 - Q_K = \sum_{i=1}^{3} |i\rangle\langle i| \tag{129}$$

with $\langle 1| = 1, |1\rangle = \Phi(\xi)$ and $\langle i| = \xi_i, |i\rangle = \xi_i \Phi(\xi)$ for $i = 2,3$. The reason for including the projection operator Q_K in the definition of this approximate memory function is to ensure that is conserves mass and the components of momentum ξ_3 and $\xi_2 = \hat{q}\cdot\boldsymbol{\xi}$, which are, respectively, normal to the interface and parallel to the propagation vector $\mathbf{q}$.

This single-relaxation-time model is an obvious generalization of the BGK[75] model, which frequently is used in the kinetic theory of gases. It is a model that invariably leads to physically sensible conclusions and, in addition, produces a theory of remarkable mathematical simplicity. Thus it is a valuable tool to use in probing unfamiliar situations.

The kinetic equation appropriate to this model may be written as

$$\left\{z\delta(1-\bar{1})-\Omega(1,\bar{1})+i\tau^{-1}Q_K\delta(1-\bar{1})\right\}\delta f(\bar{1},z)=\delta f(1,t=0) \quad (130)$$

where $\Omega=-i\hat{\Omega}$ is the sum of the singular part

$$\Omega_1(1,1')=\left\{v_0(\xi_2 q-i\xi_3\nabla_3)-iv_0[\nabla_3 \ln n_0(x_3)]\partial_3\right\}\delta(1-1') \quad (131a)$$

which characterizes the self motion, and a second, mean-field term

$$\begin{aligned}\Omega_2(1,1')&=-v_0 n_0(x_3)\Phi(\xi)(\xi_2 q-i\xi_3\nabla_3)C(\mathbf{q};x_3,x_3')\\&=-v_0 q n_0 C|2\rangle\langle 1|+iv_0 n_0(\nabla_3 C)|3\rangle\langle 1|\end{aligned} \quad (131b)$$

which is descriptive of the collective, distinct-particle response. [Here and in the remainder of this section we suppress the phase space coordinates, some of which occur as dummy variables of integration.]

We now multiply (130) by $\Gamma_S\equiv(z-\Omega_1+i\tau^{-1})^{-1}$ to obtain

$$\begin{aligned}&\delta f(z)+v_0 q\Gamma_S n_0 C|2\rangle\langle 1|\delta f(z)-iv_0 n_0\Gamma_S(\nabla_3 C)|3\rangle\langle 1|\delta f(z)\\&\quad -i\tau^{-1}\sum_{j=1}^{3}\Gamma_S|j\rangle\langle j|\delta f(z)=-\Gamma_S\delta f(t=0)\end{aligned} \quad (132)$$

This equation illustrates the remarkable simplifications that result from using the (modified) BGK model. Thus, by multiplying (132) by each of $\langle 1|,\langle 2|$, and $\langle 3|$, we obtain a closed set of three linear inhomogeneous algebraic equations for the functions $\langle i|\delta f(z), i=1,2,3$. The remaining part of δf, namely, $Q_K\delta f$, is then completely determined by these three lowest moments. The difficulty comes with the evaluation of the matrix elements of Γ_S. Approximations can be based on the power series $\Gamma_S=-i\tau-(i\tau)^2(z-\Omega_1)-\ldots$ or the alternative expansion $\Gamma_S=(z+i\tau^{-1})^{-1}+(z+i\tau^{-1})^{-2}\Omega_1+\ldots$. By using the latter it is possible to write (132) in the equivalent form,

$$\begin{aligned}&i\partial_t\langle i|\delta f(t)+v_0 n_0[qC\delta_{i2}-i(\nabla_3 C)\delta_{i3}]\langle 1|\delta f(t)\\&\quad+\sum_{l\geq 1} v_0[-iq\langle i|\Theta_l|2\rangle n_0 C-\langle i|\Theta_l|3\rangle n_0(\nabla_3 C)]\int_0^t ds\, e^{-s/\tau}s^{l-1}\langle 1|\delta f(t-s)\\&\quad+\sum_{l\geq 1}\frac{1}{\tau}\sum_{j=1}^{3}\langle i|\Theta_l|j\rangle\int_0^t ds\, e^{-s/\tau}s^{l-1}\langle j|\delta f(t-s)\\&\quad+\sum_{l\geq 1}\sum_{k\geq 1}\langle i|\Theta_l|k\rangle e^{-t/\tau}t^{k-1}\langle k|\delta f(0)=0\end{aligned} \quad (133)$$

where $\Theta_l \equiv \Omega_1^l / i^{l-1}(l-1)!$. The extended basis of bras and kets appearing in this equation is assumed to satisfy the orthonormality condition $\langle i|j\rangle = \delta_{ij}$ and so may be identified with properly normalized modifications of the generalized Hermite polynomials described in Appendix B.

D. Dispersion of Surface Waves

The results obtained in sections IV.B and IV.C provide practical means for carrying out the program outlined in section IV.A. Thus the simple projection operator defined by (104) can be used to extract from any one of the kinetic equations of section IV.C an approximate dynamics for the collective variables associated with the liquid–vapor interface. The variables that concern us most are a few of the low-order momentum moments of the phase space fluctuation δf, such as the density fluctuation δn and the momentum density $\mathbf{g}$. These and other momentum moments of δf can be dealt with systematically in terms of the generalized Hermite polynomials,

$$\psi^{(n)}_{i_1 i_2 \dots i_n}(\xi) = (-)^n \frac{1}{\Phi(\xi)} \partial_{i_1} \dots \partial_{i_2} \Phi(\xi) \tag{134}$$

whose properties are considered in Appendix B. Thus, for example,

$$\delta n(\mathbf{q}; x_3, t) = \int d^3\xi \psi^{(0)}(\xi) \delta f(\mathbf{q}; x_3, \xi, t) \tag{135a}$$

and

$$g_i(\mathbf{q}; x_3, t) = m v_0 \int d^3\xi \psi_i^{(1)}(\xi) \delta f(\mathbf{q}; x_3, \xi, t) \tag{135b}$$

Each of the kinetic equations of Section IV.C can be transformed into a set of equations for the momentum moments and these, in turn, when acted upon by the projection operator P_S, yield sets of equations for the surface variables. The construction and solution of these coupled equations is a major undertaking that has yet to be completed for any of the approximate kinetic equations. Our objective here is the much more modest one of deriving from kinetic theory the nondissipative dispersion relation characteristic of capillary waves. In Ref. 13 we accomplished this by considering the set of moment equations associated with the collisionless Vlasov equation. Here, instead, we begin with the moment equations, (133), based on the modified BGK model. As previously noted, the three lowest order members of this set form a closed system of equations for $\delta n = \langle 1|\delta f$, $g_2 = m v_0 \langle 2|\delta f$ and $g_3 = m v_0 \langle 3|\delta f$. To obtain the nondissipative forms of these three equations we consider the limit $\tau \to 0$, in which case $\tau^{-1}\exp(-t/\tau) \to \delta(t)$. Consequently, the integral occurring on the second line of (133) vanishes in this limit, as do the terms involving the initial value of δf. The final results are the equations

$$\partial_t \delta n(\mathbf{q}; x_3, t) + i\frac{q}{m} g_2(\mathbf{q}; x_3, t) + \frac{1}{m}\nabla_3 g_3(\mathbf{q}; x_3, t) = 0 \qquad (136a)$$

$$\partial_t g_2(\mathbf{q}; x_3, t) + i\beta^{-1} q\left[\delta(x_3 - \bar{x}_3) - n_0(x_3) C(\mathbf{q}; x_3, \bar{x}_3)\right]\delta n(\mathbf{q}; \bar{x}_3, t) = 0 \qquad (136b)$$

$$\partial_t g_3(\mathbf{q}; x_3, t) + \beta^{-1}\{\nabla_3 - [\nabla_3 \ln n_0(x_3)]\} \left[\delta(x_3 - \bar{x}_3) - n_0(x_3) C(\mathbf{q}; x_3, \bar{x}_3)\right]\delta n(\mathbf{q}, \bar{x}_3, t) = 0 \qquad (136c)$$

In analyzing the propagating (high-frequency sound) modes characteristic of the bulk phases one must treat the three components of momentum on equal footings, since each is associated with a separate constant of the motion. The surface (capillary wave) modes, which owe their existence to the breakdown of translational symmetry, differ fundamentally from these acoustic disturbances and we can show that for these the tangential components of momentum are of little consequence compared to g_3, the component normal to the interface. Thus, by applying the surface projection operator defined by (104), we obtain from (136a) and (136c)

$$\partial_t \delta n^{(S)}(\mathbf{q}, t) + i\frac{q}{m} g_2^{(S)}(\mathbf{q}, t) + \frac{1}{2\varepsilon m}\left[g_3(\mathbf{q}; \varepsilon, t) - g_3(\mathbf{q}; -\varepsilon, t)\right] = 0 \qquad (137a)$$

and

$$\partial_t g_3^{(S)}(\mathbf{q}, t) + \frac{\beta^{-1}}{2\varepsilon m}\int_{-\varepsilon}^{\varepsilon} dx_3 n_0(x_3)\nabla_3\left[K(\mathbf{q}; x_3, \bar{x}_3)\delta n(\mathbf{q}; \bar{x}_3, t)\right]$$

$$- \frac{\beta^{-1}}{2\varepsilon m}\int_{-\varepsilon}^{\varepsilon} dx_3 \left[\nabla_3 n_0(x_3)\right] K(\mathbf{q}; x_3, \bar{x}_3)\delta n(\mathbf{q}; \bar{x}_3, t) = 0 \qquad (137c)$$

respectively. Here use has been made of the relationship (116d) between C and K. By comparing the present notation to that of Section III [of (87)] we conclude that $\delta n^{(S)}$ and $g_3^{(S)}$ are identical to $(n_l - n_v)\zeta/2\varepsilon$ and $\phi/2\varepsilon$, respectively.

The second term of (137a) is proportional to the surface wavenumber q and so can be ignored in the long-wavelength, hydrodynamic limit. The integrand of the third term on the right-hand side of (137c) includes the almost singular, surface-specific factor $\nabla_3 n_0(x_3)$, which is lacking from the second. Consequently, for thin transition zones the second term is negligible in comparison to the third. Thus, for long wavelengths and small interfacial thicknesses, the second terms can be discarded from both of these equations.

What we must do next is eliminate the coupling between the surface and bulk phase variables from (137c). This is done by inserting a factor of $1 = P_S + Q_S$ into the integrand of the third term, between the functions $K(\mathbf{q}; x_3, \bar{x}_3)$ and $\delta n(\bar{x}_3)$, and then discarding the projection $Q_S = P_B$. The result of this manipulation is the replacement of the integral occurring in the third term of (137c) with

$$-\frac{1}{n_l - n_v}\left\{\int dx_3 \int dx_3' [\nabla_3 n_0(x_3)] K(\mathbf{q}; x_3, x_3')[\nabla_3' n_0(x_3')]\right\}\delta n^{(S)}(\mathbf{q}, t)$$

which, according to (20) and (22), is equal to $-(n_l - n_v)^{-1}\beta q^2 \sigma_D + 0(q^4)$. Consequently, in the long-wavelength limit we obtain in place of (137a) and (137c),

$$\partial_t \delta n^{(S)}(\mathbf{q}, t) + \frac{1}{2\varepsilon m}\left[g_3(\mathbf{q}; \varepsilon, t) - g_3(\mathbf{q}; -\varepsilon, t)\right] = 0 \qquad (137a')$$

and

$$\partial_t g_3^{(S)}(\mathbf{q}, t) + \frac{\sigma_D q^2}{n_l - n_v}\delta n^{(S)}(\mathbf{q}, t) = 0 \qquad (137b')$$

respectively. Attention should be drawn to the fact that these special forms of (137a) and (137c) are valid only in the limit of a thin interfacial zone and in the absence of coupling between the bulk and surface phases. In Ref. 12 [see (42) to (44)] we used quasihydrodynamic arguments to show that

$$\frac{1}{2\varepsilon}\left[g_3(\mathbf{q}; \varepsilon, t) - g_3(\mathbf{q}; -\varepsilon, t)\right] \approx -q g_3^{(S)}(\mathbf{q}, t) \qquad (138)$$

This reduces (137a') to

$$\partial_t \delta n^{(S)}(\mathbf{q}, t) \approx \frac{q}{m} g_3^{(S)}(\mathbf{q}, t) \qquad (139)$$

which, when combined with (137c′), produces for $g_3^{(S)}$ the equation of motion

$$\partial_t^2 g_3^{(S)} = -\frac{q^2\sigma_D}{n_l - n_v}\partial_t \delta_n^{(S)} \approx -\frac{\sigma_D q^3}{m(n_l - n_v)} g_3^{(S)} \tag{140}$$

The sinusoidal solution $[g_3^{(S)}(\mathbf{q}, t) = g_3^{(S)}(\mathbf{q}, 0)\exp(i\omega t)]$ of this equation then leads to the dispersion relation $\omega^2 = \sigma_D q^3/m(n_l - n_v)$, which is identical to that for surface ripples.

The one unsatisfactory feature of the last few lines of this analysis is its reliance on a quasihydrodynamic derivation of (138) and, hence, of the essential relationship (139). To repair the situation we begin with the two sum rules

$$\langle \delta\dot{n}\delta\dot{n}\rangle = (m\beta)^{-1}(q^2 + \nabla_3\nabla_3')[n_0(x_3)\delta(x_3 - x_3')]$$

and

$$\langle g_3 g_3\rangle = m\beta^{-1} n_0(x_3)\delta(x_3 - x_3')$$

By applying the surface projection operator P_S to both of the spatial arguments occurring in these sum rules we find that

$$\langle \delta\dot{n}^{(S)}\delta\dot{n}^{(S)}\rangle(\mathbf{q}) = \frac{q^2}{m^2}\langle g_3^{(S)} g_3^{(S)}\rangle(\mathbf{q}) \tag{141}$$

This implies that $|\delta\dot{n}^{(S)}|$ and $|g_3^{(S)}|$ are very nearly linearly dependent and so differ at most by a constant, complex-valued proportionality factor. Therefore, the Schwartz inequality

$$\langle \delta\dot{n}^{(S)}\,\delta\dot{n}^{(S)}\rangle\langle g_3^{(S)} g_3^{(S)}\rangle \geq |\langle \delta\dot{n}^{(S)}\, g_3^{(S)}\rangle|^2$$

can be treated as an approximate equality, which, when combined with (141), leads to the conclusion that

$$\langle \delta\dot{n}^{(S)}\, g_3^{(S)}\rangle(\mathbf{q}) \approx \frac{q}{m}\langle g_3^{(S)} g_3^{(S)}\rangle(\mathbf{q})$$

This, in turn, implies that

$$\partial_t[\delta n^{(S)}(\mathbf{q}, t)]_{t=0} \approx \frac{q}{m} g_3^{(S)}(\mathbf{q}, 0)$$

Thus, since (139) is satisfied at $t=0$, it is reasonable to assume that it remains approximately valid for times different from zero.

This completes the modest program that we set for ourselves at the beginning of this section. It has been demonstrated above that the surface projection operators and approximate kinetic equations constructed in Sections IV.B and IV.C do, indeed, provide practical means for conducting a detailed analysis of interfacial dynamics. Our considerations here, to be sure, have been limited to the nondissipative behavior of an isolated interface, but even this special and uncomplicated case is not without interest and importance. Thus, it has been clearly demonstrated that our techniques actually do permit us to separate the interfacial region from the bulk phases and so enable us to investigate the intrinsic properties of this zone. It is in this context that we encountered the dispersion relation $\omega^2=(\sigma/\rho)q^3$ for capillary waves. Its derivation was dependent on the neglect of all dissipative processes, including those specific to the transition zone itself, and so having no direct connection to the usual bulk–surface viscous interactions on which the hydrodynamic theory of surface wave damping is based. Although we have not included an analyses of this and other dissipative processes here, the tools assembled in this chapter and its appendices appear to be sufficient to the task.

Appendix A: Correlation Function Formalism

The purpose of this appendix is to provide a compendium of definitions and properties related to the correlation functions appearing in the text.

The symbol A_i is used to denote a real-valued dynamic variable that, because of its dependence on the coordinates and momenta of a classical many-body system, is an implicit function of time. A_i may depend, additionally, on a position vector $\mathbf{r}$. An example of $A_i(\mathbf{r},t)$ is the momentum density $\Sigma_\alpha \mathbf{p}^\alpha(t)\delta[\mathbf{r}-\mathbf{r}^\alpha(t)]$, with $\mathbf{r}^\alpha$ and $\mathbf{p}^\alpha$ denoting the position and momentum of molecule α. Related to this is the total momentum, $A_i(t)\equiv \int d^3r A_i(\mathbf{r},t)=\Sigma_\alpha \mathbf{p}^\alpha(t)$. In what follows, the position or time dependence of a dynamic variable may be suppressed if it is irrelevant to the issue being considered.

The time derivative of a dynamical variable is given by the expression $\dot{A}_i=[A_i,H]\equiv i\mathfrak{L}A_i$ with $[a,b]$ the Poisson bracket of two variables a and b, H the Hamiltonian of the system, and $\mathfrak{L}$ the associated (and formally self -adjoint) Liouville operator. Thus $A_i(t)=\exp(i\mathfrak{L}t)A_i$ wherein $A_i=A_i(t=0)$.

I. Correlation Functions, Transforms, and Sum Rules[85]

The correlation function of two dynamic variables A_i and A_j is the object

$$G_{ij}(\mathbf{r}, t; \mathbf{r}', t') \equiv \langle A_i(\mathbf{r}, t) A_j(\mathbf{r}', t') \rangle \tag{A.1}$$

where by $\langle X \rangle = \mathrm{Tr}\{X\rho_{\mathrm{eq}}\} = \int d\Omega\, X(\Omega) e^{-\beta H} / \int d\Omega e^{-\beta H}$ we indicate the average of the phase function X over an equilibrium ensemble. Here Ω denotes a point in the phase space of the system and $\beta = 1/k_B T$.

When the equilibrium state is isotropic, G_{ij} is translationally invariant and so depends only on the relative values of its spatial arguments $\mathbf{r}$ and $\mathbf{r}'$, that is, $G_{ij}(\mathbf{r}, t; \mathbf{r}', t') \to G_{ij}(\mathbf{r}-\mathbf{r}'; t, t')$. Under these circumstances it often proves useful to introduce the spatial Fourier transform

$$G_{ij}(\mathbf{k}) \equiv \langle A_i(\mathbf{r}) A_j(\mathbf{r}') \rangle (\mathbf{k}) \equiv \int d^3(r-r') e^{-i\mathbf{k}\cdot(\mathbf{r}-\mathbf{r}')} \langle A_i(\mathbf{r}-\mathbf{r}') A_j(\mathbf{0}) \rangle \tag{A.2}$$

Similarly, because of the stationarity of the equilibrium ensemble, G_{ij} does not depend separately on the time arguments of the variables A_i and A_j, but only on their difference. Therefore, the Laplace transform of $G_{ij}(t)$ can be written in any of the equivalent forms

$$G_{ij}(z) \equiv i \int_0^\infty dt\, e^{izt} \langle A_i(t) A_j \rangle \equiv \langle A_i(t) A_j \rangle (z) = \langle A_i A_j(-t) \rangle (z)$$

$$= \langle A_i | e^{-i\mathfrak{L}t} | A_j \rangle (z) = -\langle A_i | (z - \mathfrak{L})^{-1} | A_j \rangle \tag{A.3}$$

where $\mathrm{Im}\, z > 0$. The last two of these expressions make use of the inner product defined by $\langle A | B \rangle \equiv \langle A^* B \rangle$. Combining the notations of (A.2) and (A.3) we obtain the Fourier-Laplace transform

$$G_{ij}(\mathbf{k}, z) \equiv i \int_0^\infty dt \int d^3 r\, G_{ij}(\mathbf{r}, t) \exp[-i(\mathbf{k}\cdot\mathbf{r} - zt)] \tag{A.4}$$

Let us consider now several results specific to the time dependence of correlation functions. First of all, it easily is established that the Laplace and Fourier transforms $G_{ij}(z) = i\int_0^\infty dt\, e^{izt} G_{ij}(t)$ and $G_{ij}(\omega) = \int_{-\infty}^{+\infty} dt\, e^{i\omega t} G_{ij}(t)$ are related to one another in the manner

$$G_{ij}(z) = \int_{-\infty}^{+\infty} \frac{d\omega}{2\pi} \frac{G_{ij}(\omega)}{\omega - z} \tag{A.5}$$

The inverse of this is obtained by setting z equal to $\omega' + i0$ and using the relationship $(x \pm i0)^{-1} = \mathcal{P}(1/x) \mp i\pi\delta(x)$, where $\mathcal{P}$ denotes the principal value. This results in the relationship

$$G_{ij}(\omega) = 2\left[\operatorname{Im} G_{ij}(z)\right]_{z=\omega+i0} \tag{A.6}$$

Expanding the integrand factor of (A.5) in the manner $(\omega - z)^{-1} = -z^{-1}\Sigma_{n \geq 0}(\omega/z)^n$ leads to the expression

$$G_{ij}(z) = -\sum_{n \geq 0} \frac{\langle \omega^n \rangle_{ij}}{z^{n+1}} \tag{A.7}$$

where

$$\langle \omega^n \rangle_{ij} \equiv \int_{-\infty}^{+\infty} \frac{d\omega}{2\pi} \omega^n G_{ij}(\omega) \tag{A.8}$$

is the nth moment of the Fourier transform $G_{ij}(\omega)$. These moments are related to the derivatives of the time correlation function $G_{ij}(t) = \int_{-\infty}^{+\infty} (d\omega/2\pi) e^{-i\omega t} G_{ij}(\omega)$ by the formulas

$$\langle \omega^n \rangle_{ij} = i^n \{\partial_t^n G_{ij}(t)\}_{t=0} \tag{A.9}$$

To obtain alternative expressions for these moments or "sum rules" we note that $\langle \dot{A}B \rangle = \langle [A, H]B \rangle = \mathrm{Tr}\{[A, H]B\rho_{\mathrm{eq}}\}$, where $[A, H]\rho_{\mathrm{eq}} = -\beta^{-1}[A, \rho_{\mathrm{eq}}]$. Consequently, $[A, H]B\rho_{\mathrm{eq}} = -\beta^{-1}\{[A, B\rho_{\mathrm{eq}}] - [A, B]\rho_{\mathrm{eq}}\}$ and[60]

$$\langle \dot{A}B \rangle = \beta^{-1} \langle [A, B] \rangle; \qquad \beta = \frac{1}{k_B T} \tag{A.10}$$

From this very useful relationship it then follows that

$$\begin{aligned} \langle \omega^n \rangle_{ij} &= i^n \langle \{\partial_t^n A_i\}_{t=0} A_j \rangle = i^n \beta^{-1} \langle [\{\partial_t^{n-1} A_i\}_{t=0}, A_j] \rangle \\ &= i^n \beta^{-1} \langle \underbrace{[[..[A_i, H], H, .., H]}_{n-1 \text{ fold}}, A_j] \rangle \end{aligned} \tag{A.11}$$

II. Equations of Motion for Correlation Functions

To derive the equation of motion for G_{ij} we introduce the projection operator $P = 1 - Q \equiv |A_i\rangle \chi_{ij}^{-1} \langle A_j|$, where $\chi_{ij} = \langle A_i | A_j \rangle$. Here and henceforth the summation convention is adopted for repeated indices. Next, by applying the identity $(A\mp B)^{-1}=A^{-1}\pm A^{-1}B(A\mp B)^{-1}$ to the last expression in (A.3) we conclude that

$$G_{ij}(z)=-\langle A_i|(z-\mathfrak{L}[P+Q])^{-1}|A_j\rangle$$
$$=-\langle A_i|(z-\mathfrak{L}Q)^{-1}|A_j\rangle-\langle A_i|(z-\mathfrak{L}Q)^{-1}\mathfrak{L}P(z-\mathfrak{L})^{-1}|A_j\rangle \tag{A.12}$$

Because $Q|A_j\rangle=(1-P)|A_j\rangle=0$, the first of these two terms becomes

$$-\langle A_i|(z-\mathfrak{L}Q)^{-1}|A_j\rangle=-\langle A_i|z^{-1}\left(1-\frac{1}{z}\mathfrak{L}Q\right)^{-1}|A_j\rangle$$
$$=-\frac{1}{z}\langle A_i|\left\{1+\frac{1}{z}\mathfrak{L}Q+\cdots+(\;)Q\right\}|A_j\rangle$$
$$=-\frac{1}{z}\chi_{ij}$$

The second term on the right-hand side of (A.12) can be written as

$$-\langle A_i|(z-\mathfrak{L}Q)^{-1}\mathfrak{L}|A_k\rangle\chi_{kl}^{-1}\langle A_l|(z-\mathfrak{L})^{-1}|A_j\rangle$$
$$=\langle A_i|\frac{1}{z}(z-\mathfrak{L}Q+\mathfrak{L}Q)(z-\mathfrak{L}Q)^{-1}\mathfrak{L}|A_k\rangle\chi_{kl}^{-1}G_{lj}(z)$$
$$\equiv\frac{1}{z}\Sigma_{il}(z)G_{lj}(z)$$

with

$$\Sigma_{il}(z)=\Omega_{il}+\Sigma_{il}^{(c)} \tag{A.13}$$

and where

$$\Omega_{il}\equiv\langle A_i|\mathfrak{L}|A_k\rangle\chi_{kl}^{-1}=i\langle\dot{A}_i|A_k\rangle\chi_{kl}^{-1}=-i\langle A_i|\dot{A}_k\rangle\chi_{kl}^{-1} \tag{A.14}$$

$$\Sigma_{il}^{(c)}\equiv\langle A_i|\mathfrak{L}Q(z-\mathfrak{L}Q)^{-1}\mathfrak{L}|A_k\rangle\chi_{kl}^{-1}=\langle\dot{A}_i|Q(z-\mathfrak{L}Q)^{-1}Q|\dot{A}_k\rangle\chi_{kl}^{-1} \tag{A.15}$$

By combining these results we obtain the equations of motion

$$(z\delta_{il}-\Sigma_{il})G_{lj}=-\chi_{ij} \tag{A.16}$$

for the correlation functions $G_{ij}(z)$. These are the Laplace transforms of the equations

$$\partial_t G_{ij}(t)+\hat{\Omega}_{il}G_{lj}(t)-\int_0^t ds\,\Sigma_{il}^{(c)}(t-s)G_{lj}(s)=0 \tag{A.17}$$

with $\hat{\Omega}_{ij}=i\Omega_{ij}$ and

$$\Sigma_{ij}^{(c)}(t)=-\langle \dot{A}_i|Qe^{-iQ\mathcal{L}Qt}Q|\dot{A}_k\rangle\chi_{kj}^{-1} \tag{A.18}$$

From (A.17) and the definition $G_{ij}(t)=\langle A_i(t)A_j\rangle$ it follows that

$$\partial_t A_i(t)+\hat{\Omega}_{il}A_l(t)-\int_0^t ds\,\Sigma_{il}^{(c)}(t-s)A_l(s)+\hat{F}_i(t)=0 \tag{A.19}$$

provided that

$$\langle \hat{F}_i(t)A_j\rangle=0, \qquad \forall i,j \tag{A.20}$$

It can be shown[55] that the "fluctuating force" $\hat{F}(t)$ appearing in this generalized Langevin equation is related to the non-Markovian memory function $\Sigma^{(c)}$ by the fluctuation–dissipation theorem

$$\Sigma_{ij}^{(c)}(t)=\langle \hat{F}_i(t)\hat{F}_k\rangle\chi_{kj}^{-1} \tag{A.21}$$

All the relationships (A.12) to (A.21) can be applied to dynamic variables that depend on position as well as time, provided that $G_{ij}(z)$, $G_{ij}(t)$, and χ_{ij} are replaced with $G_{ij}(\mathbf{k},z)$, $G_{ij}(\mathbf{k},t)$, and $\chi_{ij}(\mathbf{k})=\langle A_i|A_j\rangle(\mathbf{k})$, respectively. Furthermore, the extension of the theory from discrete indices to continuous variables can be accomplished without difficulty. It is by doing this that one obtains the kinetic equation (98) of Section IV.

III. Dyson-Like Equations and Non-Markovian Memory Functions

Equations (A.16) can be cast into the somewhat simpler forms

$$(z1-\Sigma)\hat{G}=-1; \qquad \Sigma=\Omega+\Sigma^{(c)} \tag{A.22}$$

by introducing the matrix $\hat{G}\equiv G\chi^{-1}$ of "renormalized" correlation functions. Equation (A.22) then can be rewritten in the form $(G_0^{-1}+\Sigma^{(c)})\hat{G}=1$ or as the integral equation

$$\hat{G}=G_0+G_0(-\Sigma^{(c)})\hat{G} \tag{A.23}$$

where G_0 is the matrix satisfying the equation $(z1-\Omega)G_0=-1$. The Dyson -like[86] equation (A.23) can be represented diagrammatically by

$$\begin{aligned} \Rightarrow &= \rightarrow + \rightarrow\Box\Rightarrow \\ &= \rightarrow + \rightarrow\Box\rightarrow + \rightarrow\Box\rightarrow\Box\rightarrow + \cdots \end{aligned} \quad (A.23')$$

with the symbols $\Rightarrow$, $\rightarrow$, and $\Box$ in place of $\hat{G}$, G_0, and $-\Sigma^{(c)}$, respectively.

So far as we are aware, Dyson-like equations of this sort first were introduced into mathematical physics by Ornstein and Zernike.[83] Since the original work by those authors is relevant to some parts of this chapter we briefly outline the Ornstein-Zernike theory in such a way as to emphasize its similarity to (A.23). The pair correlation function $h(\mathbf{r},\mathbf{r}')\equiv n_2(\mathbf{r},\mathbf{r}')/n_0(x_3)n_0(x_3')-1$ may be relatively long-ranged, whereas the direct correlation function $C(\mathbf{r},\mathbf{r}')$ is a short-ranged function.[87] The two are connected by the integral relationship

$$h(\mathbf{r},\mathbf{r}')=C(\mathbf{r},\mathbf{r}')+\int d^3\bar{r}\,C(\mathbf{r},\bar{\mathbf{r}})n_0(\bar{x}_3)h(\bar{\mathbf{r}},\mathbf{r}') \quad (A.24)$$

which can be represented diagrammatically by (A.23′) provided that $\Rightarrow$, $\rightarrow$, and $\Box$ are identified with h, C, and n_0, respectively.

Let us return now to (A.23) and the associated equation, $(z1-\Omega)G_0=-1$, for the bare propagator G_0. Because the "frequency matrix" Ω is independent of z, it can be identified with the instantaneous or mean-field response of the system: the complex collisional dynamics of the many-body system are recorded by the frequency-dependent non-Markovian memory or collisional matrix $\Sigma^{(c)}$. The calculation of $\Sigma^{(c)}$ generally is impossible, except for a few ideal cases, such as harmonic oscillator chains.[53]

Systematic perturbation theories for $\Sigma^{(c)}$ have been described by Kadanoff and Baym.[54] For example, $\Sigma^{(c)}$ can be expanded in terms of G_0 (traditional [bare] perturbation theory) or in terms of $\hat{G}$ (renormalized perturbation theory[88, 89]). An alternative procedure is to adopt for $\Sigma^{(c)}$ the approximate model that originally was proposed by Akcasu and Duderstadt[84] and then subsequently modified by Jhon and Forster.[74] The prescription for $\Sigma^{(c)}$ given by this model can be viewed as an interpolation ansatz between the hydrodynamic (low-frequency) and high-frequency limits. The first few sum rules provide important and exact inputs because they characterize the high-frequency behavior of the system. Thus $\Sigma^{(c)}$, defined by (A.15), can be expanded in powers of z^{-1} to obtain the high-frequency limit

$$\begin{aligned} \Sigma_{ij}^{\infty} &= \lim_{z\to\infty}\{z\Sigma_{ij}^{(c)}\}=\langle\dot{A}_i|Q|\dot{A}_k\rangle\chi_{kj}^{-1} \\ &= \{\langle\dot{A}_i|\dot{A}_k\rangle-\langle\dot{A}_i|A_m\rangle\chi_{mn}^{-1}\langle A_n|\dot{A}_k\rangle\}\chi_{kj}^{-1} \end{aligned} \quad (A.25)$$

the value of which is completely determined by the static inputs contained within the three lowest order members of the set of sum rules, (A.9) or (A.11).

The collision matrix then can be equated to the product,

$$\Sigma_{ij}^{(c)}(z)=\Sigma_{ik}^{\infty}\hat{\mathfrak{F}}_{kj}(z) \tag{A.26}$$

of the quantity Σ_{ik}^{∞} defined by (A.25) and a function $\hat{\mathfrak{F}}_{kj}(z)$, whose high-frequency limit is equal to $z^{-1}\delta_{kj}$. According to the single-relaxation-time ansatz of Akcasu and Duderstadt, $\hat{\mathfrak{F}}_{kj}(z)$ is to be replaced with $(z+i\tau^{-1})^{-1}\delta_{kj}$, where τ denotes a relaxation time whose value is determined by low-frequency, hydrodynamic information. Thus the interpolation mentioned above for the memory function is the approximation,

$$\Sigma_{ij}^{(c)}(z)\doteq\frac{1}{z+i\tau^{-1}}\Sigma_{ij}^{\infty} \tag{A.27}$$

the continuous variable analogue of which is (121) of Section IV.C.

In conclusion, let use briefly consider the stability condition that must be satisfied by the approximate memory function defined by (A.27). The second law of thermodynamics dictates that the imaginary part of this function, namely,

$$\left[\operatorname{Im}\Sigma_{ij}^{(c)}\right]_{z=\omega+i0}=\frac{-\tau}{\omega^2+\tau^2}\Sigma_{ij}^{\infty}$$

must be negative semidefinite. Therefore, because Σ_{ij}^{∞}, defined by (A.25), is positive valued, we conclude that the value of the relaxation time must be positive. The negative semidefinite property of $\operatorname{Im}\Sigma_{ij}^{(c)}$ is a consequence of the irreversibility of spontaneous processes. In contrast to this behavior of the approximate memory function, the imaginary part of the exact collision matrix [defined by (A.15)] is identically zero. Thus, irreversibility is introduced by the coarse-graining procedure, which discards some of the fine details of the many-body dynamics; the dynamic information lost by the ansatz equation (A.27) leads to the production of entropy.

Appendix B: Generalized Hermite Polynomials and the Solution of Fokker-Planck-like Equations

1. *Generalized Hermite Polynomials*

In 1949 H. Grad[90] invented generalized, n-dimensional Hermite polynomials and used them to conduct a (velocity) moment analysis of the

Boltzmann equation of gas kinetic theory. Here we generate these polynomials by a procedure that is identical to the second quantization formalism for Boson systems, our purpose being to establish a bridge between these polynomials and the problem of solving the Fokker-Planck-like equation encountered in Section IV.C.2 of the text.

We define the scalar product of two real-valued functions ϕ and ψ by the integral[74]

$$\langle \phi(\xi) | M(\xi,\xi') | \psi(\xi') \rangle \equiv \int d^d\xi \int d^d\xi' \phi(\xi) M(\xi,\xi') \Phi(\xi') \psi(\xi') \tag{B.1}$$

where $\Phi(\xi) = (2\pi)^{-d/2} \exp(-\frac{1}{2}\xi\cdot\xi)$ is a d-dimensional Maxwell-Boltzmann distribution function. When $M(\xi,\xi') = M(\xi)\delta(\xi-\xi')$ this integral becomes

$$\langle \phi(\xi) | M(\xi) | \psi(\xi) \rangle \equiv \int d^d\xi \, \phi(\xi) M(\xi) \Phi(\xi) \psi(\xi) \tag{B.2}$$

The bras and kets occurring in these formulas are defined (unsymmetrically) by $\langle \phi(\xi)| = \phi(\xi)$ and $|\psi(\xi)\rangle = \Phi(\xi)\psi(\xi)$: their product is understood to involve integration over the domain of the dimensionless momentum vector ξ.

Let us now introduce the ladder operators

$$a_i = \partial_i + \xi_i, \qquad a_i^\dagger = -\partial_i; \qquad i = 1,\ldots,d \tag{B.3}$$

with $\partial_i \equiv \partial/\partial\xi_i$. This notation reflects the fact that a_i and $a_i^\dagger$ are adjoint operators with respect to the scalar product defined by (B.2), in the sense that $\langle \phi | a_i | \psi \rangle = \langle \psi | a_i^\dagger | \phi \rangle$. Furthermore, these operators satisfy the Boson commutator relationships[91]

$$[a_i, a_j] = [a_i^\dagger, a_j^\dagger] = 0 \tag{B.4a}$$

$$[a_i, a_j^\dagger] = \delta_{ij} \tag{B.4b}$$

and finally, the vacuum state, defined by

$$a_i|0\rangle = 0, \qquad \langle 0| a_i^\dagger = 0 \qquad \forall i \tag{B.4c}$$

corresponds to the function $\psi^{(0)}(\xi) = 1$.

Consider now the basis functions $|n; i_1 \ldots i_n\rangle \equiv a_{i_1}^\dagger \ldots a_{i_n}^\dagger |0\rangle$, where $i_1 \ldots i_n$ is an ordered set of integers, no more than d of which can be distinct. Then, from (B.4) it follows immediately that

$$a_i^\dagger|n; i_1 \ldots i_n\rangle = a_i^\dagger a_{i_1}^\dagger \ldots a_{i_n}^\dagger|0\rangle = |n+1; i, i_1 \ldots i_n\rangle \tag{B.5a}$$

and

$$a_i|n; i_1 \ldots i_n\rangle = a_i a_{i_1}^\dagger \ldots a_{i_n}^\dagger|0\rangle = \{\delta_{ii_1} a_{i_2}^\dagger \ldots a_{i_n}^\dagger + a_{i_1}^\dagger a_i a_{i_2}^\dagger \ldots a_{i_n}\}|0\rangle$$

$$= \begin{cases} \delta_{ii_1} a_{i_2}^\dagger \ldots a_{i_n}^\dagger|0\rangle + \delta_{ii_2} a_{i_1}^\dagger a_{i_3}^\dagger \ldots a_{i_n}^\dagger|0\rangle + \cdots + \delta_{ii_n} a_{i_1}^\dagger \ldots a_{i_{n-1}}^\dagger|0\rangle \\ + a_{i_1}^\dagger \ldots a_{i_{n-1}}^\dagger a_i|0\rangle \end{cases}$$

$$= \sum_{k=1}^{n} \delta_{ii_k} a_{i_1}^\dagger \ldots a_{i_{k-1}}^\dagger a_{i_{k+1}}^\dagger \ldots a_{i_n}^\dagger|0\rangle$$

$$= \sum_{k=1}^{n} \delta_{ii_k}|n-1; i_1 \ldots i_{k-1} i_{k+1} \ldots i_n\rangle \tag{B.5b}$$

This establishes that $a_i^\dagger$ and a_i are the creation and annihilation operators for component i.[91] By repeated applications of (B.4) it can be demonstrated that

$$\langle n; i_1 \ldots i_n | m; j_1 \ldots j_m\rangle = n!\,\delta_{nm} \Delta_{j_1 \ldots j_m}^{i_1 \ldots i_n} \tag{B.6}$$

where $\Delta_{j_1 \ldots j_m}^{i_1 \ldots i_n} = 1$ when the set $\{i_l\}$ is a permutation of $\{j_l\}$; in all other cases $\Delta_{j_1 \ldots j_m}^{i_1 \ldots i_n} = 0$. Therefore, the basis functions $|n; i_1 \ldots i_n\rangle$ are orthogonal to one another but do not have unit norms.

The functions $\psi_{i_1 \ldots i_n}^{(n)}(\xi)$ defined by

$$|n; i_1 \ldots i_n\rangle = \Phi(\xi)\psi_{i_1 \ldots i_n}^{(n)}(\xi)$$

$$= a_{i_1}^\dagger \ldots a_{i_n}^\dagger \Phi(\xi) = (-)^n \partial_{i_1} \ldots \partial_{i_n} \Phi(\xi) \tag{B.7a}$$

and equal to

$$\psi_{i_1 \ldots i_n}^{(n)}(\xi) = \frac{(-)^n}{\Phi(\xi)} \partial_{i_1} \ldots \partial_{i_n} \Phi(\xi) \tag{B.7b}$$

are Grad's generalized Hermite polynomials. The first few members of this set of functions are

$$\psi^{(0)}(\xi) = \frac{(-)^0}{\Phi(\xi)}\Phi(\xi) = 1$$

$$\psi_i^{(1)}(\xi) = \frac{(-)^1}{\Phi(\xi)}\partial_i \Phi(\xi) = \xi_i \tag{B.8}$$

and

$$\psi_{ij}^{(2)}(\xi) = \frac{(-)^2}{\Phi(\xi)}\partial_i\partial_j \Phi(\xi) = \xi_i\xi_j - \delta_{ij}$$

These functions are used in Section IV.D in conjunction with the moment analysis of kinetics equations.

This same formalism can be applied to the Gaussian model for the density gradient by identifying ξ with the single-component vector $x_3/\mu_2^{1/2}$ and $\Phi(\xi)$ with

$$\nabla_3 n_0(x_3) = -\frac{n_l - n_v}{(2\pi\mu_2)^{1/2}}\exp\left(\frac{-x_3^2}{2\mu_2}\right) \tag{B.9}$$

The analogues of the functions $\psi_{i_1\ldots i_n}^{(n)}(\xi)$, defined by (B.7b), are the functions

$$\psi_{33\ldots 3}^{(p)}(x_3) \equiv \chi_p(x_3) = \frac{(-)^p}{\nabla_3 n_0(x_3)}\left[\underbrace{\frac{\partial}{\partial(x_3/\mu_2^{1/2})}\cdots\frac{\partial}{\partial(x_3/\mu_2^{1/2})}}_{p\text{-factors}}\right]\nabla_3 n_0(x_3)$$

$$= \frac{(-\mu_2^{1/2})^p}{\nabla_3 n_0(x_3)}\nabla_3^{p+1} n_0(x_3) \tag{109}$$

the first three of which are

$$\chi_0(x_3) = 1$$

$$\chi_1(x_3) = \frac{(-\mu_2^{1/2})}{\nabla_3 n_0(x_3)}\nabla_3^2 n_0(x_3) = \mu_2^{1/2}\frac{x_3}{\mu_2} = \frac{x_3}{\mu_2^{1/2}} \tag{B.10}$$

and

$$\chi_2(x_3) = \frac{\mu_2}{\nabla_3 n_0(x_3)} \nabla_3^3 n_0(x_3) = \left(\frac{x_3}{\mu_2^{1/2}}\right)^2 - 1$$

Furthermore, the analogue of the orthogonality relationship, (B.6), is the equation

$$\begin{aligned}\langle \chi_n(x_3)|\delta(x_3 - x_3')|\chi_m(x_3')\rangle &= \langle \chi_n|\chi_m\rangle = -(n_l - n_v)n!\,\delta_{nm}\Delta^{33\ldots3}_{33\ldots3} \\ &= -(n_l - n_v)n!\,\delta_{nm} \qquad \text{(110b)}\end{aligned}$$

of Section IV.D.

2. *Solution of Fokker-Planck-like Equations*[72–74, 84]

From (B.5a) and (B.5b) we see that

$$\begin{aligned}\sum_{i=1}^{d} a_i^\dagger a_i|n; i_1\ldots i_n\rangle &= \sum_{i=1}^{d} a_i^\dagger \sum_{k=1}^{n} \delta_{i,i_k}|n-1; i_1\ldots i_{k-1}i_{k+1}\ldots i_n\rangle \\ &= \sum_{k=1}^{n} a_{i_k}^\dagger|n-1; i_1\ldots i_{k-1}i_{k+1}\ldots i_n\rangle = n|n; i_1\ldots i_n\rangle\end{aligned} \qquad \text{(B.11)}$$

This proves that $\Sigma_{i=1}^{d} a_i^\dagger a_i$ is the number operator and, as is now demonstrated, it is a Fokker-Planck operator as well. Thus, by substituting the definitions (B.3) of $a_i^\dagger$ and a_i into (B.11) we obtain the expression

$$-\sum_{i=1}^{d} \partial_i(\partial_i + \xi_i)|n; i_1\ldots i_n\rangle = n|n; i_1\ldots i_n\rangle \qquad \text{(B.12)}$$

This, when combined with (B.6), yields the result

$$\langle m; j_1\ldots j_m|K_{\mathrm{FP}}^{(d)}(\xi, \xi')|n; i_1\ldots i_n\rangle = -n\,n!\,\delta_{nm}\Delta^{i_1\ldots i_n}_{j_1\ldots j_m} \qquad \text{(B.13a)}$$

wherein $K_{\mathrm{FP}}^{(d)}$ is the d-dimensional Fokker-Planck operator,

$$K_{\mathrm{FP}}^{(d)}(\xi, \xi') = \partial\cdot(\partial + \xi)\,\delta(\xi - \xi') \qquad \text{(B.13b)}$$

This establishes the delightful fact that the multidimensional Fokker-Planck operator is diagonal in the basis of generalized Hermite polynomials. Therefore, it is a simple matter to obtain analytic expressions for the solutions of the Fokker-Planck equation. There is, however, a complication that at first glance appears to spoil the value of this nice result. The streaming terms,

$$\left[v_0(i\boldsymbol{\xi}_\perp \cdot \mathbf{q} + \xi_3 \nabla_3) + v_0(\nabla_3 \ln n_0(x_3))\partial_3 \right] \delta(\boldsymbol{\xi} - \boldsymbol{\xi}') \tag{B.14}$$

which occur along with the Fokker-Planck terms $[\alpha_\perp \boldsymbol{\partial}_\perp \cdot (\boldsymbol{\partial}_\perp + \boldsymbol{\xi}_\perp) + \alpha_\parallel \partial_3(\partial_3 + \xi_3)]\delta(\boldsymbol{\xi} - \boldsymbol{\xi}')$ as part of the operator Σ_1 defined by (126a), are not diagonal in the Hermite polynomial representation. Fortunately, this presents only minor difficulties for, since the streaming terms are linear in a_i (or $a_i^\dagger$), it is possible to construct a transformed basis set that diagonalizes the entire operator Σ_1. The recipe for performing this diagonalization is now given.

The transformed ladder operators $\tilde{a}_i^\dagger$ and $\tilde{a}_i$ are related to $a_i^\dagger$ and a_i by the formulas

$$\tilde{a}_i^\dagger = u_d a_i^\dagger u_d^{-1}, \qquad \tilde{a}_i = u_d a_i u_d^{-1} \tag{B.15}$$

where, for the present purposes, it is sufficient to choose for u_d an operator of the form

$$u_d = \exp\left[-\sum_{j=1}^{d} (\lambda_j a_j^\dagger - \mu_j a_j) \right] \tag{B.16}$$

Here, the objects λ_j and μ_j are complex-valued c-numbers. It is an immediate consequence of the definitions, (B.15) and (B.16), that the transformed operators $\tilde{a}_i^\dagger$ and $\tilde{a}_i$ obey the Boson commutation relations, (B.4a) and (B.4b), and so can be interpreted as annihilation and creation operators. What remains to be done is the construction of the transformed basis set.

From the well-known operator identity,

$$\exp(L)A\exp(-L) = A + [L, A] + \frac{1}{2!}[L, [L, A]] + \dots \tag{B.17}$$

we obtain from (B.15) and (B.16) the connections

$$\tilde{a}_i^\dagger = a_i^\dagger + \mu_i, \qquad \tilde{a}_i = a_i + \lambda_i \tag{B.18}$$

Therefore, the new vacuum state $|\tilde{0}\rangle$ is related to $|0\rangle \equiv \Phi(\xi) =$

$(2\pi)^{-d/2}\exp(-\frac{1}{2}\xi\cdot\xi)$ in the manner,

$$|\tilde{0}\rangle = u_d|0\rangle \tag{B.19}$$

An interesting feature of this (which we previously encountered in an analysis of the velocity distribution of photodissociation products[92]) is that the transformed vacuum state is a "shifted" Maxwell-Boltzmann distribution function. For example, in the one-dimensional case the situation is as follows:

Old vacuum	New vacuum
$a\|0\rangle = 0$ $(\frac{d}{d\xi}+\xi)\|0\rangle = 0$ $\|0\rangle = \text{const.}\exp(-\frac{1}{2}\xi^2)$	$\tilde{a}\|\tilde{0}\rangle = 0$ $(\frac{d}{d\xi}+\xi+\lambda)\|\tilde{0}\rangle = 0$ $\|\tilde{0}\rangle = \text{const.}\exp(-\frac{1}{2}[\xi+\lambda]^2)$

From the vacuum $|\tilde{0}\rangle$ one constructs the transformed basis functions $|\tilde{n}; i_1 \dots i_n\rangle$ as follows:

$$\begin{aligned}|\tilde{n}; i_1\dots i_n\rangle &= \tilde{a}_{i_1}^\dagger \tilde{a}_{i_2}^\dagger \dots \tilde{a}_{i_n}^\dagger|\tilde{0}\rangle = \{u_d a_{i_1}^\dagger u_d^{-1} u_d a_{i_2}^\dagger u_d^{-1} \dots u_d a_{i_n}^\dagger u_d^{-1} u_d\}|0\rangle \\ &= \{u_d a_{i_1} a_{i_2} \dots a_{i_n}\}|0\rangle = u_d|n; i_1\dots i_n\rangle \end{aligned}\tag{B.20}$$

Then since

$$\begin{aligned}\sum_{i=1}^{d} \tilde{a}_i^\dagger \tilde{a}_i &\equiv \sum_{i=1}^{d}(a_i^\dagger+\mu_i)(a_i+\lambda_i) = \sum_{i=1}^{d}(a_i^\dagger a_i + \mu_i a_i + \lambda_i a_i^\dagger + \mu_i\lambda_i) \\ &= \sum_{i=1}^{d}(a_i^\dagger a_i + \mu_i[a_i + a_i^\dagger] + [\lambda_i - \mu_i]a_i^\dagger + \mu_i\lambda_i)\end{aligned}$$

wherein $a_i + a_i^\dagger = \xi_i$, we see that

$$\sum_{i=1}^{d} \tilde{a}_i^\dagger \tilde{a}_i = -\partial\cdot(\partial+\xi) + \boldsymbol{\mu}\cdot\xi + (\boldsymbol{\mu}-\boldsymbol{\lambda})\cdot\partial + \boldsymbol{\mu}\cdot\boldsymbol{\lambda} \tag{B.21}$$

Consequently, the analogue of (B.13) is

$$\langle\tilde{m}; j_1\dots j_m|K_{\text{SFP}}^{(d)}(\xi,\xi')|\tilde{n}; i_1\dots i_n\rangle = (-n+\boldsymbol{\mu}\cdot\boldsymbol{\lambda})n!\delta_{nm}\Delta_{j_1\dots j_m}^{i_1\dots i_n} \tag{B.22a}$$

where $K_{\mathrm{SFP}}^{(d)}$ is the *shifted* Fokker-Planck operator

$$K_{\mathrm{SFP}}^{(d)}(\xi,\xi')=[\partial\cdot(\partial+\xi)+(\boldsymbol{\lambda}-\boldsymbol{\mu})\cdot\partial-\boldsymbol{\mu}\cdot\xi]\delta(\xi-\xi') \qquad \text{(B.22b)}$$

The streaming operator for a homogeneous system does not include a term proportional to ∂ [see (B.14)] and so $\boldsymbol{\lambda}=\boldsymbol{\mu}$ in that case. As is discussed in Section IV.1, the terms proportional to ∂_3 that appear in kinetics equations for inhomogeneous, two-phase systems can be interpreted in terms of an effective force, $F(x_3)\equiv\beta^{-1}\nabla_3 \ln n_0(x_3)$, due solely to the existence of the interfacial density gradient.

What we have proved is that the shifted Fokker-Planck operator, consisting of streaming terms and Fokker-Planck operators, is diagonal in the transformed basis of (B.20). Thus, it is easy to obtain solutions of closed, analytic form for the motion of the tagged particle to which this kinetic operator is specific. What really interests us, however, is a more general kinetic equation, similar to (125) of Section IV.D.2, and of the form

$$(z-\Sigma_1-\Sigma_2)G=-n_0\Phi(\xi)\delta(\xi-\xi') \qquad \text{(B.23)}$$

Here, Σ_1 is an operator of the type we have just described. Its form is that of $K_{\mathrm{SFP}}^{(d)}$, containing a Dirac delta function, and it is connected with the motion of a tagged particle. On the other hand, Σ_2 consists of a finite number of (velocity) polynomial terms (or dyads) that describe the backflow of all the other particles belonging to the many-body system. The remainder of this appendix is devoted to the construction of solutions for this more general type of kinetic equation. The spatial dependencies of the operators and correlation functions are ignored, since these are irrelevant to the problem of solving the integrodifferential equation for the momentum dependence of the correlation functions. It is shown that the problem of solving (B.23) can be reduced to that of solving a finite matrix equation: the specific case (125) corresponds to a 4×4 matrix equation.

Before proceeding further, however, let us introduce a more compact notation. Thus, the symbol $|j\rangle\equiv\Phi(\xi)\psi_j(\xi)$ is used to denote one of the suitably ordered and unit-normalized (i.e., $\langle i|j\rangle=\delta_{ij}$) basis functions $|n;i_1\ldots i_n\rangle$. For example, $|0\rangle=|0;\rangle, |1\rangle=|1;1\rangle, |2\rangle=|1;2\rangle$, and $|3\rangle=|1;3\rangle$.

The experimental observables invariably consist of a few low-order momentum moments of G. Therefore, it is natural and convenient to introduce the functions

$$G_{ij}\equiv\int d^d\xi\int d^d\xi'\,\psi_i(\xi)G(\xi,\xi')\psi_j(\xi') \qquad \text{(B.24)}$$

where, for example, G_{00} is related to the dynamical structure factor according to (A.6). From (B.23) and (B.24) and the identity $(A\pm B)^{-1}=A^{-1}\mp A^{-1}B(A\pm B)^{-1}$ it follows that

$$\begin{aligned} G_{ij} &= -\langle i|(z-\Sigma_1-\Sigma_2)^{-1}n_0|j\rangle \\ &= -\langle i|(z-\Sigma_1)^{-1}n_0|j\rangle - \langle i|(z-\Sigma_1)^{-1}\Sigma_2(z-\Sigma)^{-1}n_0|j\rangle \end{aligned}$$

Then since Σ_2 is a finite sum of polynomials, which can be written in the form

$$\Sigma_2 = \textstyle\sum_{k,l}^{N}\alpha_{kl}|k\rangle\langle l| \tag{B.25}$$

this becomes

$$G_{ij} = -\langle i|(z-\Sigma_1)^{-1}n_0|j\rangle - \sum_{k,l}^{N}\langle i|(z-\Sigma_1)^{-1}\alpha_{kl}|k\rangle\langle l|(z-\Sigma)^{-1}n_0|j\rangle$$

We now introduce the matrix elements,

$$(G_S)_{ij} = -\langle i|(z-\Sigma_1)^{-1}n_0|j\rangle \tag{B.26}$$

of the tagged-particle propagator. The equations determining the elements of G then become

$$G_{ij} = (G_S)_{ij} - \sum_{k,l}^{N}(G_S)_{ik}\frac{\alpha_{kl}}{n_0}G_{lj} \tag{B.27}$$

These are analogous to (133) of Section IV.D.3, which we associated with the moments of the modified BGK equation.

The entire set of elements G_{ij} is obtained by solving the $N\times N$ matrix equation with $1\le i\le N, 1\le j\le N$. This equation involves the quantities $(G_S)_{ij}$ which can be computed using the transformed basis. Thus

$$(G_S)_{ij} \equiv -\langle i|(z-\Sigma_1)^{-1}n_0|j\rangle = -\sum_{m,n}\langle i|\tilde{m}\rangle\langle\tilde{m}|(z-\Sigma_1)^{-1}n_0|\tilde{n}\rangle\langle\tilde{n}|j\rangle \tag{B.28}$$

where $\langle\tilde{m}|(z-\Sigma_1)^{-1}n_0|\tilde{n}\rangle$ is diagonal. Finally, algorithms, for computing the quantities $\langle i|\tilde{m}\rangle$ are readily constructed. Results for a homogeneous system have been reported by Jhon and Forster.[74, 93]

In conclusion, a systematic procedure has been developed for solving Fokker-Planck-like equations of the types represented by (B.23) and (125) of Section IV.D.2. Our method is very similar to that used by Montroll and Potts[94] (and by Lax[95]) in their studies of lattice defects. Indeed, there is a complete formal equivalence between an infinite array of harmonic oscillators with localized defects and the Fokker-Planck theory for a particle coupled to a responsive, background fluid.

Acknowledgments

Two of us, MSJ and JSD, wish to acknowledge the National Science Foundation and the Donors of the Petroleum Research Fund, administered by the American Chemical Society, for support of this research. RCD acknowledges support by the National Research Council of Canada.

References and Notes

1. I. Prigogine and R. Defay, *Surface Tension and Adsorption*, Wiley, New York, 1966.
2. L. D. Landau and E. M. Lifshitz, *Statistical Physics*, Addison-Wesley, Reading, 1970, Sect. 142.
3. A. Harasima, *Adv. Chem. Phys.*, **1**, 203 (1958).
4. J. K. Lee, J. A. Barker, and G. M. Ponnd, *J. Chem. Phys.*, **60**, 1976 (1974).
5. J. W. Gibbs, *The Scientific Papers of Josiah Williard Gibbs*, Vol. I Dover, 1961, p. 219.
6. F. P. Buff and R. A. Lovett, in *Simple Dense Fluids*, H. L. Frisch and Z. W. Salsburg, Eds., Academic, New York, 1968.
7. B. Widom, in *Phase Transitions and Critical Phenomena*, C. Domb and M. S. Green, Eds., Academic, New York, 1972, Vol. II, Chap. 3, p. 79.
8. Note that the exponent μ is related to the correlation length exponent ν, the compressibility exponent γ, and the degree $1/\beta$ of the coexisting curve by $\mu+\nu=\gamma+2\beta$; see B. Widom, *J. Chem. Phys.*, **43**, 3892 (1965); P. G. Watson, *J. Phys. C.*, **1**, 268 (1968). The classical mean field theory $[\nu=\frac{1}{2}, \gamma=1$, and $\beta=\frac{1}{2}]$ gives $\mu=\frac{3}{2}$.
9. J. M. H. Levelt Sengers, *Physica*, **82A**, 319 (1976).
10. J. K. Percus, in *The Equilibrium Theory of Classical Fluids*, H. L. Frisch and J. L. Lebowitz, Eds., Benjamin, New York, 1964, Sect. IV, pp. 11–45.
11. M. S. Jhon, R. C. Desai, and J. S. Dahler, *Chem. Phys. Lett.*, **56**, 151 (1978).
12. M. S. Jhon, R. C. Desai, and J. S. Dahler, *J. Chem. Phys.*, **68**, 5615 (1978).
13. M. S. Jhon, R. C. Desai, and J. S. Dahler, *J. Chem. Phys.*, **70**, 1544 (1979).
14. J. G. Kirkwood and F. P. Buff, *J. Chem. Phys.*, **17**, 338 (1949).
15. F. P. Buff, *Z. Elecktrochem.*, **56**, 311 (1952).
16. A. G. MacLellan, *Proc. Roy. Soc.* (*Lond.*) **A213**, 274 (1952); A. Harasima, *J. Phys. Soc. Jap.*, **8**, 343 (1953).
17. R. P. Feynman, *Statistical Mechanics*, Benjamin, Reading, MA, 1972, Sect. 2.6.
18. H. Goldstein, *Classical Mechanics*, Addison-Wesley, Reading, 1950 pp. 258–261; relevant formulas are (8.62) to (8.64) and (8.67).
19. A more symmetric form for G is

$$G=\text{const.}+\sum_{\alpha}\left\{\tfrac{1}{2}(x_1^\alpha P_1^\alpha+x_2^\alpha P_2^\alpha)-x_3^\alpha P_3^\alpha\right\}\ldots \tag{2.12}$$

This does not change the result, (13).
20. M. Toda, *J. Phys. Soc. Jap.*, **10**, 512 (1955).

21. R. H. Fowler, *Physica*, **5**, 39 (1938).
22. To illustrate the trick of performing this calculation let us consider the factor of $\theta(x_3)\theta(x_3')$ that occurs in the first term of (15.2). This factor can be rewritten as

$$\theta(x_3)\theta(x_3+X_3)=\theta(X_3)\theta(x_3+X_3)+\theta(-X_3)\theta(x_3)=[\theta(X_3)+\theta(-X_3)]\theta(x_3) + \theta(X_3)[\theta(x_3+X_3)-\theta(x_3)]$$

where $X_3 \equiv x_3' - x_3$. Because of the symmetry of the integrand the first of the two terms in this final expression makes no contribution to σ_P. The x_3 integral of the second term is equal to $X_3\theta(X_3)$. The other terms of (15.2) can be handled similarly.
23. R. W. Pastor and J. Goodisman, *J. Chem. Phys.*, **68**, 3654 (1978).
24. R. Evans, *Adv. Phys.*, **28**, 143 (1979).
25. J. Yvon, *Proceedings of IUPAP Symposium on Thermodynamics*, Brussells, 1948, p. 9; see also Ref. 6.
26. R. A. Lovett, P. W. de Haven, J. J. Vieceli, and F. P. Buff, *J. Chem. Phys.* **58**, 1880 (1973).
27. D. G. Triezenberg and R. Zwanzig, *Phys. Rev. Lett.*, **28**, 1183 (1972).
28. Compare with (5) of Ref. 26 and (14) and (15) of Ref. 27.
29. A. J. M. Yang, P.D. Fleming, and J. H. Gibbs, *J. Chem. Phys.*, **64**, 3732 (1976).
30. J. D. van der Waals, *Z. Phys. Chem.*, **13**, 657 (1894) and translation thereof by J. S. Rowlinson, *J. Stat. Phys.*, **20**, 197 (1979). See also Lord Rayleigh, *Philos. Mag.*, **33**, 209 (1892).
31. S. Fisk and B. Widom, *J. Chem. Phys.*, **50**, 3219 (1969).
32. B. U. Felderhof, *Physica*, **48**, 541 (1970).
33. V. Bongiorno and H.T. Davis, *Phys. Rev.*, **A12**, 2213 (1975): V. Bongiorno, L. E. Scriven, and H. T. Davis, *J.C.I.S.*, **57**, 462 (1976).
34. B. Widom, in *Statistical Mechanics and Statistical Methods in Theory and Application*, Uzi Landman, Ed., Plenum Press, New York, 1976.
35. C. Ebner, W. F. Saam, and P. Stroud, *Phys. Rev.*, **A14**, 2264 (1976).
36. B. S. Carey, L. E. Scriven, and H. T. Davis, *J. Chem. Phys.*, **69**, 5040 (1978); *AIChE J.*, **24**, 1076 (1978).
37. C. A. Leng, J. S. Rowlinson, and S. M. Thompson, *Proc. Roy. Soc. (Lond.)*, **A358**, 267, (1977).
38. M. S. Jhon, R. C. Desai, and J. S. Dahler, *J. Chem. Phys.*, **70**, 5228, (1979).
39. R. Lovett, C. Y. Mou, and F. P. Buff, *J. Chem. Phys.*, **65**, 570 (1976).
40. Here and elsewhere in this Chapter we use a "summation convention," according to which the occurrence of a repeated, overscored coordinate such as $\bar{x}_3$ or $\bar{\mathbf{r}}$ implies integration over the domain of the variable.
41. L. D. Landau and E. M. Lifshitz, *Fluid Mechanics*, Pergamon, London 1959, Sect. 61.
42. V. G. Levich, *Physicochemical Hydrodynamics*, Prentice Hall, Englewood Cliffs, NJ, 1962, Chap. XI.
43. R. F. Feynman, R. B. Leighton, and M. Sands *The Feynman Lectures on Physics, Vol.* 1, Addison Wesley, Reading, Pa, 1963, Chap. 51.
44. J. S. Huang and W. W. Webb, *Phys. Rev. Letts.*, **23**, 160 (1969).
45. *J. Phys. Colloq. C-1*, Suppl. au No. 2–3, Tome 33, (Feurier-Mars 1972); *J. Phys. (Paris)*, **33**, C1-77, C1-141, C1-149 (1972).
46. M. J. Stephen and J. P. Straley, *Rev. Mod. Phys.*, **46**, 617 (1974).
47. D. Langevin and M. A. Bonchiat, *J. Phys. (Paris)*, **33**, 101 (1972).
48. D. O. Edwards and W. F. Saam, in *Progress in Low Temperature Physics*, Vol. VII A, D. F. Boewer, Ed., North-Holland, Amsterdam, 1978, Chap. 4.
49. M. Nelkin and S. Ranganathan, *Phys. Rev.*, **164**, 222 (1967).

50. B. J. Alder and T. E. Wainwright, *Phys. Rev.*, **A1**, 18 (1970).
51. D. Bedeaux and P. Mazur, *Physica*, **73**, 431 (1974).
52. M. S. Jhon and J. S. Dahler, *J. Stat. Phys.*, **20**, 3 (1979); R. Kapral, Adv. Chem. Phys. (in press).
53. M. S. Jhon and J. S. Dahler, *J. Chem. Phys.*, **68**, 812 (1978).
54. L. P. Kadanoff and G. Baym, *Quantum Statistical Mechanics*, Benjamin, New York, 1962.
55. H. Mori, *Prog. Theor. Phys.*, **34**, 423 (1965).
56. L. P. Kadanoff and P. C. Martin, *Ann. Phys.*, **24**, 419 (1963).
57. D. Forster, *Hydrodynamic Fluctuations, Broken Symmetry and Correlation Functions*, Benjamin, New York, 1975.
58. B. I. Halperin and P. C. Hohenberg, *Phys. Rev.*, **188**, 898 (1969).
59. P. C. Martin, P. S. Pershan, and J. Swift, *Phys. Rev. Lett.*, **25** 844 (1970); D. Forster, T. C. Lubensky, P. C. Martin, J. Swift, and P. S. Pershan, *Phys. Rev. Lett.*, **26**, 1016 (1971).
60. The quantal version of (76) is analogous to the Bogoliulov inequality; see, for example, Ref. 57 and N. D. Mermin and H. Wagner, *Phys. Rev. Lett.* **17**, 1133 (1966).
61. R. Kubo, *Prog. Phys.*, **29**, 255 (1966), see (5.20). This relationship is derived in Appendix A of this chapter, see (A.10).
62. J. Goldstone, *Nuovo Cimento*, **19**, 154 (1961); J. Goldstone, A. Salam, and S. Weinberg, *Phys. Rev.*, **127**, 965 (1962); J. Bernstein, *Rev. Mod. Phys.*, **46**, 7 (1974).
63. R. V. Lange, *Phys. Rev.*, **146**, 301 (1966); H. Stern, *Phys. Rev.*, **147**, 94 (1966).
64. K. Kawasaki, *Ann. Phys.*, **61**, 1 (1970).
65. M. E. Ernst, E. H. Hauge, and J. M. J. van Leeuwen, *Phys. Rev. Lett.*, **25**, 1254 (1970); R. Kapral and M. Weinberg, *Phys. Rev.*, **A8**, 1008 (1973).
66. T. D. Lee and C. N. Yang, *Phys. Rev.*, **87**, 410 (1952).
67. M. E. Fisher, in *Lectures in Theoretical Physics*, Vol. 7C, University of Colorado Press, Boulder, Colorado, 1965; C. J. Thompson, *Mathematical Statistical Physics*, Macmillan, New York, 1971.
68. C. Kittel, *Introduction to Solid State Physics*, 3rd ed., Wiley, New York, 1966, p. 493.
69. Because the two-phase system is invariant with respect to translations in the plane parallel to the interface, the partial Fourier transform of $\langle A(\boldsymbol{\rho},0)B(\boldsymbol{\rho}',0)\rangle$ is given by

$$\begin{aligned}\langle A(\boldsymbol{\rho},0)B(\boldsymbol{\rho}',0)\rangle(\mathbf{q}) &= \int d^2(\rho-\rho')\langle A(\boldsymbol{\rho},0)B(\boldsymbol{\rho}',0)\rangle e^{-i\mathbf{q}\cdot(\boldsymbol{\rho}-\boldsymbol{\rho}')}\\ &= S^{-1}\int d^2\rho\int d^2(\rho-\rho')\langle A(\boldsymbol{\rho},0)B(\boldsymbol{\rho}',0)\rangle e^{-q\cdot(\boldsymbol{\rho}-\boldsymbol{\rho}')}\\ &= \int d^2\rho\int d^2\rho'\langle[S^{-1/2}e^{-\mathbf{q}\cdot\boldsymbol{\rho}}A(\boldsymbol{\rho},0)][S^{-1/2}e^{i\mathbf{q}\cdot\boldsymbol{\rho}'}B(\boldsymbol{\rho}',0)]\rangle\\ &\equiv\langle A^*(\mathbf{q})B(\mathbf{q})\rangle\end{aligned}$$

with

$$A(\mathbf{q})\equiv S^{-1/2}\int d^2\rho\, e^{i\mathbf{q}\cdot\boldsymbol{\rho}}A(\boldsymbol{\rho})$$

and where S denotes the interfacial area. Thus (88) could be written in the more conventional form of the Schwartz inequality

$$\langle\zeta^*(\mathbf{q},0)\zeta(\mathbf{q},0)\rangle\geq|\langle\dot{\phi}^*(\mathbf{q},0)\zeta(\mathbf{q},0)\rangle|^2/\langle\dot{\phi}^*(\mathbf{q},0)\dot{\phi}(\mathbf{q},0)\rangle$$

70. D. C. Montgomery and D. A. Tidman, *Plasma Kinetic Theory*, McGraw-Hill, New York, 1964.
71. Ta-You Wu, *Kinetic Equations of Gases and Plasmas*, Addison-Wesley, Reading, MA, 1966.
72. J. L. Lebowitz, J. K. Percus, and B. Sykes, *Phys. Rev.*, **188**, 487 (1969).
73. E. P. Gross, *Phys. Rev.*, **158**, 147 (1967).
74. M. S. Jhon and D. Forster, *Phys. Rev.*, **A12**, 254 (1975).
75. P. L. Bhatnager, E. P. Gross, and M. Krook, *Phys. Rev.*, **94**, 511 (1964).
76. See for examples S. A. Rice, and P. Gray, *The Statistical Mechanics of Simple Fluids*, Wiley, New York, 1965.
77. A similar projection operator and perturbation analysis of the equations of motion has been given by R. H. Terwiel, *Physica*, **74**, 248 (1974).
78. M. Bixon, *Ann. Rev. Phys. Chem*., **27**, 65 (1976).
79. K. F. Freed and S. F. Edwards, *J. Chem. Phys.*, **61**, 3626 (1974).
80. S. A. Adelman and K. F. Freed, *J. Chem. Phys.*, **67**, 1380 (1977); A. Peterlin, *Polym. Lett.* **10**, 101 (1972).
81. C. A. Croxton, *Adv. Phys.*, **22**, 385 (1973); A. C. C. Opitz, *Phys. Letts.*, **47A**, 439 (1974); M. H. Kalos, J. K. Percus, and M. Rao, *J. Stat. Phys.*, **17**, 111 (1977).
82. H. T. Davis, *J. Chem. Phys.*, **67**, 3636 (1977).
83. L. S. Ornstein and F. Zernike, *Proc. Sect. Sci. K. Med. Akad. Wet.*, **17**, 793 (1914).
84. A. Z. Akcasu and J. J. Duderstadt, *Phys. Rev.*, **188**, 479 (1969).
85. P. C. Martin, in *Many-Body Physics*, C. DeWitt and R. Balian, Eds., Gordon and Breach, New York, 1968.
86. D. Pines, *The Many-Body Problem*, Benjamin, New York, 1962.
87. H. E. Stanley, *Introduction to Phase Transitions and Critical Phenomena*, Oxford University Press, 1971.
88. G. F. Mazenko, *Phys. Rev.*, **A7**, 209 (1973); **7**, 222 (1973); **9**, 360 (1974).
89. The advantages of renormalized perturbation theory compared with the traditional theory (see Ref. 79) are described in conjunction with the hydrodynamic theory of polymer solutions by K. F. Freed and H. Metiu, *J. Chem. Phys.*, **68**, 4604 (1978).
90. H. Grad, *Commun. Pure Appl. Math.*, **2**, 325 (1949); *Encyclopedia of Physics*, Vol. VII, S. Flügge, Ed., Springer-Verlag, Berlin, 1958, p. 205.
91. R. P. Feynman, *Statistical Mechanics*, Benjamin, Reading, MA, 1972, Chap. 6.
92. M. S. Jhon and J. S. Dahler, *J. Chem. Phys.*, **69**, 819 (1978).
93. M. S. Jhon and J. S. Dahler, J. Chem. Phys. (in press).
94. E. W. Montroll and R. B. Potts, *Phys. Rev.*, **100**, 525 (1955).
95. M. Lax, *Phys. Rev.*, **94**, 1391 (1954).

STATISTICAL THERMODYNAMICS OF PROTEINS AND PROTEIN DENATURATION

AKIRA IKEGAMI

The Institute of Physical and Chemical Research, Hirosawa, Wako-shi, Saitama, Japan

CONTENTS

I. INTRODUCTION

The most significant characteristic of proteins is their highly ordered structure, which directly relates to their biological function. Recent improvements in X-ray analysis produced the detailed three-dimensional structure models of proteins in crystal and provide new information on the physical chemistry of proteins. The overall feature of many X-ray models is their tightly packed structure, in which many hydrophobic groups are found in the center, whereas most polar groups are found on the outside.

Several physicochemical measurements indicate that these structural models in crystal give at least the gross features of protein structure in solution, although there is no reason to believe such static structures in detail. The specific interaction between substrates and enzymes in their catalytic action also supports the existence of highly ordered structures in solution.

The three-dimensional structure of a native protein in solution is supposed to be determined by minimizing the Gibbs free energy of a given polypeptide chain in a given environment. This thermodynamic origin of the native structure is clearly demonstrated by renaturation experiments. The refolding of the native structure of ribonuclease, after full denaturation by reductive cleavage of its four disulfide bonds, was observed by Anfinsen et al.[1] Since the refolding of this molecule requires only 1 of the 105 possible pairings of eight sulfhydryl groups to form four disulfide linkages, thermodynamic stability of the native conformation is indicated even in the polypeptide chain before linking of the disulfide bonds.

The native ordered structure of proteins is altered to the unfolded state by denaturations induced by the change of external conditions, such as temperature, pressure, and the compositions of low-molecular-weight solutes. The unfolding of native structure is usually reversible and appears sharply under certain external conditions, suggesting the important role of cooperative interactions.

The equilibrium properties of the denaturation are commonly accounted for by the simple two-state model in which only two conformational states are assumed for the protein structure.[2–6] The kinetic properties of the denaturation, however, apparently contradict the simple two-state model, because two, or sometimes three, relaxation times are observed in the denaturation processes.[7]

Although the cooperative interaction is essential to maintain the native protein structure, it is unlikely that the cooperative effects are sufficient to cover all the secondary bonds in a protein as predicted by the simple two-state model, and local fluctuations in structure are expected even in a highly ordered native state. In fact, hydrogen exchange kinetics[8–11] and highly resolved nuclear magnetic resonance spectroscopy[12, 13] of native proteins indicate considerable structural fluctuations inside the protein molecules. Thus, to obtain a more complete understanding of the stability, fluctuation, and kinetics of protein structure, it is necessary to have more precise and realistic theories based on statistical considerations. Unfortunately, complicated and specific structures of proteins prevent us from describing them in a general but realistic model. In this chapter, however, we wish to discuss the structure of proteins by a simple statistical thermodynamic model and to elucidate the general nature of protein structure. The model can reasonably explain a number of common phenomena related to reversible denaturation, structural changes, and fluctuations in many proteins. The thermal denaturations of five globular proteins are quantitatively analyzed using the model. The molecular parameters obtained by the least-squares method reflect the average properties of proteins and are reasonably accounted for by the primary and tertiary structures of proteins.

II. PROTEIN STRUCTURE AND MODEL

A. The Nature of Protein Structure

The native structure of proteins is maintained by many secondary (noncovalent) bonds, such as hydrogen bonds, hydrophobic interactions, and electrostatic or dipole interactions between side chains of amino acids and between main chain segments, as well as between side chains and main chains. Their bond energy is perhaps a few times that of thermal energy, however, and if there are no cooperative interactions between the secondary bonds they may be broken frequently by the thermal agitation. Therefore, cooperative interaction must be essential to the stability of the native structure. The stable, native structure should appear when the total energy of the cooperative secondary bonds overcomes the entropy of the polypeptide chains.

The folding into the unique, native structure of proteins from unfolded, random polypeptide chains is an interesting problem. In the case of synthetic homopolymers, interactions between segments are the same for all possible pairs of segments. Then many different forms of folding could have almost the same total interaction energy and no special structure would be favorable. In proteins interactions between side chains are different from each other, and secondary bonds can be formed only between specific pairs of sites. Pairs of secondary bonds in a certain fold cannot be exchanged freely because many different binding sites are uniquely arranged on a chain. In the uniquely folded native structure, a large number of specific pairs exist and stabilize the structure. The probability of forming alternative folds is likely to be quite small, since tertiary arrangements of side chains would probably be less likely to form many secondary bonds than in the case of the unique fold. It is possible that polypeptide chains having such a unique stable fold may be selected by evolution on account of their stable functions. In the following model probability forming alternative folds are neglected, and the protein structure is described by the binding state of possible secondary bonds in the uniquely folded structure.

B. Statistical Thermodynamic Model[14]

1. The tertiary structure of proteins is described by the state of the system consisting of N_0 secondary bonds between unique pairs of specific sites. Every secondary bond can exist in either a bonded or unbonded state.

2. In an "ideal" solution at 0° K, all secondary bonds are in the bonded state and contribute to form a fixed unique conformation dictated by the energy minimum. There is no flexibility of chains and the entropy of the system equals zero.

3. At ordinary temperatures, the number of bonded states N_b decreases and the entropy of the system increases. At a certain "structural state," which is characterized by a specific value of $Y \equiv N_b/N_0$, a protein molecule can assume different "structures" and can fluctuate between them. The term "structure" describes hereafter a group of conformations in which the states of all the secondary bonds are specified. Therefore, many different conformations belong to a "structure" because of the flexibility of polypeptide chains (see Fig. 1).

4. A bonded state can be formed only between a unique pair of specific sites. We neglect the possibility that a bond is formed between other pairs of sites, because many kinds of binding sites are uniquely arranged on a chain and they cannot move independently to make another bonding form possible. Therefore, the number of possible bonds N_0 is independent of temperature.

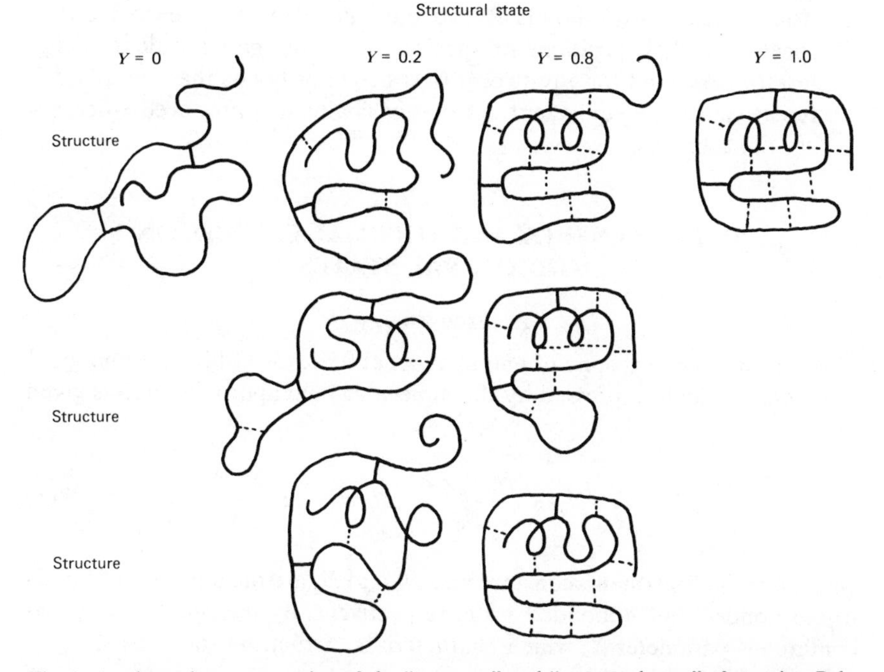

Fig. 1. A schematic representation of the "structure" and "structural state" of proteins. Polypeptide chains and secondary bonds are represented by solid and dotted lines, respectively.

5. There is a loss of energy when neighboring bonds are in different states. That is, cooperative interaction between neighboring bonds exists, since the binding sites are connected to each other by the polypeptide backbone.
6. As a result of thermal agitation, every secondary bond fluctuates in a random way between the bonded and unbonded state, but with constant average rates. Thus the number of secondary bonds in their bonded state N_b cannot be constant, but is continuously fluctuating. The thermodynamic principle of the minimum free energy cannot dictate the microscopic "structure" or conformation, but the statistical average of the "structural state" $\overline{Y}$. The kinetics of the structural change, such as the denaturation, can be described by the time course of the structural state according to the stochastic process.
7. To simplify the calculations, we make the following "equivalent and uniform" assumptions. All the possible secondary bonds N_0 are uniformly arranged on a lattice topologically similar to their distribution in the protein structure at 0°K. They have the equivalent number of

nearest neighboring bonds Z and the equivalent bond energy ε irrespective of their positions or species. Similarly, the equivalent energy loss J is assumed for any two nearest neighbor bonds that are in a different state. The equivalent entropy of chains α is produced when every bond breaks.

III. FREE ENERGY AND THERMAL EFFECTS ON STRUCTURAL STATES

A. Free Energy

When the standard state of energy is set at 0°K, the Gibbs free energy of a protein molecule expressed by the molecular field approximation is given by

$$G_0(T, N_b) = N_u\varepsilon + \frac{N_b N_u ZJ}{N_0} - N_u \alpha T - kT\ln\frac{N_0!}{N_b!N_u!} \tag{1}$$

where k is the Boltzmann constant and N_b and N_u are the numbers of bonds in the bonded and unbonded state, respectively. By the use of the quasi-continuous parameter X, which characterizes "structural state" as

$$\tfrac{1}{2}(1+X) = Y = \frac{N_b}{N_0}$$

$$\tfrac{1}{2}(1-X) = 1 - Y = \frac{N_u}{N_0} \tag{2}$$

(1) is reduced to

$$\begin{aligned} G_0(T, X) = \tfrac{1}{2}N_0 ZJ\Big[& A(1-X) + \tfrac{1}{2}(1-X^2) \\ & + BT\Big[\tfrac{1}{2}(1+X)\ln\frac{1}{2}(1+X) + \tfrac{1}{2}(1-X)\ln\tfrac{1}{2}(1-X)\Big] \\ & -(1-X)BTC\Big] \end{aligned} \tag{3}$$

where

$$A = \frac{\varepsilon}{ZJ}, \qquad B = \frac{2k}{ZJ}, \qquad C = \frac{\alpha}{2k} \tag{4}$$

depend only on the protein species considered.

B. Thermal Denaturation

By the condition of minimum free energy

$$\frac{\partial G(T, X)}{\partial X} = 0 \tag{5}$$

the most probable value of X, X_m is given by

$$\tanh^{-1} X_m = \frac{A + X_m}{BT} - C \tag{6}$$

Then, by letting

$$\frac{A + X}{BT} - C = x \tag{7}$$

the most probable structural state X_m is determined from the intersection of the straight line expressed by (7) and the hyperbolic tangent curve expressed by $\tanh x = X$ (see Fig. 2). As shown in Fig. 2, the straight lines

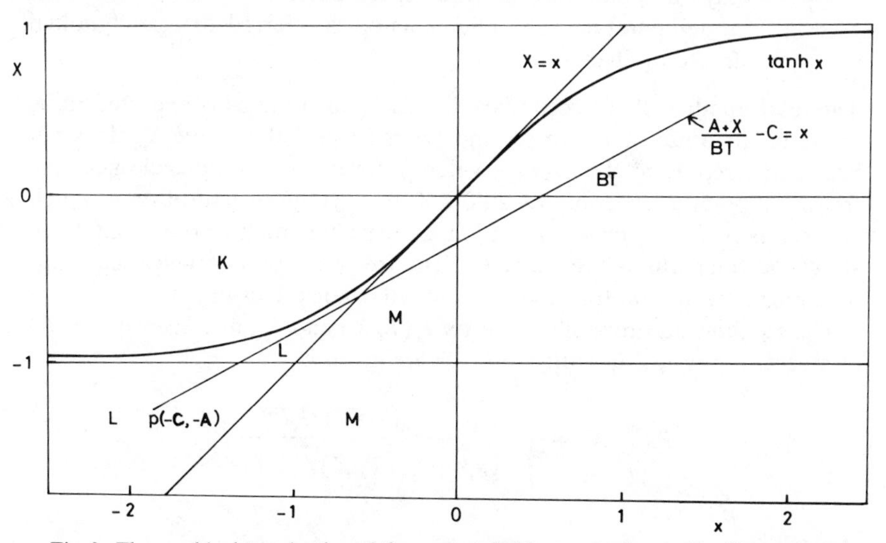

Fig. 2. The graphic determination of the most probable structural state X_m. The most probable value X_m can be determined from the intersection of the hyperbolic tangent curve $X = \tanh x$ and the straight lines expressed by (7), which pass through the characteristic point $p(-C, -A)$ and have slope BT.

expressed by (7) pass through a "characteristic point" $p(-C, -A)$ of a protein and have a slope BT varying with temperature. At the denaturation temperature $T_d \equiv A/BC$, the straight lines pass through the point o.

The changes in the most probable structural state with temperature depend on the position of the characteristic point.

Case A. If the characteristic point p is in the regions K and L in Fig. 2, where $C > A$, the free energy of the system has two minima near the denaturation temperature T_d, since the straight line intersects with the curve at three points. At the denaturation temperature T_d, the most probable structural state changes discretely from the ordered state to the disordered state and a "structural transition" occurs, which is essentially a first-order transition.

Case A′. Especially when the characteristic point is in the region K in Fig. 2, where $C > \tanh^{-1} A$, after a sharp transition from the ordered to the disordered state, to some extent ordered structure increases with a further rise in temperature.

Case B. When the characteristic point is on the line $X = x$, where $A = C$, a second-order transition of the structural state occurs at T_d.

Case C. When the characteristic point is in the region M in Fig. 2, where $A > C$, the straight line and hyperbolic tangent curve intersect at only one point and the value of X_m decreases gradually with rising temperature. The phenomenon is referred to as a "gradual structural change."

The total number of the secondary bonds N_0 of proteins is expected to be small, of the order of 100 or so, and the most probable value X_m does not necessarily represent the average value $\bar{X}$ for a protein molecule near the transition point. Hence the transition of the average structural state, which is essentially a first-order transition, is expected to be broad and large structural fluctuations are expected. The free energy of a typical case undergoing a "structural transition" is plotted against X in Fig. 3.

The equilibrium probability density $P_e(T, X)$ and the average value $\bar{S}(T)$ of a certain physical quantity $S(T, X)$ are given by

$$P_e(T, X) = \frac{\exp(-G_0(T, X)/kT)}{\int_{-1}^{1} \exp(-G_0(T, X)/kT)\,dX} \tag{8}$$

and

$$\bar{S}(T) = \int_{-1}^{1} S(T, X) P_e(T, X)\,dX \tag{9}$$

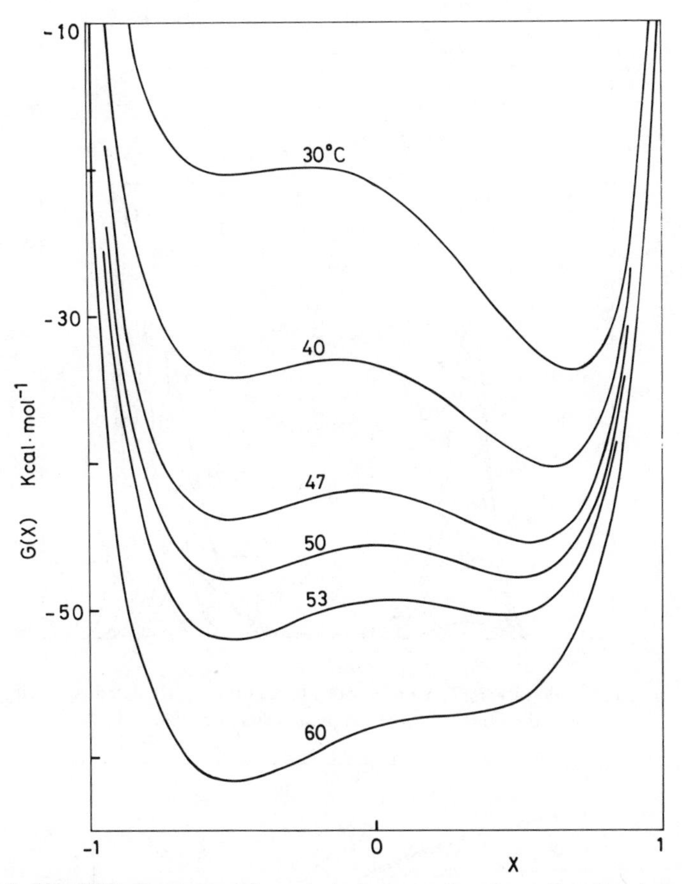

Fig. 3. The Gibbs free energy for a typical case undergoing a "structural transition" at various temperatures calculated according to (3), with $A=0.5$, $B=2.813\times10^{-3}$, $C=0.55$, $N_0=500$. These values correspond to $T_d=50°C$, $\varepsilon=706$ cal/mole, $ZJ=1412$ cal/mole, and $\alpha=2.185$ eu.

respectively. The probability density $P_e(T, X)$ and the average structural state $\overline{X}$ of the case of a structural transition is shown in Figs. 4 and 5.

Near the transition temperature, the free energy has two minima corresponding separately to the "ordered" and "disordered" state in this temperature range.

The maximum free energy difference separating two states (at $X=0$) is probably less than 10k cal, as shown in Fig. 3. The rate of the structural transition between two states may be fast, as is discussed below. That is, the free energy maximum between ordered and disordered states is high

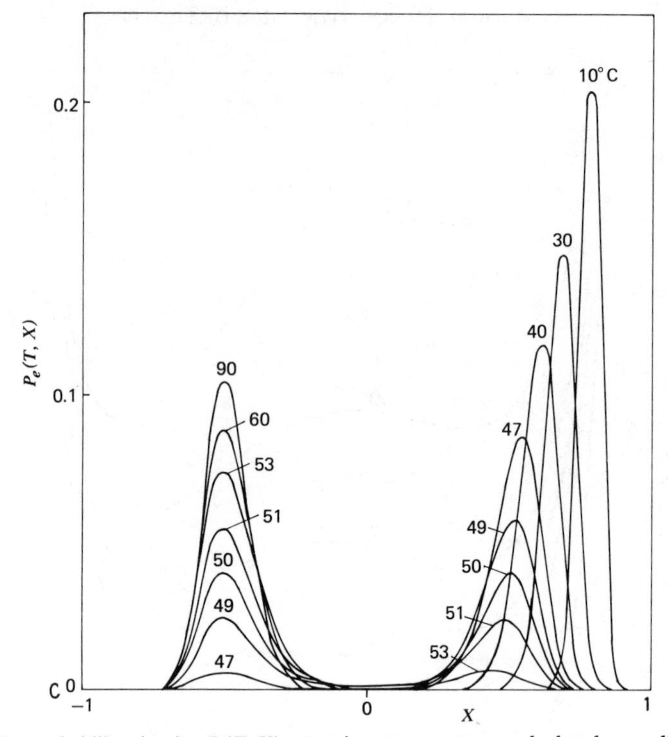

Fig. 4. The probability density $P_e(T, X)$ at various temperatures calculated according to (8) with the same values of the parameters as in Fig. 3.

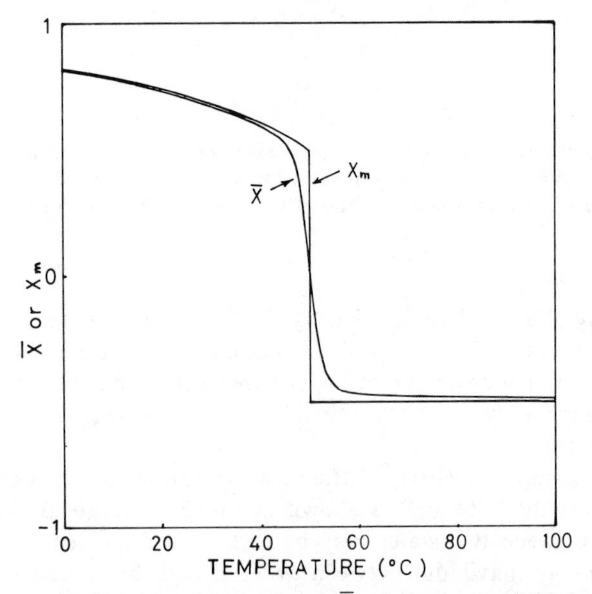

Fig. 5. The changes in the average structural state $\overline{X}$ and the most probable structural state X_m with temperature. The same values of the parameters are used as in Fig. 3.

enough to separate the "two states," but low enough to permit a fast transition rate, this being the physical condition for equilibrium between two different states without any special assumption. The origin of this condition comes from the fact that the number of secondary bonds or amino acids in protein molecules has an appropriate value.

The free energy $\underline{G_0}$, the probability density $P_e(T, X)$, and the average structural state $\overline{X}$ of a system undergoing a "gradual structural change" are plotted against X in Figs. 6, 7, and 8, respectively, In this case, the probability distribution has only one peak for all temperatures even though the average value $\overline{X}$ changes markedly in a certain range of temperature.

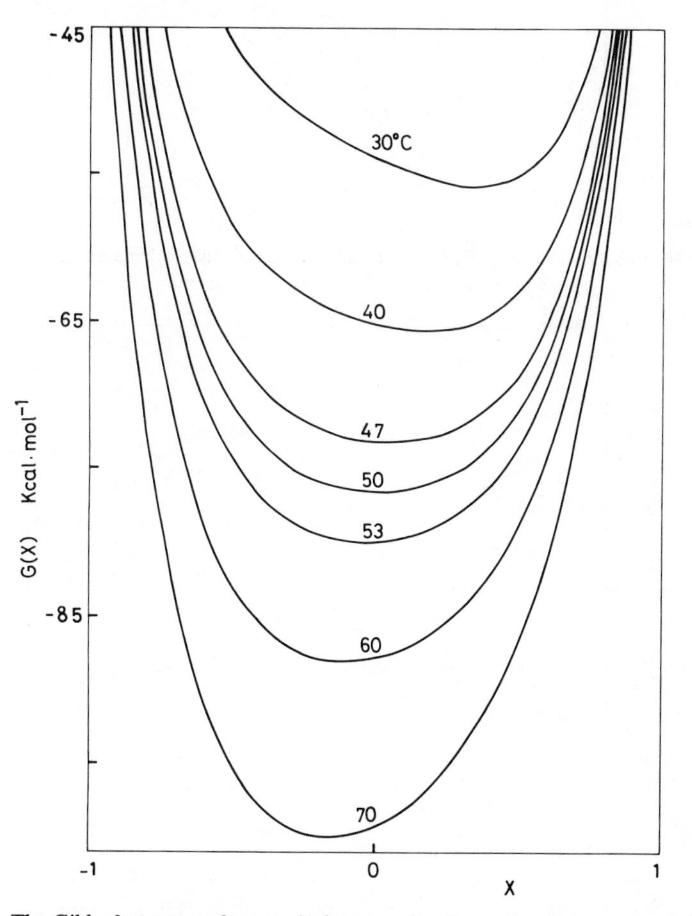

Fig. 6. The Gibbs free energy for a typical case undergoing a "gradual structural change" at various temperatures calculated according to (3), with $A=0.5$, $B=3.438\times10^{-3}$, $C=0.45$, $N_0=500$. These values correspond to $Td=50°C$, $\varepsilon=578$ cal/mol, $ZJ=1156$ cal/mole, and $\alpha=1.609$ eu.

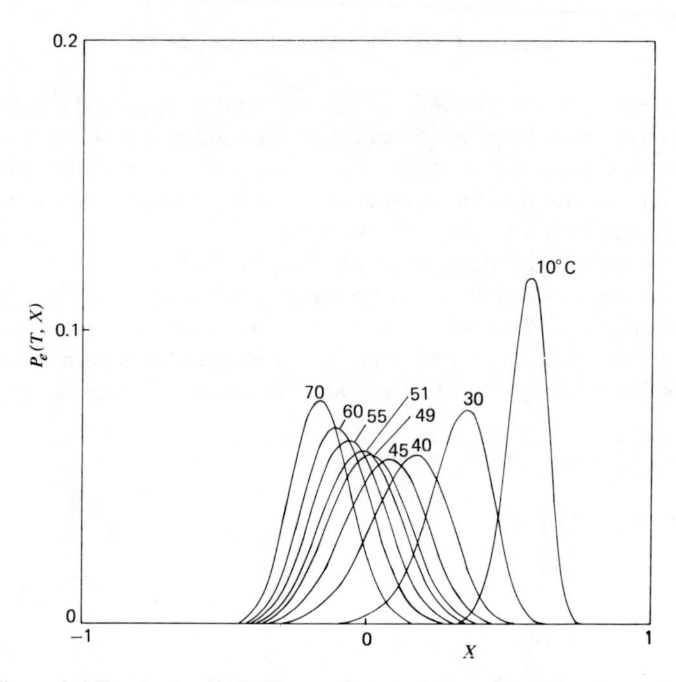

Fig. 7. The probability density $P_e(T, X)$ at various temperatures calculated according to (8) with the same values of the parameters as in Fig. 6.

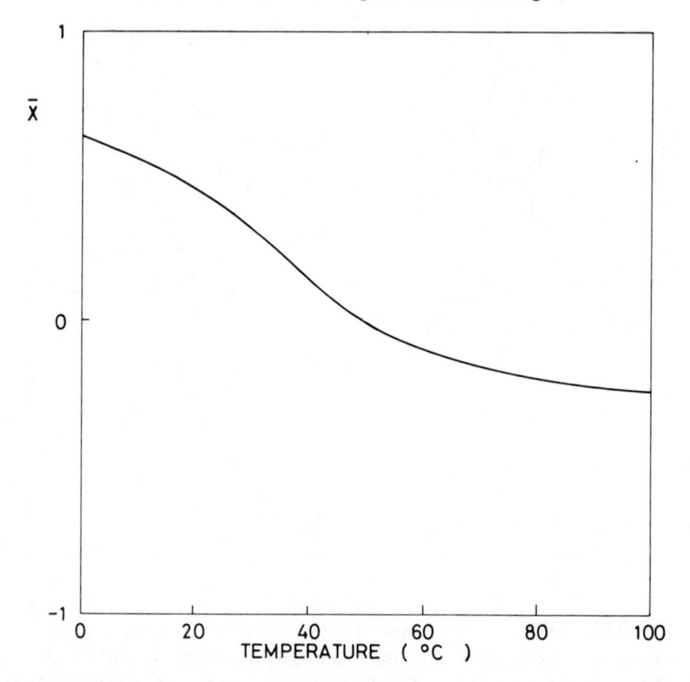

Fig. 8. The change in the average structural state $\bar{X}$ with temperature. The same values of the parameters are used as in Fig. 6.

Since enthalpy or heat capacity is directly related to the fraction of bonded states, their changes are important to characterize changes in structural state. The average enthalpy $\overline{H}(T)$ and heat capacity $\overline{C}_p$ are given by

$$\overline{H}(T)=\tfrac{1}{2}N_0\varepsilon(1-\overline{X})+\tfrac{1}{4}N_0ZJ(1-\overline{X^2}) \tag{10}$$

$$\overline{C}_p(T)=-\tfrac{1}{2}N_0\varepsilon\frac{\partial\overline{X}}{\partial T}-\tfrac{1}{4}N_0ZJ\frac{\partial\overline{X^2}}{\partial T} \tag{11}$$

C. Fluctuations

The average fluctuation of a structural state is given by

$$\overline{\Delta X^2}=\overline{(X-X_m)^2}=\frac{\int_{-1}^{1}(\Delta X)^2\exp[-G_0(T,X)]\,dX}{\int_{-1}^{1}\exp[-G_0(T,X)]\,dX} \tag{12}$$

In the case of a "gradual structural change," $\overline{\Delta X^2}$ is inversely proportional to the curvature of the free energy plot at the minimum point and is given by

$$\Delta X^2\simeq\frac{2}{N_0ZJ\left[BT/(1-X_m^2)-1\right]} \tag{13}$$

Especially when $A=C$ (case B), the average fluctuation at the transition point becomes infinite for $BT_d\equiv A/C=1$. In the case of a "structural transition," as is discussed earlier, each protein molecule fluctuates between two different states; large fluctuations in the structural state are expected from (12).

IV. INTERACTION AND STRUCTURAL STATES

A. Interaction with Small Molecules

Here we consider a multisystem in which a protein molecule interacts with small molecules, such as denaturants and substrates. The protein molecule is assumed to have n equivalent binding sites with the equivalent binding energy ΔF for a certain component. The semi-grand partition function of the system can be written as

$$Z(T, N_b, \mu) = \frac{N_0!\alpha^{N_u}}{N_b!N_u!}\exp\left[-\frac{\varepsilon N_u + ZJ(N_b N_u/N_0)}{kT}\right]$$
$$\times\left[\sum_{n_b}^{n}\frac{n!}{n_b!(n-n_b)!}\exp\left(-\frac{(\Delta F-\mu)n_b}{kT}\right)\right] \tag{14}$$

where μ is the chemical potential of the component. Then the free energy of the system is expressed by the sum of the free energy $G_0(T, N_b)$ in (1) and the interaction term:

$$G(T, P, N_b, \mu) = N_u\varepsilon + \frac{N_b N_u ZJ}{N_0} - N_u\alpha T - kT\ln\frac{N_0!}{N_b!N_u!}$$
$$-n(X)kT\ln[1+K(T, P, X)a] \tag{15}$$

where the activity of the component a is

$$a = \exp\left(\frac{\mu}{kT}\right)$$

and the binding constant K is

$$K = \exp\left(\frac{-\Delta F}{kT}\right)$$

Generally n is a function of the structural states X, and K is a function of X, temperature T, and pressure P. The average structural state of the protein molecule is affected by binding of the component through $n(X)$ and $K(T, P, X)$.

1. Nonspecific Binding and Denaturation

A sharp transition from "ordered" to "disordered" state has been observed in many proteins by denaturants such as urea and guanidine hydrochloride. These denaturants are supposed to interact directly with polypeptide chain rather than protein structure in a nonspecific manner. For such "nonspecific binding," we propose that the binding constant K does not depend on the structural state X, but the number of binding sites n is proportional to the fraction of unbonded state $\frac{1}{2}(1-X)$. Thus the free energy of the interaction between a protein and the ith component in a nonspecific manner is given by

$$\Delta G_i^{NS}(T, P, X) = -\tfrac{1}{2}N_0(1-X)l_i kT\ln[1+K_i(T, P)a_i] \tag{16}$$

where l_i is the number of sites per secondary bond. Then the most probable value of X, X_m, is given by

$$\tanh^{-1} X_m = \frac{A+X_m}{BT} - (C+D_i) \tag{17}$$

where

$$D_i(T,P,a_i) = \tfrac{1}{2} l_i \ln(1+K_i(T,P)a_i) \tag{18}$$

Equation (18) indicates that the "characteristic point" moves a distance D_i in the $-x$ direction from the point $p(-C,-A)$ with increasing activity of the ith component. The structural state always changes to a more disordered state by the addition of the "nonspecific binding" component. A "structural transition" is expected at a certain concentration of the ith component when

$$BT < 1$$

Usually heat-denatured proteins are not in a completely disordered state and further denaturation occurs when GuHCl is added after completion of the thermal denaturation.[3] Such phenomena are easily understood by the present model. The most probable structural state X_m takes the value -1 when the activity of denaturants approaches infinity, but X_m is larger than -1 even at infinite temperature, for the value of C is finite.

2. *Specific Binding*

Contrary to the case of nonspecific binding, we assume that the binding constant K_j of specific molecules such as substrates, inhibitors, and allosteric effectors depends on the structural state X, but the number of the binding sites n_j is independent of X. The free energy of the specific interaction between a protein and the jth component is

$$\Delta G_j^S(T,P,X) = -n_j kT \ln(1+K(T,P,X)a_j) \tag{19}$$

From the condition of minimum free energy,

$$\tanh^{-1} X_m = \frac{A+X_m}{BT} - (C-E_j) \tag{20}$$

$$E_j(T,P,a_j) = \frac{n_j}{N_0} \frac{a_j(\partial K_j/\partial X)_{X=X_m}}{1+K_j(T,P,X)a_j} \tag{21}$$

The shift of the "characteristic point" by the presence of the jth component is complicated, for the shift itself is a function of X. The direction of the shift and then the direction of the structural change, however, can be predicted from the value of $(\partial K_j/\partial X)$ as follows:

1. If $\partial K_j/\partial X=0$, no structural change occurs by the addition of the jth component.
2. If $\partial K_j/\partial X>0$, that is, the binding constant increases with the increase of the structural state X, the characteristic point moves toward the $+x$ direction and the structural state changes to a more ordered state.
3. If $\partial K_j/\partial X<0$, the characteristic point moves toward the $-x$ direction and the structural state changes to the direction of more disordered state.

In cases 2 and 3, drastic structural changes are expected at a certain concentration of the jth component, as is the case for nonspecific binding. Structural changes induced by the specific binding of low-molecular-weight components have been discussed by many authors. Koshland[15] proposed that protein molecules change their conformation to a more suitable one for the binding when they interact with substrates. This "induced fit" model could correspond to our case 2. Thus the interaction between protein and substrate increases ordered structures and reduces possible structures or fluctuation. In the "rack model" proposed by Lumry,[16] the structure of the substrate is assumed to be distorted by the binding to the enzyme. The model may correspond to our case 3, where the binding of the substrate is unfavorable for the ordered state of the enzyme.

B. Polymerization of Proteins

Polymers of protein molecules of the same kind are widely observed in higher order structures or organs, such as allosteric proteins, bacterial flagella, virus, and muscle. Generally they take a helical or tubular form, and sometimes a spherical or polyhedral form.

The phenomena of helical polymerization have been extensively investigated, and their similarity to crystallization or condensation has been demonstrated.[17, 18] In the polymerization of actin, a muscle protein, induced by neutral salts, a certain minimum concentration of monomers is necessary for the polymerization, and only the excess amount of monomers over this critical concentration is involved in the formation of polymers. In other words, the concentration of monomers in equilibrium with polymers is always constant, independent of the total concentration of protein. Then the relation between polymers and monomers resembles that between crystal and solute in a saturated solution. Moreover, the polymerization of G-actin at low salt concentration is accelerated by the addition of nuclei, fragments of F-actin made by sonic vibration. The effect of nuclei is clearly demonstrated in the polymerization of flagellin monomers to flagella; the presence of nuclei is essential because no polymerization takes place without the sonicated fragments of flagella.

Such a property as crystallization indicates the cooperative interaction between monomers, and further suggests the structural change associated with polymerization.

In reality a protein structure is often observed to change toward a more ordered state as a result of the polymerization. In particular, a large increase in the α-helical structure is reported for the polymerization of flagellin. Here we discuss a general treatment of the structural change associated with polymerization.

As a result of the polymerization, many secondary bonds are evolved in the contact area between protein molecules, though they decrease the entropy of chains in this area. Under our assumption, secondary bonds evolved between protein molecules cannot be distinguished from the secondary bonds originally existing within each protein. Then the free energy of an m-mer can be expressed by

$$\begin{aligned} G_p(T, X_p) = {} & \tfrac{1}{2}mN_0(1+\delta)(1-X_p) + \tfrac{1}{4}mN_0(1+\delta)ZJ\left(1-X_p^2\right) \\ & - \tfrac{1}{2}mN_0(1+\delta)\frac{\alpha}{1+\delta}T(1-X_p) \\ & + 4N_0(1+\delta)kT\left[\tfrac{1}{2}(1+X_p)\ln\tfrac{1}{2}(1+X_p) + \tfrac{1}{2}(1-X_p)\ln(1-X_p)\right] \end{aligned} \tag{22}$$

where $N_0\delta$ is the number of excess secondary bonds per monomer produced by the polymerization. Note that the entropy of the polymer chain per secondary bond α_p is equal to $\alpha/(1+\delta)$.

The most probable structural state of polymer X_{pm} is given by

$$\tanh^{-1} X_{pm} = \frac{A + X_{pm}}{BT} - \frac{C}{1+\delta} \tag{23}$$

Then the characteristic point p in Fig. 2, moves in the $+x$ direction from $(-C, -A)$ to $(-C/(1+\delta), -A)$. The structural state always changes to more ordered state with polymerization. The polymer becomes more stable to temperature than the monomer, for the denaturation temperature of polymer T_{pd} is given by

$$T_{pd} = \frac{A}{BC_p} = T_d(1+\delta) \tag{24}$$

Since the number of secondary bonds in the polymer is larger than in the monomer a steeper structural transition is expected for the polymer.

A model calculation[19] of the thermal stability of hemoglobin is shown in Fig. 9. In the calculation, the hemoglobin molecule is assumed to be tetramer of myoglobin, and the values of parameters determined for myoglobin (see Section VI) are used for the monomer. Furthermore, to account for the surface effect on Z, the effective number of the nearest neighbors,

$$Z_{\text{eff}}(N)=Z(1-N^{-\frac{1}{3}}) \tag{25}$$

is used, where $N=N_0$ for monomer and $N=4N_0(1+\delta)$ for tetramer. Here we assume that the number of nearest neighbors of a secondary bond situated on the surface is $Z-1$ and that the number of such secondary bonds is proportional to $N^{\frac{1}{3}}$.

C. Allosteric Effect

The allosteric effect is an important factor in the metabolic regulatory function of enzymes.

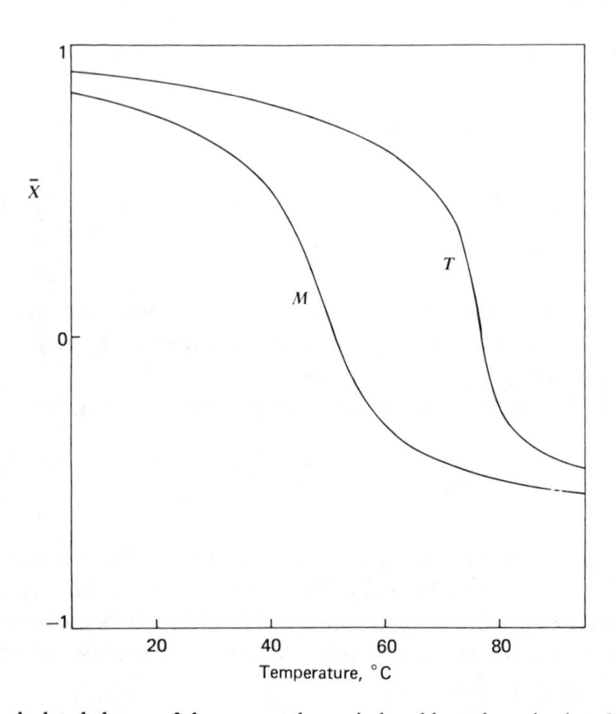

Fig. 9. The calculated change of the structural state induced by polymerization. M and T represent myoglobin and hemoglobin, a tetramer of myoglobin, respectively.

The initial rate of a simple enzymic reaction V is given by the Michaelis-Menten equation

$$V = \frac{V_m}{1 + K_m/(S)} \tag{26}$$

where V_m is the maximal velocity, K_m is the Michaelis constant, and (S) is the substrate concentration. The hyperbolic rate isotherm generated by this equation is shown in Fig. 10.

In the metabolic passways, there are two different modes of regulation both of which utilize nonhyperbolic isotherms as shown in Fig. 10. The first is the sigmoidal rate–concentration isotherm representing positive cooperativity, a small change in substrate concentration produces a large change in reaction rate over a certain concentration range. The sigmoidicity suggests the change in substrate binding with its concentration and the cooperative interactions between binding sites. Metabolic inhibitors cause the initial rate–concentration isotherm to become more sigmoidal, whereas metabolic activators make it less sigmoidal, as shown in Fig. 10. The binding sites of these activators and inhibitors are quite distinct from the catalytic sites, for their structures are dissimilar to those of substrates. Thus they are called allosteric effectors. The second mode is the regulation by negative cooperativity; the binding is strong at low substrate concentrations and becomes weaker as the concentration increases, as shown in Fig. 10. Then the essential character of an allosteric effect is that the specific binding of a small component to a protein molecule affects the affinity of another component for a sterically different bonding site.

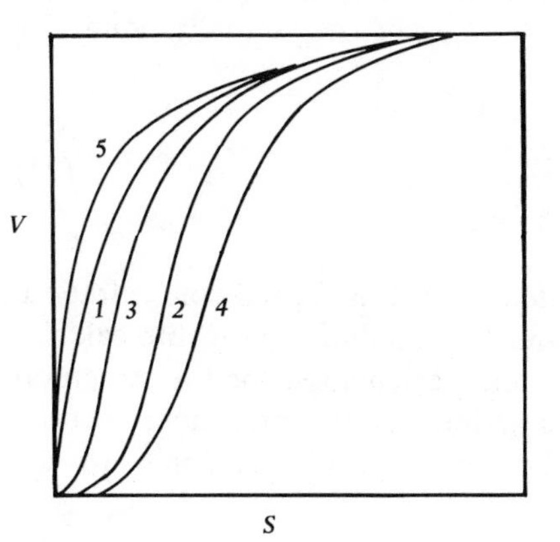

Fig. 10. Schematic plots of the steady-state initial velocity V versus the substrate concentration S, illustrating the cooperative allosteric effect. (*1*) The hyperbolic rate isotherm generated by (26). (*2*) The sigmoidal rate isotherm representing positive cooperativity. (*3*) and (*4*) The rate isotherms showing the effects of an activator and an inhibitor on the positive cooperativity. (*5*) The rate isotherm representing negative cooperativity.

Two limiting models have been proposed to describe the allosteric effect. One is the concerted model proposed by Monod et al.,[20] and the other is the sequential model by Koshland et al.[21] Both models assume that the allosteric enzyme is an oligomer consisting of several subunits that can take two conformation states. In the concerted model, the conformational changes of all subunits occur in a concerted manner. In the sequential model, however, the conformational change of a subunit is induced only when it binds a substrate, and the ligand-induced conformation change in one subunit alters its interactions with neighboring subunits.

In the present model, such an allosteric effect could be expected from the general nature of protein without any further assumptions. For the case of specific binding, the binding constant depends on the structural state X, and the binding changes the structural state itself; therefore, the binding of a specific component is affected by the binding of other components through the modification of structural state X. If a binding constant of a substrate s of an allosteric protein increases with the increase of the structural parameter, $X, i, e, \partial K_s/\partial X>0$, all the specific components that have a similar binding nature, $\partial K_a/\partial X>0$, might be allosteric activators and all the other specific components i that have an opposite binding nature, $\partial K_i/\partial X<0$, might be allosteric inhibitors. Thus an allosteric protein does not necessarily have to be an oligomer, but can be monomer, as far as the heterotropic effect is concerned.

Since most allosteric proteins exhibit the homotropic effect, which assumes several identical binding sites in a protein, allosteric proteins should be oligomers. A positive or negative homotropic effect can be explained in connection with the change of the structural state induced by the subunit association.

Results of a model calculation[19] to describe the positive homotropic effect in the oxygen binding of hemoglobin are shown in Fig. 11. The average structural state $\overline{X}$ and the average fraction of sites actually bound by the ligand

$$\overline{P}=\frac{\overline{K_a}}{1+\overline{K_a}} \tag{27}$$

of hemoglobin and myoglobin are plotted against the ligand concentration. Molecular parameters of myoglobin and hemoglobin used in the calculation are the same as those used in a model calculation for the structural change induced by polymerization described in the preceding section. Binding constants $K(T, X)$ of both hemoglobin and myoglobin are as-

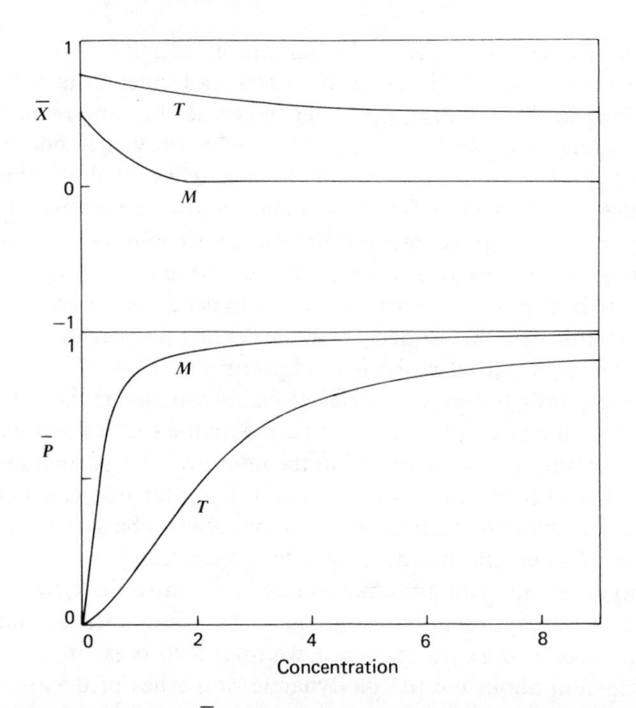

Fig. 11. The structural state $\bar{X}$ and the average fraction of ligand bound sites $\bar{P}$ are plotted against the ligand concentration to describe the positive homotropic effect of hemoglobin (see text). M and T represent myoglobin and hemoglobin, respectively.

sumed to be of the same form:

$$K(T,X)=K_0(T)\left(\frac{1-X}{2}\right)^3 \tag{28}$$

V. WATER STRUCTURE AND HYDROPHOBIC BONDS

A. Free Energy of Hydrophobic Bonds

Water has several unique physico chemical properties, indicating that there is a fundamental difference in structure between water and most other liquids. It has generally been accepted that these anomalous properties are essentially due to the existence of hydrogen-bonded quasicrystalline structures of water molecules in the water.

Frank and Evans[22] postulated the formation of such quasicrystalline structures (icebergs) around isolated, nonpolar molecules in aqueous solution. According to this concept, the water molecules become more ordered around the solute, with an increasing extent of hydrogen bonding in this region; this hypothesis can account for the negative enthalpy change and very large negative entropy change associated with the transfer of a nonpolar solute from a nonpolar medium into aqueous solution. The negative volume change associated with the transfer is also explained by the structural similarity between the iceberg around a hydrocarbon and the gas hydrate. The melting of the iceberg with increasing temperature can also account for the high partial molal heat capacities.

The X-ray crystallography of various globular proteins reveals one common feature that the polar residues are situated on the surface and that most of the nonpolar side chains are buried in the interior.[23] A hydrophobic bond in a protein is considered to be formed when nonpolar side chains tend to adhere so as to reduce the formation of unfavorable iceberg structures.[24] It is not an actual energetic bond, as is a hydrogen bond, but a statistical thermodynamic bond primarily maintained by entropic effects of the solvent. Hence the hydrophobic bond must be taken into account separately in our free energy expression if we intend to present a somewhat detailed discussion about the thermodynamic properties of denaturation.

Here we add the following assumption in our model.[26] When one secondary bond changes from the bonded to the unbonded state, γ water molecules are brought into new contact with nonpolar groups. The free energy difference per mole of water between the first layer around a nonpolar group and the bulk water is Δg_w. Note that the contributions to hydrophobic bonding from van der Waals interactions and internal rotations of bonds are included in our model through the bond energy ε and the chain entropy α, so that the term Δg_w represents only changes in the water structure.

Nemethy and Scheraga estimated[27] the thermodynamic parameters of hydrophobic bonds on the basis of their statistical thermodynamic treatments of water.[28, 29] As their calculations are in fairly good agreement with experimental data, we make use of their expression as,

$$\Delta g_w = a + bT + cT^2 \tag{29}$$

Then the Gibbs free energy of a protein molecule in aqueous solution takes the form

$$\begin{aligned} G_a(T, X) &= G_0 + N_u \gamma \Delta g_w \\ &= \tfrac{1}{2} N_0 [h(T) - s(T)T](1 - X) + \tfrac{1}{4} N_0 Z J (1 - X^2) + N_0 kT \\ &\quad \times \left[\tfrac{1}{2}(1+X) \ln \tfrac{1}{2}(1+X) + \tfrac{1}{2}(1-X) \ln \tfrac{1}{2}(1-X) \right] \end{aligned} \tag{30}$$

where

$$h(T)=\varepsilon+\gamma a-\gamma cT^2 \tag{31}$$

$$s(T)=\alpha-\gamma b+2\gamma cT \tag{32}$$

From this free energy, the average enthalpy and heat capacity are calculated

$$\overline{H}(T)=\tfrac{1}{2}N_0h(T)(1-\overline{X})+\tfrac{1}{4}N_0ZJ(1-\overline{X^2}) \tag{33}$$

$$\overline{C_p}(T)=-\tfrac{1}{2}N_0h(T)\frac{\partial\overline{X}}{\partial T}-\tfrac{1}{4}N_0ZJ\frac{\partial\overline{X^2}}{\partial T}-N_0\gamma cT(1-\overline{X}) \tag{34}$$

The thermodynamic parameters ε and α in (1) correspond to $h(T_d)$ and $s(T_d)$ respectively, and they must be regarded, respectively, as an "effective" bond energy and chain entropy containing the contribution from the water structure.

B. Water Structure and Denaturants

The introduction of the effect of water structure into our free energy expression contributes to the detailed analysis of the denaturation through the temperature dependence of the thermodynamic functions, $\overline{X}$, $\overline{H}$, and $\overline{C_p}$. The structural state $\overline{X}$ and the specific heat $\overline{C_p}$ calculated from the free energy with and without consideration of the water structure are shown in Fig. 12. Large discrepancies are observed between two calculations in the specific heat above the transition temperature.

Experimental results show that the denatured state of any protein has a much higher value of $\overline{C_p}$ than the native one, and generally agree with the calculations taking into consideration the water structure. The much higher heat capacity of the disorder state can be accounted for by the melting of the iceberg structures around the nonpolar groups exposed by the denaturation.

Since many protein denaturants, such as urea, GuHCl, and LiCl, are known as the "breakers" of water structure, their main action can also be accounted for by their effect on the iceberg, though it is discussed as their nonspecific binding effect (see Section IV.A.1). Therefore, we assume that the free energy difference of a water molecule, Δg_w, is simply reduced by the presence of denaturants. In practice, we assume that the presence of the denaturants effectively reduces the value of γ, the number of water molecules in the first layer around a nonpolar groups, without any alteration of Δg_w.

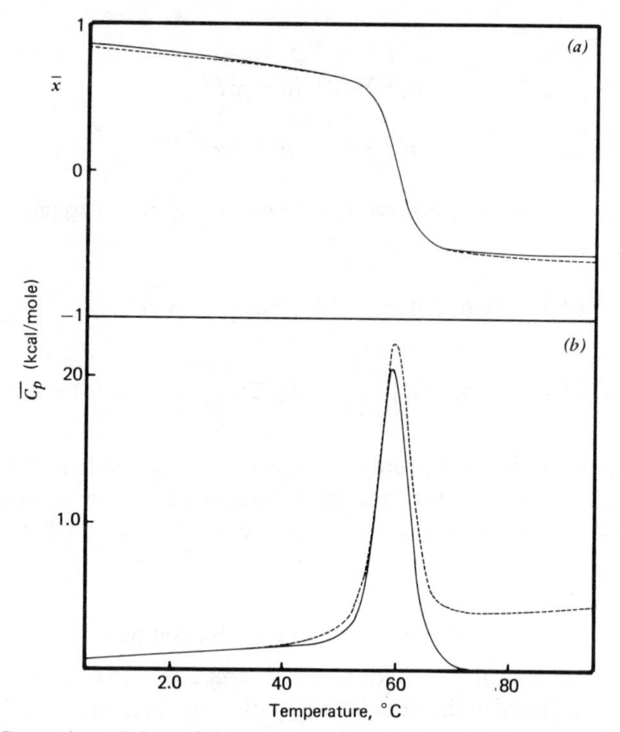

Fig. 12. Comparison of thermal denaturation curves with and without consideration of the water structure: (a) structural state $\overline{X}$; (b) specific heat $\overline{C_p}$. (...) Calculated from the free energy of (30), with $N_0=250$, $T_d=60°\text{C}$, $\varepsilon+\gamma a=-2$ kcal/mole, $\gamma c T_d^2=-3$ kcal/mol, and $ZJ=1.5$ kcal/mol [hence $h(T_d)=\varepsilon+\gamma a-\gamma c T_d^2=1$ kcal/mol]; (________) corresponding curves without consideration of the water structure calculated from (3) with $N_0=250$, $T_d=60°\text{C}$, $\varepsilon=1$ kcal/mole $ZJ=1.5$ kcal/mol.

The temperature dependence of the average structural state $\overline{X}$ at various values of γ, which represents the thermal denaturation of protein at various values of denaturant concentration c, are shown in Fig. 13. The same data are replotted in Fig. 14 as the isothermal denaturation at various values of temperature. The function $\gamma(c)$ is treated as a linearly decreasing function of c: $\gamma(c)=3.2-0.1c$. The sample calculation shown in Fig. 13 qualitatively agrees with what actually has been observed in the presence of denaturants by many authors, such as Pace and Tanford,[30] Tanford and Aune,[31] Salahuddin and Tanford,[32] and Brandts and Hunt,[33] as shown in Figs. 15a to 15d. The overall structural changes, including parabolic curves as shown in Fig. 13, would be common to all proteins, although the experimental observations do not always cover sufficient ranges of temperature.

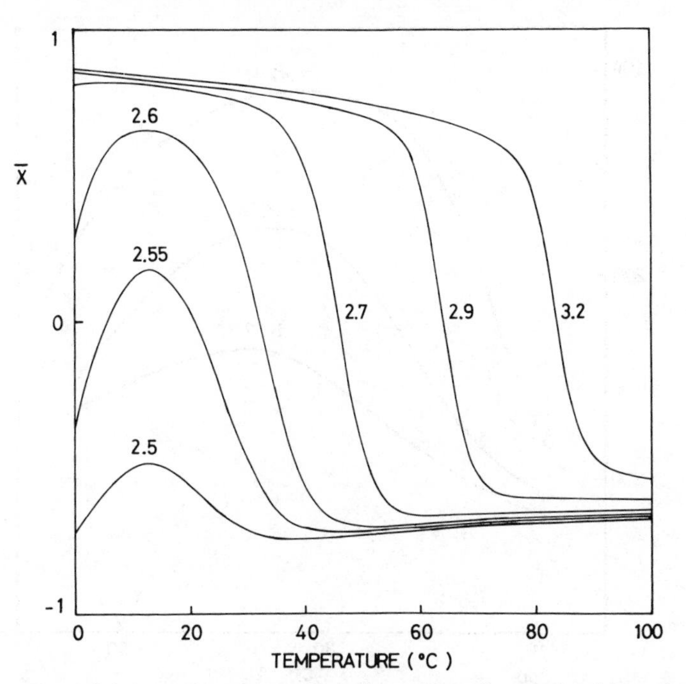

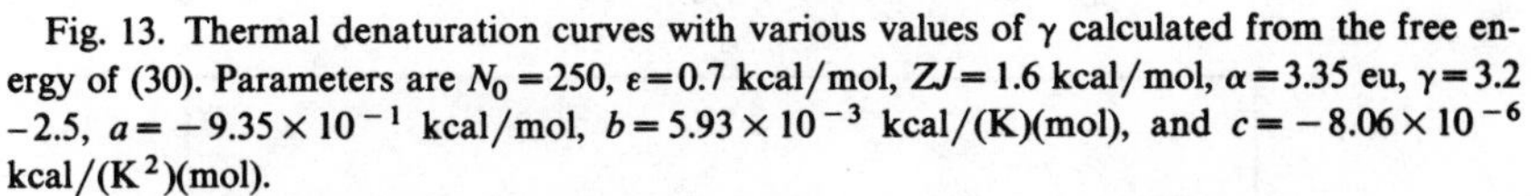
Fig. 13. Thermal denaturation curves with various values of γ calculated from the free energy of (30). Parameters are $N_0=250$, $\varepsilon=0.7$ kcal/mol, $ZJ=1.6$ kcal/mol, $\alpha=3.35$ eu, $\gamma=3.2-2.5$, $a=-9.35\times10^{-1}$ kcal/mol, $b=5.93\times10^{-3}$ kcal/(K)(mol), and $c=-8.06\times10^{-6}$ kcal/(K^2)(mol).

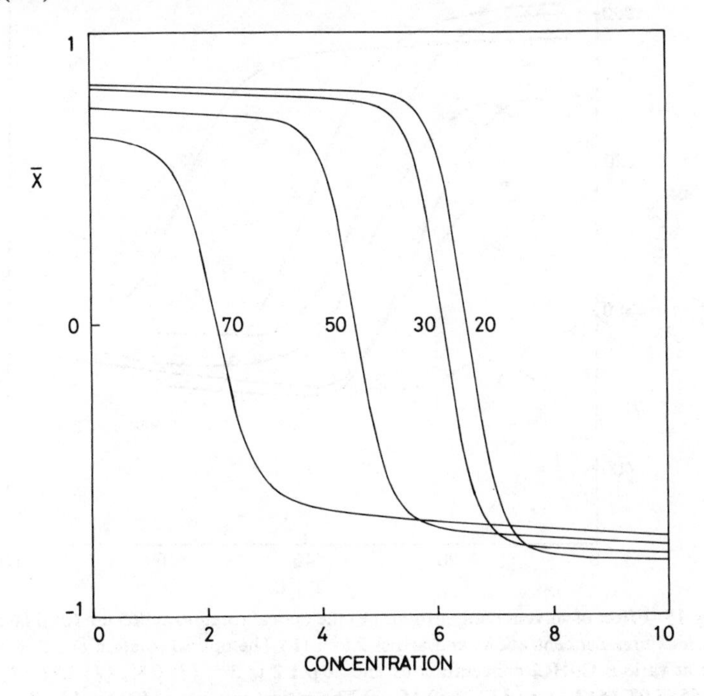

Fig. 14. Isothermal denaturation curves by protein denaturant at various values of temperature. The parameters are the same as those in Fig. 13. Denaturant concentration is arbitrarily scaled and is transformed to γ as $\gamma(c)=3.2-0.1c$, where c is the concentration of the denaturant.

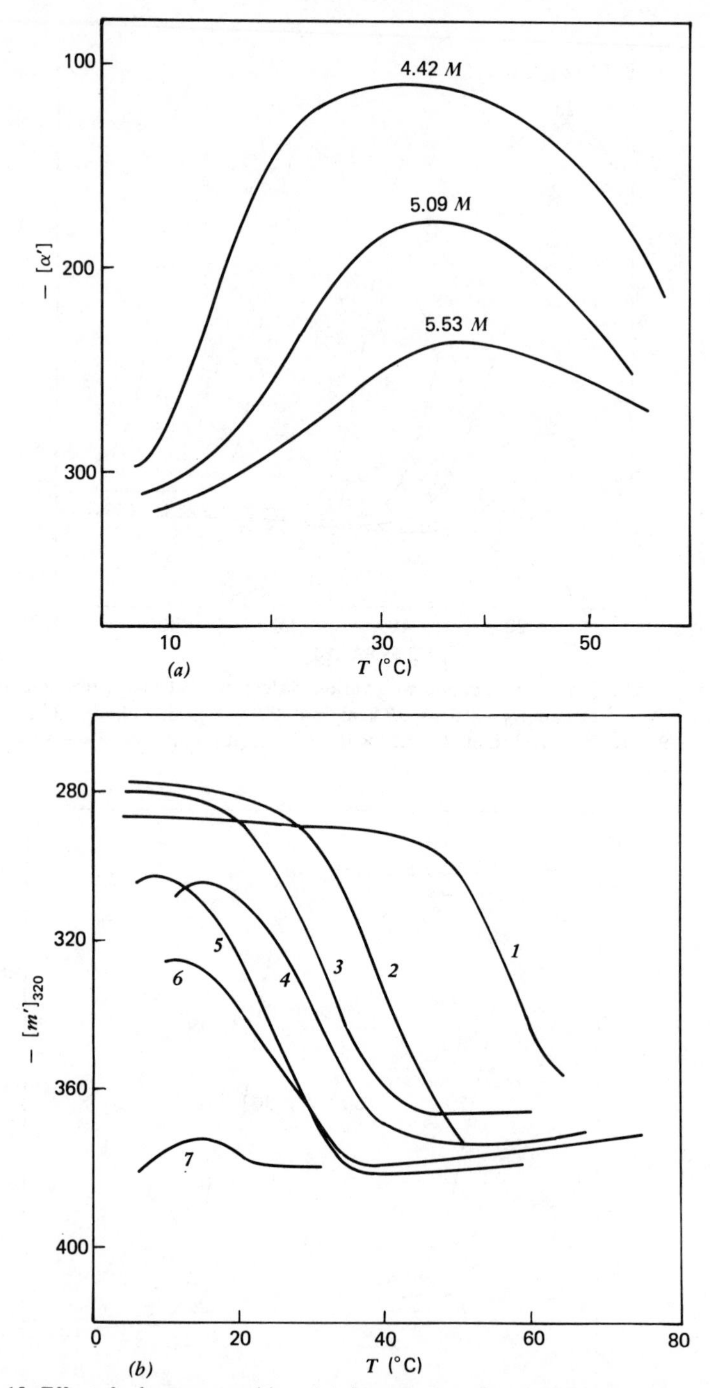

Fig. 15. Effect of solvent composition: (*a*) the optical rotation at 365 nm for β-lactoglobulin at various urea concentrations and at pH 2.5[30]: (*b*) The optical rotation at 320 nm for lysozyme at various GuHCl concentrations and at pH 2 to 3[31]: (*1*) 0.85, (*2*) 1.92, (*3*) 2.23, (*4*) 2.79, (*5*) 3.08, (*6*) 3.14, and (*7*) 3.50 *M*. (*c*) The optical rotation at 400 nm for ribonuclease at various GuHCl concentration and at pH 6[32]: (*A*) 4.31, (*B*) 3.44, (*C*) 3.24, (*D*) 3.10, (*E*) 2.79, and (*F*) 2.29 *M*. (*d*) The change in the extiction coefficient at 365 nm for ribonuclease A at various urea concentrations and at pH 3[33].

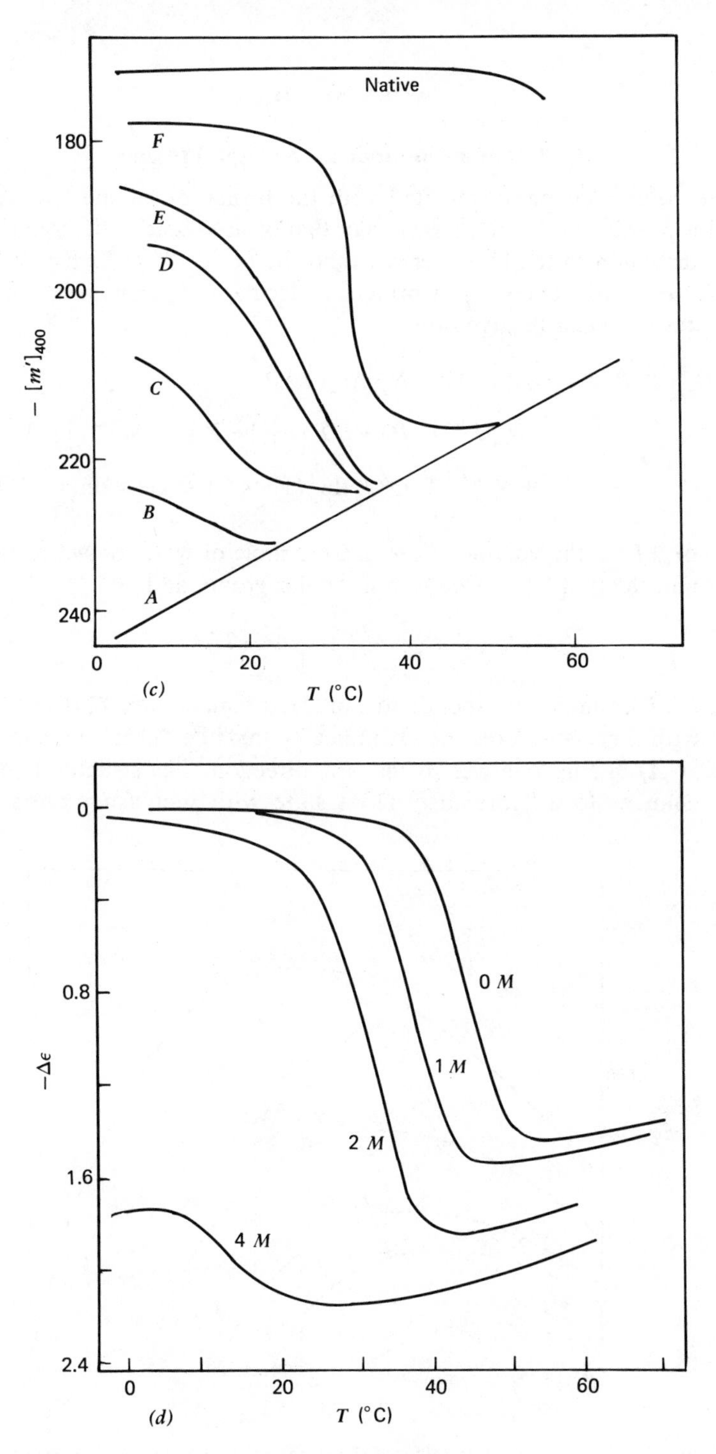

Fig. 15 (*continued*)

C. Denaturation Induced by High Pressure

The volume change associated with the formation of the hydrophobic bond is positive and much larger than that of usual energetic bonds. Thus, its contribution to the free energy cannot be neglected at high pressure.

The Gibbs free energy of a protein molecule in aqueous solution under high pressure must be given by

$$\begin{aligned}G_{ap}(T,P,X)&=G_a(T,X)-N_u\gamma\Delta v_w(T)P\\&=\tfrac{1}{2}N_0[h^p(T,P)-s(T)T](1-X)+\tfrac{1}{4}N_0ZJ(1-X^2)\\&\quad+N_0kT\left[\tfrac{1}{2}(1+X)\ln\tfrac{1}{2}(1+X)+\tfrac{1}{2}(1-X)\ln(1-X)\right]\end{aligned}\tag{35}$$

Here, $\Delta v_w(T)$ is the volume difference per mole of water between the bulk water and the first layer around a nonpolar group, and

$$h^p(T,P)=h(T)-\gamma\Delta v_w(T)P\tag{36}$$

Then $h^p(T)$, which corresponds to ε in (1) at temperature T, decreases linearly with increasing pressure. It indicates that the "characteristic point" $p(-C,-A)$ in Fig. 2 moves in the $+X$ direction. The structural state always changes to a more disordered state with increasing pressure. A

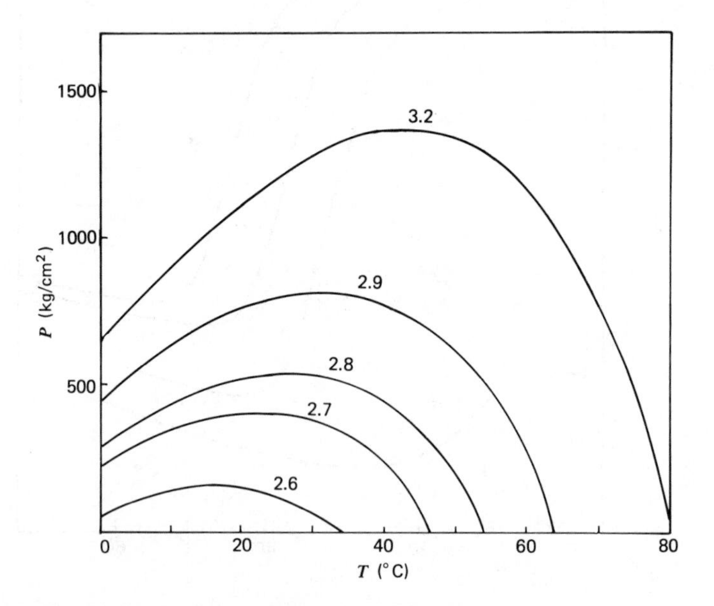

Fig. 16. The calculated relation between the denaturation pressure and temperature. The parameters are the same as those in Fig. 13.

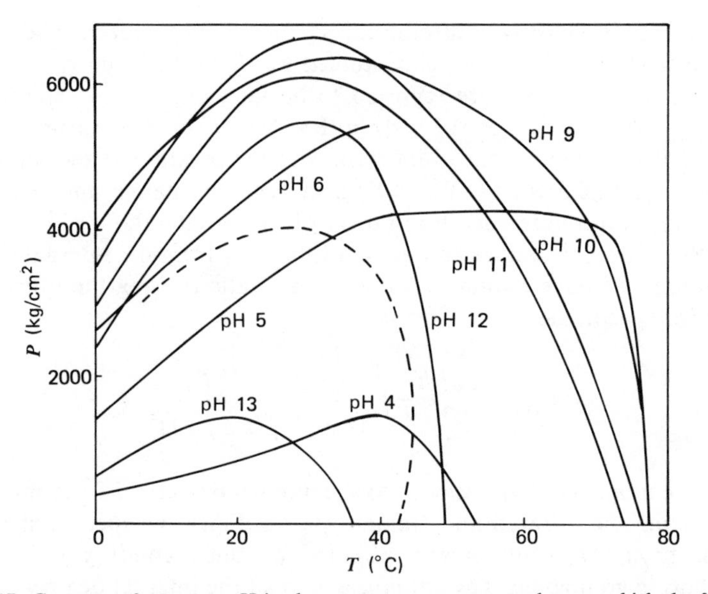

Fig. 17. Contours of constant pH in the pressure–temperature plane at which the free energy of denaturation $\Delta G = 0$. The native state is more stable than the denatured state inside each contour. (________) Metmyoglobin at various pH values[34]. (– – – –) Chymotrypsinogen at pH 2.07[35].

"structural transition" is expected at a certain pressure when $BT < 1$. Since the values of $v_w(T)$ estimated by Nemethy and Scheraga[27] decrease sharply with increasing temperature,[27] the denaturation pressure, at which the structural transition takes place, is expected to increase with increasing temperature when $T < T_d$. Figure 16 shows the temperature dependence of the denaturation pressure calculated with the same values of parameters as in Fig. 13. The sample calculation qualitatively agrees with the experimental results observed by Zipp and Kauzmann,[34] and Hawley,[35] as shown in Fig. 17.

VI. NUMERICAL ANALYSIS OF THERMAL DENATURATION

A. Methods

1. Specific Heat of Denaturation

Privalov and Khechinashvili[36] have investigated the thermal denaturation of five globular proteins with known three-dimensional structures, ribonuclease A, lysozyme, α-chymotrypsin, cytochrome c, and myoglobin, using a highly sensitive differential scanning calorimeter. The partial specific heats of all the proteins studied at 20°C, sufficiently below the transition temperatures, are reported to be the same within experimental error; C_p (20°C) = 0.31 ± 0.02 cal/(g)(K). Moreover, the partial specific heats of five proteins seem to increase with temperature at the same rate, $dC_p/dT = (2.2 \pm 0.5) \times 10^3$ cal/(g)(K^2), in both the native and denatured states, except in the temperature ranges of the denaturation.

As these absolute values contain contributions from the internal degrees of freedom, we have estimated[26] these contributions using the Einstein relation for specific heat,

$$C_{\text{vib}} = \sum_i \left(\frac{h\nu_i}{kT}\right)^2 \frac{\exp(h\nu_i/kT)}{(\exp(h\nu_i/kT)-1)^2} \tag{37}$$

The wave numbers of vibration ν_i to the internal degrees of freedom are assigned from the infrared and Raman spectra. Since the spectra of certain atomic groups are not greatly affected by the secondary bonds with neighboring groups, the wave numbers ν_i to all the internal degrees of freedom of a protein are determined from the corresponding values of amino acids assigned using spectra for small molecules.[37] Some unknown degrees of freedom are interpreted as those of backbone bending, of which the wave number is assumed to be 450 cm^{-1}. The internal rotations are expected to be frozen-in almost completely in a native state by the secondary bonds, but become free on denaturation. Therefore, corresponding degrees of backbone bending are altered to those of internal rotations, the wave number of which is assumed to be 100 cm^{-1}. The calculated results shown in Table I are nearly the same for these five globular proteins: C_p(20°C) = 0.25

TABLE I
Calculated Specific Heat per Gram of Protein [cal/(g)(K)]

	Native		Denatured	
Protein	300°K	350°K	300°K	350°K
Cytochrome c	0.252	0.304	0.277	0.323
Ribonuclease	0.246	0.295	0.270	0.314
Lysozyme	0.249	0.299	0.273	0.317
Myoglobin	0.254	0.306	0.279	0.325
α-Chymotrypsin	0.250	0.301	0.274	0.320
Average	0.250	0.301	0.275	0.320

cal/(g)(K) and $dC_p/dT = 1\times10^{-3}$cal/(g)(K^2). Thus the estimated value in the native state is in fairly good agreement with the expreimental result. By the subtraction of these values from the reported specific heat, the specific heat of denaturation is obtained.

2. *Least-Squares Method*

The thermodynamic parameters of our model have been determined by the least-squares fit of the specific heat of denaturation. Since the denaturation temperature T_d is known experimentally, five parameters to be determined, N_0, ε, α, ZJ, and γ, can be reduced to four, N_0, $\varepsilon'(=\varepsilon+\gamma a)$, $\beta'(=-\gamma c)$, and ZJ. The optimization function is

$$I(P) = \sum_{i=1}^{\text{no. of data}} \left[\overline{C_p}(P, T_i) - C_{p_i}^{\text{exp}}\right]^2 \tag{38}$$

where P represents the parameters to be fitted and the superscript exp denotes an experimental value. The average specific heat $\overline{C_p}$ is given by (34). The optimized point (local minimum of the function I) is determined by the variable metric method of Fletcher and Powell.[38]

On account of the very weak dependence of the optimization function on N_0, the value of N_0 determined in our optimization routine is rather ambiguous. The value of three energies ε', $\beta' T_d^2$, and ZJ, have been obtained by the least-squares method using the values of N_0 given in Table II. These

TABLE II
Calculated Thermodynamic Parameters for Five Globular Proteins

Protein	pH	T_d	N_0	ZJ	ε	α	γ	Variance
Myoglobin	12.0	54	269	1.45	0.846	4.18	3.73	2.72
	11.5	63		1.49	0.789	3.90	3.61	4.97
	11.1	70.5		1.54	0.788	3.78	3.46	5.16
	10.7	76		1.56	0.724	3.89	4.24	2.45
Cytochrome c	3.0	59.5	186	1.48	0.851	4.06	3.58	2.64
	3.7	70		1.57	0.872	4.24	4.04	4.08
	4.6	77.5		1.66	0.733	3.55	3.49	3.00
α-Chymotrypsin	2.6	44	378	1.76	0.702	3.29	2.51	6.05
	3.0	52		1.56	0.812	3.80	2.99	3.43
	4.0	57.5		1.48	0.965	4.46	3.52	9.19
Lysozyme	2.0	56	215	1.54	0.898	3.93	2.75	2.36
	2.5	66.5		1.60	0.901	3.93	2.90	3.04
	4.5	78.5		1.65	0.854	3.73	3.01	13.3
Ribonuclease A	2.7	43	204	1.50	0.904	3.75	2.11	10.1
	3.7	55		1.55	1.02	4.35	2.87	4.10
	5.4	63.5		1.62	0.937	3.94	2.65	4.53

N_0 values are tentatively adopted by assuming that the number of secondary bonds N_0 is proportional to the molecular weight and that the value for ribonuclease A is around 200. For ribonuclease A, three parameters have been determined from a similar analysis of the enthalpy by changing values of N_0, and we have concluded that the value of 200 seems to be most suitable to reproduce the experiments for all pH values.[26] The parameters a, b, and c in (29) are determined for each protein from its amino acid composition, averaging the numerical values calculated[27] for aliphatic and aromatic side chains.

B. Thermodynamic Parameters of Globular Proteins

The calculated thermodynamic parameters and the structural parameters of five globular proteins are shown in Tables II to IV. Here we discuss the thermodynamic parameters of these proteins referring to their primary and tertiary structures.

1. Number of Secondary Bonds

On account of the high cooperativity between secondary bonds, exact determination of the value of N_0 is rather difficult. If we take two actual

TABLE III
Structural Parameters of Five Globular Proteins[a]

	Mgl	Cyr	Chm	Lys	RNA
Molecular weight	17,900	12,400	25,200	14,300	13,600
Number of residues	153 + heme	104 + heme	241	129	124
Number of disulfide bonds	—	—	5	4	4
Number of polar contacts (b–b)[36]	98	37	105	54	53
(b–s)	22 133	23 70	54 173	20 89	19 81
(s–s)	13	10	14	15	9
Number of nonpolar contacts (4.0 Å)[36]	213	136	238	108	90
(4.5 Å)	456	298	586	302	209
Percentage of α-helix content[26]	65–77		12–13	28–42	6–18
β-Structure content	0		23–32	10	36
Random coil content	32–23		66–55	62–48	58–46
Number of single bonds about which rotation can occur[b]	725	487	1053	578	563
Partial specific volume[40]	0.743	0.72	0.721	0.703	0.692

[a]Abbreviations: Mgl, myoglobin; Cyr, cytochrome c; Chm, α-chymotrypsin; Lys, lysozyme; RNA, ribonuclease A; b–b, backbone–backbone; b–s, backbone–side chain; s–s, side chain–side chain.

[b]Our computation.

TABLE IV
Molecular Parameter of Five Globular Proteins

	Mgl	Cyr	Chm	Lys	RNA
Value of N_0	269	186	378	215	204
Average value of ε	0.787	0.819	0.826	0.884	0.952
Average value of ZJ	1.51	1.57	1.60	1.59	1.56
Number of polar contacts per N_0	0.494	0.376	0.458	0.414	0.397
Number of nonpolar contacts (4.0 Å) per N_0	0.792	0.731	0.630	0.502	0.441
Number of total contacts per N_0	1.29	1.11	1.09	0.916	0.838
Relative fraction of polar contacts n_{H}	0.384	0.340	0.421	0.452	0.474
Relative fraction of nonpolar contacts $n_{\mathrm{H}\phi}$	0.616	0.660	0.579	0.548	0.526
Average value of α	3.94	3.95	3.85	3.87	4.01
Value of α per residue, $m\alpha$	6.93	7.06	6.04	6.45	6.60
Number of internal rotations per residue, w	4.74	4.68	4.37	4.48	4.54
Number of minima of rotational potential, $x(x^w = e^{m\alpha/R})$	2.09	2.14	2.00	2.06	2.08
Average value of γ	3.76	3.70	3.00	2.89	2.54
Number of nonpolar contacts within 4.0 Å/100 g	1.19	1.10	0.944	0.755	0.662
Number of nonpolar groups per residue[a]	1.38	1.11	1.26	1.06	1.00
Average value of $\gamma/n_{\mathrm{H}\phi}$	6.10	5.61	5.18	5.27	4.82

[a] Calculated from the amino acid composition. Number of nonpolar groups: Ala, 1; Val, 3; Leu, 4; Ile, 4; Met, 4; Cys, 2; Pro, 3; Phe, 7.

secondary bonds, instead of one, as a unit, the values of ε, α, and γ may be roughly doubled. Nevertheless, our estimation of N_0 seems to be fairly good in relation to the sum of numbers of polar contacts (hydrogen bonds) and nonpolar contacts within 4.0 Å (hydrophobic bonds) in Table IV. Hence our secondary bond unit can be regarded as one of these actual bonds.

2. *Bond Energy*

The value of ε averaged over all pH values in Table IV seems to be around 0.8 kcal/mole, a little higher than RT at room temperature. This energy value must be related mainly to the hydrogen bond energy and the van der Waals interaction energy of the hydrophobic bonds. If we assume that the average bond energy ε is expressed by

$$\varepsilon = \varepsilon_{\mathrm{H}} n_{\mathrm{H}} + \varepsilon_{\mathrm{H}\phi} n_{\mathrm{H}\phi} \tag{39}$$

where n_{H} and $n_{\mathrm{H}\phi}$ are the relative fraction of polar and nonpolar contacts, respectively, the energy of a hydrogen bond ε_{H} and that of van der Waals interaction of a hydrophobic bond $\varepsilon_{\mathrm{H}\phi}$ can be estimated from the differences in ε among five proteins. The least-squares fitted values are: $\varepsilon_{\mathrm{H}} =$ 1.4 kcal/mole and $\varepsilon_{\mathrm{H}\phi} = 0.44$ kcal/mole.

The van der Waals contribution to the hydrophobic bonds can be estimated according to the Lennard-Jones potential function. It gives, for example, the values of 0.15 kcal/mole for a pair of aliphatic groups and 0.50 kcal/mole for a pair of aromatic groups.[27] A number of estimates of the enthalpy change involved in breaking one hydrogen bond in aqueous environment spread from 0 to 2 kcal/mole.[6] Lower estimations may arise from the presence of nonpolar groups around the interacting donor and acceptor groups.[39] In the absence of nonpolar groups, the enthalpy change on formation of a hydrogen bond seems to be around 1.5 kcal. Thus the agreement between these estimates and our results is good. As the free energy contribution from the water structure to a hydrophobic bond can be estimated as about 0.7 kcal/mole at room temperature, the total free energy of a hydrophobic bond is not so different from that of a hydrogen bond. This fact seems to support our "equivalent and uniform" assumption of the secondary bonds in the protein structure.

3. *Cooperative Energy*

The cooperative energy ZJ, which mainly reflects the connectivity of the chain, is nearly constant for all the proteins studied, in the range 1.5 to 1.6 kcal/mole.

4. *Chain Entropy*

The chain entropy α of a completely disordered state ($X = -1$) is around 4 eu per secondary bond. Since the rotations around single bonds in a protein molecule are supposed to be completely frozen at $X = 1$ and to be released at $X = -1$, we can estimate the average number of minima of the rotational potential, or the number of possible configurations taken by one single bond from the value of α. The actual number of single bonds about which rotation can occur, such as C—C, C—O, C—N, and C—S, is found from the known primary structures and is shown in Table III. The results show almost the same values, a little higher value than 2 for all the proteins considered (see Table IV).

The rotational potential function of the C—C or C—N usually has three minima. In a protein molecule, however, the number of possible configurations must be considerably reduced by the steric restrictions. Such a small number of possible configurations is estimated from the energy calculations of several oligopeptides. The conformational entropy also must be restricted by the cross-linked S—S bridges. Our results in Table IV, however, do not show any significant difference between the proteins with and without S—S bonds. They suggest that the hemegroup in myoglobin and cytochrome c would restrict the conformational entropy to the same extent as about four S—S bridges.

5. *Hydrophobicity*

The value of γ, or the number of water molecules brought into new contact with nonpolar groups, is better related to the number of nonpolar contacts than to the number of nonpolar groups themselves, such as CH_3, CH_2, and CH, as shown in Table IV. Thus the average number of water molecules affected by one hydrophobic bond, $\gamma/n_{H\phi}$, is roughly constant, being between 5 and 6. The result is in good agreement with the estimation reported by Nemethy and Scheraga.[27] They calculated such a number from a molecular model on forming a hydrophobic bond of maximum strength between two isolated side chains and obtained a value ranging from 4 to 12.

C. Individuality of Protein Molecules and the Model

So far as we have analyzed the thermal denaturation of five globular proteins, the difference in the parameters among different proteins is not observed for the chain entropy α and the cooperative energy ZJ. The average number of rotational minima around a single bond in the completely disordered state should reflect the steric restrictions by the neighboring residues on a chain. As no definite correlation between two nearest neighboring amino acids in the primary structure is observed in usual proteins, the result is reasonably understood. Similarly, the cooperative energy ZJ, which reflects mainly the connectivity of the chain, is expected to have the same value for each protein.

The individuality of a protein, as far as its structural stability is concerned, is mainly represented by the average bond energy ε and the parameter γ specifying the hydrophobic nature of protein. The bond energy ε is well related to the numbers of hydrogen bonds and van der Waals interactions in the tertiary structure of each protein.

Two different physical properties or concepts are included in the parameter γ. One is the original concept of the average number of water molecules that establish new contacts with the nonpolar groups of a protein resulting from the breakage of hydrophobic bonds. Therefore, it is directly related to the number of hydrophobic bonds in the tertiary structure. In addition, the change in free energy of the water structure, Δg_w, due to the change in the solvent composition is conveniently expressed by the change in the parameter γ. Many factors, such as the addition of denaturants or salts and, even the change of pH, are expected to affect the water structure and hence to change γ.

In the present model, the structure of a globular protein has been expressed by the state of equivalent secondary bonds uniformly distributed in a lattice using the essential molecular parameters at the chain level. The "equivalent and uniform" assumption of the model is extensively supported by the fact that the average values of the molecular parameters ε, α, and γ determined are quantitatively explained by the molecular properties and the tertiary structures of proteins.

This fact strongly suggests that the values of the parameters can be reasonably determined from the nature of the bonds, the primary structure, and the tertiary structure at 0°K or in the crystal, and that the theory could predict the structural stability of individual proteins.

VII. DYNAMICS

A. Dynamic Aspects of the Protein Structure

As is described earlier, equilibrium properties of the reversible denaturation of globular proteins have been well analyzed by a simple two-state model. Early kinetic studies of the denaturation indicated one single relaxation time and confirmed the simple two-state model. By the recent fast relaxation measurements of the denaturation, however, the defect of the simple two-state model was revealed. The kinetics of the denaturation are generally biphasic and sometimes triphasic; the major reaction is the slow reaction for which relaxation time is in the second to minute range, and the other one or two minor reactions are fast reactions having relaxation times in the millisecond range. Apparently the phenomena are not consistent with the simple two-state model, and many kinetic models have been proposed.

By the temperature and pH jump methods, the kinetics of ribonuclease A have been extensively studied by Baldwin et al. To explain the observed fast and slow reactions, they assumed[4] the native form (N) and two denatured forms (D_1 and D_2). They attributed the fast reaction to the main folding and unfolding process ($N \leftrightarrow D_1$) and the slow reaction to the interconversion of two unfolded forms ($D_1 \leftrightarrow D_2$). Thus the reaction mechanism is

$$N \overset{\text{fast}}{\leftrightarrow} D_1 \overset{\text{slow}}{\leftrightarrow} D_2 \tag{40}$$

Ikei et al.[42] reported equilibrium and kinetic data for the reversible denaturation of horse heart ferricytochrome c by guanidine hydrochloride. They tried to analyze the results by three or four species models and proposed the following mechanism as the most plausible:

$$N \rightleftarrows X_1 \rightleftarrows D \rightleftarrows X_2 \tag{41}$$

where X_1 is an intermediate on the pathway between the native (N) and denatured (D) states, whereas X_2 is a relatively highly ordered state on a dead-end pathway. A cooperative sequential model was proposed by Tsong et al.,[43] in which kinetic behavior is accounted for by two processes: the cooperative nucleation process, which has a very slow rate, and the sequential process involving many intermediate states. Unfortunately, there is no direct evidence for the several intermediate states proposed by these models, and there is no direct reason to select a certain reaction scheme. The general explanation of phenomena is still open to further discussion.

Highly resolved NMR spectra observed in the reversible unfolding of ribonuclease A[12] showed the fast interconversion between different native conformations in addition to the large cooperative transition with slow interconversion rate. Similarly, the local fluctuation in the native state and large interconversion between the native and the denatured state are indicated by hydrogen–deuterium exchange measurements[9] or by the dynamic accessibility to proteolytic enzymes.[44]

All these phenomena, together with the kinetics of the reversible denaturation, could be summarized by the two different structural changes. The first is the major cooperative transition between the native and denatured states, and the second is the minor structural changes or local fluctuations within the native or denatured states. Generally the rate of the first major process is much slower than the rate of the second, and its temperature dependence is much larger than that of the second.

Here we present a stochastic approach based on our statistical thermodynamic model to protein dynamics.[45] The theory predicts the relaxation times of the two structural changes mentioned above. It can account for the dynamic phenomena in a unified manner.

B. Stochastic Theory

1. Master Equation

The time course of the structural state, X or N_b, is assumed to obey the Markovian master equation

$$\begin{aligned}\frac{\partial P(T, N_b, t)}{\partial t} &= -[W_{ub}(N_b \to N_b + 1) + W_{bu}(N_b \to N_b - 1)]P(T, N_b, t)\\ &\quad + W_{ub}(N_b - 1 \to N_b)P(T, N_b - 1, t)\\ &\quad + W_{bu}(N_b + 1 \to N_b)P(T, N_b + 1, t) \qquad (42)\end{aligned}$$

where $P(T, N_b, t)$ is the probability of finding the structural state N_b at time t and temperature T.

The transition probability $W_{ub}(N_b \to N_b + 1)$ is assigned for the formation of one secondary bond from N_b to $N_b + 1$, and $W_{bu}(N_b \to N_b - 1)$ is for the breakage of one secondary bond from N_b to $N_b - 1$. These transition probabilities must be chosen so as to satisfy the equilibrium probability density

$$P_e(T, N_b) = Z^{-1} \exp\left[-\frac{G_a(T, N_b)}{kT} \right]$$
$$= Z^{-1} \frac{N_0!}{N_b! N_u!} \exp\left[-\frac{N_u(h(T) - s(T)T) + (N_u N_b / N_0) ZJ}{kT} \right] \tag{43}$$

where Z is the partition function. The transition probabilities can generally be a complex function of local structures, but the simplest possible assumption is that the rate factor depends only on the structural state N_b:

$$W_{ub}(N_b \to N_b + 1) = \kappa(T) N_u \exp\left[\frac{h(T) - s(T)T + (2N_b / N_0) ZJ}{2kT} \right] \tag{44}$$

$$W_{bu}(N_b \to N_b - 1) = \kappa(T) N_b \exp\left[-\frac{h(T) - s(T)T + (2N_u / N_0) ZJ}{2kT} \right] \tag{45}$$

where $\kappa(T)$ is the micro transition probability between the bonded and unbonded states of a single isolated secondary bond. By the absolute rate theory of Eyring, $\kappa(T)$ should be given by

$$\kappa(T) = \frac{kT}{h} \exp\left(-\frac{\Delta E^* - T\Delta S^*}{kT} \right) \tag{46}$$

where ΔE^* and ΔS^* are the activation energy and entropy of the isolated secondary bond.

2. Two Relaxation Processes

By the change of thermodynamic variables of a solvent with which a protein molecule is in thermal equilibrium, Gibbs free energy of the protein is altered and the initial probability density is redistributed into the final equilibrium distribution through two different relaxation processes. The first is the small shift of a peak in the initial probability density to a new stable position corresponding to the shift of a minimum in the free energy. The second is the flow of the probability density from one stable state to the more stable state passing over the maximum of the free energy. Thus it is the large structural change corresponding to the cooperative structural transition.

Apparently, the first process is the relaxation around the new stable state and is equivalent to the relaxation associated with fluctuation near the stable state. The relaxation time of this process in the molecular field Ising model was treated by Griffiths et al.[46] More general treatment has been reported by Kubo et al.[47] As the transition probabilities are written in the form proportional to the total number of the secondary bonds N_0, they are expressed in terms of the quasicontinuous variable X and time t, and the master equation (42) can be replaced by a corresponding partial differential equation. As we discuss the realxation process under the same temperature, we omit a variable T in functions hereafter. For large N_0, the probability $P(X,t)$ is a rapidly varying function of X, but $\phi(X,t)$, defined by

$$P(X,t)=\exp[N_0\phi(X,t)] \tag{47}$$

should be a less rapidly varying function. Suppose that $\phi(X,t)$ has a smooth maximum in X at $X_m(t)$ and expand $\phi(X,t)$ as

$$\phi\left(X+\frac{1}{2N_0},t\right)\simeq\phi+\frac{2}{N_0}\phi_x$$

$$\phi\left(X,t+\frac{1}{N_0}\right)\simeq\phi+\frac{1}{N_0}\phi_t \tag{48}$$

Here subscripts denote partial derivatives. Inserting this approximate expression into the master equation, we get

$$\kappa^{-1}(e^{\phi_t}-1)=\frac{1}{N_0}W_{ub}(X)(e^{-2\phi_x}-1)+\frac{1}{N_0}W_{bu}(X)(e^{2\phi_x}-1) \tag{49}$$

Near the maximum point $X_m(t)$, (49) is reduced to

$$\phi_t=-\kappa C_1(X)\phi_x \tag{50}$$

where

$$C_1=\frac{2}{N_0}[W_{ub}(X)-W_{bu}(X)] \tag{51}$$

Then $X_m(t)$ satisfies the equation

$$\frac{dX_m}{dt}=+\kappa C_1(X_m) \tag{52}$$

As C_1 vanishes at the equilibrium point X_e the structural change around the equilibrium point is expressed by

$$\Delta \bar{X}(t) = \Delta \bar{X}(0) \exp(k_e t) \tag{53}$$

$$k_e = \kappa \left(\frac{\partial C_1}{\partial X} \right)_{X_e} = 2\kappa(T) \left[1 - \frac{ZJ}{2kT}(1 - X_e^2) \right]$$
$$\times \exp\left(\frac{ZJ}{2kT} \right) \cosh \left\{ \frac{1}{2kT} [h(T) - s(T)T + ZJX_e] \right\} \tag{54}$$

The rate of the second process corresponds to the relaxation time from the metastable to stable states in the molecular field Ising model, which is also considered by Griffiths et al.[46] To discuss longer time behavior, define the function $q(X, t)$ by

$$P(X, t) = q(X, t) P_e(X) \tag{55}$$

so as to approach 1 as t become very large. Then from the master equation, we get

$$q_t = k(T) \left[C_1(X) q_X + \frac{1}{N_0} C_2(X) q_{XX} \right] \tag{56}$$

where

$$C_2(X) = \frac{2}{N_0} [W_{ub}(X) + W_{bu}(X)] \tag{57}$$

The second term in (56) should be important near extreme points of the free energy where $C_1 = 0$. Near the maximum point of the free energy, X_u, a static solution $q = f(X)$ to (56) satisfies the approximate equation

$$\frac{1}{N_0} C_2(X_u) f_{xx}(X) = -(X - X_u) C_{1x}(X_u) f_x(X) \tag{58}$$

The general solution to (58) is given by an error function as

$$f(X) = A + B \operatorname{erf}\left[(N_0 r)^{1/2} (X - X_u) \right] \tag{59}$$

$$r = \frac{C_{1x}(X_u)}{2C_2(X_u)} > 0 \tag{60}$$

where A and B are arbitrary constants. Roughly speaking, $f(X)$ is a step function at $X=X_u$. Except for the immediate vicinity of X_u, that is, $|X-X_0|\geq 0(N_0^{-1/2})$, it has a value of $A+B$ on the metastable side of X_u, and a value of $A-B$ on the stable side of X_u. Thus, at the stationary condition after a large values of t, an approximate solution (56) should be expressed by the very slow time dependence of A and B as

$$q(x,t)=A(t)+B(t)\mathrm{erf}\left[(Nr)^{1/2}(X-X_u)\right] \tag{61}$$

The time dependence of A or B can be determined from the rate of the "probability flow" path through X_u. This flow is expressed by the transition probability $W(X)$ and the stationary solution $q(X,t)$ in the immediate vicinity of X_u. According to the calculation by Griffiths et al.,[46] the time dependence of B is given by

$$B(t)=B_0 e^{-k_m t} \tag{62}$$

In our case of proteins, the change in the peak height of the probability density at X_e is given by

$$P(T,X_e,t)=P(T,X_e,0)e^{-k_m t} \tag{63}$$

The rate k_m is calculated as

$$k_m=\frac{\kappa(T)}{\pi}\left[\frac{ZJ}{2kT}(1-X_u^2)-1\right]^{1/2}\left\{\frac{1}{1-X_e}-\frac{ZJ}{2kT}\right\}^{1/2} \times\exp\left(\frac{ZJ}{2kT}\right)\exp\left\{-\frac{1}{2kT}\left[G(X_u)-G(X_e)\right]\right\} \tag{64}$$

using a Gaussian approximation to P_e near X_e.

C. Kinetics of Protein Denaturation

1. Relaxation Times and Molecular Parameters

The rate of the fast relaxation, k_e, is independent of the size of proteins, N_0. The rate of the relaxation from the metastable state, k_m, however, depends on N_0 because the free energy difference $G(X_u)-G(X_e)$ in (14) is proportional to N_0. The size dependence of k_m near the transition temperature is shown in Fig. 18. The same values of the parameters for models of structural transition used in Fig. 3 are used for the calculation, but with varying values of N_0. As is shown in Fig. 18, the theory predicts that the

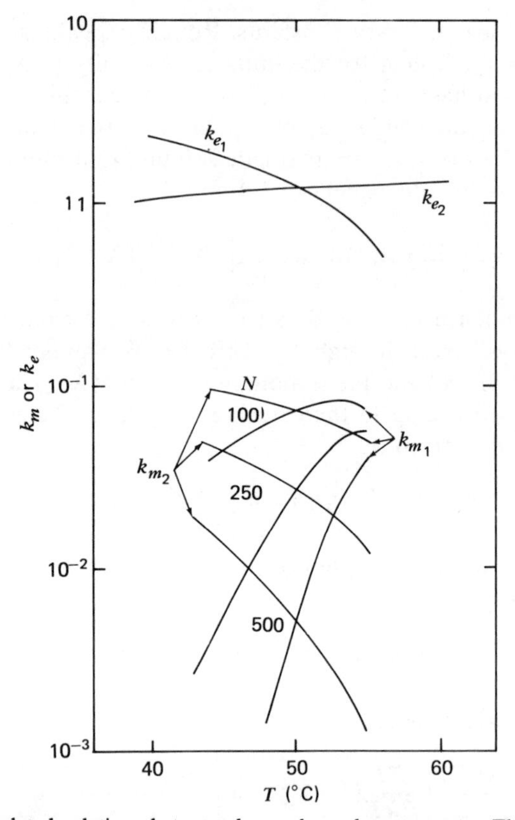

Fig. 18. The calculated relations between k_m or k_e and temperature. The same values of parameters are used as in Fig. 3, but the values of N_0 are varied.

ratios k_e/k_m are of the order of 10 or 100 for most globular proteins, which have a few hundred secondary bonds. Moreover, temperature dependence of the slower rate k_m is greater than that of the faster rate k_e. These results agree well with the experimental results for several proteins.

The relations between the ratio k_e/k_m and other molecular parameters can be estimated generally by (65). The ratio k_e/k_m at the denaturation temperature T_d, which must be the most typical value obtained form kinetic experiments, is given by

$$\frac{k_e(T_d)}{k_m(T_d)} = \frac{-1}{2\pi}\left[\frac{\tanh^{-1}X_d - X_d}{(1-X_{d^2})\tanh^{-1}X_d - X_d}\right]^{1/2} \times\left[(1+X_d)^{(2+X_d)/4}(1-X_d)^{(2+X_d)/4}\right]^{N_0} \tag{65}$$

depending only on N_0 and X_d, the most probable value of X at T_d. Equation (65) indicates that the ratio is the increasing function of X_d and N_0. Since the value of X_d expresses the extent of the two-state-like nature of the transition, that is, the height of the free energy barrier between two stable states, the dependence given by (65) reflects mainly the strong decrease of k_m with the increase of X_d.

Dependence of k_e or k_m on various values of ε and α is shown in Figs. 19 and 20 respectively. As X_d is the monotonously increasing function of C/A

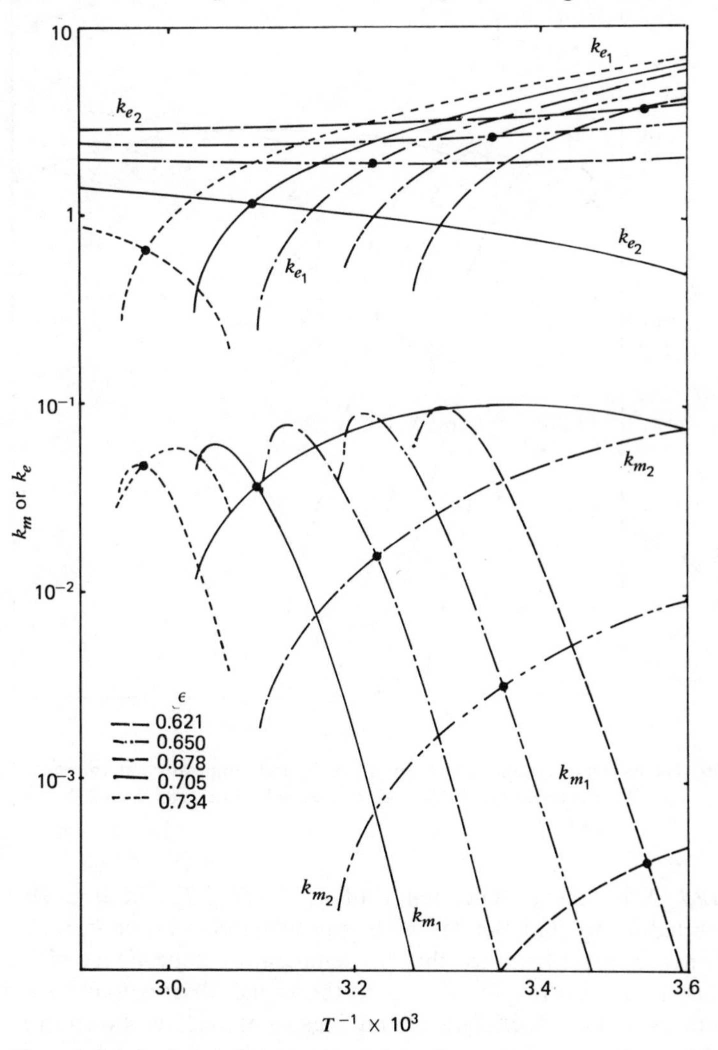

Fig. 19. The calculated relations between k_m or k_e and temperature at various values of ε. Parameters are $ZJ = 1412$ cal/mol, $\alpha = 2.185$ eu, $N_0 = 200$.

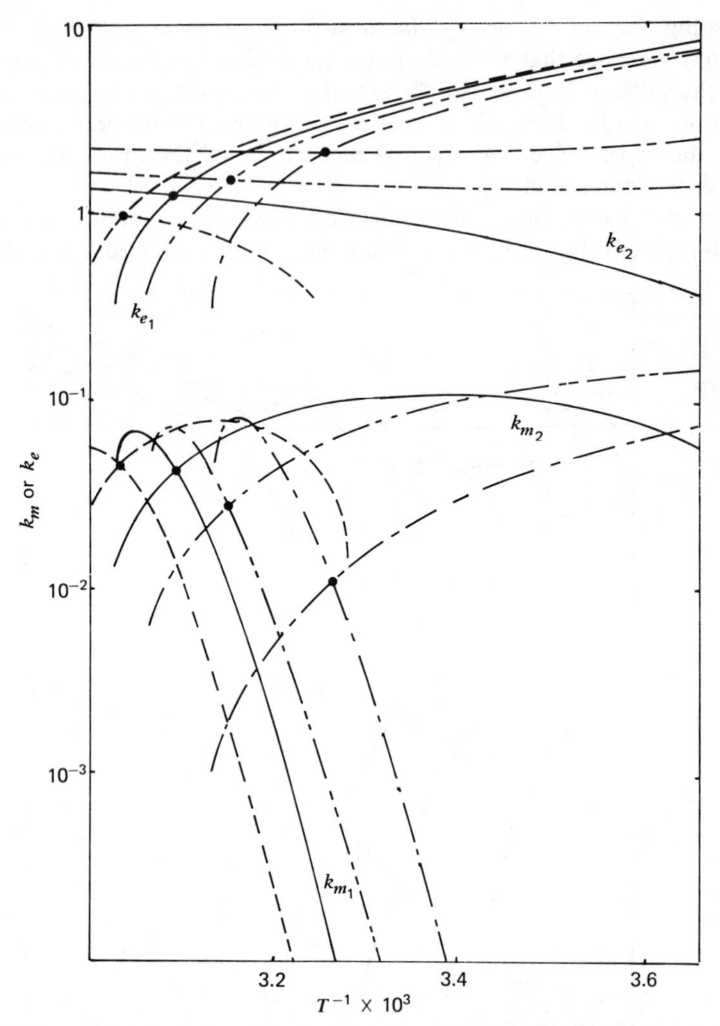

Fig. 20. The calculated relations between k_m or k_e and temperature at various values of α. Parameters are $\varepsilon = 0.706$ kcal/mol, $ZJ = 1412$ cal/mol, $N_0 = 200$.

($\equiv \alpha ZJ/2k\varepsilon$), the ε dependence of $k_e(T_d)/k_m(T_d)$ in Fig. 19 can be accounted for by (65). Similarly, an opposite effect of α on k_e or k_m can be explained. It must be noted that the temperature dependence of k_m shown in Figs. 19 and 20 is quite similar to the actual observations by a number of authors, Pohl,[48, 49] McPhie,[50] and Segawa et al.,[51] as shown in Figs. 21*a* and 21*b*.

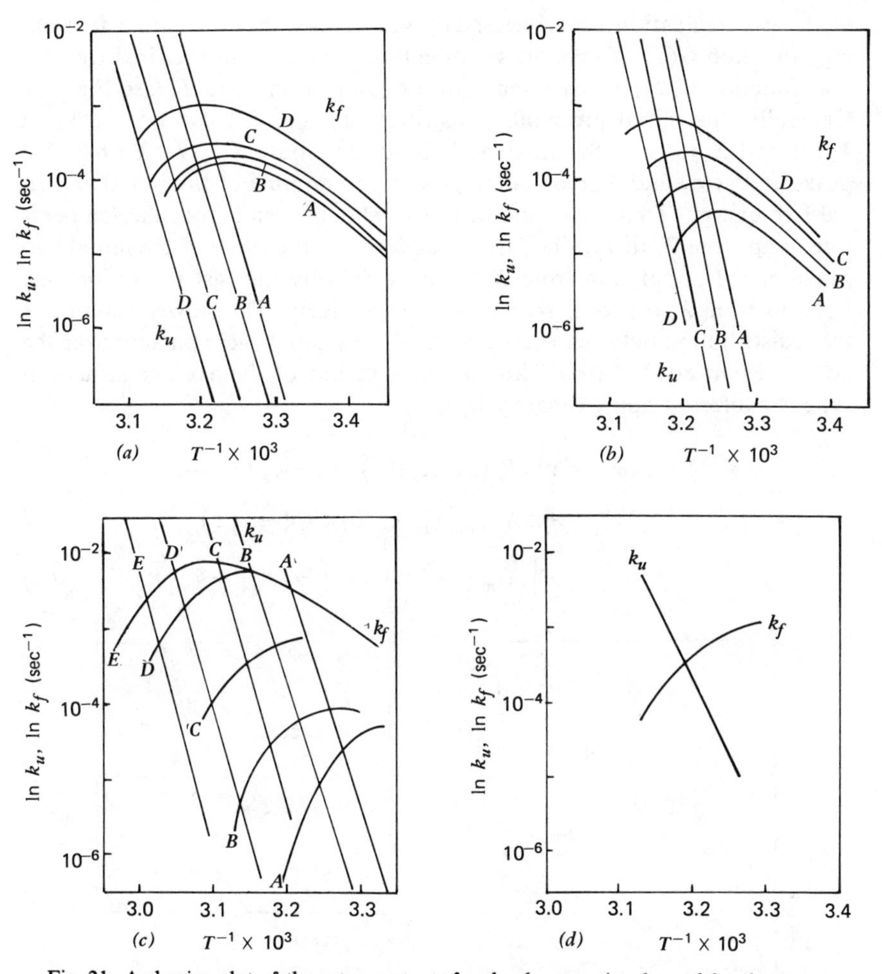

Fig. 21. Arrhenius plot of the rate constants for the denaturation k_u and for the renaturation k_f: (a) trypsin at various pH values[48]: (A) 1.59 and 1.80, (B) 2.01, (C) 2.22, (D) 2.40: (b) Chymotrypsin at various pH values: (A) 2.11, (B) 2.33, (C) 2.51 and chymotrypsinogen at pH 2.31, (D) 49. (c) Rebonuclease A at various pH values[50]: (A) 1.24, (B) 2.24, (C) 2.94, (D) 3.95, (E) 7.55, respectively. (d) Lysozyme in the presence of 4.5 M lithium bromide[51].

The pH dependence of these observed proteins may be accounted for by the change in either ε or α, or in both.

2. *Simulations of Temperature-Jump Experiments*

The change in the structural state of a protein by a temperature jump is somewhat complicated, depending on the initial distribution and the rates of the two relaxation processes. Here we suppose that the initial free energy function $G(T_1, X)$ changes suddenly at time $t=0$ to the final free energy function $G(T_2, X)$ by a temperature jump from T_1 to T_2 (see Fig. 22). Generally, the initial probability distribution $P_e(T_1, X)$ has two peaks at $X_{e_1}(0)$ and $X_{e_2}(0)$ and the final probability distribution $P_e(T_2, X)$ has two peaks at $X_{e_1}(\infty)$ and $X_{e_2}(\infty)$, corresponding to minima of the initial and final free energy. Two relaxation processes should result from the temperature jump from T_1 to T_2. The faster reactions are the shifts of the initial two peaks of the population from $X_{e_1}(0)$ to $X_{e_1}(\infty)$ by the fast relaxation rate k_{e_1}, and from $X_{e_2}(0)$ to $X_{e_2}(\infty)$ by k_{e_2}, respectively. The slower reaction is the redistribution between two peaks at $X_{e_1}(\infty)$ and $X_{e_2}(\infty)$ passing over the potential barrier at $X_u(\infty)$. Thus the time course of the average structural state is expressed approximately by

$$\begin{aligned}\bar{X}(t)-\bar{X}(\infty)=&\Delta X_1 P_{e_1}(T_1, X_{e_1}(0))\exp(-k_{e_1}t)\\&+\Delta X_2 P_{e_2}(T_1, X_{e_2}(0))\exp(-k_{e_2}t)\\&+\left[\bar{X}_{\text{int}}-\bar{X}(\infty)\right]\exp\left[-(K_{m_1}+k_{m_2})t\right]\end{aligned} \quad (66)$$

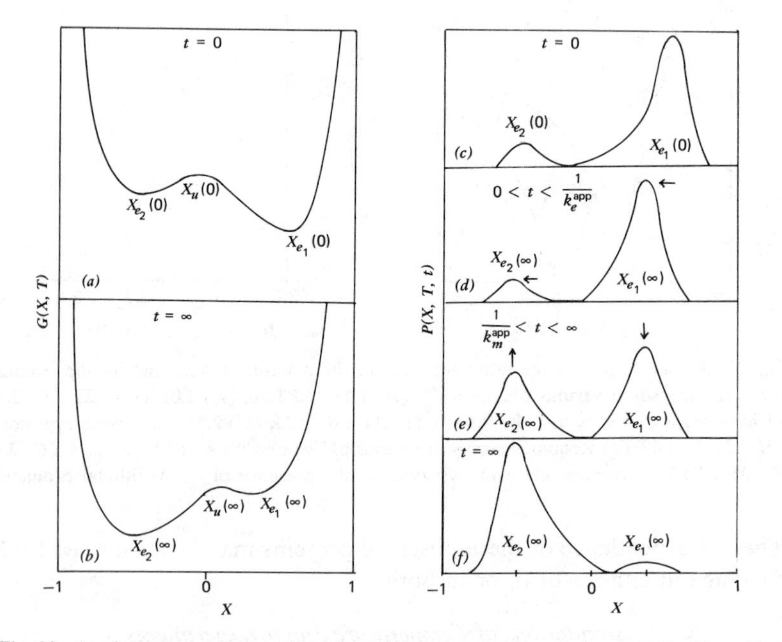

Fig. 22. A schematic representation of a temperature-jump experiment from T_1 to T_2 (see text): (a and b) Gibbs free energy of the initial and final temperatures; (c to f): time cause of the probability density $P(T_2, X, t)$.

where

$$\Delta X_i = X_i(0) - X_i(\infty) \qquad i=1,2$$

$$\overline{X}_{\text{int}} = X_{e_1}(\infty)P_e(T_1, X_{e_1}(0)) + X_{e_2}(\infty)P_e(T_1, X_{e_2}(0)) \tag{67}$$

Since the separation between two fast reactions in (66) is difficult in the usual relaxation experiments, the following approximate expression was used in our simulations:

$$\overline{X}(t) - \overline{X}(\infty) = \Delta\overline{X}_e \exp(-k_e^{\text{app}}t) + \Delta\overline{X}_m \exp(-k_m^{\text{app}}t) \tag{68}$$

where

$$k_e^{\text{app}} = P_e(T_1, X_{e_1}(0))k_{e_1} + P_e(T_1, X_{e_2}(0))k_{e_2}$$

$$k_m^{\text{app}} = k_{m_1} + k_{m_2}$$

$$\Delta\overline{X}_e = \overline{X}(0) - \overline{X}_{\text{int}}$$

$$\Delta\overline{X}_m = \overline{X}_{\text{int}} - \overline{X}(\infty) \tag{69}$$

The result of the temperature-jump simulations[45] from the various initial temperatures to a fixed final temperature, and those from a fixed temperature to various final temperatures are shown in Figs. 23 and 24, respectively. (See pp. 410 and 411) These simulations are supposed to represent the observed kinetics of the protein denaturation. Actually the kinetic simulations[45] using the values of parameters determined from the calorimetric measurements of ribonuclease A agree fairly well with its temperature- and pH-jump experiments.

VII. CONCLUDING REMARKS

Two fundamental factors are combined in the present theory. One is the highly ordered, crystal structure of proteins, and the other is the flexible and dynamic nature of proteins. Our lattice model is not the lattice approximation often used for high polymer solutions or sometimes used for proteins, but a crystal approximation. The flexibility is introduced directly into each lattice site as the entropy of the chain. Thus the model implies that the molecular motion of an amino acid is confined within a certain space surrounded by several secondary bonds in their bonded state.

The essential idea of the model is supported by a recent computer simulation by McCammon et al.,[52] who calculated the molecular motion of all

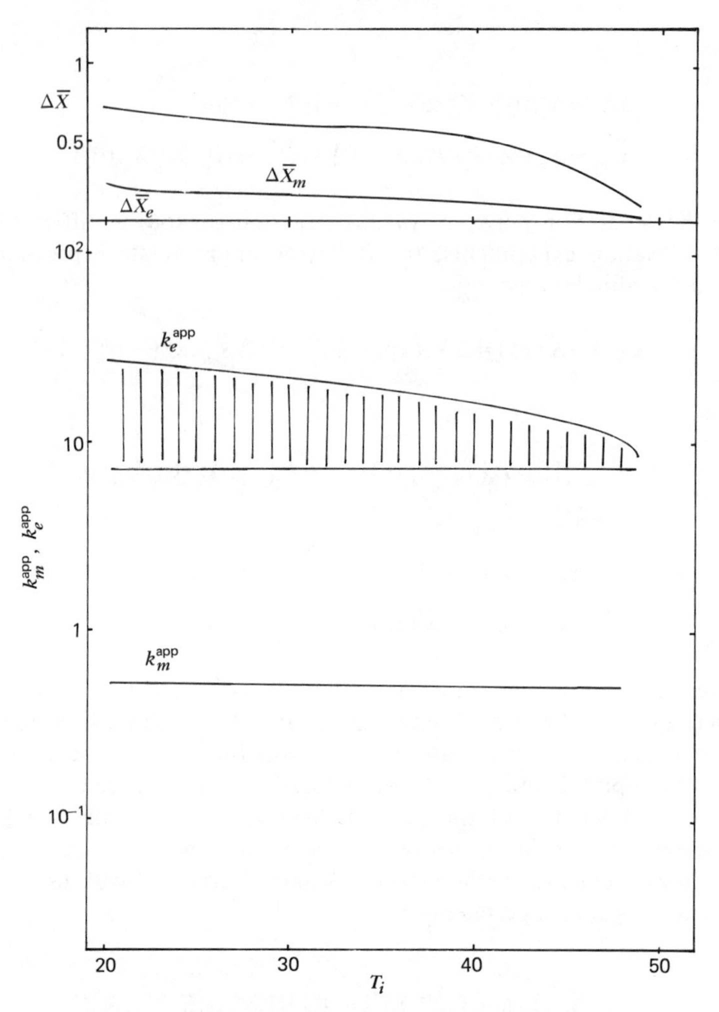

Fig. 23. The temperature-jump simulations from various initial temperatures T_i to the final temperature of 50°C.

the atoms in the small protein pancreatic trypsin inhibitor. Their results revealed considerable fluctuations, though the value for the time-averaged structure is near that of the X-ray structure. The internal motion of the protein is governed by fluidlike collisions between neighboring nonbonded atoms rather than by rotational potentials of covalent bonds.

By the use of this crystal approximation, we can, in principle at least, infer the structural stability and dynamics of each protein from the knowledge of its specific tertiary structure at 0°K. All theoretical calculations

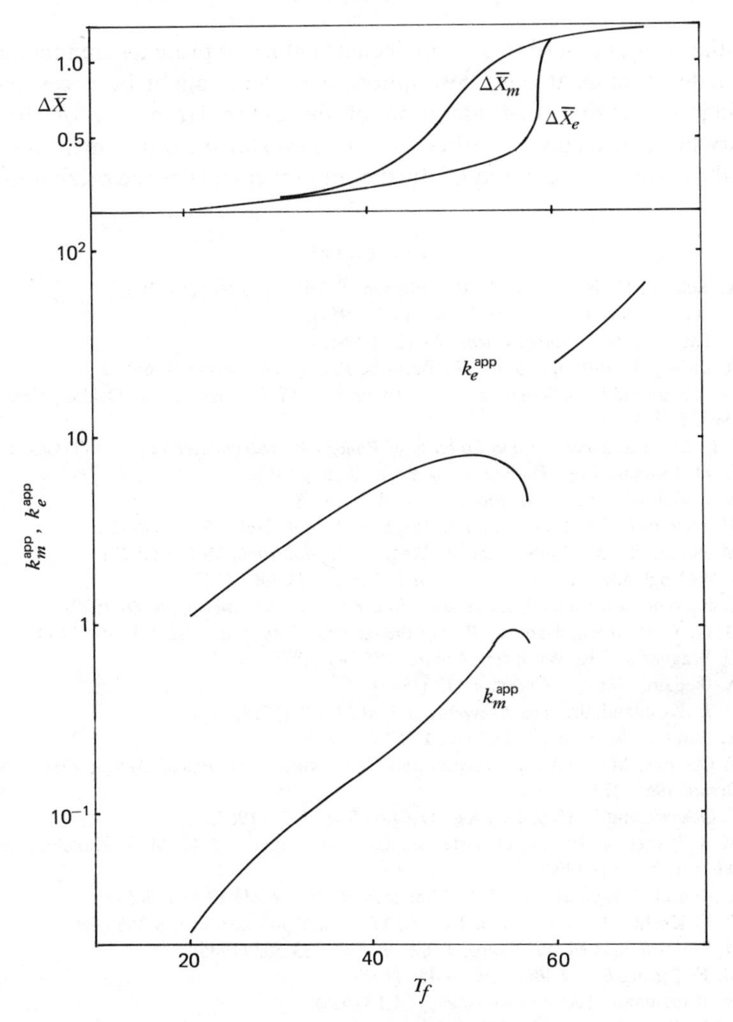

Fig. 24. The temperature-jump simulation from the initial temperature of 10°C to various final temperatures T_f.

have been carried out under the simplest assumption based on the fundamental idea. Apparently the "equivalent and unform" assumption has oversimplified the real situation. The results have shown, however, the essential features of the equilibrium and kinetic properties of protein denaturations induced by temperature, pressure, and denaturants. Furthermore, structural changes induced by the specific binding, polymerization, and allosteric effects can be explained in a unified manner. Although the essential factors describing the molecular nature of proteins are included in the model, a more precise description of proteins might be necessary for making predictions. Consideration of the dissimilar nature of the secondary bond or chains, together with the new statistical treatment for nonperiodic systems, is required for further improvements of the present model.

References

1. M. Sela, F. H. White, and C. B. Anfinsen, *Science*, **125** 691 (1957).
2. W. Kauzmann, *Adv. Protein Chem.*, **14** 1 (1959).
3. C. Tanford, *Adv. Protein Chem.*, **23** 121 (1968).
4. R. Lumry, R. Biltonen, and J. F. Brandts, *Biopolymers*, **4** 917 (1966).
5. D. Poland and H. S. Scheraga, *Poly-αAmino Acid*, G. D. Fasman Ed., Dekker, New York, 1967, p. 391.
6. J. F. Brandts, *Structure and Stability of Biological Macromolecules*, S. N. Timasheff and G. D. Fasman, Eds., Dekker, New York, 1969, p. 213.
7. R. L. Baldwin, *Ann. Rev. Biochem.*, **44** 453 (1975).
8. M. Nakanishi, M. Tsuboi, and A. Ikegami, *J. Mol. Biol.*, **70** 351 (1972).
9. M. Nakanishi, M. Tsuboi, and A. Ikegaim, *J. Mol. Biol.*, **75** 673 (1973).
10. S. W. Englander and A. Rolfe, *J. Biol. Chem.*, **248** 4852 (1973).
11. C. K. Woodward and B. D. Hilton, *Ann. Rev. Biophys. Bioeng.*, **8** 99 (1979).
12. D. G. Westmoreland and C. R. Matthews, *Proc. Nat. Acad. Sci. US*, **70** 914 (1973).
13. G. Wagner and K. Wüthrich, *Nature*, **275** 247 (1978).
14. A. Ikegami, *Biophys. Chem.*, **6** 117 (1977).
15. D. E. Koshland, Jr., *The Enzymes*, 2nd ed., **1** 105 (1959).
16. R. Lumry, *The Enzymes*, 2nd ed., **1** 157 (1959).
17. F. Oosawa, M. Kasai, S. Hatano, and S. Asakura, *Ciba Found. Symp., Princ. Biomol. Organ.* **1965**, 273.
18. F. Oosawa and S. Higashi, *Prog. Theoret. Biol.*, **1** 79 (1967).
19. M. I. Kanehisa, Ph. D. Dissertation. Univ. of Tokyo (1974). M. I. Kanehisa and A. Ikegami. to be published.
20. J. Monod, J. Wyman, and J. P. Changeux, *J. Mol. Biol.*, **12** 88 (1965).
21. D. E. Koshland, Jr., G. Nemethy, and D. Filmer, *Biochemistry*, **5** 365 (1966).
22. H. S. Frank and M. W. Evans, *J. Chem. Phys.*, **13** 507 (1945).
23. M. F. Perutz, *Eur. J. Biochem.*, **8** 455 (1969).
24. W. Kauzmann, *Adv. Protein Chem.*, **14** 1 (1959).
25. I. M. Klotz, *Science*, **128** 815 (1958).
26. M. I. Kanehisa and A. Ikegami, *Biophys. Chem.*, **6** 131 (1977).
27. G. Nemethy and H. A. Scheraga, *J. Phys. Chem.*, **66** 1773 (1962).
28. G. Nemethy and H. A. Scheraga, *J. Chem. Phys.*, **15**, 3382 (1962).
29. G. Nemethy and H. A. Scheraga, *J. Chem. Phys.*, **15** 3401 (1962).
30. N. C. Pace and C. Tanford, *Biochem.*, **7** 198 (1968).
31. C. Tanford and K. C. Aune, *Biochem.*, **9** 206 (1970).
32. A. Salahuddin and C. Tanford, *Biochem.*, **9** 1342 (1970).
33. J. F. Brandts and L. Hunt, *J. Am. Chem. Soc.*, **89** 4826 (1967).
34. A. Zipp and W. Kautzmann, *Biochem.*, **12** 4217 (1973).
35. S. A. Hawley, *Biochem.*, **10** 2436 (1971).
36. P. L. Privalov and N. N. Khechinashvili, *J. Mol. Biol.*, **86** 665 (1974).
37. S. Mizushima and T. Shimanouchi, *Infrared Absorption and Raman Effect*, Kyoritsu Shuppan, Tokyo, 1958 [in Japanese].

38. R. Fletcher and M. J. D. Powell, *Comput. J.*, **6** 163 (1963).
39. H. A. Scheraga, in *The Proteins*, Vol. 1, H. Neurath, Ed., Academic, New York, 1963, p. 477.
40. I. D. Kuntz and W. Kauzmann, *Adv. Protein Chem.*, 239 (1974).
41. J. R. Garel and R. L. Baldwin, *J. Mol. Biol.*, **94** 611 (1975).
42. A. Ikai, W. W. Fish, and C. Tanford, *J. Mol. Biol.*, **73** 165 (1973).
43. T. Y. Tsong, R. L. Baldwin, P. McPhie, and E. L. Elson, *J. Mol. Biol.*, **63** 453 (1972).
44. A. W. Burgess, L. I. Weinstein, D. Gabel, and H. A. Scheraga, *Biochemistry*, **14** 197 (1975).
45. A. Ikegami, A. Nasuda, and M. I. Kanehisa, in preparation.
46. R. B. Griffiths, C. Y. Weng, and J. S. Langer, *Phys. Rev.*, **149**, 301 (1966).
47. R. Kubo, K. Matsuo, and K. Kitahara, *J. Stat. Phys.*, **9** 51 (1973).
48. F. M. Pohl, *Eur. J. Biochem.*, **7** 146 (1968).
49. F. M. Pohl, *Eur. J. Biochem.*, **4** 373 (1968).
50. P. McPhie, *Biochem.*, **11** 879 (1972).
51. S. Segawa, Y. Hushimi, and A. Wada, *Biopolymers*, **12** 2521 (1973).
52. J. A. McCammon, B. R. Gelin, and M. Karplus, *Nature*, **267** 585 (1977).

AUTHOR INDEX

SUBJECT INDEX